RECENT DEVELOPMENTS IN DURABILITY ANALYSIS OF COMPOSITE SYSTEMS

PROCEEDINGS OF THE FOURTH INTERNATIONAL CONFERENCE ON DURABILITY ANALYSIS OF COMPOSITE SYSTEMS – DURACOSYS 99/BRUSSELS/BELGIUM/11 – 14 JULY 1999

Recent Developments in Durability Analysis of Composite Systems

Edited by

Albert H. Cardon
Free University, Brussels, Belgium

Hiroshi Fukuda
Science University Tokyo, Japan

Ken L. Reifsnider
Virginia Polytechnic Institute and State University, Blacksburg, Virginia, USA

Georges Verchery
ISAT, University of Burgundy, Nevers, France

A.A. BALKEMA/ROTTERDAM/BROOKFIELD/2000

This conference was sponsored by the Free University of Brussels (V.U.B), the Foundation for Scientific Research of the Flemish Region (FWO-Vlaanderen), Virginia Polytechnic Institute and State University, the Office of Naval Research, the European Research Office of the US Army, the Science University Tokyo and the University of Bourgogne (ISAT-Nevers).

The texts of the various papers in this volume were set individually by typists under the supervision of each of the authors concerned.

Published by
A.A. Balkema, P.O. Box 1675, 3000 BR Rotterdam, Netherlands
Fax: +31.10.413.5947; E-mail: balkema@balkema.nl; Internet site: www.balkema.nl
A.A. Balkema Publishers, Old Post Road, Brookfield, VT 05036-9704, USA
Fax: 802.276.3837; E-mail: info@ashgate.com

ISBN 90 5809 103 1
© 2000 A.A. Balkema, Rotterdam
Printed in the Netherlands

*Recent Developments in Durability Analysis of Composite Systems, Cardon, Fukuda, Reifsnider & Verchery (eds) ©
2000 Balkema, Rotterdam, ISBN 90 5809 103 1*

Table of contents

[*] Invited General Lectures
[**] Special Invited Lectures

Infrastructure applications

Experimental aspects

Viscoelasticity – Creep analysis

Applications

*Recent Developments in Durability Analysis of Composite Systems, Cardon, Fukuda, Reifsnider & Verchery (eds) ©
2000 Balkema, Rotterdam, ISBN 90 5809 103 1*

Preface

With the increasing number of mechanical, aerospace and civil engineering applications of composite systems, including joining aspects, reliability of those constructions become more and more important.

As a basis for a complete reliability analysis we need a good understanding of the long term behaviour of the composite, his internal transformations, including the interphase properties and the damage initiation, development and control. Durability analysis is the prediction of the residual structural integrity of a composite system after an imposed life time under a complex mechanical loading history in interaction with environmental variations.

After the meeting on Durability Analysis of Polymer Based Composite Systems (Brussels, August 27-31, 1990), an Informal Invited International Workshop on the Subject in July 1993 in Porto, we had the International Conference on Progress in Durability Analysis of Composite Systems (Duracosys 95, Brussels, July 16-21, 1995), followed by the Duracosys 97 Meeting in Virginia Tech on September 24-27, 1997.

Different organisations such as the ASTM, AMAC and many industrial groups become more and more concerned with this problem as soon as the composite structural component, to be designed or repaired, has to perform safely for a long time.

All composite systems are considered with focus on polymeric, cement and ceramic matrix composites.

Applications in civil engineering (bridges, buildings, off shores, pipe lines and infrastructure in general) and in mechanical engineering (transport on land, sea and in the air and robotics), were discussed.

Theoretical, fundamental, computational and experimental aspects were presented.

Organisation

Conference co-chairmen
Albert H.Cardon, V.U.B., Brussels, Belgium
Hiroshi Fukuda, Science University Tokyo, Japan
Ken L.Reifsnider, Virginia Tech, Blacksburg, Va., USA
Georges Verchery, ISAT, University of Bourgogne, Nevers, France

International Scientific Board
R.Adams, Bristol, United Kingdom
S.Aivazzadeh, Nevers, France
L.Berglund, Luleå, Sweden
O.Bergsma, Delft, Netherlands
T.Błaszczyński, Poznan, Poland
H.Brinson, Houston, USA
O.Brüller, Munich, Germany
I.Emri, Ljubljana, Slovenia
P.Gudmundson, Stockholm, Sweden
Z.Hashin, Tel Aviv, Israel
N.Himmel, Kaiserslautern, Germany
P.Hogg, London, United Kingdom
S.Johnson, Georgia Tech, Atlanta, USA
W.Knauss, Caltech, Pasadena, USA
P.Lagace, MIT, Cambridge, USA

J.A.Manson, EPfL, Lausanne, Switzer
Tiangxiang Mao, Beijing, China
A.Miravete, Zaragoza, Spain
Y.Miyano, Kanazawa, Japan
G.Papanicolaou, Patras, Greece
T.Peijs, Eindhoven, Netherlands
D.Perreux, Besançon, France
R.Pyrz, Aalborg, Denmark
E.Sancaktar, Akron, USA
K.Schulte, Hamburg-Harburg, Germa
R.Talreja, Georgia Tech, Atlanta, USA
V.Tamusz, Riga, Latvia
A.Vautrin, St.Etienne, France
I.Verpoest, K.U.Leuven, Belgium
J.Weitsman, Knoxville, USA

Belgian Host Committee
A.Cardon, V.U.B., Brussels
R.Bourgois, Royal Military School, Brussels
J.Degrieck, University of Ghent

P.Halleux, U.L.Brussels
R.Keunings, U.C.Louvain-La-Neuve
I.Verpoest, K.U.Leuven

Local Organizing Committee
A.Cardon
M.Bourlau
A.Vanaeken
P.Bouquet
K.Hoes
Chr.Van Vossole
S.Harou-Kouka

K.De Proft
A.Vrijdag
F.Boulpaep
R.Heremans
D.Debondt
G.Van Den Nest

Strength criteria and failure mechanisms

Recent Developments in Durability Analysis of Composite Systems, Cardon, Fukuda, Reifsnider & Verchery (eds)
© *2000 Balkema, Rotterdam, ISBN 90 5809 103 1*

Progressive failure of 3-D-stressed laminates: Multiple nonlinearity treated by the failure mode concept (FMC)

R.G.Cuntze
Main Department Analysis, MAN Technologie AG, Karlsfeld/Munich, Germany

ABSTRACT:
This contribution focuses the two aspects: The derivation of failure conditions for a unidirectional (UD) lamina with the prediction of *initial failure of the embedded laminae* and secondly, the treatment of nonlinear, progressive failure of 3-dimensionally stressed laminates until *final failure*. The failure conditions are based on the so-called Failure Mode Concept (FMC) which takes the symmetries of the UD-lamina homogenized to a 'material' into account. For the solution procedure of the multi-nonlinear analysis the secant modulus of the activated failure modes was employed. Each mode is associated to a stress-strain curve, summing up to five in the case of the transversally-isotropic UD-material. The test cases were taken from the '*World-wide Failure Exercise*' in the UK involving strength criteria *and* nonlinear analysis in order to obtain accurate stresses. Furthermore, the reserve factor necessary for the *Proof of Design* is defined and outlined for the cases 'linear elastic analysis permitted' and 'nonlinear analysis mandatory'. An extensive conclusion with outlooks terminates the contribution. This includes a brief comparison of Puck's and Cuntze's theory.

1 GENERAL

Progressive failure analysis of laminates or the prediction of laminate behaviour up to fracture is an old challenge and again, was the focus of the *'Failure (theory) Exercise'* (failure theory: = nonlinear laminate analysis + failure conditions) arranged by M. Hinton and P. Soden, published in (Hin97). The results of this competition were at least 'surprising', if viewing their scatter. A comparison with test results will be published soon. The author hopes to add a lamina stresses-based *engineering approach* being a 'physically'-based 3D phenomenological model, that will improve the quality of predicting *'Progressive failure of laminates subjected to 2D/3D-states of stress'*. Basis is the Failure Mode Concept (FMC) and macromechanical laminate analysis, which is in industry the level of the finite element analysis for composite structures.

The idea of thinking in strength failure modes is not a new idea, e.g., A. Puck in 1968 still distinguished fibre failure (FF) and inter fibre-failure (IFF). Further, Z. Hashin in (Has80) based his investigations on the same understanding and put forward the idea that brittle failure occurs in the plane of the highest Mohr-stress based effort. And in 1992, A. Puck worked out the mode-related Mohr-stresses based Hashin/Puck 'Action plane strength criteria' (see Puc 92, 96; VDI 97 annex AVI).

Also, the idea of applying invariants was intensively pursued, e.g. by J. Boehler's group since 1980 and by Z. Hashin (Has80). The results unfortunately attracted too little attention. In 1996 the author influenced R.Jeltsch-Fricker to pick up the invariant-based idea [Jel99]. S. Tsai's polynomial failure conditions may be transformed into formulations of invariant terms, too.

The failure mode *concept* just more strictly applies the mode thinking and more consequently uses the advantage of formulating the failure conditions (*interaction of stresses*) by the material symmetries respecting invariants, which contain the lamina stresses of the FEM output. The author's strong connections to structural reliability, where failure mode thinking is one basement, helped to simply model the *interaction of modes*.

The paper addresses several topics:
1) A brief introduction into the FMC; 2) Modelling

of nonlinearity and programming; 3) Derivation of initial and final failure curves for two test cases, based on 1) and 2); 4) Integration of design aspects with the objective to obtain a positive margin of safety or a reserve factor >1, respectively.

In order to correctly assess the following contribution: It is a private and non-funded work.

2 MAIN FEATURES OF THE FAILURE MODE CONCEPT (FMC)

The features of the FMC are briefly summarised in table 1. Additional aspects are collected in table 2.

Table 1. Main features of the FMC

> • Each *mode* represents one failure mechanism and one piece of the complete *failure surface* (surface of the failure body or *limit surface*)
> • Each failure *mechanism* is represented by one failure *condition*. One failure mechanism is governed by one basic *strength* and therefore has a clearly defined equivalent stress σ_{eq}
> • Curve-fitting of the course of test data is only permitted in the pure failure mode's regime
> • Different, however, similar behaving materials obey the same function as failure condition but have *different curve parameters*
> • Rounding-off in mode interaction zones is performed by probabilistics-based spring models. This rounding-off of adjacent *mode failure curves* (*partial* surfaces) in their interaction zone is leading again to a *global* failure curve (surface) or to a 'single surface failure description' (such as with Tsai/Wu).

Table 2. Additional FMC aspects/informations

• An invariant formulation of a failure condition in order to achive a *scalar* potential considering the material's symmetries (Chr97) is possible
• Each invariant term of the failure function shall be related to a physical mechanism observed in the solid
• Hypotheses applied:
  Hashin/Puck with Beltrami (choice of invariants),
  Mohr-Coulomb (friction, thinking in Mohr's stresses)
* Proof of Design and *Strength* analysis:
  - For each *mode* one *reserve factor* $f_{Res}^{(mode)}$ or one stress effort (if nonlinear) is to be determined, displaying, where the design key has to be turned
  - The probabilistics-based 'rounding-off' approach delivers the *resultant reserve factor*
    linked to the *margin of safety* by $MS = f_{Res}^{(res)} - 1$.
* *Stress* analysis with Degradation:
  - Equivalent stresses and stress efforts are used in the (nonlinear) progressive damage description.

3 BASICS

• State of stress:
For the unidirectional (UD) material element <u>figure 1</u> depicts the prevailing 3D-state of stress. Additionally, with respect to the symmetries of this transversally-isotropic material (modelled an ideal crystal), the 5 basic strengths and 5 elasticities are given. A UD-lamina is a low-scale structure with the constituents fibre, matrix and interphase (not interface). After homogenization (smearing) it is called material.

• *Invariants:*
Strength criteria or failure conditions may be formulated by invariants based on the UD-stresses:

$$I_1 = \sigma_1; \quad I_2 = \sigma_2 + \sigma_3 \quad ; \quad I_4 = (\sigma_2 - \sigma_3)^2 + 4\tau_{23}^2;$$

$$I_3 = \tau_{31}^2 + \tau_{21}^2 \quad ; \quad \text{(Has80, Boe85)}$$

$$I_5 = (\sigma_2 - \sigma_3)(\tau_{31}^2 - \tau_{21}^2) - 4\tau_{23}\tau_{31}\tau_{21} \ .$$

The shear stress product $\tau_{23}\tau_{31}\tau_{21}$ in I_5 has a sign effect. However, this does not matter because by a transformation $(\sigma_2, \sigma_3, \tau_{23}) \rightarrow (\sigma_{II}, \sigma_{III}, 0)$ the term may be dropped.

• *Strengths*
The characterisation of the strength of transversally-isotropic composites requires the measurement of five independent basic strengths: $R_{\parallel}^t$, $R_{\parallel}^c$ (fibre parallel tensile and compressive strength) as well $R_{\perp}^t$, $R_{\perp}^c$ (tensile, compressive strength perpendicular to the fibre direction) and $R_{\perp\parallel}$ (fibre parallel shear strength). $R_{\parallel}^t$ is determined by the strength of the constituent fibre and $R_{\parallel}^c$ by '*shear*

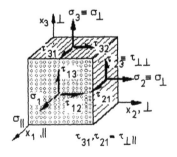

$R_{\parallel}^t (= X^t), R_{\parallel}^c (= X^c), R_{\perp\parallel} (= S), R_{\perp}^t (= Y^t), R_{\perp}^c (= Y^c);$
$E_{\parallel}, E_{\perp}, G_{\parallel\perp}, \nu_{\parallel\perp}, \nu_{\perp\perp}$

Fig 1. UD lamina (t: = tension, c: = compression. X, Y, S : American denotations)

4

stability'. The latter includes different microfailure mechanisms: The matrix may shear under loading and does not stabilise the generally somewhat misaligned fibres embedded within. Hence it comes to bending and "kinking" (structural behaviour). Also, the load "grasping" fibre as stiffer constituent may shear (this is material behaviour) under $\sigma_{\parallel}^c$ and $\tau_{\perp\parallel}$. $R_{\perp}^t$ is determined as well by the relatively low strength properties of the matrix (cohesive failure) and the interphase material in the interface fibre-matrix (adhesive failure caused by a weak fibre-matrix bond), as by the fibres acting as embedded stress raisers.

• *Rounding-off in the Interaction Zones:*
Of further interest is the rounding-off of the fracture curve in the mixed failure domain (MiFD) or interaction or transition zone of adjacent failure modes, respectively. In (Cun97) a simple formula - the 'Series Spring Model' - as *engineering approach* for the resultant reserve factor was proposed which approximates the results of a time-consuming probabilistic calculation on the safe side. In the case of residual stresses and nonlinearity instead of the reserve factor the stress effort has to be employed.

• *Classical Laminate Theory (CLT):*
Assuming transversal isotropy and the state of plane stress ('in-plane stressing' $\sigma_3 = 0$, situation of the case studies investigated) the linear stress-strain relations for the k'th lamina of a multilayered laminate are (using matrix notation; $1 = \parallel$, $2 = \perp$, $12 = \parallel\perp$; [Q], [S]: = stiffness, compliance matrix of the lamina)

$$\{\varepsilon\}_k = (\varepsilon_1, \varepsilon_2, \gamma_{12})_k^T \qquad = [S]_k \{\sigma\}_k \ ,$$
$$\{\sigma\}_k = (\sigma_1, \sigma_2, \tau_{12})_k^T \qquad = [Q]_k \{\varepsilon\}_k \ .$$

The definitions for the lamina (often called ply if prepreg and layer if winding) stresses, angles and thicknesses are illustrated in <u>figure 2</u>.

The symmetric *elasticity matrix of stiffness* (stiffness matrix) of the lamina reads:

$$[Q]_k = \begin{bmatrix} Q_{11} & Q_{12} & Q_{16} \\ Q_{21} & Q_{22} & Q_{26} \\ Q_{61} & Q_{62} & Q_{66} \end{bmatrix}_k = \begin{bmatrix} \dfrac{E_{\parallel}}{1-\nu_{\parallel\perp}\nu_{\perp\parallel}} & \dfrac{\nu_{\parallel\perp}E_{\parallel}}{1-\nu_{\parallel\perp}\nu_{\perp\parallel}} & 0 \\[2mm] \dfrac{\nu_{\perp\parallel}E_{\perp}}{1-\nu_{\parallel\perp}\nu_{\perp\parallel}} & \dfrac{E_{\perp}}{1-\nu_{\parallel\perp}\nu_{\perp\parallel}} & 0 \\[2mm] 0 & 0 & G_{\parallel\perp} \end{bmatrix}_k ,$$

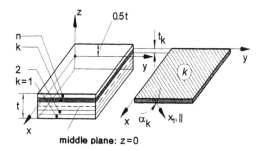

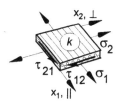

Fig 2. Laminate and k' th lamina subjected to a plane state of stress (midplane z = 0)

with $\quad [Q]_k^{-1} = [S]_k, \quad \nu_{\perp\parallel} \cdot E_{\perp} = \nu_{\parallel\perp} \cdot E_{\parallel}$
and $\nu_{\perp\parallel}$ as the *major* Poisson's ratio!
Thus, for the application of CLT the knowledge of only four constants is essential: $E_{\parallel}$, $E_{\perp}$, $G_{\parallel\perp}$, $\nu_{\perp\parallel}$. The 2D-strength analysis employs all 5 strengths.

In the case of mechanical loading the following load-strain equations are obtained in the cross section for the load fluxes {n} and the moment fluxes {m°}

$$\begin{Bmatrix} n \\ m^\circ \end{Bmatrix} = \begin{bmatrix} A & B \\ B & D \end{bmatrix} \begin{Bmatrix} \varepsilon^\circ \\ \chi \end{Bmatrix} = [K] \begin{Bmatrix} \varepsilon^\circ \\ \chi \end{Bmatrix}$$

with [K] being the *stiffness matrix of the laminate*
and $\quad [A] = \sum\limits_{k=1}^{n} [Q']_k \cdot t_k, \quad [Q'] = [T_\sigma][Q][T_\sigma]^T,$
and transformation matrices (s = sin α, c = cos α)

$$[T_\sigma] = \begin{bmatrix} c^2 & s^2 & -2sc \\ s^2 & c^2 & 2sc \\ sc & -sc & c^2-s^2 \end{bmatrix}, [T_\varepsilon] = \begin{bmatrix} c^2 & s^2 & -sc \\ s^2 & c^2 & sc \\ 2sc & -2sc & c^2-s^2 \end{bmatrix}.$$

Having determined the strain vector {ε°} and the curvature vector {χ} for the middle plane of the laminate, the natural strains $\{\varepsilon\}_k$ and stresses $\{\sigma\}_k$ in each lamina may be calculated according to

$$\{\varepsilon\}_k = [T_\sigma]_k^{-1}(\{\varepsilon^\circ\}+z\{\chi\}), \quad \{\sigma\}_k = [Q]_k\{\varepsilon\}_k \ .$$

The equations above decouple for a symmetric lay-up to

$$\{\varepsilon^\circ\} = [A]^{-1}\{n\}.$$

5

If curing stresses have to be considered the equations read

$$\{\varepsilon^\circ\} = [A]^{-1}(\{n\} + \{n_T\}) \quad \text{with}$$

$$\{n_T\} = \sum_{k=1}^{n} \Delta T [Q']_k t_k \{\alpha'_T\}_k \quad \text{and}$$

$$\{\alpha'_T\}_k = [T_\varepsilon]_k \{\alpha_T\}_k, \quad \{\alpha_T\} = (\alpha_{T\|}, \alpha_{T\perp}, 0)^T.$$

In the case of symmetrical lay-ups (test cases of the 'failure exercise'), for the treatment of *material nonlinearity* and of *degradation,* the lamina stresses $\{\sigma\}_k$ have to be computed considering

$$\{\varepsilon'\}_k = \{\varepsilon^\circ\} \qquad \text{.... compatibility}$$

$$\{\sigma'\}_k = [Q']_k \cdot (\{\varepsilon'\}_k - \{\varepsilon'_T\}_k) \quad \text{.... Hooke}$$

$$\{\sigma\}_k = [T_\sigma]_k^{-1} \{\sigma'\}_k, \quad \{\varepsilon\}_k = [T_\varepsilon]_k^{-1} \{\varepsilon'\}_k$$

and applying

$$[T_\sigma]^{-1} = [T_\varepsilon]^T, \quad [T_\varepsilon]^{-1} = [T_\sigma]^T.$$

4 STRENGTH CRITERIA (failure conditions)

Failure conditions should exhibit – besides a sound mechanical basis – the numerical advantages: homogeneity in the stress terms, stress terms of the lowest degree, simplicity, numerically robust and rapid computation.

4.1 *Failure modes (types)*

A designer has to dimension a laminate versus inter-fibre-failure (IFF) and fibre-failure (FF). IFF normally indicates the *onset of failure* whereas the appearance of FF in a single lamina of the laminate usually marks *final failure.* In the case of brittle behaving FRP, the failure is fracture. The IFF incorporates cohesive fracture of the matrix and adhesive fracture of the fibre-matrix interface.

Further, the 'explosive' effect of a so-called *wedge shape* failure (a σ_2^c caused IFF) of an embedded lamina of the laminate may directly lead (Puc96) to failure (see a torsion spring) or via local delaminations to buckling of the adjacent laminae and therefore to final failure, too.

Fracture is understood in this article as a separation of material, which is free of damage such as technical cracks and delaminations but not free of small defects/flaws prior to loading.

Figure 3 informs about the *types of fracture* which are recognised in the case of *dense* transversally-isotropic *ideal materials.*

Whether a failure may be called a shear stress induced shear failure ,SF, or a normal stress induced

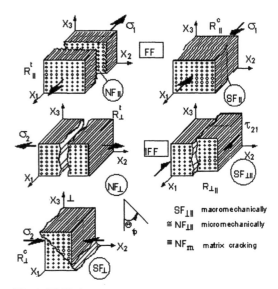

Fig 3. FMC view of the fracture types ($\equiv$ failure modes) of brittle transversally-isotropic material. (The physical fracture "planes" are pointed out in the figure (Cun98b) Θ_{fp}: = fracture plane angle)

normal failure ('cleavage') ,NF, depends on the size scale applied.

4.2 *Strain energy density basis*

Beltrami, Schleicher et al. assume at initiation of yield that the strain energy density will consist of two portions. Thus, the strain energy (denoted by W) in a cubic element of a material reads

$$W = \int \{\sigma\} \{\varepsilon\} d\{\varepsilon\} = W_{Vol} + W_{shape}.$$

Including Hooke's law in the case of a transversally-isotropic body the expression will take the form (see Lechnitski, s_{ik} = compliance coefficients):

$$\begin{aligned} W &= [s_{11}\sigma_1^2 + s_{22}\sigma_2^2 + s_{33}\sigma_3^2 + s_{44}\tau_{23}^2 + \\ &\quad + s_{55}(\tau_{12}^2 + \tau_{13}^2)]/2 + s_{12}(\sigma_1\sigma_2 + \sigma_1\sigma_3) + \\ &\quad + s_{23}\sigma_2\sigma_3 \\ &= \underset{\text{volume}}{\frac{I_1^2}{2E_\|}} + \underset{\text{volume}}{\frac{I_2(1-\nu_{\perp\perp})}{4E_\perp}} - \underset{\text{volume}}{\frac{\nu_{\perp\|}I_1I_2}{E_\|}} + \underset{\text{shape}}{\frac{I_3}{2G_{\|\perp}}} + \underset{\text{shape}}{\frac{I_4(1+\nu_{\perp\perp})}{4E_\perp}}. \end{aligned}$$

Some of the terms above describe the *volume* change of the material cube and others the change of its *shape.*

All portions may be used to formulate a failure condition, however respecting, whether the cubic material element will experience a volume change or a shape change or even both.

4.3 *Failure conditions achieved*

In engineering application due to property scatter the simplest strength criteria which still describe the physical effects should be applied. This always reduces the number of curve parameters to be determined and, besides this, additionally the numerical effort. At maximum one (statistically-based) calibration point in each failure domain should have to be experimentally determined besides the basic strength as anchor point.

Based on the idea above the following *failure conditions*, $F(\{\sigma\}) = 1$, have been derived ($\overline{R}$ marks mean strength value, $v_f \cdot \sigma_{1f} = v_f \cdot \varepsilon_1 \cdot E_\parallel^f = \varepsilon_1 \cdot E_\parallel^t$.

$F \begin{smallmatrix} > \\ = \\ < \end{smallmatrix} 1$ is called criterion, $F = 1$ is a condition):

$$F_\parallel^\sigma = \frac{I_1^2}{\overline{R}_\parallel^{t2}} = 1 \quad \rightarrow F_\parallel^\sigma = \frac{v_f \sigma_{1f}}{\overline{R}_\parallel^t} = 1,$$
$$\qquad \qquad \qquad \qquad \qquad \qquad ...... \text{ FF}$$
$$F_\parallel^\tau(I_1^2, I_3) \quad \rightarrow F_\parallel^\tau = \frac{-I_1}{\overline{R}_\parallel^c} = 1,$$

$$F_{\perp\parallel} = \frac{I_3^{3/2}}{\overline{R}_{\perp\parallel}^3} + b_{\perp\parallel}\frac{I_2 I_3 - I_5}{\overline{R}_{\perp\parallel}^3} = 1, \quad F_\perp^\sigma = \frac{I_2 + \sqrt{I_4}}{2\overline{R}_\perp^t} = 1,$$

$$F_\perp^\tau = (b_\perp^\tau - 1)\frac{I_2}{\overline{R}_\perp^c} + \frac{b_\perp^\tau I_4 + b_{\perp\parallel}^\tau I_3}{\overline{R}_\perp^{c2}} = 1 \quad .... \text{ IFF}$$

with three free curve parameters $(b_{\perp\parallel}, b_\perp^\tau, b_{\perp\parallel}^\tau)$ to be determined from multiaxial test data. Each of them has to be calculated from a test point (several measurements) or by curve fitting of the course of test data in the associated pure domain. This delivers (calibration points in the figures 4, A1)

$$b_{\perp\parallel} = \frac{1 - \left(\tau_{21}^{\perp\parallel}/\overline{R}_{\perp\parallel}\right)^2}{2\sigma_2^c \cdot \tau_{21}^{\perp\parallel 2}/\overline{R}_{\perp\parallel}^3} \quad \text{from } \left(\sigma_2^c, \tau_{21}^{\perp\parallel}\right)$$

$$b_\perp^\tau = \frac{1 + (\sigma_2^{c\tau} + \sigma_3^{c\tau})/\overline{R}_\perp^c}{(\sigma_2^{c\tau} + \sigma_3^{c\tau})/\overline{R}_\perp^c + (\sigma_2^{c\tau} - \sigma_3^{c\tau})^2/\overline{R}_\perp^{c2}}$$

$$b_{\perp\parallel}^\tau = 1 - (b_\perp^\tau - 1)\sigma_{2\perp\parallel}^{c\tau}/\overline{R}_{\perp\parallel} - b_\perp^\tau(\sigma_{2\perp\parallel}^{c\tau}/\overline{R}_{\perp\parallel})^2 .$$

Depending on the material behaviour, bounds on the safe side should be

$$0.05 < b_{\perp\parallel} < 0.15, \ 1.0 < b_\perp^\tau < 1.6, \ 0 < b_{\perp\parallel}^\tau < 0.4 .$$

The extreme value $b_{\perp\parallel} = 0$ means 'no bulge effect' and $b_\perp^\tau = 1$ means 'no friction' (see figure A1). The mapping of the failure curves shown later is based on multiaxial test data cited in literature or carried out at MAN.

According to the FMC, $F_\parallel^\sigma$ originally consists of a quadratic term in stresses, however, can be replaced by a numerically simpler linear term which regards that the fibre tensile stress and not σ_1 (the 'material' model does not hold here) has to be applied if formulating a failure condition. For reasons of simplicity and due to lacking of test data, the shear addressing invariant I_3 was not considered in $F_\parallel^\tau$.

The choice of the failure condition is strongly affected by the 'easy to be used' wish and by an easy determination of f_{Res}, which is simplified if $F(\{\sigma\})$ is a so-called homogeneous function wherein the stress terms are of the same power (grade). Therefore the $I_3^{3/2}/\overline{R}_{\perp\parallel}^3$, instead of a quadratic formulation which was used in the past, was applied leading to homogeneity of $F_{\perp\parallel}$. The term $I_2 I_3 - I_5$ respects the different inteaction of the stress combinations (σ_2, τ_{21}) and (σ_2, τ_{31}), which is described in the various papers of Puck et.al. By employing $I_2 + \sqrt{I_4}$ in $F_\perp^\sigma$ the straight line (known from isotropy) in the quasi-isotropic (σ_2, σ_3)-plane can be mapped. $F_\perp^\tau$ can not be simplified in order to consider friction, by the linear term I_2. The term I_3 was additionally taken aiming at a simpler numerical rounding-off procedure in the $(F_\perp^\tau, F_{\perp\parallel})$-interaction domain.

With respect to the 3D character of the failure conditions they may serve also as criteria for the *onset of delamination*.

One or two modes will be the design driving ones in a local 'material' point of a composite.
The basic strength value of the mode-related linear/nonlinear stress-strain curve controls the (size) *volume* of the partial failure surface (body)
Additional curve parameters, representing an effect, such as friction or fracture mechanics of defects in the material, control the *shape* of the partial failure surface

5 RESERVE FACTORS $f_{Res}^{(mode)}$, $f_{Res}^{(res)}$

5.1 *General*

Reserve factors which have to be determined for the *Proof of Design* of the laminate are *load*-related defined. There are:

- for the *initial failure*, indicated by the so-called knee in the laminate's stress-strain curve and originated by $F_\perp^\sigma, F_{\perp\parallel}$

$$f_{Res}^{initial} = \frac{\text{initial failure load}}{j_{p0.2} \cdot DLL},$$

- for the *final failure*, indicated by $F_\parallel^\sigma, F_\parallel^\tau$ or $F_\perp^\tau$,

$$f_{Res}^{final} = \frac{\text{final failure load}}{j_{ult} \cdot DLL} \quad .$$

with DLL : = Design Limit Load and
$j_{p0.2}, j_{ult}$: = design factors of safety (FoS) .

Often the reserve factor f_{Res} is defined that factor all mechanical *load*-induced stresses applied have to be multiplied with to generate failure. Geometrically it means that the stress vector $\{\sigma\}_{(L)}$ has to be stretched in its original direction by this factor in order to cause failure (see A III). This definition is valid as far as *linear* modelling can be applied. If there are no residual stresses and high design factors of safety (FoS), j, then a linear elastic modelling is permissible and a stress-based f_{Res} can be predicted.

In the case of *nonlinear* material behaviour accurate reserve factors have to be referred to loads again. Then, the procedure at first has to determine modal *equivalent stresses* and *stress efforts* outlining the remaining load capacity for the computation of the resultant reserve factor.

5.2 Determination of mode Reserve Factors

- Case "No residual stresses"

$$\{\sigma\}_{failure} = f_{Res} \cdot \{\sigma\}_{(L)} = \{\sigma\}_{(L)} + MS \cdot \{\sigma\}_{(L)}$$

with the margin of safety $MS = f_{Res} - 1$.
Inserting above definition into the failure condition

$$F = F(\{\sigma\}_{Failure}) = F(f_{Res} \cdot \{\sigma\}_{(L)}) = 1$$

yields an equation for the *stress-based* f_{Res}

$$f_{Res} \cdot \ell_{(L)} + f_{Res}^2 \cdot q_{(L)} + f_{Res}^3 \cdot c_{(L)} + \ldots = 1.$$

  * In case the failure condition will only have linear and quadratic stress terms the reserve factor can be calculated [Cun96] by resolving for f_{Res} as a root of a polynomial. This delivers $(\overline{R} \rightarrow R!)$:

$$f_{Res} = 1/\ell_{(L)} \quad (e.g. = 1/(v_f \cdot I_{1f}/R_\parallel^t) \ldots \text{linear}$$

$$f_{Res} = \left(-\ell_{(L)} + \sqrt{\ell^2_{(L)} + 4q_{(L)}} \right) / 2q_{(L)} \ldots \text{quadr.}$$

with $\ell_{(L)} = \Sigma$ linear terms, $q_{(L)} = \Sigma$ quadratic terms.

- Case "With residual stresses" (linear modelling)

$$\{\sigma\}_{failure} = f_{Res} \cdot \{\sigma\}_{(L)} + \{\sigma\}_{(R)}$$

In the case of linear terms, after substitution of the failure causing state of stress one yields

$$F = F(\{\sigma\}_{Failure}) = F(f_{Res} \cdot \{\sigma\}_{(L)} + \{\sigma\}_{(R)}) = 1$$

and $\{\sigma\}_{(R)}$ from curing stresses computation etc.
This procedure can be applied as long as the stresses have not caused an essential amount of damage which would lead to stress-redistribution and a reduction of the size of the residual stresses.

5.3 Determination of resultant reserve factor or rounding-off of failure modes

The *resultant Reserve Factor* (superscript res) takes account of the interactions of the modes. In the case of *linearity* it may be estimated by (figure A1)

$$(1/f_{Res}^{(res)})^{\dot{m}} = (1/f_{Res}^{\perp\sigma})^{\dot{m}} + (1/f_{Res}^{\perp\parallel})^{\dot{m}} + \ldots$$

$$+ \ldots + (1/f_{Res}^{\perp\tau})^{\dot{m}} = f(f_{Res}^{(modes)})$$

with $\dot{m} \leq \min m \approx 1.2$ /max cov (safe side) causing maximum smoothing, by taking the maximum coefficient of variation, cov, of the 5 basic strengths. The value of $\dot{m}$ has to be lowered further by fitting experience, due to the fact that in the interaction zones micromechanical and probabilistic effects will commonly occur and cannot be discriminated (simplifying *assumption*: $\dot{m}$ is the same for each interaction zone).

Figure 4 visualizes in the upper part the evaluation of test data and in the bottom part as well the rounding-off (by the spring model) in the multifold (MfFd) and mixed failure domains (MiFD) as the shrunk design spaces, to be used by the designer in the 'dimensioning' and in the 'proof of design'.
Additionally to the FMC-based Mode Fit the Global Fit (e.g. Tsai-Wu criterion) is pointed out. The Global Fit interacts the UD-stresses *and* the *independent* failure modes in one equation achieving a description of the complete failure surface. This procedure is simple, however error-prone, due to its physical drawbacks.

In order to consider the twofold failure in the (σ_2^t, σ_3^t)-domain the term $(1/f_{Res}^{\perp\sigma})^{\dot{m}}$ has to be made effective (see figure 4) twice.
The numerically simplest way is by including the multifold failure term

$$f_\perp^{MfFd} = 2R_\perp^t /(\sigma_2^t + \sigma_3^t) \quad .$$

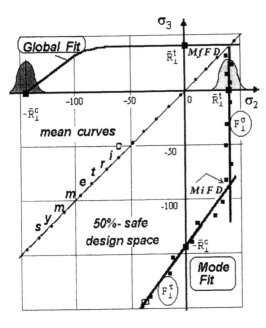

* Test data evaluation

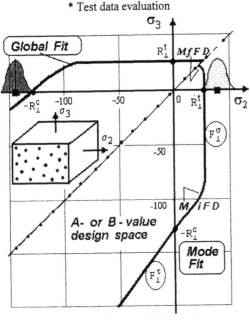

* Rounding-off in the MfFD, MiFD

Fig 4. "Global Fit" and "Mode Fit".
Example: CFRP-IFF-curve of UD-material.
(MiFD: = mixed failure domain = fracture due to 2 modes. MfFD: = multi-fold failure domain of the same mode "Normal Fracture", $NF_\perp$ twice.
A-curve: 99% reliability, 95 % confidence [Mil-Hdbk]
B-curve: 90% / 95%, mean-curve: 50% / 50%)
In A-, B-design space: $\overline{R} \rightarrow$ strength allowable R

5.4 *Application to the UD-lamina (3D-condits.)*

The *mode Reserve Factors* explicitly read

$$\bullet f_{Res}^{\|\sigma} = R_\|^t / v_f \cdot \sigma_{1f} \ \hat{=} \ R_\|^t / \sigma_{eq}^{\|\sigma} = R_\|^t / \varepsilon_1 \cdot E_\|^t$$

or generally $f_{Res}^{(mode)} = R^{(mode)} / \sigma_{eq}^{(mode)}$,

$$\bullet f_{Res}^{\|\tau} = \frac{-R_\|^c}{v_f \sigma_{1f}}; \qquad \bullet f_{Res}^{\perp\sigma} = \frac{2R_\perp^t}{I_2 + \sqrt{I_4}} = \frac{R_\perp^t}{\sigma_{eq}^{\perp\sigma}}$$

$$\bullet f_{Res}^{\perp\|} = R_{\perp\|} / (I_3^{3/2} + b_{\perp\|}(I_2 I_3 - I_5))^{1/3}$$

$$\bullet f_{Res}^{\perp\|} = \frac{R_\perp^c}{2} \frac{(-b_\perp^\tau - 1)I_2 + \sqrt{(b_\perp^\tau - 1)^2 I_2^2 + 4 b_\perp^\tau I_4 + 4 b_{\perp\|}^\tau I_3}}{b_\perp^\tau I_4 + b_{\perp\|}^\tau I_3} .$$

If a $f_{Res}^{(mode)}$ will become negative for numerical reasons a value of +100 shall replace the negative value.

6 EQUIVALENT STRESS, MODE EFFORT

In the case of small FoS (e.g. in spacecraft) *nonlinear* analyses, only, will enable the stress man to predict the stress effort and then the load-based $f_{Res}^{(mode)}$. The actual stress effort of a mode $Eff^{(mode)}$ is the actual portion of the maximum 100 % achieved at mode failure. The procedure of determining the resultant stress effort $Eff^{(res)}$ in each lamina of the laminate is similar to that of $f_{Res}^{(res)}$. The stress effort is related to the reserve factor in the case of *linear* behaviour and zero residual stresses by

$$Eff^{(res)} = 1 / f_{Res}^{(res)} .$$

Also similar to the $f_{Res}^{(res)}$ procedure at first the vector of the modes' equivalent stresses

$$\{\sigma_{equiv.}^{(mode)}\} = (\sigma_{eq}^{\|\sigma}, \sigma_{eq}^{\|\tau}, \sigma_{eq}^{\perp\sigma}; \sigma_{eq}^{\perp\tau}; \sigma_{eq}^{\perp\|})^T$$

will be computed which now includes the nonlinearly load dependent load stresses $\{\sigma\}_{(L)}$ and the equally nonlinearity dependent residual (curing stresses etc.) stresses $\{\sigma\}_{(R)}$.

Then for the resultant stress effort holds

$$Eff^{(res)\bar{m}} = (\sigma_{eq}^{\|\sigma} / \overline{R}_\|^t)^{\bar{m}} + (\sigma_{eq}^{\|\tau} / \overline{R}_\|^c)^{\bar{m}} + (\sigma_{eq}^{\perp\sigma} / \overline{R}_\perp^t)^{\bar{m}} +$$
$$+ (\sigma_{eq}^{\perp\tau} / R_\perp^c)^{\bar{m}} + (\sigma_{eq}^{\perp\|} / \overline{R}_{\perp\|})^{\bar{m}} .$$

In order to consider the interaction of the modes as triggering approach is recommended:

9

- $\sigma_{eq}^{>(mode)} = \sigma_{eq}^{(mode)} \cdot \overset{(res)}{Eff} / \max \overset{(mode)}{Eff} .. ..\Delta\sigma > 0$

  being an influence increase

- $\sigma_{eq}^{<(mode)} = \sigma_{eq}^{(mode)} \cdot \max \overset{(mode)}{Eff} / \overset{(res)}{Eff} .....\Delta\sigma < 0$

  being an influence decrease .

This approach has to be verified - before general acceptance - for all stress combinations possible, of course.

7 DESCRIPTION OF NON-LINEARITY

Nonlinear behaviour of well-designed composites is physically (laminae behaviour) but rarely geometrically (laminate behaviour) caused.

A full 3D-input in stress analysis demands for 5 elastic properties in case of Fibre Reinforced Plastics (FRP) and the strength analysis for 5 strengths. In the 2D-case the required input is 4 elastic properties and 5 strength properties. Further, for the nonlinear stress analysis additionally the relevant nonlinear stress-strain curves are to be provided.

Material *hardening* is defined until the stress reaches its strength value R_m and thereby an initial failure level of IFF type. From that level on, that means for the *progressive failure* or damage regime, the term *softening* is used. Of course damaging still begins with material hardening.

7.1 *Mapping of hardening*

The degree of nonlinearity mainly depends on the nonlinearly behaving matrix material which mainly affects $E_\perp^c$ and $G_{\perp\parallel}$.

For the nonlinear stress analysis the relevant nonlinear stress-strain curves are to be provided. Data for the secant moduli of $E_\perp$, $G_{\parallel\perp}$ applied in the multi-nonlinear analysis may be derived from Ramberg/Osgood mapping of the course of test data which reads (denotations see figure 5)

$$\varepsilon = \sigma / E_{(o)} + 0.002(\sigma / R_{p0.2})^n$$

with the Ramberg/Osgood exponent

$$n = \ell n\big(\varepsilon_{pl}(R_m)\big) / \ell n\big(R_m / R_{p0.2}\big)$$

estimated from the strength point $\big(R_m, \varepsilon_{pl}(R_m)\big)$.

7.2 *Mapping of softening*

Above the *Initial Failure* level an appropriate progressive failure analysis method has to be employed *(a Successive Degradation Model* for the description of post initial failure) by using a failure mode condition that indicates failure type and damage size.

Final Failure occurs after the structure has degraded to a level where it is no longer capable of carrying additional load.

Figure 5 depicts a) for an isolated e.g. tensile coupon specimen in the usual load controlled test, b) in a strain controlled test. This is possible at the institute BAM, Berlin, which has a test rig of a very high frame stiffness. The curve b) is assumed here.

$E_\perp^c$ and $G_{\parallel\perp}$ are reducing gradually rather than being suddenly annihilated. A rapid collapse (ply discount method) of $E_\perp^t$ is unrealistic and leads to convergence problems.

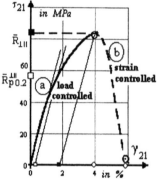

Fig 5. Mapping of measured stress-strain curves of an isolated UD-specimen. Example τ_{21} (γ_{21})

Modelling of *Post Initial Failure* behaviour of a laminate requires that assumptions have to be made regarding the decaying properties of the actuallydegrading lamina or laminae. A simple function was used to map this softening, in order to later derive the secant moduli, (the suffix s denotes softening)

$$\sigma_s = R_m / (1+\exp[(\bar{a} + \varepsilon) / \bar{b}])$$

with two curve parameters $\bar{a}$, $\bar{b}$ to be estimated by the data of two calibration points, e.g.

$(R_m, \varepsilon (R_m))$ and $(R_m \cdot 0.5, \varepsilon (R_m \cdot 0.5))$.

This function practically not has to model the stress-strain curve of an isolated lamina in a strain-controlled test b) but that of a lamina embedded in a laminate c), and thus regarding the altering microcrack density up to the critical damage state (CDS). Figure 6 includes this curve c for the embedded lamina. The curve is an effective curve.

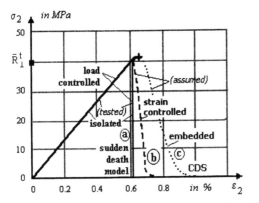

*Fig 6.*The differences in the stress-strain behaviour of isolated and embedded UD-laminae

7.3 Constraint effect of embedded laminae

If applying test data from tensile coupons to embedded laminae, one has to consider that tensile coupon tests deliver test results of *weakest link type* (series model). An embedded or even only one-sided constraint lamina, however, is belonging to the class of *redundant types* (parallel spring model). Therefore, besides being strain-controlled also all the material flaws in a *thin* lamina cannot grow freely up to microcrack size in thickness direction because the neighbouring laminae will act as microcrack-stoppers (fracture mechanics problem, energy release).

In order to consider the constraint effect on a "thin" embedded lamina following (Fla 82) and some own investigations Cuntze formerly recommended e.g. for CFRP ($t_\perp$ = thickness of lateral lamina, $t_{thr} \approx 0{,}35$ mm) as correction for the design allowable as an effective strength value in linear analysis (VDI97)

$$R_\perp{}^t \rightarrow R_\perp{}^t (t_\perp < t_{thr}) = R_\perp{}^t \sqrt{t_{thr}/t_\perp} \ .$$

For the execution of nonlinear analysis the application of an *effective* stress-strain curve is necessary which estimates the behaviour of the lamina in the laminate regarding the stack, its position, and the thickness (figure 7).

In the nonlinear analysis normally mean values are regarded in order to perform stress analysis that corresponds to an average *structural behaviour*. Therefore, when executing a nonlinear *stress* analysis the secant moduli to be utilized are mean values, too (in the *strength* analysis within the *Proof of Design* 'A' or 'B' design allowables have to be regarded).

For simply deriving unique data for the secant moduli two regimes have to be distinguished the regime below and that above $\varepsilon(\overline{R}_m)$.

In order to provide the nonlinear analysis with the input needed normalized stress-strain curves have been constructed with a hardening part measured and a softening part assumed (figure 7).

The peak value of the embedded curve is higher than the strength point $\overline{R}$ of the isolated specimen due to the change from the 'weakest link' behaviour to a redundant behaviour.

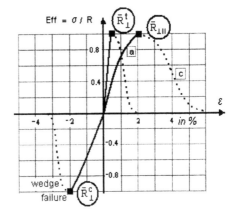

Fig 7. Normalized effective stress-stain curves

For the sake of simplicity this peak effect is flattened in the analytical description of softening.

7.4 Variation of Poisson's ratio

The alteration of Poisson's ratio $v_{\perp\|}$ is linked to the associated failure mode. E.g. in the case of shear failure under compressive lateral stresses the value for $v_{\perp\|}$ will be higher than for tensile lateral stresses. Respecting the low effect Poisson's ratios have if using FRP with stiff fibres the following estimation will be a good approach before mode failure occurs:

$$F_\perp^\sigma: \quad v_{\perp\|} = v_{\perp\|(0)} \cdot E_{\perp(sec)}/E_{\perp(0)}$$

7.5 Choice of different ṁ values

For the reduced Weibull modulus ṁ a constant value was recommended. This might not work if the interaction effects covered by the refined conditions (Cun98a) are replaced by simpler formulations, and the resultant reserve factor formula should be split into several mode interaction formulae replacing the single formula (loss of advantage). Interaction will address two or at maximum three modes.

The advantage of this measure would be the possibility of accounting for different values in respect of the different interaction effects in the different mode interaction domains.

7.6 Remarks on design and modelling

In composite structures composed of stiff fibres and well-designed by netting theory the fibre net controls the strain behaviour.

The FMC considers the interlaminar stresses and classifies the failure modes. Therefore, associated degradation models are inherent making gradual degradation of the affected property possible.

In order to properly design a laminate not only verified failure conditions have to be available but also proper stresses have to be analytically provided. Therefore, analogous to isotropic materials the nonlinear stress-strain curves have to be taken into account below reaching initial failure.

Above the initial failure level an appropriate progressive failure analysis method has to be employed by taking a *Successive Degradation Model* and by using a failure mode condition that indicates failure type and damage size.

Final failure occurs after the structure has degraded to a level where it is no longer capable of carrying additional load. This is most often caused by FF, however in specific cases by IFF, too. An inclined wedge-shaped inter-fibre crack caused by $F_\perp^\tau$ can lead to final failure (Puc96).

Multidirectional laminates usually are still capable of carrying load beyond *initial failure* which usually is determined by IFF.

8 CALCULATION PROCEDURE

Figure 8 presents the scheme of the nonlinear calculation. The solution procedure of the nonlinear analysis is to establish static equilibrium on each load step after material properties have been changed. For each iteration the procedure is repeated until convergence or total failure is reached.

By employing the equivalent stress reached in each failure mode the associated secant modulus of each mode was determined for the hardening and the

Input:
- Geometry, temp. load, boundary conditions
- Nonlinearity descriptions

Load step 0:
- Initial mechanical properties, increment 0
- $\{\sigma\}_o, \{\varepsilon\}_o \rightarrow \left\{\sigma_{equiv}^{(modes)}\right\}_o$,
- determination of degraded secant moduli
- secant moduli from load step 0
- determination of associated Poisson's ratios

Load step 1
- $\{\sigma\}_1, \{\varepsilon\}_1 \rightarrow \left\{\sigma_{equiv}^{(modes)}\right\}_1$,
- determination of degraded secant moduli
- iterations of static equilibrium $\quad\uparrow$

Load step 2: And so further

Figure 8. Nonlinear calculation scheme chosen

softening regime.

According to a consistent stress concept for all $\sigma_{eq}^{(modes)}$ an explicit dependency $E_{sec}(\sigma_{eq}^{(mode)})$ has to be provided. For reasons of achieving such an explicit formulation two separate formulae are discriminated which are linked in the strength point. This automatically respects that the nonlinear calculation procedure chosen demands for the dependencies of the secant moduli on the corresponding equivalent stress for, (example $F_\perp^\tau$):

- $\Delta\sigma > 0$ (increasing stress for hardening regime)

$$E_{\perp(sec)}^t = E_{\perp(o)}^t$$

$$E_{\perp(sec)}^c = E_{\perp(o)}^c / [1+0.002 \cdot (E_{\perp(o)}^c / R_{p0.2}^{\perp c}) \cdot$$
$$\cdot (\sigma_{eq}^{\perp\tau} / R_{p0.2}^{\perp c})^{n_\perp^c - 1}$$

$$G_{\|\perp(sec)} = G_{\|\perp(o)} / [1+0.002(G_{\|\perp(o)} / R_{p0.2}^{\perp\|}) \cdot$$
$$[\sigma_{eq}^{\perp\|} / R_{p0.2}^{\perp\|})^{n_{\perp\|} - 1}]$$

- $\Delta\sigma < 0$ (softening regime)

$$E_{\perp(sec)}^t = \sigma_{eq}^{\perp\sigma} / \varepsilon(\sigma_{eq}^{\perp\sigma})$$
$$= (\sigma_{eq}^{\perp\sigma} / b_s^{\perp t}) / \left[\ell n(\frac{R_\perp^t - \sigma_{eq}^{\perp\sigma}}{\sigma_{eq}^{\perp\sigma}}) - \frac{a_s^{\perp t}}{b_s^{\perp t}} \right].$$

For the further modes the same formula is valid, however, the mode parameters are different.

If the laminate's stiffness matrix is recomputed after each step of damage increase the laminate's damage evolution may be continuously computed.

This includes the decrease of the residual stresses versus increasing damage or nonlinearity.

9 APPLICATION TO A LAMINATE

9.1 Definition of test cases

In table 3 the properties for a CFRP and a GFRP laminate are outlined. Table 4 provides with the initial and final failure envelopes to be computed .

9.2 Assumptions and remarks for the plots

- Thermal stresses as a result of the curing process are included
- *Post-initial failure* considered by gradually degraded properties of embedded laminae (no Sudden Death of the failed lamina). The course of the *softening* (suffix d) is assumed.
- First FF is *final failure*. The two FF $F_{\parallel}^{\sigma}$ (tensile fibre failure) and $F_{\parallel}^{\tau}$ (shear instability, buckling), and further, the IFF $F_{\perp}^{\tau}$ are defined to cause *final failure*.
- Failure mode identification is inherent to the Failure Mode Concept
- Parameters $\dot{m}$, $b_{\perp}^{\tau}$, $b_{\perp\parallel}$, and $b_{\perp\parallel}^{\tau}$ are roughly assumed for the examples.
- Due to the lack of information on the moisture content (see also annex AIV) instead of the difference *stress free* temperature to *room* temperature 22°C an *effective* temperature difference (table 1) is applied in order to consider the combined effect of all residual stresses.
- The stress-strain curves of the UD-lamina are interpreted load-based macro-mechanical stresses.
  The micromechanical curing stresses (residual stresses of the 2nd kind at filament/matrix level) are not considered. It is assumed that the stress-strain curves are *mean* curves, the type one needs for test data mapping (see figures 4)
- Edge effect not considered (3D state of stress), ('closed' structure assumed)
- The progressive behaviour of $E_{\parallel}^{t}$ in the case of C-fibres was not regarded.
- The loading is monotonic.

For the computation of the two test cases the following failure conditions may be employed:

$$\frac{v_f \cdot \sigma_{1f}}{\text{Eff} \cdot \overline{R}_{\parallel}^{t}} = 1; \qquad \frac{-\sigma_1}{\text{Eff} \cdot \overline{R}_{\parallel}^{c}} = 1,$$

$$\frac{\tau_{21}^{3} + b_{\perp\parallel} 2\sigma_2\tau_{21}^{2}}{(\text{Eff} \cdot \overline{R}_{\perp\parallel})^{3}} = 1; \qquad \frac{\sigma_2}{\text{Eff} \cdot \overline{R}_{\perp}^{t}} = 1$$

$$\frac{(b_{\perp}^{\tau} - 1)(\sigma_2 + \sigma_3)}{\text{Eff} \cdot \overline{R}_{\perp}^{c}} + \frac{b_{\perp}^{\tau}(\sigma_2 - \sigma_3)^{2} + b_{\perp\parallel}^{\tau}\tau_{21}^{2}}{(\text{Eff} \cdot \overline{R}_{\perp}^{c})^{2}} = 1$$

wherein $\sigma_3 = -p_{ex}$ is to be inserted in the case of tube specimens loaded by external pressure p_{ex} and for flat specimens $\sigma_3 = 0$

The consideration of $\sigma_3 = -p_{ex}$ shifts the biaxial strength capacity to higher negative values.
The equivalent stress, building up the denominators, was defined by

$$\text{Eff} \cdot R = \sigma_{eq} .$$

The residual stress is included in σ_{eq}. It is decaying with decreasing stiffness and has to be superimposed to the load stress according to

$$\{\sigma\} = \{\sigma\}_{(L)} + \{\sigma\}_{(R)} .$$

9.3 Stress-strain curves of the UD-lamina

On the following figures 10 through 14 as well the course of the test data (solid lines) is depicted as the assumed curves for the embedded UD-lamina (dotted curve):

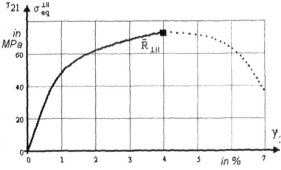

Fig. 9. In-plane shear stress-strain curve $\tau_{21}(\gamma_{21})$; UD-lamina (softening parameters assumed) GFRP: E-glass/MY750/HY917/DY063.
$\overline{R}_{\perp\parallel} = 73$ MPa, $\overline{G}_{\parallel\perp 0} = 5.83$ GPa;
$n^{\perp\parallel} = 6.6$; $a_s^{\perp\parallel} = -7.0\%$, $b_s^{\perp\parallel} = 0.53\%$

Table 3: Mechanical and thermal properties of two UD-laminae of the 'failure exercise' [Sod98]

Fibre		AS4	Silenka E-Glass 1200tex
Matrix		3501-6 epoxy	MY750/HY917/DY063 epoxy
Specification		Prepeg	Filament winding
Manufacturer		Hercules	DRA
Fibre volume fraction,	V_f	0.60	0.60
Longitudinal modulus, [GPa]	$E_{\parallel}$	126[a]	45.6
Transverse modulus, [GPa]	$E_{\perp}$	11	16.2
In-plane shear modulus, [GPa]	$G_{\parallel\perp}$	6.6[a]	5.83[a]
Major Poisson's ratio	$\nu_{\perp\parallel}$	0.28	0.278
Through thickness Poisson's ratio	$\nu_{\perp\perp}$	0.4	0.4
Longitudinal tensile strength [MPa]	$R_{\parallel}{}^t$	1950[b]	1280
Longitudinal compressive strength, [MPa]	$R_{\parallel}{}^c$	1480	800
Transverse tensile strength, [MPa]	$R_{\perp}{}^t$	48	40
Transverse compressive strength, [MPa]	$R_{\perp}{}^c$	200[b]	145[b]
In-plane shear strength, [MPa]	$R_{\perp\parallel}$	79[b]	73[b]
Longitudinal tensile failure strain, [%]	$e_{\parallel}{}^t$	1.38	2.807
Longitudinal compressive failure strain [%]	$e_{\parallel}{}^c$	1.175	1.754
Transverse tensile failure strain [%]	$e_{\perp}{}^t$	0.436	0.246
Transverse compressive failure strain [%]	$e_{\perp}{}^c$	2.0	1.2
In-plane shear failure strain [%]	$\gamma_{\perp\parallel}$	2	4
Strain energy release rate, [Jm^{-2}]	G_{IC}	220[c]	165
Longitudinal thermal coefficient, [10^{-6}/°C]	$\alpha_{T\parallel}$	-1	8.6
Transverse thermal coefficient, [10^{-6}/°C]	$\alpha_{T\perp}$	26	26.4
Curing: Stress free temperature [°C]		177	120
effective temperature difference [°C]		-125[e]	-68

[a] Initial modulus; [c] Double cantilever specimen; [d] assumption: linearized, reference temperature = RT = 22°C;
[b] Nonlinear behaviour, stress/strain curves and data points are provided; [e] -177 + RT + 30 (moisture effect) = -125°C

Table 4: Summary of laminate types, material types and plots required from contributors ($\hat{\sigma}_x = n_x / t$; $\hat{\sigma}_y = n_y / t$)

Laminate type	Material type	Plots required, loading conditions
• 0° unidirectional lamina (isolated)	E-glass/MY750/HY917/DY063 AS-4 /3501-6	σ_2 vs τ_{21} failure stress envelopes
		σ_1 vs τ_{21} failure stress envelopes
		σ_2 vs σ_1 failure stress envelopes
• [90/45/-45/0]$_s$ laminate t = 1.1 mm, t_k = t / 8	AS4/3501-6 (quasi-isotropic, widely used)	$\hat{\sigma}_y$ vs $\hat{\sigma}_x$ failure stress envelope
		Stress/strain curve $\hat{\sigma}_y / \hat{\sigma}_x = 1/0$
		Stress/strain c. for $\hat{\sigma}_y / \hat{\sigma}_x = 2/1$
• [+55/-55]$_s$ angle-ply laminate t = 1.0 mm, t_k = t / 4 (angle measured from x-axis)	E-glass/MY750/HY917/DY063 (piping, pressure vessels)	$\hat{\sigma}_y$ vs $\hat{\sigma}_x$ failure stress envelope
		Stress/strain curve $\hat{\sigma}_y / \hat{\sigma}_x = 0/1$
		Stress/strain c. for $\hat{\sigma}_y / \hat{\sigma}_x = 2/1$

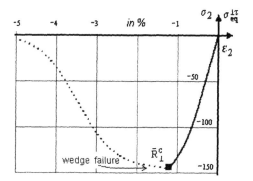

Fig. 10. Transv. compr. stress/strain curve $\sigma_2^c(\varepsilon_2)$;
UD-lamina (softening assumed).
GFRP: E-glass/MY750/HY917/DY063.
$\overline{R}_\perp^c = 145$ MPa, $\overline{E}^c{}_{\perp 0} = 16.2$ GPa;
$n^{\perp c} = 6.6; a_s^{\perp c} = -3.47\%, b_s^{\perp c} = 0.47\%$

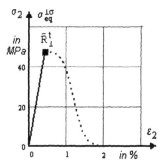

Fig. 11. Transv.tensile stress/strain curve $\sigma_2^t(\varepsilon_2^t)$;
UD-lamina (softening assumed).
CFRP: AS4/3501-6 epoxy.
$\overline{R}_\perp^t = 48$MPa, $\overline{E}_{\perp 0}^t = 11$GPa; $a_s^{\perp t} = -1.18\%, b_s^{\perp t} = 0.15\%$

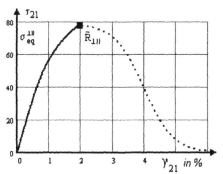

Fig. 12. In-plane shear stress-strain curve $\tau_{21}(\gamma_{21})$;
UD-lamina (softening assumed)
CFRP: AS4/3501-6 epoxy.
$\overline{R}_{\perp \parallel} = 79$ MPa, $\overline{G}_{\parallel \perp 0} = 6.6$ GPa;
$n^{\perp \parallel} = 5; a_s^{\perp \parallel} = -4.0\%, b_s^{\perp \parallel} = 0.46\%$

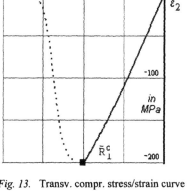

Fig. 13. Transv. compr. stress/strain curve $\sigma_2^c(\varepsilon_2)$;
UD-lamina (softening assumed).
CFRP: AS4/3501-6 epoxy.
$\overline{R}_\perp^c \approx 200$ MPa, $\overline{E}^c{}_{\perp 0} = 11$ GPa;
$n^{\perp c} = 5; a_d^{\perp c} = -2.74\%, b_d^{\perp c} = 0.12\%$

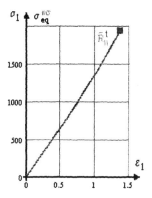

Fig. 14. Longit. tensile stress/strain curve $\sigma_1^t(\varepsilon_1)$;
UD-lamina.
CFRP: AS4/3501-6 epoxy.
$\overline{R}_\parallel^t = 1950$ MPa, $\overline{E}^t{}_{\parallel 0} = 126$ GPa;

9.4 Biaxial failure envelopes for the UD-lamina

In the following UD failure envelopes (figures 15 through 18) the *residual stresses* are not regarded (due to a lack of information), only the *load stresses* from the mechanical load test are considered. Ramberg/ Osgood exponent and assumed softening parameters are added. The course of the tested curves shown has been verified by own tests and tests cited in the literature (ZTL80, VDI97, Kna72, ...).

15

Depicted are several cross-sections of the five-dimensional IFF-body: The graph (τ_{21}, σ_2) represents the IFF-responsible stresses in the plane of the lamina, the graph (τ_{31}, σ_2) outlines that τ_{31} has not the same action plane as σ_2^t. In the graphs (σ_2, σ_3) and (τ_{23}, σ_2) fracture excellently is described by the homogenized stresses.

The graph (σ_2, σ_1) shows the limited applicability of the homogenized lamina stresses, because σ_1 or I_1 is not the fracture

stress. This is the fibre stress σ_{1f}. In order to remain on composite level in the graph σ_{1f} has to be multiplied by the fibre colume fraction (approach: $\sigma_{1f} \cdot v_f \approx \sigma_1$).

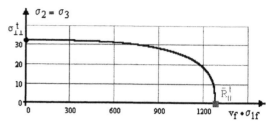

Fig. 17. Zoom of Fig. 18

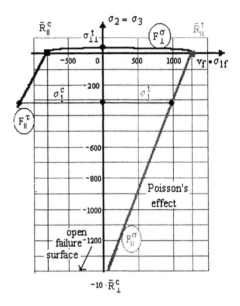

Fig. 18. Biaxial failure stress envelope $(\sigma_2 = \sigma_3, v_f \cdot \sigma_{1f})$; $\sigma_{\perp\perp}^t \approx R_\perp^t / \sqrt[m]{2}$ (Cun96) in MPa. UD-lamina E-glass / MY750 epoxy.

$b_\perp^\tau = 1.56, b_{\perp\parallel} = 0.12, \ b_{\perp\parallel}^\tau = 0.4, \ \dot{m} = 3.1$

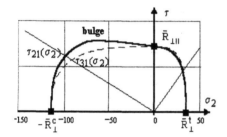

Fig. 15. Biaxial failure stress envelope (τ_{21}, σ_2) and (τ_{31}, σ_2) in MPa. UD-lamina .
GFRP: E-glass MY 556 epoxy (VDI97, Kna72)

$b_\perp^\tau = 1.5, b_{\perp\parallel} = 0.13, \ b_{\perp\parallel}^\tau = 0.4, \ \dot{m} = 3.1$

9.5 Initial and final biaxial failure envelopes

For the determination of the failure envelopes (see figures 19, 20) nonlinear CLT and softening were applied.

In order to simplify the progressive failure computations (otherwise a higher programming effort is necessary) $b_{\perp\parallel}$ may be set zero with the effect that the bulge is not considered. Furthermore, in $F_\perp^\tau$ the first term (friction) and $b_{\perp\parallel}^\tau$ may be deleted.

Fig. 19 incorporates the initial and the final failure envelope of this GFRP-laminate. In the positive

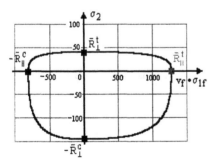

Fig. 16. Biaxial failure stress envelope $(\sigma_2, v_f \cdot \sigma_{1f})$ in MPa. UD-lamina.
E-glass / MY750 epoxy (ZTL80)

$b_\perp^\tau = 1.56, b_{\perp\parallel} = 0.12, \ b_{\perp\parallel}^\tau = 0.4, \ \dot{m} = 3.1$

In the domain $\sigma_2^c = \sigma_3^c > -10 R_\perp^c$ the IFF-curves due to Poisson's effect will be closed by $F_\parallel^\sigma$.

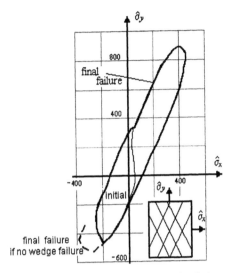

Fig. 19. Initial and final failure envel. $\hat{\sigma}_y(\hat{\sigma}_x)$.
$[+55/-55]_s$-laminate, E-glass / MY750 epoxy
$b^\tau_\perp = 1.5, b_{\perp\parallel} = 0.13, b^\tau_{\perp\parallel} = 0.4, \dot{m} = 3.1, \hat{\sigma}_y :$ = average
hoop stress of the laminate

quadrant there are corners. They become smoothed
due to the effect of high interaction of $F^\sigma_\parallel$ in
adjacent laminae which was not performed here.

In the negative quadrant wedge failure may occur.
The event of a wedge failure is equal to the onset of
delamination damage. In the case of a plane
specimen with an antibuckling device, applied when
testing in the compression regime, the wedge will
slide and then cause a compressive reaction σ_3^c
normal to the lamina's plane onto the adjacent
laminae. This will induce delamination or might
increase the original delamination size. However, in
the case of tube specimens loaded by external
pressure p as well the multiaxial strength is
increased (σ_3 = -p is acting in a favourable manner)
as – after fracture – the sliding friction due to p is
increased similarly until its maximum will be
exceeded.
A correct analysis of the boundary conditions and
the stress state of the test specimen is mandatory.

Fig. 20 depicts the symmetrical failure envelopes of
this CFRP laminae. Also here, the sharp corners still
have to be rounded-off in a refined procedure taking
the joint failure probability of the laminate (Cun87)
into account. In the neg. quadrant IFF covers FF.

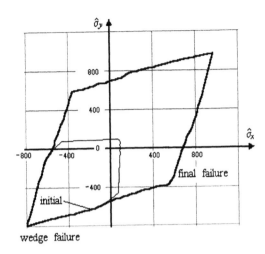

wedge failure

Fig.20. Initial and final failure envel. $\hat{\sigma}_y(\hat{\sigma}_x)$.
$[90/+45/-45/0]_s$-laminate, AS4/3501-6
$b^\tau_\perp = 1.5, b_{\perp\parallel} = 0.13, b^\tau_{\perp\parallel} = 0.4, \dot{m} = 3.1$

9.6 Stress-strain curves of the laminates

Figure 21 outlines the deformation behaviour of a
pressure vessel, which usually is designed for one
special load case inner pressure that means for
$\hat{\sigma}_y / \hat{\sigma}_x = 2:1$. Load combinations outside of this
ratio may lead to high shear strains and to a limit of
usage. This *design limit* was assumed here to be 4 %
shear strain.
Due to curing/moisture stresses the graphs do not
begin in the origin.

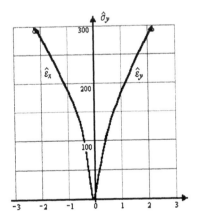

Fig. 21. Stress-strain curves $\hat{\sigma}_y : \hat{\sigma}_x = 1:0$.
$[+55/-55]_s$-laminate, E-glass / MY750;
$b^\tau_\perp = 1.5; b_{\perp\parallel} = 0.13; b^\tau_{\perp\parallel} = 0.4, \dot{m} = 3.1$

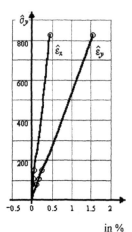

in %

Figure 22. Stress-strain curves $\hat{\sigma}_y : \hat{\sigma}_x = 2:1$.
$[90/+45/-45/0]_s$-laminate AS4/3501-6 epoxy

Figure 22 depicts a quasi-isotropic laminate, the stress-strain relationship of which is fibre-dominated and therefore linear after the 'knee' caused by initiation of IFF.

10 SOME CONCLUSIONS, OUTLOOKS

10.1 *Regarding the FMC-based conditions*

- A general concept was highlighted for the establishment of Failure Conditions (F = 1) for *Initial Failure* (IFF) of the lamina and Final Failure of the laminate
- The *complete* failure surface consists of piecewise smooth regimes (partial failure surfaces). Each regime represents *1 failure mode* and is governed by *1 basic strength*
- Sufficient for pre-dimensioning are the basic strengths R.
  The remaining unknown *curve parameters*
  $b_{\perp\|}$, $b_\perp^\tau$, $b_{\perp\|}^\tau$ can be estimated
- The *interaction* (rounding-off) *of adjacent failure modes* is automatically considered when calculating the stress effort $\mathrm{Eff}^{(res)}$ as function of the mode efforts $\mathrm{Eff}^{(modes)}$
- The concept enables to correctly turn the design screw respecting the most critical mode and location (Cun98b)
- The solution procedure worked, both, in the failure exercise code [Mathcad] and in the FEM implementation [Marc/Mentat, (Cun98)]
- Homogenization of the UD-material comes to its

limit if a *constituent* stress governs the failure. This is the case for $F_\|^\sigma$ where the macromechanical stress σ_1 has to be replaced by the actual fibre stress σ_{1f}. A fibre stress may be not zero even for zero σ_1. Therefore, it has to be estimated by $\sigma_{1f} = \varepsilon_1 \cdot E_{1f}$. In order to remain on *composite* stress visualization level σ_{1f} will be multiplied by the fibre volume fraction v_f

- The 'mode fit' avoids the shortcomings of the 'global fit' which maps the course of test data by mathematically linking failure modes which are mechanically not linked. One typical shortcoming is that a reduction of the strength of one mode will increase the multiaxial strength in another (independent) mode.
- For the prediction of final failure the initial failure approach is not of that high concern, if wedge failure, caused by $F_\perp^\tau < 1$, followed by delamination failure, will not occur (see A. Pucks drive shaft or torsion spring)
- Each failure condition describes the interaction of stresses affecting the same failure mode and assesses the actual state of stress in a 'material point'
- For (σ_2, σ_3) states of stress as well Mohr's stresses, Mohr's envelope curve as the inclined fracture angle may be determined
- Damage mechanics is captured in the FMC conditions so far as the stiffness reduction is determinable via the σ_{eq}-strain curve, and by the predictability of delamination initiation, applying $F_\perp^\tau$
- Regarding the investigations in theory and test carried out in Germany in the last years (still going on) on the *lamina* material level the understanding has improved a lot and seems to be a good basis to tackle *laminates* stacked-up of UD-lamina and fabric lamina. For other textile preforms (3D, stitched etc.) engineering models have to be developed.
- The transferability to rhombically-orthotropic composites (fabrics) works.

10.2 *Regarding progressive failure analysis*

- Degradation modelling looks promising, however, research has to verify the effective softening curves for embedded laminae. Appropriate test specimens and test evaluation have to be discussed
- *Probabilistic* tools should be applied in order to improve the *deterministic* procedures and to smooth the sharp corners of a *laminate's* failure envelope.

- In order to implement the multifold nonlinearity approach into a commercial FEM code and taking advantage of its solution architecture model flow rules have to be provided.
- For stress *concentration* loci in the laminate such as bolt holes an engineering approach has to be provided
- In the case of stress *intensity* (after delamination onset) fracture mechanics has to be employed
- The definition of the terms damage, failure, failure modes, flaws, defect, imperfection, etc. has to be urgently worked out in order to generate a common understanding in the composite community (ICCM11 panel).

10.3 *Industrial needs, areas of further work*

- Failure conditions which are independent of the laminate configuration
- Improved world-wide standardization encouraged by manufacturers, technical associations and authorities
- Industry has to cope with damage and the Proof of Design (justification) of damaged structures or damaged laminates, respectively.
- Industry asks to replace the expensive 'Make and Test' design development procedure by verified failure conditions working as engineering models.
- The limitations of the failure conditions are to be indicated
- Good multiaxial tests in the frame of a coordinated research programme with an exchange of results have to be focussed. Probabilistic effects of the lamina stack have to be included in the analysis (Cun93, Rac97)
- Verification of engineering approaches on FEA-output level is necessary
- A practical 'progressive failure analysis procedure' has to be provided for the designer
- RAMS (reliability, availability, maintainability, safety of human beings) needs the many failure modes for its various analyses.
- An accurate failure prediction firstly consists of a physically nonlinear stress analysis if the corresponding lamina stresses are high and of a geometrically nonlinear analysis if the laminate extremely will deform. If the computed lamina stresses do not repect this, the second part, the strength analysis cannot work well.
- Recommended NDI methods for damage detection
- Criteria for the assessment of damage size and criticality of delamination
- Design guideline for improving damage tolerance.

The treatment of fatigue and damage growth has to be enhanced on failure mode basis
- The lamina is the general building brick (basic computational element) for the prediction of laminate behaviour
- Not only 2D/3D-*strength* analysis by failure conditions are not verified as they should be, but also the 3D-*stress* analysis of laminate shells by commercial FEA codes are still too time-consuming. First steps (e.g. Rol97) indicate industry applicability.

10.4 *Regarding the 'failure exercise'*

The work to understand and predict failure of laminates must be continued, however mind in the failure exercise, both parts of a failure theory are commonly compared: the physically and/or geometrically nonlinear stress analysis together with the failure conditions. Therefore it is not a competition of the predictive capabilities of the failure conditions, only. And further, the judging of the failure theories without viewing the nonlinear stress analysis is just the half story.

- If a part of the predicted initial or final failure envelope might exceed the test results this may be caused by the many assumptions to be made, also
- Leakage of fluid taken as test measure for indicating IFF may overpredict
- The application of external pressure to tubes (Cun 93) asks for a 3D treatment of the strength problem. A 2D-treatment underpredicts the multiaxial strength capacity.
- The resulting curves generated for the failure exercise' are similar to Puck's curves.

10.5 *Regarding the discussions at DURACOSYS 99*

In reaction to the discussions at DURACOSYS 99 and at the 'ICCM11 panel on composite failure conditions 1997' has to be noted:
- The joint failure probability of highly stressed adjacent laminae needs to be investigated by means of system reliability of the laminate consisting of the system components laminae
- *Stress*-based failure conditions which include the residual stresses can predict failure, only. A stress may be acting even in the case of zero strain (temperature loading).
Simple *strain*-based failure conditions based on long practical experience are useful tools, e.g. the 0.4% strain limit approach for dimensioning.

- The failure conditions generated are models, only, which approximate the material behaviour on the macro-mechanical level. Micro-mechanics has to be applied in order to gain a better mechanical understanding.

ACKNOWLEDGEMENT

The author wishes to thank Mr. A Freund for the programming in MATHCAD and the time-consuming nonlinear calculations of the test cases. As well Dr. J. Broede's support for the mapping of the degradation is gratefully acknowledged as Dr. R. Beck-Teran's comments on nonlinear calculations. Prof. A. Puck is thanked for the commonly formulated comparison of the coincidences and differences of Puck's and Cuntze's UD-strength conditions.

LITERATURE

(Awa78) Awaji, H. and Sato, S.: A Statistical Theory for the Fracture of Brittle Solids under Multiaxial Stresses. Intern. *Journal of Fracture 14 (1978), R13-16*

(Boe85) Boehler, J.P.: Failure Criteria for Glass-Fiber Reinforced Composites under Confining Pressure. *J. Struct. Mechanics 13, 371*

(Boe95) Boehler, J.P.: Personal note to author on Fabric Invariants, 1995

(Chr97) Christensen, R.M.: Yield Functions/Failure Criteria for Isotropic Materials. *Proc. R. Soc. Lond. A (1997) 453, 1473-1491*

(Chr97a) Christensen, R.M.: Stress based Yield/Failure Criteria for Fiber Composites. *Int. J. Solids Structures 34. (1997), no. 5, 529-543*

(Chr98) Christensen, R.M.: The Numbers of Elastic Properties and Failure Parameters for Fiber Composites. *Transactions of the ASME, Vol. 120 (1998), 110-113*

(Cun87) Cuntze, R.B.: "Failure Path Analysis of Multilayered Fibre Reinforced Plastic Components with the Reliability Calculation Programme FRPREL". *Noordwijk, Oct. 1987*

(Cun93) Cuntze, R.G.: Deterministic and Probabilistic Prediction of the Distribution of Inter-Fibre Failure Test Data of Prestrained CFRP Tubes composed of Thin Layers and loaded by radial pressure. Wollongong. *Advanced Composites '93, 579-585. The Minerals, Metals & Materials Society, 1993*

(Cun96) Cuntze, R.G.: "Fracture-type Strength Criteria" formulated by Invariants which consider the Materials Symmetries of the Isotropic/Anisotropic Material used. Conf. on Spacecraft Structures, Materials and Mechanical Testing. ESA-CNES-DARA: Noordwijk, March 1996 (*Conf. Hdbk*)

(Cun97) Cuntze, R.G.: Evaluation of Multiaxial Test Data of UD-laminae by so-called "Fracture Type Strength Criteria" and by supporting Probabilistic Means. *ICCM-11*, Gold Coast, Australia, 1997

(Cun98) Cuntze, R.G. and Sukarie, G.: Effective Dimensioning of 3D-stressed UD-laminae on Basis of Fracture-type Strength Criteria. Int. conf. on Mechanics of Composite Materials. Riga, April 20-23, 1998. *Conference handbook*, Presentation

(Cun98a) Cuntze, R.G.: The Failure Mode Concept - A new comprehensive 3D-strength Analysis Concept for Any Brittle and Ductile behaving Material. Europ. Conf. on Spacecraft Structures, Materials and Mechanical Testing. ESA-CNES-DGLR-DLR; Braunschweig, Nov. 1998, *ESA SP-428*, 269-287

(Cun98b) Cuntze, R.G.: Strength Prediction for Multiaxially Loaded CMC-Materials. Conf. on Spacecraft Structures, Materials and Mech. Testing. ESA-ESTEC: Noordwijk, March 1998 (*Conf. Hdbk*)

(Cun98c) Cuntze, R.G.: Application of 3D-strength criteria, based on the so-called "Failure Mode Concept", to multiaxial test data of sandwich foam, concrete, epoxide, CFRP-UD lamina, CMC-Fabric Lamina. *ICCE/5*, Las Vegas, July 1998 (presentation)

(Cun99) *Cuntze, R.G.:* Progressive Failure of 3D-stressed Laminates: Multiple Nonlinearity treated by the Failure Mode Concept (FMC). DURACOSYS99, Brussels, July 1999 (to be published by Balkema)

(Fla82) Flaggs, D.L. and Kural, M.H.: "Experimental Determination of the In Situ Transverse Lamina Strength in Graphite Epoxy Laminates". *J. Comp. Mat. Vol 16 (1982), S. 103-116*

(Gol66) Goldenblat, I.I., Kopnov, V.A.: Strength of Glass-reinforced Plastics in the complex stress state. *Polymer Mechanics of Mechanical Polimerov, Vol. 1 1966, 54-59*

(Gri89) Grimmelt, M. and Cuntze, R.G.: Probabilistic Prediction of Structural Test Results as a Tool for the Performance Estimation in Composite Structures Design. Beuth Verlag, *VDI-Bericht 771 (1989), 191-200*

(Har93) Hart-Smith, L.J.: An Inherent Fallacy in Composite Interaction Failure Curves. *Designers Corner, Composites 24 (1993), 523-524*

(Has80) *Hashin, Z.:* Failure Criteria for Unidirectional Fibre Composites. *J. of Appl. Mech. 47 (1980), 329-334*

(Hin97) Hinton, M.J., Soden, P.D. and Kaddour, A.S.: Comparison of Failure Prediction Methods for Glass/Epoxy and Carbon/Epoxy Laminates under Biaxial Stress. *ICCM11, Vol. V, 672-682 (1997)*

(Hin98) Hinton, M.J., Soden, P.D. and Kaddour, A.S.: Failure Criteria in Fibre-Reinforced-Polymer Composites. Special Issue, *Composites Science and Technology 58 (1998)*

(Huf95) Hufenbach, W. and Kroll, L.: A New Failure Criterion Based on the Mechanics of 3-Dimensional Composite Materials. *ICCM-10*, Whistler, Canada, 1995

(Jel96a) Jeltsch-Fricker, R.: Bruchbedingungen vom Mohrschen Typ für transversal-isotrope Werkstoffe am Beispiel der Faser-Kunststoff-Verbunde. *ZAMM 76 (1996), 505-520*

(Jel96b) Jeltsch-Fricker, R. und Meckbach, S.: "Fast Solver of a Fracture Condition According to Mohr for Unidirectional Fibre-Polymer Composite" (German*), Scripts of the University of Kassel on Appl.Math.*, Pre-print No. 1/96

(Jel99) Jeltsch-Fricker, R. and Meckbach, S.: A parabolic Mohr Fracture Condition in Invariant Formulation for Brittle Isotropic Materials (in German). *ZAMM, 79 (1999), 465-471*

(Kna72) Knappe, W., Schneider, W.: "Bruchkriterien für unidirektionalen Glasfaser/Kunststoff unter ebener Kurzzeit- und Langzeitbeanspruchung". *Kunststoffe, Bd. 62, 1972, 864-868*

(Kop96) Kopp, J. and Michaeli, W.: Dimensioning of Thick Laminates using New IFF Strength Criteria and some Experiments for their Verification. Proceedings "Conf. on Spacecraft Structures Materials and Mechanical Testing", ESA, 27-29 March 1996

(Kop99) Kopp, J. and Michaeli, W.: The New Action Plane related Strength Criterion in comparison with Common Strength Criteria, Proceedings of ICCM-12, Paris, France, July 1999

(Mal67) Malmeister, A.: Geometry of Theories of Strength. *Polymer Mechanics, 2, 1967, 324-331*

(Mec98) Meckbach, S.: Invariants of Cloth-reinforced Fibre Reinforced Plastics. *Kasseler Schriften zur angewandten Mathematik, Nr. 1/1998 (in German)*

(MIL17) Plastics for Aerospace Vehicles. Vol I "Reinforced Plastics"; Vol. II; Vol. III "Utilization of Data." Dep. of Defence (DOD), USA

(Moh00) Mohr, O.: Welche Umstände bedingen die Elastizitätsgrenze und den Bruch eines Materials? *Civilingenieur XXXXIV (1900), 1524-1530, 1572-1577*

(Pau68) Paul, B.: Generalized Pyramidal Fracture and Yield Criteria. *Int. J. Solids Structures 1968, 175-196*

(Puc69a) Puck, A.: Calculating the strength of glass fibre/plastic laminates under combined load. *Kunststoffe, German Plastics*, 1969, 59, 18-19 (German test pp. 780-787).

(Puc69b) Puck, A. and Schneider, W.: On failure mechanisms and failure criteria of filament-wound glass-fibre/resin composites. *Plast. Polym.*, 1969, Feb., 33-43.

(Puc92a) Puck, A.: Praxisgerechte Bruchkriterien für hochbeanspruchte Faser-Kunststoffverbunde. *Kunststoffe 82 (1992) 2, S. 149-155* (Fracture Criteria for highly Stressed Fibre Plastics Composites which Meet Requirements of Design Practice. *Kunststoffe German Plastics 82 (1992) 2, p.36-38*)

(Puc92b) Puck. A.: Faser-Kunststoff-Verbunde mit Dehnungs-oder Spannungs-Kriterien auslegen? *Kunststoffe 82 (1992) 5, S. 431-434* (Should fibre-Plastics Composites be Designed with Strain or Stress Criteria? *Kunststoffe German Plastics (1992) 5, P. 34-36*)

(Puc92c) Puck, A.: Ein Bruchkriterium gibt die Richtung an. *Kunststoffe 82 (1992) 7, S. 607-610* (A failure criterion shows the Direction – Further Thoughts on the Design of Laminates-. *Kunststoffe German Plastics 82 (1992) 7, p. 29-32*)

(Puc96) Puck, A.: Festigkeitsanalyse von Faser-Matrix-Laminaten - Modelle für die Praxis -. München: *Carl Hanser Verlag*, 1996

(Puc97) Puck, A.: "Physically based IFF-criteria allow realistic strength analysis of fibre-matrix-laminates" (German), *Proceedings of the DGLR-Conference 1996*, Ottobrunn, Germany, 1997, pp. 315-352

(Puc98) Puck, A. and Schürmann, H.: Failure Analysis of FRP Laminates by Means of Physically based Phenomenological Models. Special issue of *"Composite Science and Technology" 58 (1998), part A of the 'failure exercise'*

(Rac97) Rackwitz, R. and Gollwitzer, S.: A New Model for Inter-Fibre-Failure of high strength Unidirectionally Reinforced Plastics and its Reliability Implications. *NATO-workshop PROBAMAT-21.* Century, Perm, Russia, Sept. 10-12, 1997

(Rac97) Rackwitz, R. and Cuntze, R.G.: System Reliability Aspects in Composite Structures. Eng.' *Opt., 1987, Vol. 11*, pp. 69-76

(Rol97) Rolfes, R., Noor, A.H. and Rohwer, K.: Efficient Calculation of Transverse Stresses in Composite Plates. *MSC-NASTRAN User Conference*, 1997

(Row85) Rowlands, R.E.: Strength (Failure) Theories and their Experimental Correlation. In Sih, G.C. Skudra, A.M., Editioren, *Handbook of Composites, Band III, Kapitel 2,* Elsevier Science Publisher B.V., Madison, WI, U.S.A., 1985, 71-125

(Sli97) Slight, D.W., Knight, N.F. and Wang, J.T.: Evaluation of a Progressive Failure Analysis Methodology for Laminated Composite Structures. 38th Structure, Structure Dynamic and Material Conference, April 1997. *AIAA Paper 97-1187*

(Sod98) Soden, P.D., Hinton, M.J. and Kaddour, A.S.: Lamina Properties, Lay-up Configurations and Loading Conditions for a Range of Fibre-reinforced Composite Laminates. Special Issue, *Composite Science and Technology 58 (1998), 1011-1022*

(Sod98a) Soden, P.D., Hinton, M.J. and Kaddour, A.S.: A Comparison of the Predictive Capabilities of Current Failure Theories for Composite Laminates. Special Issue, *Composites Science and Technology 58 (1998), 1225-1254*

(Suk96) Sukarie, G.: Einsatz der FE-Methode bei der Simulation des progressiven Schichtversagens in laminierten Faserverbundstrukturen. *Symposium* "Berechnung von Faserverbundstrukturen unter Anwendung numerischer Verfahren". München, Techn. Univ., 13./14. März 1996

(Tho98) Thom, H.: A Review of the Biaxial Strength of Fibre-Reinforced Plastics. *Composites Part A 29* (1998), 869-886

(Tsa71) Tsai, S.W. and Wu, E.M.: A General Theory of Strength for Anisotropic Materials. *Journal Comp. Mater, Vol. 5 (1971), 58-80*

(VDI97) Cuntze, R.G., et.al.: Neue Bruchkriterien und Festigkeitsnachweise für unidirektionalen Faserkunststoffverbund unter mehrachsiger Beanspruchung – Modellbildung und Experimente -. VDI-Fortschrittbericht, Reihe 5, Nr. 506 (1997, 250 pages)

(Wan93) Wang, J.Z.: Failure Strength and Mechanism of Composite Laminates under Multiaxial Loading Conditions. Dissertation, Univ. of Illinois at Urbana, 1993

(Yeh98) Yeh, H.Y. and Kilfoy, L.T.: A Simple Comparison of Macroscopic Failure Criteria for Advanced Fiber Reinforced Composites. *J. of Reinforced Plastics and Composites, Vol. 17 (1998), 406-445*

(ZTL80) Dornier, Fokker, MBB, DLR: Investigations of Fracture Criteria of Laminae. 1975-1980, Grant from BMVg. (multiaxial testing, reports in German)

ANNEXES ON DESIGN ASPECTS

AI Design safety factors

As well in the stress analysis as in the failure criterion of the strength analysis there are scattering design parameters. The uncertainty of these

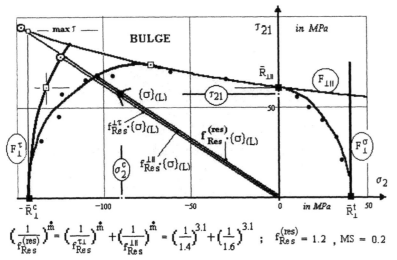

$$\left(\frac{1}{f_{Res}^{(res)}}\right)^{\dot m}=\left(\frac{1}{f_{Res}^{\tau\perp}}\right)^{\dot m}+\left(\frac{1}{f_{Res}^{\perp\parallel}}\right)^{\dot m}=\left(\frac{1}{1.4}\right)^{3.1}+\left(\frac{1}{1.6}\right)^{3.1}\ ;\quad f_{Res}^{(res)}=1.2\ ,\ MS=0.2$$

Figure A1. Visualization of the reserve factor $\{\sigma\}_{(L)}$: = load stress vector, $\dot m$ = rounding-off exponent. ☐ calibration point, O max τ limitation

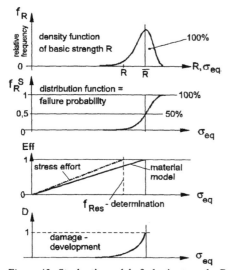

Figure A2. Stochastic model of a basic strength R (σ_{eq}: = equivalent stress, linear elastic)

parameters (loads, strengths, geometrical quantities, Young's modulus, ...) is of *physical nature* or of *statistical nature* (shortage of information due to a too small sample size of a certain design parameter measured). Besides this there is always some *uncertainty in the calculation model* and in the *testing*.

Scatter is usually considered in the design by the use of fixed, deterministic *factors of safety*, better called *design* factors of safety (FoS), which are based on long experience with structural tests.

In aerospace industry structures are dimensioned by using FoS, j, which *increase* ('load factors') the so-called *design limit load* (DLL) up to the design ultimate load (DUL) according to DUL = j_{ult} · DLL. This procedure includes the idea of *proportional loading* and that the reserve factor f_{Res} is related to the external load. The load factors above distinguish the *onset of yield* (e.g. $j_{p0.2}$ = 1.1) of the material applied and *ultimate fracture* (e.g. j_{ult} = 1.5). In practice j is applied to the maximum anticipated load or combinations of loads which the structure is expected to experience which results in the DLL.

Structures are dimensioned and later to be proven for each *single* failure mode discriminating e.g. for the strength failure modes *onset of yield* and *ultimate fracture* in the case of isotropy. The failure surface for onset of yield is inside of that for fracture. Including the different design factors it depends on the actual material behaviour which failure type will be the design driver.

AII Design allowables

The design allowables to be used as strength values are statistics-based minimum values. This might be a so-called Mil-Hdbk 17 "B-value" (or "A"-) with 90 % (99 %) reliability and 95 % confidence probability, the latter number regarding the confidence in the transfer of the finite number of sample test data to the parent distribution with its theoretically infinite number of data.

AIII 'Proof of design' and margin of safety

In the following <u>table A1</u> all the Proofs of Design are listed in in order to gain a general view of which proofs are necessarily to be made in the case of composite parts.

Table A1: 'Proofs of design' of parts

loading	damage free	with damage
static	MS_y , MS_{ult}	residual strengths
'dynamic'	fatigue lifetime	damage growth

The structure shall exhibit positive margins of safety MS: $= f_{Res}$ -1 (figure A1) with respect to any design condition such as for the failure modes ($\equiv$ limit states) buckling, yield, fracture, etc. This usually is demanded for the worst combination of simultaneously occurring loads and associated environmental conditions. And this should be seen in respect of the following facts:

All design parameters scatter, they are stochastic. The size of a stochastic design parameter (strength, load, E-modulus, thickness) is called an uncertain basic variable before its realization and random variable after its realization. Uncertainties can be treated by probabilistic means. By applying probabilistics one can conclude: *An analysis (also a failure condition) represents a model of the reality. A test result represents one realization of the reality.* Test results therefore are random.

Proof of Design requires "No relevant *limit state* is exceeded" and "All *dimensioning load cases* are considered".

a) If $F = \sigma / R = 1$ then failure is achieved. As special case of the failure criterion $F \gtrless 1$ the failure condition $F = 1$ mathematically represents the so-called failure surface or limit surface as limit state description F, respectively. This failure surface envelopes all states of stress which are not yet leading to failure.

b) If $F < 1$ then, by regarding $|\{\sigma\}_{(L)}| < |\{\sigma\}_{fr}|$ fracture stress, reserves or margins are still there. The stress state vectors $\{\sigma\}_{(L)}$ and $\{\sigma\}_{fr}$ are assumed to be parallel under the precondition 'linear behaviour and proportional loading'. All load stresses (σ, τ) in $\{\sigma\}_{(L)}$ consequently have to be increased by the same factor to obtain failure or to meet the failure surface while $\{\sigma\}_{(L)}$ represents all load stresses resulting from the design limit load

multiplied by the load increasing (design) factor of safety j.
• Residual stresses have to be respected vector-like.
• Correct margins cannot be computed without considering the *residual stresses* on the *lamina level* such as the curing stresses in the laminae of a laminate or without considering the moisture or the operational thermal stresses (see annex AIV).
• The still used *failure index* does not deliver practical informations to the designer.

In the (interaction) transition zones of the failure modes respectively in the domains of mixed fracture or interaction, respectively, an increased number of chances to fracture exists. This probabilistic/mechanistic fracture behaviour leads to a smooth complete failure surface. It can be approximately considered on lamina level within the determination of a *resultant reserve factor* being a function of the reserve factors of all activated failure modes.

AIV Remark on residual stresses

Thermal stresses:

In the FRP material there are residual stresses of different levels. Whereas the residual stresses of the 1st kind (*lamina level*) are generally considered in the computation of the laminate only little effort is put at the residual stresses of the 2nd kind on the constituents level (*filament/matrix*) of the isolated lamina generated during the same curing process. These residual stresses are caused by the different coefficients of thermal elongation between fibre and matrix (thermal thrinking) and due to the volume reduction level (chemical shrinking) of the matrix at curing in the 'glassy' state.

Of course, these residual stresses of the 2nd kind are still incorporated in the cured UD tensile coupon specimen for strength testing. They eventually determine the maximum stress achievable which one calls basic strength ($R_{\perp}^t, R_{\perp}^c, R_{\perp\|}$). Often, in the case of UD-specimen it is assumed that the *lamina* curing stresses of the 2nd kind will be the same in the cured *laminate* structure and therefore by applying the lamina test data in the strength analysis the residual stresses of the 2nd kind can be 'forgotten' (as has been done in section 9.4).

Moisture stresses:

According to the moisture absorption level and if saturation through the laminate is achieved (a gradient causes an additional bending effect) the uptaking of moisture equilibrates parts of the curing stress level.

It should be noted for the computations:
- The constraining of the *laminae* in the *laminate* caused by thermal shrinking of each homogenized UD-lamina is generally taken into account. The output are the curing stresses immediately after cooling down and before onset of loading.
- Relaxation effects in the cured laminate are small and tempering has only marginal effect if applying thermosets.
- Moisture can be considered by a reduction of the stress free temperature to an effective temperature.
- Residual stresses diminish with increasing nonlinearity and degradation, respectively.

AV   Properties to be used in analysis

- In practice the number of computations has to be minimized. Therefore, it is highly appreciated in the *Design* that for the dimensioning of structures in *stress* analysis normally the mean properties have to be utilized. By this procedure the most probable (50 %) deformation behaviour of the structure is determined. This is also valid in the case of test data prediction if properties for the test article are not known. Later, in the *Proof of Design* (*strength* analysis) at the Hot Spots minimum properties have to be taken.
- In the evaluation of structural test results the actual data of the test article's properties have to be applied.
- In the case of stability-critical structures, of pressure tanks (in the membrane area these are *single load path structures* which lack of redundancy) instead of the nominal (drawing mean) thickness the minimum thickness shall be used in stress analysis. This adds to the demanded worst case choice for load and environmental conditions another seldom case resulting in a combination of minimum probability.
It should be noted here further: The application of minimum thickness is not always on the safe side.

AVI   Comparison of Puck's fracture plane based criteria and Cuntze's FMC-invariant formulations

(This section is a common formulation of Puck and Cuntze, because both authors have often been asked for an explanation of the coincidences and differences of the two approaches.)

As early as 1968/69 Puck concluded from experimental observations, that two completely different types of fracture should be distinguished and theoretically treated by separate failure criteria: Fibre Failure (FF) and Interfibre Failure (IFF) [Puc69a,69b]. In the early seventies the discrimination of these two fracture types became common practice in the German aerospace industry [ZTL80]. In all later papers of Puck and Cuntze the separate treatment of FF and IFF has been maintained. Both authors use simple maximum stress criteria for FF, based on the consideration, that the composite fails, when the fibres reach a certain critical stress.

Both authors feel that for the new anisotropic fibres a better approach for FF prediction may be necessary.

Since another fundamental paper of 1992 [Puc92], research in Germany concentrated on the improvement of IFF criteria. This appeared to be of higher importance after it had been learned from experience on torsional tube springs, that the wedge effect of oblique fractures under transverse compression can cause destruction of the whole composite part [Puc98].

Common foundations:

Both failure theories are based on the same fundamental assumptions:

- The UD-layer is transversally isotropic and failure occurs by brittle fracture.
- Mohr's statement is valid:
   The strengths of a material are determined by the stresses on the fracture plane.
- The fracture plane may be inclined with respect to the plane on which the external stresses are acting. (This is for instance true for uniaxial transverse compression)
- For states of stress without longitudinal shear (τ_{31}, τ_{21}), that means plane stress conditions consisting of a stress state (σ_2, σ_3, τ_{23}) which can be replaced by the so called principal stresses (σ_{II}, σ_{III}) of the transversally isotropic plane, both authors make the same assumption:
   Paul's modification of the Coulomb-Mohr theory of fracture [Pau61] is valid. This is based on the assumption, that two different modes of fracture can occur, which leads to the following fracture hypothesis (analogous formulation to that for isotropic material): "An intrinsically brittle material will fracture

in either that plane where the shear stress τ_{nt} reaches a critical value which is given by the shear resistance $R_{\perp\perp}^A$ of a fibre parallel plane increased by a certain amount of friction caused by the simultaneously acting compressive stress σ_n on that plane. Or, it will fracture in that plane, where the maximum principal stress (σ_{II} or σ_{III}) reaches the transverse tensile strength $R_\perp^t$ ".

Results for plane stress σ_{II}, σ_{III}

For this state of stress without any longitudinal shear (τ_{31}, τ_{21}) there is a complete coincidence of the procedures of Puck and Cuntze.
The treatment of this problem by Mohr's circle (representing the state of stress (σ_n, τ_{nt}) on any plane) and Mohr's envelope (representing the fracture limit for combined (σ_n, τ_{nt})-stresses) is well known.
Puck starts with the assumption of a parabolic fracture envelope $\tau_{nt} = \tau_{nt}(\sigma_n)$ for $\sigma_n < 0$ [Puc98]:

$$\tau_{nt}^2 = (R_{\perp\perp}^A)^2 - 2p_{\perp\perp}^{(-)} R_{\perp\perp}^A \sigma_n$$

wherein $R_{\perp\perp}^A$ is the shear fracture resistance of a fibre parallel plane and $p_{\perp\perp}^{(-)}$ is a friction coefficient for $\sigma_n < 0$.

At fracture Mohr's circle and envelope have a common point of contact, that means the same inclination $d\tau_{nt}/d\sigma_n$. From this condition the varying angle Θ_{fp} between the action plane of σ_{II} and the fracture plane can be calculated:

$$\cos 2\Theta_{fp} = -\cos 2\Theta_{fp}^c \, \frac{R_\perp^c}{\sigma_{II} - \sigma_{III}}$$

with Θ_{fp}^c = fracture angle under uniaxial transverse compression and $R_\perp^c$ = transverse compression strength. σ_{II} and σ_{III} are stresses at fracture.
With this result a closed form in σ_{II}, σ_{III} for the fracture condition is found which is parabolic and invariant in the transversal plane.

In contrast to Puck Cuntze starts with this invariant formulation

$$F_\perp^\tau = \frac{a_\perp^\tau}{R_\perp^c}(\sigma_{II} + \sigma_{III}) + \frac{b_\perp^\tau}{(R_\perp^c)^2}(\sigma_{II} - \sigma_{III})^2 = 1.$$

The adaptation to the uniaxial compression results (strength $R_\perp^c$ and fracture angle Θ_{fp}^c) gives

$$a_\perp^\tau = b_\perp^\tau - 1 \quad \text{and} \quad b_\perp^\tau = 1/(2\cos 2\Theta_{fp}^c + 1)$$

(or see chapter 4).
Puck's and Cuntze's approaches are connected by the relation for the fracture angle Θ_{fp}. For the calculation of the fracture stresses Cuntze's invariant formulation is of course the more convenient one.

Results for states of stress with additional longitudinal shear (τ_{31}, τ_{21})

In this field the two authors use rather different approaches:

• Puck stays with the more or less physically based consideration of the mechanical interaction of the stresses σ_n, τ_{nt}, τ_{n1} on the fracture plane. He uses simple polynomials (parabolic or elliptic) to formulate a (master-)fracture body in the (σ_n, τ_{nt}, τ_{n1})-space.
Starting from this (master-)fracture body no analytical solutions can be found for the fracture angle (with the exception of (σ_1, σ_2, τ_{21})-states of stress) and with this also no analytical solutions for the fracture bodies in $\sigma_1, \sigma_2, \sigma_3, \tau_{23}, \tau_{31}, \tau_{21}$.
Therefore, the necessary search for the fracture plane, that means for the plane with the lowest reserve factor $f_{Res}(\Theta)$, has to be done numerically (using the fracture condition in σ_n, τ_{nt}, τ_{n1}) in an angle range between $-90° \le \Theta \le +90°$. With the lowest $f_{Res}(\Theta)$, found by the numerical procedure, the stresses ($\sigma_1, \sigma_2, \dots \tau_{21}$) at facture can be calculated.

The numerical search for the fracture plane is an inconvenience, but on the other hand the user of this approach automatically gets an information on the fracture angle and on the "fracture mode". Puck defines the fracture mode as the stress combination (σ_n, τ_{nt}, τ_{n1}) or ($\sigma_\perp$, $\tau_{\perp\perp}$, $\tau_{\perp\parallel}$) on the fracture plane.

The results can be visualized by fracture bodies in a 3-dimensional (σ_{II}, σ_{III}, $\tau_{\omega1}$)-space, where $\tau_{\omega1}$ is the "resultant" of τ_{31} and τ_{21}.
These fracture bodies are not symmetric in respect of the ($\sigma_{II} = \sigma_{III}$)-plane [Kop99].

• Cuntze uses three simple invariant formulations (1 linear, 1 quadratic and 1 cubic polynomial) which lead to fracture bodies similar to those of Puck.

25

He feels that (micro) mechanical and probabilistic interactions can not be clearly distinguished and therefore he models the interaction mainly by a probabilistic series model ("rounding-off " by the calculation of $f_{\mathrm{Res}}^{\mathrm{res}}$).

Attention has to be paid to the fact that the expression "mode" has different meanings in the papers of Puck and Cuntze. Puck differentiates between 7 interfibre fracture modes (possible stress combinations on the fracture plane):

1.) $(\sigma_\perp^t, \tau_{\perp\perp}, \tau_{\perp\parallel})$, 2.) $(\sigma_\perp^t, 0, 0)$,

3.) $(\sigma_\perp^t, 0, \tau_{\perp\parallel})$, 4.) $(0, 0, \tau_{\perp\parallel})$, 5.) $(\sigma_\perp^c, 0, \tau_{\perp\parallel})$,

6.) $(\sigma_\perp^c, \tau_{\perp\perp}, \tau_{\perp\parallel})$, 7.) $(\sigma_\perp^c, \tau_{\perp\perp}, 0)$.

- Cuntze uses the expression "mode" to adress his three different invariant fracture conditions, based on the idea that for each of these fracture conditions either the $\sigma_\perp$- , or the $\tau_{\perp\perp}$-, or the $\tau_{\perp\parallel}$-stress is "dominant".

An absolute aggreement between Puck's (2. or 7.) and Cuntze's modes is to be found again in the (σ_{II}, σ_{III})-plane.

Of course, one has to pay for the higher convenience of the invariant approach with a certain loss of "physical correctness", but this may be acceptable in many cases of design practice.

AVII Modern failure mode based safety concept (format)

The reliability assessment of structures may be described by the following classes of measures for the Proof of Design:
- failure index: = value of failure function F
- reserve factor $f_{res}^{(res)}$ or margin of safety MS
- reliability $\Re$ of the laminate:
  - the simple semi-probabilistic procedure for just two stochastic design parameters, for 'stress' and strength.
  - the 'partial safety factor concept' for a bunch of stochastic design parameters.

Structural Integrity (SI) plays a big role in design (since Hammurabi's code). It is the characteristic of a structure that enables it to withstand the loads and other environmental loadings imposed during life time.

In order to achieve the certification for the structural product the designer has to provide it with a high reliability. This includes the reliability of all functions with SI being one of them. SI covers as well the 'strength of damage free structures' as the 'damage tolerance incl. NDI-techniques of damaged structures'.

Reliability is the aptitude of a product to perform the required functions at certain performance levels under specific conditions and for a given period of time (expressed in terms of a reliability value).

Reliability assessment requires a 'Functional Analysis' of the product *structure*, a 'Failure Mode Effect and Criticality Analysis (FMECA)' and the control of the 'Critical Points'.

A damage tolerant structure can endure the specified design limit load even in case of a predamage or a damage caused by fatigue or impact as long as the damage will be discovered by inspection or malfunction and will be repaired.

The designer of a structural part has to demonstrate to the customer and to the authorities compliance with the design requirements concerning SI of the hardware and to verify this by analysis and tests.

A standard supporting tool to ensure of SI is to follow a so-called *RAMS achievement process* including Risk Control, FMECA, Critical Points listing and probabilistic predictions in order to accept the design if e.g. mass and residual risk is acceptable.

With reducing the FoS a structural reliability analysis as part of the (e.g. launcher) system reliability analysis was becoming necessary, due to the fact that the reliability $\Re$ = 1- failure probability p_f of the subsystem structure will not be extremely high anymore (practically it was $\Re$ = 1 or $p_f \leq 10^{-10}$ and now $p_f \leq 10^{-4}$) in order to minimize mass.

Structure reliability analysis includes the 'logical modelling of the failure ystem' (e.g., of the system laminate consisting of the components laminae), the 'modelling of the stochastic design parameters', and the 'mechanical modelling'. The latter includes the definition of the structural system, of the failure modes and corresponding failure conditions.

The modern tools above will be more and more introduced in spacecraft, aircraft and automotive industry. They indicate that 'Think failure modes' is a must for the future.

In aerospace engineering damage tolerance design still now involves the provision with the 'catalogue of acceptable damages for composite parts'. It has to include the different types of damages (most often called 'defects'), their acceptable size and classifies,

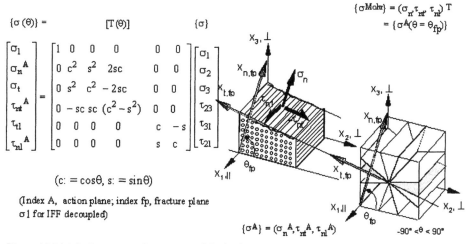

$$\{\sigma(\theta)\} = \qquad [T(\theta)] \qquad \{\sigma\}$$

$$\begin{bmatrix} \sigma_1 \\ \sigma_n^{\ A} \\ \sigma_t \\ \tau_{nt}^{\ A} \\ \tau_{tl} \\ \tau_{nl}^{\ A} \end{bmatrix} = \begin{bmatrix} 1 & 0 & 0 & 0 & 0 & 0 \\ 0 & c^2 & s^2 & 2sc & 0 & 0 \\ 0 & s^2 & c^2 & -2sc & 0 & 0 \\ 0 & -sc & sc & (c^2-s^2) & 0 & 0 \\ 0 & 0 & 0 & 0 & c & -s \\ 0 & 0 & 0 & 0 & s & c \end{bmatrix} \begin{bmatrix} \sigma_1 \\ \sigma_2 \\ \sigma_3 \\ \tau_{23} \\ \tau_{31} \\ \tau_{21} \end{bmatrix}$$

$$(c: = \cos\theta, \ s: = \sin\theta)$$

(Index A, action plane; index fp, fracture plane
$\sigma 1$ for IFF decoupled)

$$\{\sigma^{Mohr}\} = (\sigma_n, \tau_{nt}, \tau_{nl})^T$$
$$= \{\sigma^A(\theta = \theta_{fp})\}$$

$$\{\sigma^A\} = (\sigma_n^{\ A}, \tau_{nt}^{\ A}, \tau_{nl}^{\ A})$$

$$-90° < \theta < 90°$$

Figure A3. Mohr's fracture causing stresses of the lamina

how this damage has to be detected and which influence it has at SI.

For the Proof of Design all functional requirements of the structure (and the laminate) have to be fulfilled such as deformation, leakage limits, stability and strength. E.g. for stability this means all sub-failure modes (general buckling, local buckling, wrinkling, etc.) have to be accounted for structural integrity. This also has to hold for FF and IFF as sub-failure modes of strength.

A Probabilistic Safety Concept executed at minimum in the so-called critical points:
• considers the mutual dependencies of the 'failure modes and of correlations between the design parameters and is consequently an improvement of the physical understanding (the traditional Deterministic Safety Concept does not look at the behaviour of the system laminate or even structure)
• gives a quantitative measure for the different influence of the stochastic design parameters, which enables by the computation of the so-called *design sensitivities* to take the cheapest measure in the case of an alteration possible and to base test concepts better.

The German investigators on composite strength criteria are convinced that the transfer of the mode ideas to a prediction of fatigue life of composite materials is much more promising for this usually brittle material than taking a correction effect of the mean stress into account (which is also doubtful in the case of metals, however, not questioned due to the long experience with this material family).

Damage accumulation and finally, the prediction of lifetime might be better based upon a separate accumulation of NF- and SF-portions in the cycles, respecting the action plane of the failure active Mohr stresses, σ_n and τ_n (figure A3).

27

Recent Developments in Durability Analysis of Composite Systems, Cardon, Fukuda, Reifsnider & Verchery (eds)
© *2000 Balkema, Rotterdam, ISBN 90 5809 103 1*

Failure mechanics of thin coatings under multiaxial loading

Y. Leterrier, D. Pellaton & J.-A. E. Månson
Laboratoire de Technologie des Composites et Polymères, Ecole Polytechnique Fédérale de Lausanne, Switzerland

J. Andersons
Institute of Polymer Mechanics, Riga, Latvia

ABSTRACT: The initiation of cracking in nanosized oxide coatings on polyethylene terephthalate (PET) films under uniaxial and equibiaxial tension is modeled as a function of coating thickness and residual strain. Several theoretical derivations are tested against experimental data, using the coating fracture toughness as a fitting parameter. All models reproduce with reasonable accuracy the measured crack onset strain of coatings of thickness in the range 30 nm to 156 nm. However, shear lag and variational mechanics analyses of the stress transfer phenomenon based on linear elasticity overestimate the coating toughness. By contrast, a novel hypothesis of a perfectly plastic interface introduced to account for dissipative effects yields a low coating toughness, found to be equal to 3 J/m^2.

1 INTRODUCTION

Thin coatings adhering onto polymer substrates find increasing interest in applications as diverse as multilayered optical lenses, microelectronic devices, and gas barriers for pharmaceutical and food packaging (see for instance the proceedings of the annual SVC conference, 1998). The thickness of the coating and its adhesion to the substrate are among the key characteristics to be tailored for both performance and cost optimization. Reduced thickness without impaired reliability enables cost savings. However, coatings with thickness down in the nanometer range are often associated with growth heterogeneities and high residual stresses generated during the deposition process (Henry et al. 1998). The influence of these factors on the mechanical properties of the coating and adhesion to the substrate, and therefore on the functional reliability of the coated parts is a major concern. The failure mechanisms of elastic thin coatings under multiaxial loading are investigated in this work to simulate the complex stress loadings present during manufacture and service.

Particular attention is paid to modeling crack onset strain in thin, brittle coatings, which is also relevant to off-axis ply cracking in composite laminates. A novel hypothesis of a perfectly plastic interface is introduced, and the effect of uniaxial and equibiaxial tension is considered. The analysis of the stress transfer and durability of such multiphase materials is further expected to refine the understanding of the interfacial behavior in polymer composites. A promising output of this approach is the generation of tailored interfaces for improved durability, which is still a challenge for this class of materials if one aims at increasing their use in advanced applications.

2 THEORETICAL DERIVATION OF CRACK ONSET STRAIN

Cracking of brittle coatings shows close analogies with off-axis ply cracking in brittle-matrix composite laminates subjected to tensile load. In both cases a crack, once initiated, usually propagates instantaneously, spanning the whole width of the specimen, stopping at the interface with adjacent plies in the laminate case. An important difference concerning the direction of intralaminar crack propagation is in that it is governed by reinforcing fiber direction in laminates, while in isotropic coatings crack trajectory is determined by the stress state in the coating, i.e. by applied loading. Crack onset strain is a key feature in durability analysis, and its modeling has motivated various approaches, which are briefly reviewed.

2.1 *Linear elastic approaches*

Under uniaxial tension, the weakest link model with two-parameter Weibull strength distribution has been extensively used, where the fracture of a brittle material results from flaws located both on the surface and within the bulk. The Weibull strength model was reported to underpredict the measured increase of crack onset strain with decreasing coating thickness (Leterrier et al. 1997a). This result was also observed in the case of fibers, for which, although both length and diameter effects on strength agree with Weibull model, the shape parameter associated with diameter is substantially lower than that related to fiber length (Wagner 1989).

Alternatively, linear elastic fracture mechanics (LEFM) models have been successfully employed for crack onset and crack density prediction in laminated composites, and also extended to coating cracking analysis under uniaxial loading (Nairn & Kim 1992). They differ by the accuracy of energy release rate calculation, which depends on the modeling of the stress state perturbation due to the presence of crack. Since the random nature of fracture is not considered, fracture strain depends on geometrical parameters, that is coating and substrate thickness. The corresponding energy balance equation for crack propagation implies that the strain energy released due to crack growth equals coating fracture toughness of either the transverse ply or the coating. The simplest analytical approach for stress redistribution calculation is the shear lag model. Applying the energy release rate relation derived in the work of Laws & Dvorak (1988) to the coating-substrate system in consideration and neglecting residual stresses leads to the following expression for crack onset strain:

$$\varepsilon_{onset} = \sqrt{\frac{2 G_c \xi h_s E_s}{h_c (h_s + h_c) E_c E}} \qquad (1)$$

where G_c is the coating fracture toughness, E_s, E_c, and E are the substrate, coating and film moduli, respectively, and h_s and h_c are the corresponding thickness. The factor ξ is a non-dimensional shear lag parameter determined by fitting the above equation to experimental data.

Another approach was proposed recently by Yanaka et al. (1998) to calculate the crack onset stress, with the assumption that the additional displacement in the substrate due to cracking varies linearly in the thickness direction. The corresponding crack onset strain is found inversely proportional to $h_c^{1/4}$, where the residual strain ε_{rc} is accounted for:

$$\varepsilon_{onset} + \varepsilon_{rc} = \sqrt{\frac{2 G_c}{3 E_c^{3/2} \sqrt{(1+v_c)} h_c h_s / E_s}} \qquad (2)$$

A further expression for strain energy release rate was derived by Nairn and Kim (1992, 1999) by minimizing the complementary energy of the film, assuming that the axial stress in coating and substrate does not depend on the thickness coordinate. Due to complexity, the resulting expression for first crack

strain is not given here. Both shear-lag and variation mechanics solutions can be applied to extrapolate crack onset strain vs. coating thickness, providing that the parameter ξ or the coating toughness G_c are fitted by testing one film with a given coating thickness.

2.2 Case of a plastic interface

The above approaches assume linear elastic behavior of the material system. For a ductile substrate, crack propagation can also involve small-scale yielding at the coating-substrate interface (Hu & Evans 1989) caused by shear stress, and part of the released strain energy is dissipated. Accounting for that, the energy balance takes the form:

$$G_c h_c + W_d = \Delta A - \Delta W \qquad (3)$$

where W_d, ΔA and ΔW are accordingly the energy dissipated by deforming the plastic interface, the work done by the applied load, and the strain energy released due to crack extension through the coating (all per unit width of the film). Crack onset strain for a thin film under plane-stress conditions is derived assuming that the coating and substrate are elastic in the strain range of interest, and that the coating-substrate interface is perfectly plastic with a shear strength τ. The energy terms in Equation (3) are derived from the expressions for stresses and displacements in the presence of the crack, and calculated as follows, first under uniaxial loading, and then under biaxial loading.

Uniaxial loading

Consider a film with a straight crack normal to the loading direction and spanning whole thickness and width of the coating, under applied stress $\sigma_x = \sigma$, as shown schematically in Figure 1. Stresses normal to the crack within the stress transfer zone ($x<\delta$) are given by shear-lag approximation as:

$$\sigma_{xc} = \frac{\tau x}{h_c} \quad \text{and} \quad \sigma_{xs} = \sigma \frac{H}{h_s} - \frac{\tau x}{h_s} \qquad (4)$$

where indices s and c denote substrate and coating, respectively, $H = h_s + h_c$ is the film thickness, and τ is interface yield strength in shear.

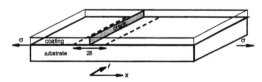

Figure 1. Schematics of the coated film under uniaxial tension with one coating crack.

30

The stress transfer zone length, δ, is proportional to the far-field stress in the coating, σ^0_{xc}: $\delta = \sigma^0_{xc} h_c / \tau$. Elementary laminated plate theory yields the far-field stress in the coating:

$$\sigma^0_{xc} = E_c \left(\varepsilon \frac{1 - \nu \nu_c}{1 - \nu_c^2} + \frac{\varepsilon_{rc}}{1 - \nu_c} \right) \qquad (5)$$

where $\varepsilon = \sigma/E$, ε_{rc} is the residual strain in the coating, ν and E denote the Poisson's ratio and modulus of the film assembly. For thin coating, perturbation of the transverse strain within the stress transfer zone can be neglected:

$$\varepsilon_{yc} = \varepsilon_{ys} = \varepsilon_y \qquad (6)$$

Knowing longitudinal stresses and transverse strain allows determining transverse stresses and longitudinal strains according to Hook's law. Expressions for coating and substrate displacements, u_{xc} and u_{xs}, within stress transfer zone are obtained by integrating ε_{xc} and ε_{xs}, correspondingly with the boundary conditions: $u_{xs}(0) = 0$ and $u_{xc}(\delta) = u_{xs}(\delta)$.

Having thus established expressions for stresses and displacements in the presence of the crack, the energy dissipated by deforming the plastic interface, W_d, the work done by the applied load, ΔA, and the change of strain energy, ΔW, caused by crack propagation, are determined as follows:

$$W_d = 2 \int_0^{\delta} \tau \left(u_{xc} - u_{xs} \right) dx$$

$$\Delta A = 2 \sigma H \Delta u_x \qquad (7)$$

$$\Delta W = \sum_{i=c,s} \frac{h_i}{E_i} \int_0^{\delta} \left(\sigma_{xi}^2 + \sigma_{yi}^2 - 2 \nu_i \sigma_{xi} \sigma_{yi} \right) dx$$
$$- \sum_{i=c,s} \frac{h_i}{E_i} \int_0^{\delta} \left[\left(\sigma_{xi}^0 \right)^2 + \left(\sigma_{yi}^0 \right)^2 - 2 \nu_i \sigma_{xi}^0 \sigma_{yi}^0 \right] dx$$

Coating crack onset strain is obtained upon substituting Equations (7) into Equation (3) by routine but tedious calculations, yielding:

$$\varepsilon_{onset} + \varepsilon_{rc} \frac{1 + \nu_c}{1 - \nu \nu_c} = \frac{1 - \nu_c^2}{1 - \nu \nu_c} \sqrt[3]{\frac{3 G_c \tau}{E_c^2 h_c \left(1 - \nu_c^2 + \frac{E_c h_c}{E_s h_s} \left(1 - \nu_s^2 \right) \right)}} \qquad (8)$$

Equibiaxial loading

For equibiaxial loading case, it is convenient to retain the coordinate system (Fig. 1) and notation introduced above. Now stress is applied to the film in both x and y directions, and $\sigma_x = \sigma_y = \sigma$. The basic modeling assumptions expressed by Equations (4)

and (6) also remain the same. The presence of the applied load component parallel to the straight crack is reflected in the model via the appropriate far-field stress and strain values. Thus, unperturbed coating stress under equibiaxial loading is:

$$\sigma^0_{xc} = \sigma^0_{yc} = \frac{E_c}{1 - \nu_c} \left(\varepsilon + \varepsilon_{rc} \right) \qquad (9)$$

where the biaxial strain $\varepsilon = \sigma(1-\nu)/E$. It follows from Equation (6) that strains and displacements in y-axis direction are not perturbed by the presence of the crack. Therefore neither plastic slip occurs along the crack, nor additional work is done by load component σ_y, and Equations (7) remain valid for biaxial load. Finally, crack onset strain under equibiaxial tension is given by:

$$\varepsilon_{onset} + \varepsilon_{rc} = \left(1 - \nu_c \right) \sqrt[3]{\frac{3 G_c \tau}{E_c^2 h_c \left(1 - \nu_c^2 + \frac{E_c h_c}{E_s h_s} \left(1 - \nu_s^2 \right) \right)}} \qquad (10)$$

3 EXPERIMENTAL TECHNIQUES

The materials investigated are 12 μm thick biaxially stretched PET films coated by physical vapor deposition (PVD) with SiO$_x$ layers, of thickness ranging from 30 nm to 156 nm (Leterrier et al. 1997b). The elastic properties of the materials are $E_s = 3920$ MPa; $\nu_s = 0.44$; $E_c = 79500$ MPa; $\nu_c = 0.2$; $\nu = 0.4$; and E was derived using the rule of mixtures.

The residual coating strain generated during the deposition process was determined from the measured curvature of the coated films, as detailed by Yanaka et al. (1998).

The techniques developed to analyze the fragmentation process of the thin coating under uniaxial and biaxial loads were detailed in separate publications (Leterrier et al. 1997a, 1999b). Fragmentation tests under uniaxial tension were performed at a constant strain rate of 2.1×10^{-4} s^{-1} by means of a Polymers Labs Minimat tensile tester mounted on a Olympus SH-2 optical microscope stage, and the crack development during straining was recorded via a CCD camera connected to the microscope. Fragmentation tests under equibiaxial plane stress conditions were performed using a bulging cell installed on the optical microscope stage. Circular specimens were pressurized stepwise by means of a precision pressure controller. Theoretical modeling of film deformation process was carried out to derive stresses and strains in the membrane as a function of applied pressure and displacement of the top of the inflated calotte, and is detailed as follows.

31

3.1 Theoretical analysis of the pressure dependence of strains and stresses in the bulge geometry

Theoretical modeling of film deformation process in the bulge geometry is complicated by both geometrical nonlinearity due to large displacements and material nonlinearity caused by substrate plasticity.

During the initial loading stage, up to the onset of material nonlinearity, only the geometrical nonlinearity has to be accounted for. A closed-form solution was derived by Hencky for a circular membrane loaded by an uniform pressure P (Hencky 1915). Expressions for stresses and displacement of the membrane in polar coordinates with the origin in membrane center where obtained as power series of the radial coordinate r. Truncating the series yields the maximum displacement d and equibiaxial stress σ in the center of membrane:

$$d = C_d R_0 \sqrt[3]{\frac{PR_0}{e_0 E}} \tag{11}$$

and

$$\sigma = C_\sigma \sqrt[3]{E\left(\frac{PR_0}{e_0}\right)^2} \tag{12}$$

where R_0 and e_0 are the initial radius of the free surface of the specimen and its initial thickness, respectively, equal to 18 mm and 12 μm in the present case, and E is the Young modulus of the membrane. Using Hook's law for equibiaxial tension, the strain in the center of membrane, ε is then given by:

$$\varepsilon = \frac{1-\nu}{E}\sigma = C_\sigma(1-\nu)\sqrt[3]{\left(\frac{PR_0}{e_0 E}\right)^2} \tag{13}$$

The prefactors in Equations (11), (12) and (13) depend on the Poisson's ratio of the film and on the number of terms retained in series.

In order to assess the accuracy of the truncated analytical solution, FEM calculations were performed using the NISA2 software (1994). 3-D general shell elements accounting for membrane, bending and transverse shear stresses were applied to model one quarter of the circular membrane with appropriate symmetry and boundary conditions. 510 nodes were used. Figure 2 shows the comparison of analytical and FEM results for a PET film with a 103 nm thick coating. The displacement of the film center vs. applied pressure, as well as membrane stress distribution in the film, are in good agreement with the analytical results.

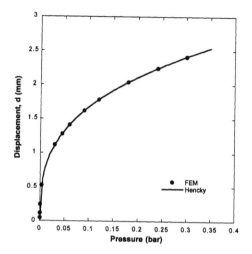

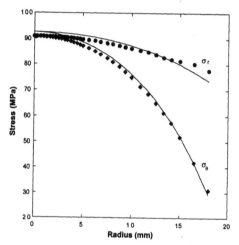

Figure 2. Comparison of FEM results (dots) and Hencky's analytical model prediction (lines) for the bulge test of PET film with a 103 nm thick SiO$_x$ coating. Top: dependence of calotte top displacement on applied pressure; bottom: radial and circumferential stress distribution at a pressure equal to 0.3 bar.

Upon plastic yield of the polymer substrate, the above elastic derivation is not applicable. A simplified geometrical calculation is proposed in this case, which assumes that for a circular specimen of thickness-to-diameter ratio smaller than $4 \; 10^{-4}$, the deformed film forms a spherical cap. This assumption leads to homogeneous equibiaxial strain. The biaxial strain ε is derived from initial and final film surface area ratio and expressed via the measured displacement d of the top of the inflated calotte:

$$\varepsilon = \frac{1}{2} Ln\left(\frac{d^2 + R_0^2}{R_0^2}\right) \tag{14}$$

4 RESULTS AND DISCUSSION

4.1 *Residual strains*

Compressive residual strains are generated in the oxide coating during the deposition process, as shown in Figure 3, where the strains are compared with values reported in the literature for a similar system (Yanaka et al. 1998). In both cases, very similar compressive strains are determined, and found to increase when the coating thickness decreases. These strains and are of the order of -0.4 %, which is far from being negligible compared to the crack onset strains values, as detailed in the following. Neglecting such residual strains would therefore lead to important overestimation of the coating toughness. The thickness dependence of the residual compression was approximated by a power-law, which was then used in the various models described in the preceding section, following the expression found in the left-hand-side of Equation (8).

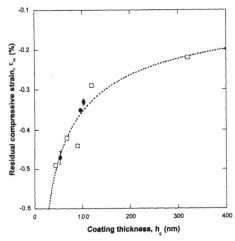

Figure 3. Residual compressive strain in the oxide coating vs. coating thickness determined in the present work (filled circles), and compared with values (open squares) determined by Yanaka et al. (1998). The dotted line is a power-law fit to the data 53-320 nm.

4.2 *Failure of the coating under uniaxial tension*

The initial cracks observed during tensile testing of the film were straight cracks, running parallel one to the other through the whole sample width, perpendicularly to the loading direction. Figure 4 reproduces an optical micrograph of the 103 nm coating, where the first two cracks were identified. It should be pointed out that an accurate measurement of the coating crack onset strain, within 0.1% error, is required to use the models descibed in the preceding section. Due to elastic recovery of the

polymer substrate, any cracks present on samples strained to less than 4 % strain will close and will not be visible if the sample is unloaded. In-situ tests, where the material is maintained under stress, resolve this drawback. Experimental values and theoretical crack onset strain of the SiO_x coatings under uniaxial loading are compared in Figure 5. All models account for the residual compression measured in the previous section. Since a direct measurement of the coating fracture toughness G_c is not possible, curve fit to experimental data was done with adjustable ξ (Eq. 1) or adjustable G_c (Eq. 2 and 8, and variational mechanics approach) values.

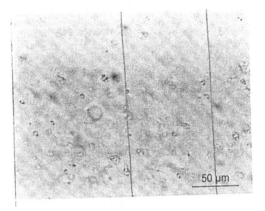

Figure 4. Optical micrograph of a 103 nm thick SiO_x coating on PET under uniaxial tension at ca. 1.4 % strain, showing the first two cracks observed during in-situ testing, perpendicular to the loading direction. The cracks were hardly visible on the printed image, and were carefully hand-traced for improved visibility.

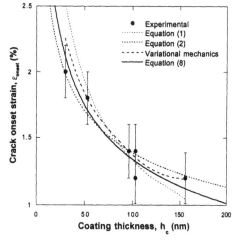

Figure 5. Dependence of the SiO_x coating crack onset strain under uniaxial load on coating thickness, and comparison of experimental data with models.

All models fit the experimental findings with reasonable accuracy, however with considerably different values. The first shear lag model (Equation 1) is hampered by the presence of the empirical parameter ξ. The apparent coating toughness G_c is found to be equal to 60 J/m^2 in the case of the variational model, and to 76 J/m^2 in the case of the shear lag model including the linear variation of the additional displacement in the substrate due to cracking (Equation 2). Both values are very high, compared to the fracture toughness of bulk glass that, depending on the composition, is of the order of 10 J/m^2 (Wiederhorn 1969). This discrepancy is likely to arise from neglecting plastic effects in the substrate, at the interface vicinity.

The strength τ of the perfectly plastic interface introduced to model the dissipative term in Equation (7) was assumed to be equal to the VonMises yield shear stress of the polymer substrate. The latter was calculated from the tensile yield stress, σ_Y, as: $\tau = \sigma_Y/\sqrt{3}$, and found to be equal to 49 MPa. This assumption is based on the strong specific interactions such as hydrogen bonds and Si-O-C and Si-C bonds resulting from the deposition process of the SiO$_x$ on PET (Rotger et al. 1995; Leterrier et al. 1999a), and leads to a low fracture toughness equal to 2.9 J/m^2.

4.3 Failure of the coating under biaxial tension

The first two cracks observed in the biaxial experiment are shown in Figure 6 for the 103 nm coating. These cracks, contrary to the uniaxial case, are not straight. Rather, they follow a complex trajectory, determined by the local stress

perturbation at the crack tip and the defect structure of the coating.

Modeling of the crack onset strain under equibiaxial load (Equation 10) is reported in Figure 7, using the fracture toughness determined in the uniaxial case. In the biaxial geometry, the model is only accurate for the thicker coating, whereas it underestimates crack onset strains of thinner coatings. The deviation of the prediction from experimental data with coating thickness reduction is likely to be caused by plastic yield of the substrate, which is assumed to be elastic in the model.

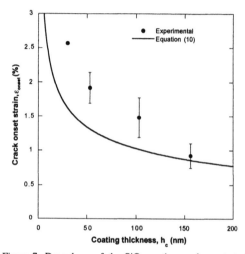

Figure 7. Dependence of the SiO$_x$ coating crack onset strain under biaxial load on coating thickness. The dots are experimental data, and the line represents Equation (10).

Estimation of substrate stress in the center of the inflated film, shown in Figure 2 for the 103 nm coated film, revealed that only in the 156 nm coating case the equibiaxial stress in the PET substrate at crack onset was well below the nominal PET tensile yield stress of 85 MPa. For the 103 nm coating case, the calculated stress in PET at coating crack onset was close to plastic limit, while for thinner coatings – exceeded it. The presence of substrate plasticity in the pressure range of interest for $h_c \leq 103$ nm was also indirectly corroborated by the measured calotte top displacement exceeding considerably the elastic analysis prediction.

The comparison of these various approaches to fracture mechanics of thin coatings highlights the importance of accounting for the behavior of the polymer/coating interface for reliable durability predictions.

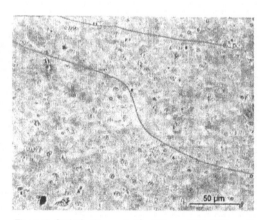

Figure 6. Optical micrograph of a 103 nm thick SiO$_x$ coating on PET at the top of the inflated calotte under equibiaxial tension at ca. 1.5 % strain. The first two cracks observed during in-situ testing were hardly visible on the printed image, and were carefully hand-traced for improved visibility.

5 CONCLUSIONS

The failure mechanisms of SiO_x coatings of thicknesses ranging from 30 nm to 156 nm on PET substrates developed for packaging applications were modeled under uniaxial and equibiaxial tension, to simulate relevant stress conditions achieved during the manufacture and service of the bilayer material system.

Compressive residual strains generated during the deposition of the coating were found to slightly increase in absolute values, and the crack onset strain was found to increase, with decreasing coating thickness. This evolution was analyzed using several fracture mechanics-based models, also relevant for transverse cracking in composite laminates.

In the uniaxial loading case, all models show a good agreement with experimental data. Elastic shear-lag and variational mechanics approaches lead to oxide fracture toughness values beyond practical limits for silicon oxide. On the contrary, the shear-lag model with a perfectly plastic interface developed in this study leads to a low fracture toughness, equal to approx. 3 J/m^2. In the biaxial loading case, substrate plasticity leads to an underestimation of the crack onset strain at small coating thickness.

ACKNOWLEDGMENTS

The authors acknowledge the Swiss National Science Foundation for financial support, and thank Lawson Mardon Packaging for supplying film samples.

REFERENCES

Hencky, H. 1915. *Z. Math. Physik* 63: 311.
Henry, B.M., et al. 1998. Microstructural characterisation of transparent silicon oxide permeation barrier coatings on PET. *Proc. 41st SVC Ann. Tech. Conf*, Boston, Apr. 18-23.
Hu, M.S. & A.G. Evans 1989. The cracking and decohesion of thin films on ductile substrates. *Acta Metall.* 37: 917.
Laws, N. & G.J. Dvorak 1988. Progressive transverse cracking in composite laminates. *J. Compos. Mater.* 22: 900.
Leterrier, Y., L. Boogh, J. Andersons & J.-A.E. Månson 1997a. Adhesion of silicon oxide layers on poly(ethylene terephthalate). I: effect of substrate properties on coating's fragmentation kinetics. *J. Polym. Sci. B: Polym. Phys.* 35: 1449.
Leterrier, Y., J. Andersons, Y. Pitton & J.-A.E. Månson 1997b. Adhesion of silicon oxide layers on poly(ethylene terephthalate). II: effect of coating thickness on adhesive and cohesive strengths. *J. Polym. Sci. B: Polym. Phys.* 35: 1463.
Leterrier, Y., P. Sutter & J.-A.E. Månson 1999a. Thermodynamic and micromechanical approaches to the adhesion between polyethylene terephthalate and silicon oxide. *J. Adhesion* 69: 13.
Leterrier, Y., et al. 1999b. Biaxial fragmentation of thin silicon oxide coatings on polyethylene terephthalate. *To be submitted.*
Nairn, J.A. & S.-R. Kim 1992. A fracture mechanics analysis of multiple cracking in coatings. *Eng. Fract. Mech.* 42: 195.
Nairn, J.A. & S.-R. Kim 1999. Fracture mechanics analysis of coating/substrate systems subjected to tension or bending loads I: theory. Submitted to *Eng. Fract. Mech.*
NISA 1994. Users Manual for NISA2. Version 94.0.
Rotger, J.C., et al. 1995. Deposition of silicon oxide onto polyethylene and polyethyleneterephthalate: An X-ray photoelectron spectroscopy interfacial study. *J. Vac. Sci. Technol.* A13: 260.
Society of Vacuum Coaters (SVC) 1998. *Proc. 41st Ann. Tech. Conf.*, Boston, Apr. 18-23.
Wagner, H.D. 1989. *Composite Material Series. Vol. 6. Application of fracture mechanics to composite materials*: 39-77, Elsevier Science Publishers, New-York.
Wiederhorn, S.M. 1969. *J. Amer. Ceram. Soc.* 52: 99.
Yanaka, M., T. Miyamoto, T. Tsukahara & N. Takeda 1998. In-situ observation and analysis of multiple cracking phenomena in thin glass layers deposited on polymer films. *Proc. ICCI7*, Shonan Inst. Technol., Japan, May 9-13. Also submitted to *Compos. Interfaces.*

Damage analysis

Recent Developments in Durability Analysis of Composite Systems, Cardon, Fukuda, Reifsnider & Verchery (eds)
© *2000 Balkema, Rotterdam, ISBN 90 5809 103 1*

Modelling and computation until final fracture of laminate composites

P. Ladevèze
LMT-Cachan, ENS Cachan / CNRS / Université Paris 6, France

ABSTRACT: One main challenge in composite design is to compute the damage state of a composite structure subjected to complex loading at any point and at any time until final fracture. Damage refers to the more or less gradual developments of microcracks which lead to macrocracks and then to rupture ; macrocracks are simulated as completely damage zones.Our solution for composites and especially laminate composites is based on what we call a damage mesomodel. It is a semi-discrete modelling for which the damage state is locally uniform within the mesoconstituents. For laminates, it is uniform throughout the thickness of each single layer ; as a complement, continuum damage models with delay effects are introduced. Attention is focused herein on latest developments which concern the interface modelling and the localization phenomena description. Last comparisons between simulations and tests for delamination tests are shown.

1. INTRODUCTION

An initial step, which has been achieved in other studies, is to define what we call a laminate mesomodel (Ladevèze 1986, 1989). At the mesoscale, characterized by the thickness of the ply, the laminates structure is described as a stacking sequence of homogeneous layers throughout the thickness and interlaminar interfaces. The main damage mechanisms are described as: fiber breaking, matrix micro-cracking and adjacent layers debonding. The single-layer model includes both damage and inelasticity. The interlaminar interface is defined as a two-dimensional mechanical model which ensures traction and displacement transfer from on ply to another. Its mechanical behavior depends on the angle between the fibers of two adjacent layers. This theory is named "mesoscale composite damage theory" by Herakovich in his book (1998).

It is well-known that fracture simulation using a classical continuum damage model leads to severe theoretical and numerical difficulties (Bazant et al. 1994). A second step which has also been achieved, is to overcome theses difficulties (Bazant & Belytschko 1985) (Bazant & Pijaudier-Cabot 1988) (Belytschko & Lasry 1988) (Slyuis & De Borst 1992) (Perzyna 1998) (Dubé et al. 1996). For laminates and, more generally, for composites, we propose the concept of the mesomodel: the state of damage is uniform within each meso-constituent. For laminates, it is uniform throughout the thickness of each single layer; as a complement, continuum damage models with delay effects are introduced (Ladevèze 1989).

Two models have to be identified: the single layer model (Ladevèze & Le Dantec 1992) and the interface model (Allix & Ladevèze 1992) (Allix et al. 1999). The appropriate tests used consist of: tension, bending, delamination. Each composite specimen, which contains several layers and interfaces, is computed in order to derive the material quantities intrinsic to the single layer or to the interlaminar interface.

The single layer model and its identification procedure is detailed in the book (Herakovich 1998). Complements concerning compression behaviour and temperature-dependent behaviour can be founded in (Allix et al. 1994, 1996).

The proposed procedure is rather simple and has been applied to various materials. Various comparisons with experimental results have been performed to show the possibilities and the limits of our proposed computational damage mechanics approach for laminates (Ladevèze 1992, 1995) (Daudeville & Ladevèze 1993) (Allix 1992).

Other contributions to damage mechanics for laminates can be founded particularly in (Highsmith & Reifsnider 1982) (Talreja 1985) (Allen 1994) (Chaboche et al. 1998) and in the book (Voyiadjis et al. 1998).

In this paper, we seek to outline the current state-of-the-art. Attention is also focused on latest developments which concern the interface modelling and the localization phenomena description. Last comparisons between simulations and tests for delamination tests are shown.

2. MESOMODELLING OF LAMINATES

In our pragmatic approach, the characteristic length is the thickness of the plies. The meso-model is defined by means of two meso-constituents:
• the single layer,
• the interface, which is a mechanical surface connecting two adjacent layers and depending on the relative orientation of their fibres (Figure 2). A priori 0°/0° interfaces are not introduced.

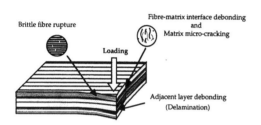

Figure 1. Damage and failure mechanisms

The damage mechanisms are taken into account by means of internal damage variables. A meso-model is then defined by adding another property: a uniform damage state is prescribed throughout the thickness of the elementary ply. This point plays a major role when trying to simulate a crack with a damage model. As a complement, delayed damage models are introduced.

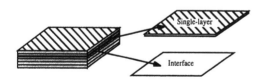

Figure 2. Laminate modelling

One limitation of the proposed meso-model is that the fracture of the material is described by means of only two types of macrocracks:
• delamination cracks within the interfaces,
• cracks, orthogonal to the laminate mid-plane, with each cracked layer being completely cracked throughout its thickness.

Another limitation is that very severe dynamic loading cannot be studied; the dynamic wavelength must be larger than the thickness of the plies.

The single-layer model is briefly given here. A similar model is used for the interface.

2.1 Damage kinematics of the single-layer model

The composite materials (e.g.: carbon-fibre/epoxy-resin) under consideration in this study has only one reinforced direction. In what follows, the subscripts 1, 2 and 3 designate respectively the fibre direction, the transverse direction inside the layer and the normal direction. The energy of the damaged material defines the damage kinematics. Using common notations, this energy is:

$$E_D = \frac{1}{2(1-d_F)}\left[\frac{\langle\sigma_{11}\rangle^2}{E_1^0} + \frac{\varnothing(\langle-\sigma_{11}\rangle)}{E_1^0} - \left(\frac{v_{21}^0}{E_2^0} + \frac{v_{12}^0}{E_1^0}\right)\sigma_{11}\sigma_{22}\right.$$

$$\left. - \left(\frac{v_{31}^0}{E_3^0} + \frac{v_{13}^0}{E_1^0}\right)\sigma_{11}\sigma_{33} - \left(\frac{v_{32}^0}{E_3^0} + \frac{v_{23}^0}{E_2^0}\right)\sigma_{22}\sigma_{33} + \frac{\langle-\sigma_{22}\rangle^2}{E_2^0} + \frac{\langle-\sigma_{33}\rangle^2}{E_3^0}\right]$$

$$+ \frac{1}{2}\left[\frac{1}{(1-d')}\left(\frac{\langle\sigma_{22}\rangle^2}{E_2^0} + \frac{\langle\sigma_{33}\rangle^2}{E_3^0}\right) + \frac{1}{(1-d)}\left(\frac{\sigma_{12}^2}{G_{12}^0} + \frac{\sigma_{23}^2}{G_{23}^0} + \frac{\sigma_{31}^2}{G_{31}^0}\right)\right]$$

$\varnothing$ is a material function which takes into account the non-linear response in compression (Allix et al. 1994). d_F, d and d' are three scalar internal variables which remain constant within the thickness of each single-layer and serve to describe the damage mechanisms inside. The unilateral aspect of microcracking is taken into account by splitting the energy into a "tension" energy and a "compression" energy; $\langle.\rangle$ denotes the positive part. The thermodynamic forces associated with the mechanical dissipation are:

$$Y_d = \frac{\partial}{\partial d}\langle\langle E_D \rangle\rangle\big|_{\sigma:cst} = \frac{1}{2(1-d)^2}\langle\langle\frac{\sigma_{12}^2}{G_{12}^0} + \frac{\sigma_{23}^2}{G_{23}^0} + \frac{\sigma_{31}^2}{G_{31}^0}\rangle\rangle$$

$$Y_{d'} = \frac{\partial}{\partial d'}\langle\langle E_D \rangle\rangle\big|_{\sigma:cst} = \frac{1}{2(1-d')^2}\langle\langle\frac{\langle\sigma_{22}\rangle^2}{E_2^0} + \frac{\langle\sigma_{33}\rangle^2}{E_3^0}\rangle\rangle$$

$$Y_F = \frac{\partial}{\partial d_F}\langle\langle E_D \rangle\rangle\big|_{\sigma:cst} = \frac{1}{2(1-d_F)^2}\langle\langle\frac{\langle\sigma_{11}\rangle^2}{E_1^0} + \frac{\varnothing(\langle-\sigma_{11}\rangle)}{E_1^0}$$

$$-\left(\frac{v_{12}^0}{E_1^0}+\frac{v_{21}^0}{E_2^0}\right)\sigma_{11}\,\sigma_{22}-\left(\frac{v_{13}^0}{E_1^0}+\frac{v_{31}^0}{E_3^0}\right)\sigma_{11}\,\sigma_{33}-\left(\frac{v_{32}^0}{E_3^0}+\frac{v_{23}^0}{E_2^0}\right)\sigma_{22}\,\sigma_{33}\rangle\rangle$$

$\langle\langle.\rangle\rangle$ denotes here the integral value within the thickness and contrary to other previous papers not the mean value.

2.2 Damage evolution law of the single-layer model

From experimental results, it follows that the governing forces of damage evolution are:

$$Y=\left[Y_d+bY_{d'}\right]\ ,\ \ Y'=\left[Y_{d'}+b'Y_d\right]\ ,\ \ Y_F$$

where b and b' are material constants which balance the transverse energy's influence and the shear energy's influence. For small damage rates, we get :

$$d=f_d\left(\underline{Y}^{1/2}\right) for\ d\le 1$$

$$d'=f_{d'}\left(\underline{Y}'^{1/2}\right) for\ d'\le 1$$

$$d_F=f_F\left(\underline{Y}_F^{1/2}\right) for\ d_F\le 1$$

where:

$$\underline{r}|_t= \sup_{\tau\le t} r|_\tau$$

$f_d, f_{d'}$ and f_F are material functions; both progressive and brittle damage evolution are present. For large damage rates, we have introduced a damage model with delay effects:

$$\dot d=\frac{1}{\tau_c}\left[1-exp\left(-a\left\langle f_d\left(Y^{1/2}\right)-d\right\rangle\right)\right]\ \ if\ d<1\ \ ,\ \ d=1\ otherwise$$

$$\dot d'=\frac{1}{\tau_c}\left[1-exp\left(-a\left\langle f_{d'}\left(Y'^{1/2}\right)-d'\right\rangle\right)\right]\ \ if\ d'<1\ \ ,\ \ d'=1\ otherwise$$

$$\dot d_F=\frac{1}{\tau_c}\left[1-exp\left(-a\left\langle f_F\left(Y_F^{1/2}\right)-d_F\right\rangle\right)\right]\ \ if\ d_F<1\ \ ,\ \ d_F=1\ otherwise$$

The same material constants, τ_c and a, are taken for the three damage evolution laws. For this damage model with delay effects, the variations of the forces Y, Y' and Y_F do not lead to instantaneous variations of the damage variables d, d' and d_F. There is a certain delay, defined by the characteristic time τ_c. Moreover, a maximum damage rate, which is $1/\tau_c$,

does exist. Let us also note herein that a clear distinction can be made between this damage model with delay effects and viscoelastic or viscoplastic models: the characteristic time introduced in the damage model with delay effects is of several orders of magnitude less than in the viscous case. This characteristic time is, in fact, related to the fracture process.

Remarks

• Two damage variables are used to described the damage associated to the matrix microcraking and the fiber-matrix debonding. It seems to contain all the proposed damage kinematics including those starting from an analysis of the microcraks. Many works have derived experimentally or theoretically a relation between the microcrak density and our damage variable d which can be very useful for the identification of a damage fatigue model.

• What we call the single-layer is the assemblage of adjacent usual elementary plies of same direction. The damage forces being integral values over the thickness of the single-layer can be interpreted as energy release rates. It follows that the damage evolution law of the single-layer is thickness-dependent. For single-layers which are not too thick, such damage evolution laws include results coming from shear lag analyses. Such a connection will be detailed in a companion paper. It results that the size effects - observed for example in tension - are produced by both the single-layer model and the interface model through a structure problem. Such a theory, which is very simple, works very well for most engineering laminates; however for rather thick layers, it can not be satisfying. A first solution is to modify the damage evolution law, the thickness being a parameter.

• The damage variables are active for $[0°, 90°]_n$ laminates even if the apparent modulus does not change. The model predicts this hidden damage (Ladevèze 1992).

2.3 Damage/plasticity (or viscoplasticity) coupling of the single-layer model

The microcracks, i.e. the damage, lead to sliding with friction and thus to inelastic strains. Such models are reported in (Ladevèze & Le Dantec 1992).

3 IDENTIFICATION

The single-layer model and the interface model have been identified for various materials. The single-layer model identification is based on three canonic tests:

41

$\left| 0°, 90° \right|_s$, $\left| +45°, -45° \right|_s$, $\left| +67,5°, -67,5° \right|_{ls}$. For the interface model, 3D-calculations are combined with the classical delamination tests DCB and NEF. A review of delamination tests can be founded in (Bathias 1995). Quite a lot of experimental works have been done in particular (Whithney 1989) (Albertsen et al. 1995) (Kim 1989) (Davis 1990) (O'Brien 1982) (Crossman & Wang 1982) (Robinson & Song 1992).

Figure 3 gives the main damage material function for the M55J/M18 material; it is valid for any loading and any stacking sequence.

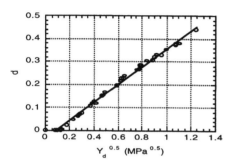

Figure 3. Shear damage material function $Y_d^{1/2} \rightarrow f_d\left(Y_d^{1/2}\right)$ of the single-layer for the M55J/M18 material

Concerning the interface model, $\pm \theta$-interface have been studied (Allix et al. 1999). Figure 4 shows the corrected critical energy release rate obtained for classical delamination tests and for different stacking sequences; the damages inside the plies have been taken into account. These values do not depend on θ excepted for the 0°/0° interface which appears to be something special. Consequently, we use the hypothesis: the interface model does not depend on θ.

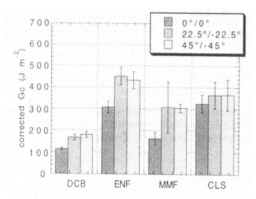

Figure 4. : Critical energy release rates at propagation

4 QUALITATIVE ANALYSIS OF THE MESODAMAGE MODEL WITH DELAY EFFECTS

In order to investigate the performance of the damage model with delay effects, we consider the classical example of a bar in Dynamics. More details can be founded in (Allix & Deu 1997) (Ladevèze et al. 1999).

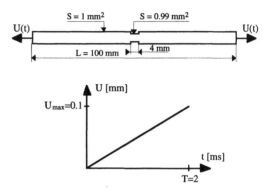

Figure 5. One-dimensional bar problem

The analysis is based on a simple one-dimensional damage model with only one scalar damage variable. The model is defined by its strain energy, E_D, which is split into two parts according to whether the cracks are closed or open.

$$\sigma = E^0 \left(1-d\right) \langle \epsilon \rangle + E^0 \langle -\epsilon \rangle$$

$$E_D = \frac{1}{2}\left[\frac{\langle \sigma \rangle^2}{E^0\left(1-d\right)} + \frac{\langle -\sigma \rangle^2}{E^0} \right]$$

$$Y = \frac{\partial E_D}{\partial d}\Big|_{\sigma : cst} = \frac{\langle \sigma \rangle^2}{2E^0\left(1-d\right)^2} = \frac{E^0\langle \epsilon \rangle^2}{2}$$

where Y is the damage energy release rate. Y is assumed to drive the damage evolution. In fact, for many long-fibre composites and for a progressive damage mode, a typical quasi-static damage evolution law, for slow damage rates, is:

$$\begin{cases} d = f\left(\underline{Y}\right) & \text{if } d < 1 \\ d = 1 & \text{otherwise} \end{cases} \quad \text{with} \quad \begin{cases} \underline{Y}\big|_t = \sup_{\tau \le t} Y\big|_\tau \\ f\left(\underline{Y}\right) = \left\langle \frac{\sqrt{\underline{Y}} - \sqrt{Y_0}}{\sqrt{Y_c}} \right\rangle \end{cases}$$

The corresponding damage model with delay effects is:

$$\dot{d} = \frac{1}{\tau_c} \left| 1 - exp\left(-a\left\langle f(Y) - d\right\rangle\right)\right| \quad if\, d < 1 \ , \ d = 1 \ otherwise$$

The results obtained for different meshed are plotted in figure 6. It can be seen that the numerical results do not depend on the mesh size. Finally, figure 7 displays the global load versus the prescribed end displacement. All of these results demonstrate that fracture phenomena are well described by a damage model with delay effects.

Remarks

• The delay effects introduce two more material constants which have to be identified. A first link is given by the value of the critical energy release rate: damage mechanics must contain the usual fracture mechanics for cracks. To find another relation, let us note that the proposed damage model with delay effects introduces a maximum value for the damage rate which is $1/\tau_c$. Figure 8 shows what happens at the bar's centre: just before the material is completely destroyed the damage rate reaches its maximum value. A first identification which seems to be consistent in regard to experimental results is to take for $1/\tau_c$ half of the matrix Rayleigh wave speed.

It is to say that the values of these material constants do not play an important role in final fracture prediction for most engineering problems for which the precritical state is generally not uniform. At the contrary, the fact that the damage state is prescribed locally constant within the thickness of the layers is essential.

• For most fracture engineering problems with quasi-static loading, dynamic effects are negligible; it is then not necessary to introduce them.

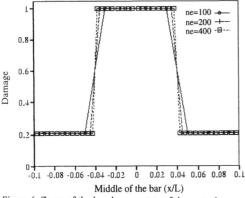

Figure 6. Zoom of the bar damage state of the central zone at time 2ms

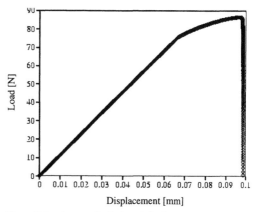

Figure 7. Load versus the bar's end displacement

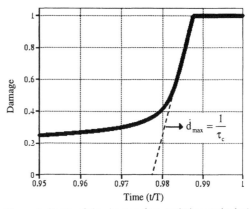

Figure 8. Zoom of the damage-time evolution at the bar's centre

5 EXAMPLES OF LAMINATED STRUCTURE COMPUTATIONS

Various calculations and comparisons with tests have been done for quite a lot of materials.

5.1 Comparison calculations/tests for tension and compression loadings - Case without delamination

Different stacking sequences and different materials have been studied in uniaxial tension (Ladevèze & Le Dantec 1992) (Herakovich 1998) (Ladevèze 1992). Compression loading is investigated in (Allix et al. 1994) and temperature effects in (Allix 1996). All the model predictions are satisfying for loading directions and stacking sequences for which delamination is negligible.

5.2 Comparison calculations/tests for Edge delamination tension specimens

The complete damage mesomodel is active. Comparisons are reported in (Daudeville & Ladevèze 1993) (Ladevèze et al. 1998).

5.3 Comparison calculations/tests for classical delamination tests

As in the previous case, the interface model is active. It is identified using DCB and for a part ENF delamination tests. The calculations take also into account the damage inside the plies which can be important. Here, the results concerning MMF and ENF tests are reported on figures 9 and 10 [see also (Ladevèze et al. 1998)] for M55J/M18 material.

5.4 Comparison calculations/tests for low velocity impacts

The damage mesomodel has been used; calculations have been done using the code DSDM (Allix 1992). Data are given figure 11. The reported results have been obtained in (Allix & Guinard 1999). Figure 12

gives the final damage state which is rather well described in regard to experiments. It is not simple: transverse cracks and delamination cracks are both present.

5.5 Comparison calculations/tests for an holed plate submitted to quasi-static loadings

Such results are reported in (Allix 1992) and (Ladevèze et al. 1999).

5.6 Delamination computation in Dynamics

An example of a 3D finite element computation is presented in order to demonstrate the ability of the damage mesomodel to predict the response of a composite structure in dynamics until its ultimate fracture. This response is computed using the explicit dynamic code LS-DYNA3D. Figure 13 defines the studied structure and its loading. It is a $[+22.5°,-22.5°]_s$ holed laminated plate; the material is a SiC/MAS-L composite with silicon carbide fibres and a glass matrix made by Aerospatiale. The fibre stiffness (200 GPa) is higher than the matrix stiffness (75 GPa), and cracks first appear in the matrix. Let us note that reasonable values have been chosen for the material constants of the interlaminar interface model. In particular, the values of the

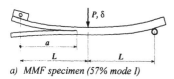

a) MMF specimen (57% mode I)

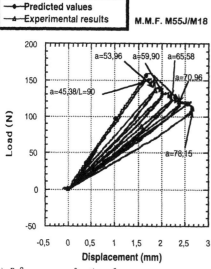

b) P-δ curves as a function of a

Figure 9. Prediction of an MMF test. Comparison between experimental results and predicted values. The initial crack closure is a = 45 mm, the crack length at the end of the test is 32.77 mm

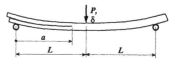

a) ENF specimen (0% mode I)

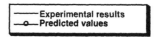

b) P-δ curve as a function of a

Figure 10. Comparison between experimental results on an ENF test and 3D F.E. prediction

44

critical times τ_c and τ'_c and the constants a and a' are:

$$\tau_c = \tau'_c = 2\,\mu s$$

$$a = a' = 1$$

Several computations have been performed especially for the stacking sequence $[n\,0°]_s$. These

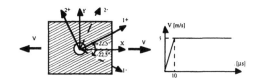

Figure 13. Holed laminate submitted to dynamic tension loading

Experimental datas

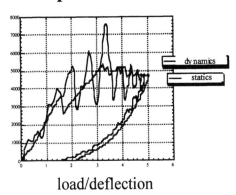

load/deflection

Figure 11. Low velocity impact data

material :	T300-914
stacking sequence :	[45°, 0°]
contact surface :	1 mm^2
force contact :	220 N

Damage in 45° bottom ply

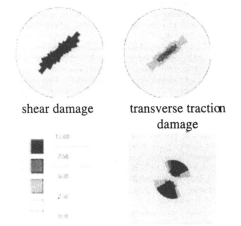

shear damage   transverse traction damage

Damage levels   Delamination

Figure 12. Final damage state showing transverse cracks and delamination cracks

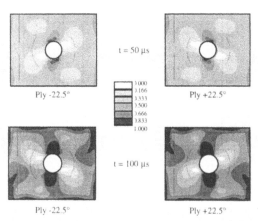

Figure 14. Interface damage map at several times

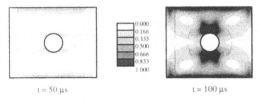

Figure 15. Shear damage maps for the plies at several times

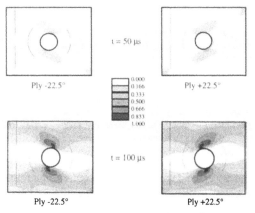

Figure 16. Longitudinal damage maps for the plies at several times

45

reasonable constant values correspond to a fracture zone size whose order of magnitude corresponds to the ply's thickness.

Figure 14 reveals the degradation of the [±22.5°] interface; the dark area represents the completely destroyed zone and then the delamination crack.

Figures 15 and 16 present the microcracking intensity maps and the fibre-direction damage maps at different times. It is clear that a transverse crack orthogonal to the fibres appears and then grows inside each ply. One can consider that the final fracture occurs around t = 100 μs; the size of the transverse cracks is about 2 mm.

Last, the global load versus the prescribed displacement is plotted in figure 17. No particular numerical difficulty with respect to time discretisation and mesh sensitivity has appeared.

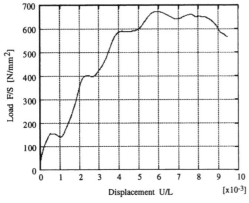

Figure 17. Global load versus the prescribed displacement

CONCLUSION

The laminate mesomodel proposed herein is able to compute the intensities of the damage mechanisms inside both the plies and the interfaces at any time, up until final fracture. Simulations have shown the macrocracks' initiation and propagation. Comparisons with experimental results have proved to be very satisfactory.

However, the computations performed with such a mesomodel do generate very large computational times. One present challenge is to develop a more effective and robust computational strategy and, in particular, to use parallel computers. Another challenge is to extend what has been carried out for laminates to other composite structures. A first attempt is the damage model derived in (Ladevèze 1995) for CMCs which seems to be valid for a larger class of composite materials.

REFERENCES

Albertsen, H.J., J. Ivens, P. Peters, M. Wevers & I. Verpoest 1995. Interlaminar fracture toughness of CFRP influenced by fibre surface treatment: Part 1: Experimental results. *Composite Science and Technology.* 54: 133-145.

Allen, D.H. 1994. Damage evolution in laminates. In R. Talreja (ed.), *Damage Mechanics of Composites Materials*: 79-114. Amsterdam: Elsevier.

Allix, O. 1992. Damage analysis of delamination around a hole. In P. Ladevèze & O.C. Zienkiewicz (eds.), *New Advances in Computational Structural Mechanics*: 411-421. Elsevier Science Publishers B.V.

Allix, O. & P. Ladevèze 1992. Interlaminar interface modelling for the prediction of laminate delamination. *Composite Structures.* 22: 235-242.

Allix, O., P. Ladevèze & E. Vittecoq 1994. Modelling and identification of the mechanical behaviour of composite laminates in compression. *Composite Science and Technology.* 51: 35-42.

Allix; O., N. Bahlouli, C. Cluzel & L. Perret 1996. Modelling and identification of temperature-dependent mechanical behaviour of the elementary ply in carbon/epoxy laminates. *Composites Science and Technology.* 56: 883-888.

Allix, O. & J.F. Deü 1997. Delay-damage modeling for fracture prediction of laminated composites under dynamic loading. *Engineering Transactions.* 45: 29-46.

Allix, O., D. Guedra-Degeorges, S. Guinard & A. Vinet 1999. 3D analysis applied to low-energy impacts on composite laminates. *Proceedings ICCM12.* T. Massard and A. Vautrin (eds.): 282-283.

Allix, O., D. Leveque & L. Perret in press. Interlaminar interface model identification and forecast of delamination in composite laminates. *Composite Science and Technology.*

Bathias, C. 1995. Une revue des méthodes de caractérisation du délaminage des matériaux composites. In O. Allix & M.L. Benzeggah (eds.), *Délaminage: bilan et perspectives* - Journée AMAC/CSMA, Cachan.

Bazant, Z.P. & T.B. Belytschko 1985. Wave propagation in a strain softening bar: Exact solution. *Journal of Engineering Mechanics.* 111: 381-389.

Bazant, Z.P. & G. Pijaudier-Cabot 1988. Non local damage: continuum model and localisation instability. *J. of Appl. Mech.*: 287-294. ASME 55.

Bazant, Z.P., Z. Bittnar & M. Jirasek 1994. *Fracture and Damage in Quasibrittle Structures* (eds.), Amsterdam, Elsevier.

Belytschko, T. & D. Lasry 1988. *Localisation limiters and numerical strategies softening materials.* In J. Mazars and Z.P. Bazant (eds.), 349-362. Amsterdam, Elsevier.

Chaboche, J.L., O. Lesné & T. Pottier 1998. Continuum Damage Mechanics of Composites: Towards a Unified Approach. In: G.Z. Voyiadjis, J.W. Wu and J.L. Chaboche (eds.), *Damage Mechanics in Engineering Materials.* 3-26. Amsterdam, Elsevier.

Crossman, F.W. & A.S.D. Wang 1982. The dependence of transverse cracking and delamination on ply thickness in graphite/epoxy laminates. In K.L. Reifsnider (ed.), *Damage in Composite Materials*: 118-139. ASTM STP 775.

Daudeville, L. & P. Ladevèze 1993. A damage mechanics tool for laminate delamination. *Journal of Composite Structures.* 25: 547-555.

Davis, P. 1990. Measurement of GIc and GIIc in Carbon/Epoxy composites. *Composite Science and Technology.* 39: 193-205.

Dubé, J.F., G. Pijaudier-Cabot & C. La Borderie 1996. Rate dependent damage model for concrete in dynamics. *Journal of Engineering Mechanics.* 122: 939-947.

Herakovich, C.T. 1998. *Mechanics of fibrous composites*. J. Wiley.

Highsmith, A. & K.L. Reifsnider 1982. Stiffness reduction mechanism in composite material. In: *Damage in Composite Materials*: 103-107. ASTM-STP 775.

Kim, R.Y. 1989. Experimental observations of free-edge delamination. In: N.J. Pagano (ed.), *Interlaminar response of composite materials*: 111-160.

Ladevèze, P. 1986. Sur la mécanique de l'endommagement des composites. In: C. Bathias & D. Menkès (eds): *Comptes-Rendus des JNC5*: 667-683. Paris, Pluralis Publication.

Ladevèze, P. 1989. About a damage mechanics approach. In D. Baptiste (ed.), *Mechanics and Mechanisms of Damage in Composite and Multimaterials*:119-142. MEP.

Ladevèze, P. 1992. A damage computational method for composite structures. *J. Computer and Structure*. 44(1/2): 79-87.

Ladevèze. P. 1992. Towards a fracture theory. In D.R.J. Owen, E. Onate & E. Hinton (eds.), *Proceedings of the Third International Conference on Computational Plasticity Part II*: 1369-1400. Cambridge, Pineridge Press.

Ladevèze, P. & E. Le Dantec 1992. Damage modeling of the elementary ply for laminated composites. *Composite Science and Technology*. 43-3: 257-267.Ladevèze, P., O. Allix, J.F. Deu and D. Leveque to appear. A mesomodel for localisation and damage computation in laminates. Comput. Meth. Appl. Mech. Engrg.

Ladevèze, P. 1995. A damage computational approach for composites: Basic aspects and micromechanical relations. *Computational Mechanics*. 17: 142-150.

Ladevèze, P. 1995. Modeling and simulation of the mechanical behavior of CMCs. In A.G. Evans & R. Naslain (eds.), *High-temperature Ceramic-matrix composites*: 53-63. Ceramic Transaction.

Ladevèze, P., O. Allix, L. Gornet, D. Leveque and L. Perret 1998. A computational damage mechanics approach for laminates: Identification and comparison with experimental results. In G.Z. Voyiadjis, J.W. Wu & J.L. Chaboche (eds.), *Damage Mechanics in Engineering Materials*: 481-500. Amsterdam, Elsevier.

Ladevèze, P., O. Allix, J.F. Deu & D. Leveque in press. A mesomodel for localisation and damage computation in laminates. *Comput. Meth. Appl. Mech. Engrg*.

O'Brien, T.K. 1982. Characterisation of delamination onset and growth in a composite laminate. In K.L. Reifsnider (ed.), D*amage in Composite Material*: 140-167. ASTM STP 775.

Perzyna, P. 1998. Dynamic localized fracture in inelastic solids. In G.Z. Voyiadjis, J.W. Wu and J.L. Chaboche (eds.), *Damage Mechanics in Engineering Materials*: 183-202. Amsterdam, Elsevier.

Robinson, P. & D.Q. Song 1992. A modified DCB specimen for mode I testing of multidirectional laminates. *Composite Science and Technology*. 26: 1554-1577.

Russell, A.J. & K.N. Street 1985. Moisture and temperature effects on the mixed-mode delamination fracture of unidirectional graphite/epoxy. In W.S. Johnson (ed.), *Delamination and debonding of materials*: 349-370. Philadelphia, ASTM STP 876.

Slyuis, L.J. & R. de Borst 1992. Wave propagation and localisation in a rate-dependent cracked medium: Model formulation and one dimensional examples. *International Journal of Solids and Structures*. 29: 2945-2958.

Talreja, R. 1985. Transverse cracking and stiffness reduction in composite laminate. *Journal of Composite Materials*. 19: 355-375.

Voyiadjis, G.Z., J.W. Wu & J.L. Chaboche 1998. *Damage Mechanics in Engineering Materials*.Amsterdam, Elsevier.

Whithney, J.M. 1989. Experimental characterization of delamination fracture. In N.J. Nagano (ed.): Interlaminar response of composite materials. *Comp. Mat. Series*. 5: 111-239.

Recent Developments in Durability Analysis of Composite Systems, Cardon, Fukuda, Reifsnider & Verchery (eds)
© 2000 Balkema, Rotterdam, ISBN 90 5809 103 1

Kinetic modelling of weight changes during the isothermal oxidative ageing of bismaleimide matrix

X. Colin, C. Marais & J. L. Cochon
ONERA, Département Matériaux et Systèmes Composites, Châtillon, France

J. Verdu
ENSAM, Laboratoire de Transformation et de Vieillissement des Plastiques, Paris, France

ABSTRACT: With the intention of being used for the next generation supersonic aircraft, bismaleimide polymer used as a matrix for the T800H/F655-2 composite system was put into context by subjecting it to accelerated isothermal ageing conditions. Since the oxidation process is controlled by oxygen diffusion over the glass temperature range of this thermoset, a physical approach coupling oxygen diffusion to chemical reaction is applied in order to modelise weight loss and thickness of oxidised layer versus ageing time. The kinetic model shows a good agreement with experimental data and allows to well understand kinetic mechanisms involved in the thermo-oxidation of the bismaleimide polymer.

1 INTRODUCTION

Carbon/polymer composites are likely to be used as structural materials for the next generation supersonic aircraft and thus will have to face severe service conditions. Previous studies (Favre et al. 1996, Nam & Seferis 1992) showed up a gradual structural modification of the organic matrix of composite laminates during accelerated thermal ageing conditions. Damage is a combination both of oxidation and thermally induced micro-cracking of the matrix. Therefore, suitable sampling and testing as well as adequate modelling are required to predict the long-term behaviour of composite materials.

The present paper focuses on matrix oxidation after isothermal ageing. Since carbon fibres are thermally stable over a wide temperature range from 150°C to 240°C, the study will be restricted to the bismaleimide polymer used as a matrix for the T800H/F655-2 composite system. Besides, since the microscopic analysis (Colin et al. 1999) displays an heterogeneous oxidation of the polymer such as an ageing-induced « skin-core » structure representative of an oxidation process controlled by oxygen diffusion, the methodology, set for classical linear polymers (Audouin et al. 1994), will be applied on the bismaleimide thermoset. This methodology is summarised in figure 1.

First, an one-dimensional kinetic model is build-up for the isotropic bismaleimide polymer. This model predicts profiles both of oxygen concentration and quantity of consumed oxygen within the sample by coupling an oxygen diffusion law to the chemical reaction kinetics. The oxygen diffusion is modelised by the classical Fick's second law where coefficients of oxygen diffusion D and oxygen solubility S are determined by permeation experiments. Analytical expressions for chemical oxygen consumption are deduced from the standard oxidation scheme for classical linear polymers (Bolland & Gee 1946) and from thermogravimetry measurements carried out on very thin polymer films subjected to various oxygen partial pressures.

Then, the validity range of the kinetic model is discussed. With this intention, a theoretical criterion is proposed to stand for the thickness of the oxidised layer and compared to experimental data coming from an investigation of the weight loss dependence on the specimen thickness.

2 MATERIALS AND EXPERIMENTAL DETAILS

2.1 *Materials*

The composite system was selected for this study by an previous assessment of the oxidative stability of fibre/resin systems currently used in aircraft composite structures. It is made of carbon fibres/bismaleimide resin (Hexcel T800H/F655-2).

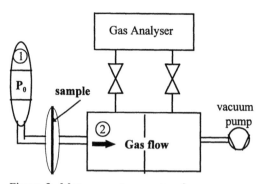

Figure 2. Measurement apparatus for oxygen permeation experiments.

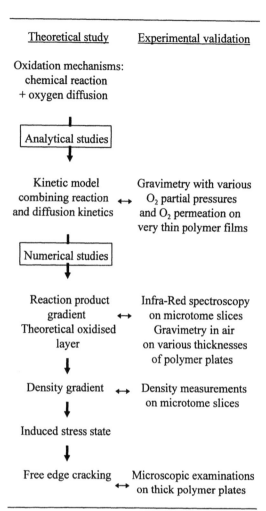

Theoretical study	Experimental validation
Oxidation mechanisms: chemical reaction + oxygen diffusion	
Analytical studies	
Kinetic model combining reaction and diffusion kinetics ↔	Gravimetry with various O₂ partial pressures and O₂ permeation on very thin polymer films
Numerical studies	
Reaction product gradient ↔ Theoretical oxidised layer	Infra-Red spectroscopy on microtome slices Gravimetry in air on various thicknesses of polymer plates
Density gradient ↔	Density measurements on microtome slices
Induced stress state	
Free edge cracking ↔	Microscopic examinations on thick polymer plates

Figure 1. Methodology for studying the neat polymer oxidation.

Neat bismaleimide films and plates were processed by press-moulding and then post-cured under primary vacuum in accordance with the recommended temperature cure cycle. Then, all the materials were inspected prior to testing. Particularly, their glass transition temperature was measured by means of Differential Scanning Calorimetry as the temperature of the onset of the endothermal increase in heat capacity on the thermogram. For a heating rate of 20°C/min, T_g is located around 255°C.

2.2 *Experimental procedures*

All the specimens were conditioned under vacuum (10^{-3} Pa) before testing. All the thermal tests were performed over the glass temperature range of the bismaleimide polymer.

Coefficients of oxygen diffusion D and oxygen solubility S were measured experimentally at various temperatures from 20°C to 120°C. With this intention, a polymer film is put between two volumes (Figure 2), the first one containing oxygen at a pressure P_0 whereas the second one being initially under vacuum (10^{-5} Pa). The pressure difference induces a gas flow through the sample. The gas, which penetrates in the second volume, is analysed by means of a mass spectrometer which gives access to the oxygen partial pressure versus test time.

The influence of oxygen partial pressure on the chemical reaction kinetics was determined by subjecting very thin polymer films to isothermal ageing from 150°C to 240°C. The polymer sample is placed on the plateau of a microbalance METTLER AM50 under various nitrogen /oxygen atmospheres of 1 bar pressure and the weight of the sample is recorded versus ageing time continuously.

The coupling oxygen diffusion/chemical reaction was investigated by exposing various thicknesses of polymer samples to isothermal ageing conditions from 150°C to 240°C in air-circulating. Samples are removed from ovens intermittently, cooled at room temperature in a dessicator, weighted and returned to ovens.

3 KINETIC EQUATIONS

Most of the authors modelise oxygen diffusion throughout a bulk polymer by the classical Fick's second law and boundary conditions, on specimen free surfaces, by the Henry's law. Thus the kinetic models only differ from one to another by selecting of an analytical expression of the local oxygen consumption rate R(C) :

$$\frac{\partial C}{\partial t} = D\frac{\partial^2 C}{\partial x^2} - R(C) \qquad (1.1)$$

$$C^S = SP_{O2} \qquad (1.2)$$

where C is the oxygen concentration, t is the ageing time, x is the depth of penetration for reactive species, D is the coefficient of oxygen diffusion into the material, S is the coefficient of oxygen solubility and P_{O2} is the oxygen partial pressure of the oxidising environment.

In the steady-state, the equation (1.1) becomes :

$$D\frac{\partial^2 C}{\partial x^2} - R(C) = 0 \qquad (1.3)$$

The analytical expression for R(C) is deduced from the standard oxidation scheme. This scheme composed of three main stages can be presented as following (Bolland & Gee 1946):

Initiation:	$POOH \rightarrow 2P^\circ + H_2O + \upsilon V$	k_1
Propagation:	$P^\circ + O_2 \rightarrow PO_2^\circ$	k_2
	$PO_2^\circ + PH \rightarrow POOH + P^\circ$	k_3
Termination:	$P^\circ + P^\circ \rightarrow$ inactive species	k_4
	$P^\circ + PO_2^\circ \rightarrow$ inactive species	k_5
	$PO_2^\circ + PO_2^\circ \rightarrow O_2 +$ inactive species	k_6

$$\qquad (1.4)$$

where $k_{i=1,\ldots,6}$ are rate constants and V are volatile species, formed with a yield υ, and distinguished from H_2O molecules coming from OH$^\circ$ radicals.

In the steady state, this scheme leads to complex analytical expressions for each chemical species concentration $[P^\circ]$, $[PO_2^\circ]$ and $[POOH]$.

After putting $\psi = 4\frac{k_4 k_6}{k_5^2}$ and $\beta = \frac{k_2 k_6}{2k_5 k_3 [PH]}$, and achieving some approximations such as $\psi < 1$ (Gillen et al. 1995) and other suitable transformations, the simplified expressions may be obtained :

$$[P^\circ] = \frac{k_3 [PH]}{k_5}\frac{1}{1+\beta C} \qquad (1.5)$$

$$[PO_2^\circ] = \frac{k_3 [PH]}{k_6}\frac{\beta C}{1+\beta C} \qquad (1.6)$$

and $$[POOH] = \frac{k_3^2 [PH]^2}{k_1 k_6}\frac{\beta C}{1+\beta C} \qquad (1.7)$$

From these equations, the local oxygen chemical consumption rate R(C) may be expressed as :

$$R(C) = -\frac{dC}{dt} = k_2 C[P^\circ] - k_6 [PO_2^\circ]$$
$$= 2\frac{\alpha C}{1+\beta C}\left(1 - \frac{\beta C}{2(1+\beta C)}\right) \qquad (1.8)$$

with $\alpha = \frac{k_2 k_3 [PH]}{k_5}$.

From this equation, the relationship between the local weight loss rate and oxygen consumption rate can be expressed as :

$$\frac{\left(\frac{dM}{M_0}\right)}{dt} = \frac{32}{\rho}R(C) - \frac{18}{\rho}\frac{d[H_2O]}{dt} - M_V\frac{d[V]}{dt}$$

$$= \frac{1}{\rho}\frac{\alpha C}{1+\beta C}\left((46 - \upsilon M_V) - 32\frac{\beta C}{1+\beta C}\right) \qquad (1.9)$$

where M_0 is the initial polymer weight, M is the polymer weight for the ageing time t, M_V is the molar weight of volatile species V, and ρ is the initial polymer density.

4 DETERMINATION OF KINETIC PARAMETERS

The resolution of the equations (1.1) and (1.2), based on the above assumptions, requires the knowledge of the kinetic parameters D and S as well as α and β at ageing temperature.

4.1 Experimental determination of D and S

The coefficients D and S were measured at various temperatures from 20°C to 120°C in order to estimate their variation with the temperature. The coefficient of oxygen diffusion obeys to the following Arrhenius law :

$$D(T) = D_0 \exp\left(\frac{-E_a}{RT}\right) \qquad (2.0)$$

where the activation energy E_a is equal to 16 kJ/mol and the pre-exponential factor D_0 to 1,95 10^{-10} m^2/s which gives D (240°C) = 4,6 10^{-12} m^2/s for example.

At the opposite, no clear tread of variation with the temperature for the coefficient of oxygen solubility was observed. Thus, its mean value was took :

$$S = 3,6 \cdot 10^{-4} \text{ mol/m}^3/\text{Pa} \qquad (2.1)$$

4.2 Experimental determination of the thickness of the oxidised layer

Before determining the reaction kinetic parameters α and β, adequate sampling must be achieved in order to avoid any diffusion-controlled effect on the reaction kinetics.

The boundary between non-diffusion and diffusion-controlled regimes can be simply evaluated by comparing mass loss rates for various thicknesses of polymer samples (Figure 3). For each sample thickness, weight loss rate was thus determined from the linear part of the curve corresponding to the steady state. Then, weight loss rate was plotted versus the reciprocal thickness. As an example, data obtained at 240°C are given in the Figure 4.

It appears clearly that weight loss strongly depends on the thickness of the sample. For the thick samples, oxygen diffusion controls the oxidation process kinetics and slows down weight loss rate. Whereas, for the thin samples, oxygen diffusion has no effects on oxidation process and weight loss rate reaches a maximum value.

One expects to a critical thickness corresponding to the double of the oxidised layer thickness. Its value is about 35μm at 240°C for instance.

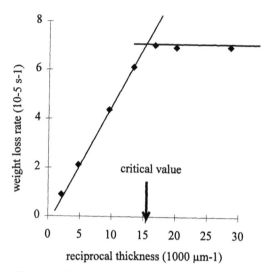

Figure 4. Weight loss rates measured for various sample thicknesses in air at 240°C.

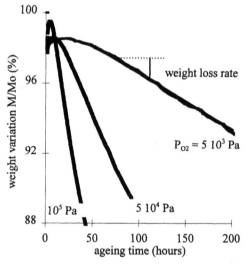

Figure 5. Example of thermogravimetry curves obtained for polymer films of 50μm thick at 240°C.

4.3 Experimental determination of α and β

Thus coefficients α and β were deduced by comparing the analytical expression for local weight loss rate (1.9) to experimental data measured for very thin polymer films.

With this intention, thermogravimetry experiments were performed on 50 μm thick films subjected to various oxygen partial pressures (Figure 5).

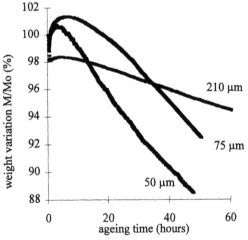

Figure 3. Example of thermogravimetry curves obtained for various sample thicknesses in air at 240°C.

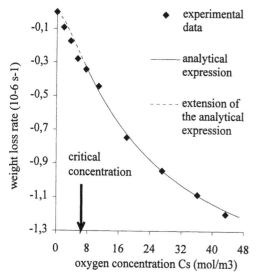

Figure 6. Weight loss rate measured on 50μm thick polymers films exposed to various oxygen partial pressures at 240°C. Comparison to theoretical values calculated with analytical expression (1.9).

For each oxygen partial pressure, oxygen concentration within the film C^S and weight loss rate in the steady state were calculated. Then, weight loss rate was plotted versus oxygen concentration C^S and fitted by the analytical expression (1.9). As an example, data obtained at 240°C are given in the Figure 6.

For $C^S > 7,2$ mole/m³, analytical expression (1.9) shows a good agreement with experimental data. Thus, values of α, β and M_v can be calculated over this concentration range :

$\alpha(240°C) = 6,42 \cdot 10^{-3}$ s⁻¹,
$\beta(240°C) = 9,17 \cdot 10^{-2}$ m³/mol
and $M_v = 47,36$ g/mol   (2.2)

At the opposite, below 7,2 mole/m³, a slight divergence is observed. In fact, it can be demonstrated that the analytical approximations made above are valid only over the concentration range $C > \dfrac{2\psi}{\beta}$.

Our result could be then explained by the fact that concentrations lower than 7,2 mole/m³ are out of this range, which would correspond to $\psi(240°C) \sim 0,35$.

But, since the above divergence remains reasonably small, we will consider, in a first approximation, that the analytical expression is valid over the whole partial pressure range.

5 NUMERICAL STUDIES OF THE KINETIC MODEL

5.1 *Numerical resolution*

A finite element analysis was achieved from a Matlab programming in order to compute the general trends of profiles both of oxygen concentration $C(x,t)$ and total quantity of consumed oxygen $Q(x,t) = \int_0^t R(C)dt$ within the thick polymer plates. Examples of profiles obtained, by using the general equations (1.1) and (1.2), are given in Figures 5 and 6.

Despite all the above approximations, profiles both of oxygen concentration and consumed oxygen become rapidly independent of ageing time which confirms the hypothesis of a steady-state. Moreover, this indicates that oxygen attack first affects specimen free surfaces and then progresses as a decreasing function of ageing time within the specimen. Thus, the thickness of oxidised layer tends towards an asymptotic value as experimentally observed previously by means of optical microscopy on polished cross-sections (Colin et al. 1999).

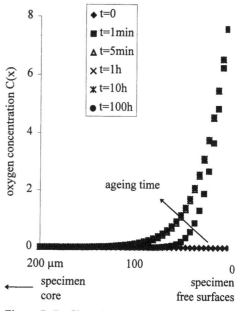

Figure 7. Profiles of oxygen concentration in air at 240°C.

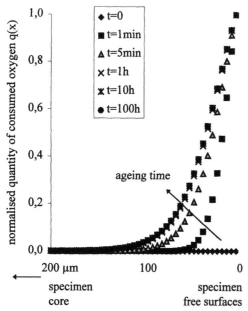

Figure 8. Profiles of normalised quantity $q(x,t) = \dfrac{Q(x,t)}{Q_s(t)}$ of consumed oxygen profiles in air at 240°C.

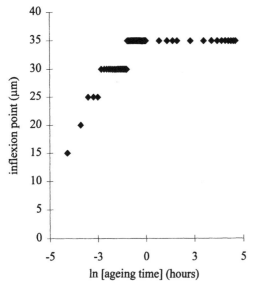

Figure 9. Evolution of the inflexion point of the curves $q(x,t) = f$(ageing time) in air at 240°C.

5.2 Experimental validation

A very simply way to check these tendencies consists in looking at the evolution of the inflexion point of the curves $q(x,t) = f$(ageing time) and comparing

this arbitrary criterion to experimental data. Since first calculations show that this inflexion point takes values which corresponds to experimental thicknesses of oxidised layers (see §4.2), we choose it to stand for the evolution of the thickness of the oxidised layer.

Thus, our kinetic model appears to be built-up on solid basis.

6 CONCLUSION

Thus, a methodology derived from studies on classical linear polymers (Audouin et al. 1994) allows us to well understand kinetic mechanisms involved in the thermo-oxidation of bismaleimide polymer matrix. We established a kinetic model leading to theoretical results in agreement with the experimental observations such as :

- oxidation process is diffusion-controlled;
- chemical reaction rate is a pseudo-hyperbolic function of the oxygen concentration within the specimen;
- oxygen attack first affects the specimen free surfaces and then progresses as a decreasing function of ageing time within the specimen ;
- the thickness of the oxidised layer tends towards an asymptotic value;
- a steady state appears rapidly.

REFERENCES

Audouin, L., V. Langlois & J. Verdu 1994. Review. Role of oxygen diffusion in polymer ageing: kinetic and mechanical aspects. *J. of Mat. Sci.* 29: 569-583.
Bolland, J.L. & G. Gee 1946. Kinetics studies in the chemistry of rubber and related materials. II. Thermochemistry and mechanisms of olefin oxidation. *Trans. Far. Soc.* 42: 244.
Colin, X., C. Marais & J-.P. Favre 1999. Damage/weight loss relationship of polymer matrix composites under thermal ageing. *Proceedings of the 12th Internat. Conf. on Comp. Mat. (ICCM-12), Paris, July 5-9th 1999.* (to be published).
Favre, J-.P., H. Levadoux, T. Ochin & J. Cinquin 1996. Ageing of organic matrix composites at moderate temperature - A first assessment. In D. Baptiste & A. Vautrin (eds), *10th Nat. Conf. on Comp. Mat. (JNC-10)* 1: 205-214. Paris: AMAC.
Gillen, K.T., J. Wise & R.L. Clough 1995. General solution for the basic auto-oxidation scheme. *Polym. Deg. Stab.* 47: 149-161.
Nam, J.D. & J.C. Seferis 1992. Anisotropic thermo-oxidative stability of carbon fiber reinforced polymeric composites. *SAMPE Quaterly.* 24(1): 10-18.

Method for identification of elastic and damage properties of laminates

R. Rikards
CAD Institute, Riga Technical University, Latvia

ABSTRACT: A numerical-experimental method for the identification of mechanical properties of laminated composites from the experimental results is developed. It is proposed to use the method of experiment design and the response surface approach to solve the identification (inverse) problems. The response surface approximations are obtained by using the information on the behavior of a structure in the reference points of the experiment design. The finite element modeling of the structure is performed only in the reference points. Therefore, a significant reduction (about 50-100 times) in calculations of the identification functional can be achieved in comparison with the conventional methods of minimization. The functional to be minimized describes the difference between the measured and numerically calculated parameters of the response of structure. By minimizing the functional the identification parameters are obtained. The method is employed to identify the elastic and damage properties of the laminates from the measured eigenfrequencies of the plates.

1 INTRODUCTION

During the last years investigations for developing a new technique for material identification, the so-called mixed numerical-experimental technique, have started (Sol 1986, Pedersen 1989, Frederiksen 1992, Mota Soares et al. 1993). The determination of stiffness parameters for complex materials such as fiber reinforced composites is much more complicated than for isotropic materials since composites are anisotropic and non-homogeneous. Conventional methods for determining stiffness parameters of the composite materials are based on direct measurements of strain fields. Boundary effects, sample size dependencies and difficulties in obtaining homogeneous stress and strain fields are some of the most serious problems. Because of this, indirect methods have recently received increasing attention. One of such indirect methods is based on measurements of the structure response and application of the numerical-experimental identification technique.

Numerical-experimental identification methods are mainly used in structural applications. For example, elastic properties of laminated composites have been identified by using experimental eigenfrequencies (Mota Soares et al. 1993). The stiffness parameters were identified from the measured natural frequencies of the laminated composite plate by direct minimization of the identification functional. Similar approach in order to identify the stiffness properties of the laminated composites was used also by other au-

thors (Frederiksen 1997, Araújo 1995). It was shown (De Visscher et al. 1997) that the mixed numerical-experimental method can be used also for identification of damping properties of polymeric composites.

In the present study a numerical-experimental method for the identification of mechanical properties of laminated polymeric composites from the experimental results of the structure response has been further developed. The difference between conventional (Mota Soares et al. 1993, Frederiksen 1997) and present approach is that instead of direct minimization of identification functional the experiment design is used, by which response surfaces of the functional to be minimized are obtained. The response surface approximations are obtained by using the information on the behavior of a structure in the reference points of the experiment design. The finite element modeling of the structure is performed only in the reference points. The functional to be minimized describes the difference between the measured and numerically calculated parameters of the response of structure. By minimizing the functional the identification parameters are obtained. The method is employed to identify the elastic properties of the cross-ply laminates from the measured eigenfrequencies of the plates. The elastic constants of a single transversely isotropic layer have been determined. The example of identification of the location and size of damage zone of transversely isotropic plate also is presented. The main advantage of the present method is a significant reduction of the computational efforts. Previously this method was used for the solution of the optimum design problems of laminated composite and sandwich plates (Rikards & Chate 1995).

2 PARAMETERS OF IDENTIFICATION AND CRITERION

The numerical-experimental method proposed in the present study consists the following stages. In the first stage the physical experiments are performed. Also the parameters to be identified, the domain of search and criterion containing experimental data are selected. In the second stage the finite element method is used in order to model the response of the structure and calculations are performed in reference points of the variables to be identified. The reference points are determined by using the method of experiment design. In the third stage the numerical data obtained by the finite element solution in the reference points are used in order to determine simple functions (response surfaces) for a calculation of the structure response. In the fourth stage, on the basis of the simple models and experimental data of the measured values of the structure response, the identification of the material properties is performed. For this the corresponding functional is minimized by using a method of non-linear programing.

The present numerical-experimental approach is used for the identification of the elastic properties of laminated composite plates. For this the experimental data of the measured eigenfrequencies are used. It is assumed that the plate dimensions (see Fig. 1), plate mass and the layer stacking sequence are known. The parameters to be identified are five elastic constants of a single transversely isotropic layer in the laminated composite plate:

- two Young's moduli, E_1, $E_2 = E_3$

- shear moduli, $G_{12} = G_{13}$

- Poisson's ratio, $\nu_{12} = \nu_{13}$

- shear modulus, $G_{23} = \dfrac{E_2}{2(1 + \nu_{23})}$

The plate is composed of unidirectionally reinforced layers. The material directions are denoted 1-2-3, where 1 is the fiber direction and 2,3 are the transverse directions. The unidirectional layer is assumed as homogeneous and transversely isotropic with respect to the fiber direction. In general, the ith layer of the laminated plate is oriented at an arbitrary angle β_i. The angles of the layers are assumed to be fixed. In the present study a cross-ply laminates are investigated, i.e. a composite with the layer angles $\beta_i = 0^0$ and $\beta_i = 90^0$. The five material parameters of the single layer can be expressed in terms of dimensionless

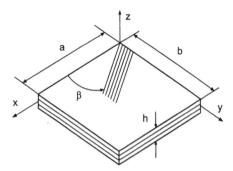

Figure 1: Geometry of the laminated plate

constants α_i (Mota Soares et al. 1993). The vector of parameters $\mathbf{x}$ to be identified is defined through these dimensionless quantities α_i

$$\mathbf{x} = [x_1, x_2, x_3, x_4] = [\alpha_2, \alpha_3, \alpha_4, \alpha_5] \qquad (1)$$

These parameters can be evaluated through the identification procedure using the experimental eigenfrequencies of the laminated composite rectangular plate of constant thickness h, length a and width b (see Fig. 1). Although plate consisting of N layers can be made with an arbitrary layer stacking sequence and fiber reinforcement angles, in the present paper we have investigated only symmetric cross-ply laminates.

Let the experimental angular eigenfrequencies be designated by $\bar{\omega}_1, \bar{\omega}_2, \ldots, \bar{\omega}_I$, where I is the number of measured eigenfrequencies $\bar{f}_i (\bar{\omega}_i = 2\pi \bar{f}_i)$. The corresponding numerical eigenfrequencies f_i $(\tilde{\omega}_i = 2\pi f_i)$ for the set of material parameters α_i are represented by $\tilde{\omega}_1, \tilde{\omega}_2, \ldots, \tilde{\omega}_I$. Let us consider the scaling parameter C which is chosen according to the relation (Mota Soares et al. 1993)

$$C = \frac{\bar{\omega}_1^2}{\tilde{\omega}_1^2(E_1^0)} \qquad (2)$$

where $\tilde{\omega}_1$ is the first numerical eigenfrequency calculated with the prior selected longitudinal Young's modulus E_1^0 of the layer.

The functional to be minimized describes deviation between the measured $\bar{\omega}_i$ and numerically calculated $\tilde{\omega}_i(\mathbf{x})$ frequencies (Mota Soares et al. 1993)

$$\Phi(\mathbf{x}) = \sum_{i=2}^{I} \frac{(\bar{\omega}_i^2 - C\tilde{\omega}_i(\mathbf{x})^2)^2}{\bar{\omega}_i^4} \qquad (3)$$

It is seen that criterion (3) is a non-linear function of the identification vector $\mathbf{x}$. The identification of the elastic constants $\mathbf{x}$ is performed on the basis of information obtained from the measurements of the I lowest frequencies. The identification problem is formulated as follows

56

Table 1: Geometric parameters and density of the cross-ply plates

Sample	a, m	b, m	h, mm	ρ, kg/m^3
PU10	0.1401	0.1401	2.011	1884
PU11	0.1401	0.1401	1.981	1908

Table 2: Experimental plate flexural frequencies $\bar{f}_i$ [Hz]

Mode	Mode shape	Specimens	
i	m, n	PU10	PU11
1	1,1	*166	*159
2	2,0	*341	*332
3	0,2	–	–
4	2,1	*484	*464
5	1,2	*542	*529
6	2,2	*902	*869
7	3,0	*971	938
8	3,1	*1090	*1050
9	0,3	*1155	*1143
10	1,3	*1273	*1240
11	3,2	*1523	*1470
12	2,3	*1643	–
13	4,0	*1898	*1838
14	4,1	*2003	*1940
15	3,3	2290	2180
16	0,4	–	–
17	4,2	2418	2338
18	1,4	–	–
19	2,4	*2733	*2665
20	4,3	–	–
21	5,0	3108	3023
22	5,1	3233	3135
23	3,4	–	–
24	5,2	3605	–

$$\min_{\mathbf{x}} \Phi(\mathbf{x}) \qquad (4)$$

subject to constraints, which define the domain of interest and conditions that the elasticity matrix of the material should be positve definite.

3 VIBRATION TEST OF PLATES

The glass/epoxy cross-ply laminated plates consisting of 8 unidirectional layers with the layer stacking sequence $[90/0/90/0]_S$ were tested. The geometric dimensions and density of plates are presented in Table 1. The eigenfrequencies of the test plates were measured by a real-time television (TV) holography. The samples were hung upon two threads in order to simulate free-free boundary conditions. The sample was located in front of the holographic testing device. A piezoelectric resonator (in the following called "sensor") was glued in one corner to exite the sample plate with increasing frequency. The sensor is of circular shape with a diameter 25 mm located at the coordinates $x = a - 12.5$ mm and $y = b - 12.5$ mm. Mass of the sensor is $m_s = 3.5$ g.

The plate is iluminated by laser light and imaged by CCD (Charged Couple Device) array, resulting in speckled image on the PC monitor. When the plate is deformed (exited), this interference pattern is slightly modified. Digital substraction of two consecutive interference patterns yields a fringe pattern depicting the surface displacements of the plate. The nodal lines of the vibration modes can be easily identified on the monitor in form of white lines on the speckled image. The digital substraction of two consecutive pictures helps to minimize noises, such as rigid body motion of the hung plate. The measurement technique more detailed was described by Yang et al. (1995) and Rikards et al. (1999). Experiments were performed for both plates considered (see Table 1) and about 20 flexural eigenfrequencies were measured. The mode shapes also were recognized in the experiment. In Table 2 the experimental plate flexural frequencies are presented. The frequencies marked with the sign (*) were taken into account in the identification. In Table 2 the quantity m denotes the wave number in the y direction and n denotes the wave number in the x direction (see Fig. 1).

4 FINITE ELEMENT SOLUTION

The eigenvalue problem for harmonic vibrations of the plate can be represented by

$$\mathbf{K}\mathbf{u} = \omega^2 \mathbf{M}\mathbf{u} \qquad (5)$$

Here $\mathbf{K}$ is the stiffness matrix of the plate, $\mathbf{M}$ is the mass matrix and $\mathbf{u}$ is the displacement vector. The eigenvalue relation (5) for the mode $\mathbf{u}_1$ which corresponds to the first experimental eigenfrequency $\bar{\omega}_1$ can be written in an equivalent form placing E_1 in evidence

$$E_1 \mathbf{K}^* \mathbf{u}_1 = \bar{\omega}_1^2 \mathbf{M}\mathbf{u}_1 \qquad (6)$$

Here $E_1 \mathbf{K}^* = \mathbf{K}$ is the stiffness matrix. Taking into account relation (2) this equation can be written as

$$C E_1^0 \mathbf{K}^* \mathbf{u}_1 = C \tilde{\omega}_1^2 \mathbf{M}\mathbf{u}_1 \qquad (7)$$

hence

$$E_1 = C E_1^0 \qquad (8)$$

where E_1^0 is the initial guess value given to the Young's modulus in the fiber direction of the layer and E_1 is the corresponding identified mechanical property. After evaluation of the optimum value of $\mathbf{x}$ the remaining mechanical properties are calculated by inverse relations (Mota Soares et al. 1993).

The eigenvalue problem (5) was solved by the subspace iteration method (Bathe 1992) and using a triangular finite element of laminated thick plate with a shear correction (Rikards et al. 1995). In order to avoid 'shear locking' a selective integration technique was applied. A 22×22 regular mesh (968 finite elements) was considered in order to achieve the necessary accuracy for at least 20 first eigenvalues of the laminated plate with FFFF (all edges free) boundary conditions.

5 METHOD OF EXPERIMENT DESIGN

Let us consider a criterion for the evaluation of the experiment designs which is independent on the mathematical model of the object. The number of variables n and the number of experiments k are the only initial information. The main principles in the approach proposed are as follows

1. The number of levels in the design space for each variable is equal to the number of the experiments, and for each level only one experiment is performed.

2. The reference points (experiment points) in the design space are distributed as regular as possible.

To realize the second principle it is suggested to use a criterion

$$\Psi = \sum_{i=1}^{k-1} \sum_{j=i+1}^{k} \frac{1}{L_{ij}^2} \Rightarrow \min \qquad (9)$$

where L_{ij} is the distance between the reference points having numbers i and j ($i \neq j$). The problem of minimizing the criterion (9) together with the first principle leads to the non-linear integer programming problem. The experiment designs were determined for different number of the design variables n and different number of the experiments k. The experiment design is characterized by the matrix B_{ij}. The following matrices were calculated for $n = 2, 3, \ldots, 15$ and for $k = 2, 3, \ldots, 25$. For example, the experiment design with nine reference points ($k = 9$) and two variables ($n = 2$) is as follows

$$\mathbf{B}^T = \begin{bmatrix} 7 & 1 & 2 & 5 & 4 & 9 & 6 & 8 & 3 \\ 2 & 6 & 3 & 5 & 1 & 4 & 9 & 7 & 8 \end{bmatrix} \qquad (10)$$

The experiment points (reference points) for this matrix are presented in Fig. 2. The domain of interest is determined as $x_j \in [x_j^{\min}, x_j^{\max}]$, where $x_j^{\min}$ and $x_j^{\max}$ are respectively the lower and the upper bound on the design variables. Thus, in this domain the reference points, where the experiments (or computer simulation) must be performed, are calculated by the expression

$$x_j^{(i)} = x_j^{\min} + \frac{1}{k-1}(x_j^{\max} - x_j^{\min})(B_{ij} - 1) \qquad (11)$$

Here $i = 1, 2, \ldots, k$ and $j = 1, 2, \ldots, n$. Since the matrices B_{ij} of the experiment design are universal, they may be used for the various identification or optimum design problems.

6 APPROXIMATION OF RESPONSE SURFACES

Techniques from experiment design and response surface methodology (Box & Draper 1987) are used to build the approximate models from the data in the reference points. Similar method based on the response surface approach was used in (Roux et al. 1998). Information on the behavior of an object can be obtained from the physical experiment or the computer solution in the reference points of the experiment design. The information can be represented as a data table, where the response function $y(\mathbf{x})$ of the object is to be in relationship with the variables $\mathbf{x}$. In our case there are four identification variables representing the elastic constants of the material. The goal is, by using the data only in the reference points (in our case these data are obtained by the finite element solution of the eigenvalue problem (5) in the reference points), to obtain the relation $y(\mathbf{x})$ in the mathematical form or the so called response surface. Here such mathematical models (response surfaces) have been obtained for the first I eigenfrequencies of the laminated plate.

The selection of the 'best' regression equation (response surface) in the subregion defined by the lower and upper bound on the design variables is performed by the following procedure. The response surface is built by using data obtained by computer simulation (or physical experiment) in all points of experiment design. First, consider the approximating function (model) of the following form

$$y(\mathbf{x}) = \sum_{i=1}^{m} A_i \phi_i(\mathbf{x}) \qquad (12)$$

where A_i are unknown coefficients and $\phi_i(\mathbf{x})$ are

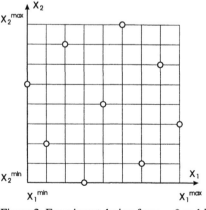

Figure 2: Experiment design for $n = 2$ and $k = 9$

the basis functions that constitute the model. These functions are built from the set of simple functions $\varphi_1, \varphi_2, \ldots, \varphi_R$. The functions φ_r ($r = 1, 2, \ldots, R$) are assumed to be in the form

$$\varphi_r(\mathbf{x}) = \prod_{j=1}^{n} x_j^{\alpha_{rj}} \tag{13}$$

where n is the number of variables and α_{rj} are positive or negative including zero integers. The form of the approximating function (12) is determined in two steps. First, the perspective functions $\phi_i(\mathbf{x})$ are selected by using the least-squares estimation. Then a step-by-step reduction procedure of the number of terms in the model is applied and further reduction of the selected functions is performed. Details of this procedure and the corresponding RESINT program were described in (Rikards 1993). Note that there is no general rules for the procedure of reduction of terms in the model (response surface function) and it is necessary to acquire some experience to obtain appropriate function. Other possibility to build a model is using engineering knowledge of the true functional form of the response (Vanderplaats 1984).

7 DETERMINATION OF ELASTIC CONSTANTS

Let us consider the procedure of identification for the cross-ply plates PU10 and PU11. The initial value of Young's modulus is taken E_1^0=50 GPa. The experiment design with four variables ($n = 4$) and 35 reference points ($k = 35$) was selected. The upper and lower limits (domain of interest) of the identification parameters are taken as follows

$$\begin{array}{ccccc}
5\,\text{GPa} & \leq & E_2 & \leq & 40\,\text{GPa}, \\
5\,\text{GPa} & \leq & G_{12} & \leq & 30\,\text{GPa}, \\
5\,\text{GPa} & \leq & G_{23} & \leq & 30\,\text{GPa}, \\
0.2 & \leq & \nu_{12} & \leq & 0.4
\end{array} \tag{14}$$

By using the matrix B_{ij} of the experiment design for $n = 4$ and $k = 35$, and the expression (11) the values of all the four identification parameters are calculated in 35 reference points. In each reference point the finite element analysis of the eigenvalue problem was performed and the first 20 natural frequencies were obtained. The finite element mesh of the plate is shown in Fig. 3. It should be noted that there is some originality in calculation of the mass matrix $\mathbf{M}$ in equation (5). In order to represent more accurate an inertia forces of the plate, the mass of sensor m_s should be taken into account. In the finite element modeling it is assumed that the finite elements where the sensor is located (see Fig. 3) have the same thickness h as the plate, but for these finite elements an equivalent density ρ_{eqv} is calculated

$$\rho_{\text{eqv}} = \rho + \frac{m_s}{F_s h}$$

Here F_s is the area of the sensor.

The data of the numerical experiment were used to determine the response surfaces (12). For this the RESINT program (Rikards 1993) was employed and approximating functions for all frequencies were obtained. These functions were used in the functional (3). In the identification for the plate PU10 at all 14 experimental eigenfrequencies and for the plate PU11 12 eigenfrequencies (see Table 2) were used. Minimization of the functional (3) subject to constraints was performed by the random search method, outlined previously (Rikards 1993). Results of identification of the layer stiffness properties of both plates are presented in Table 3. It is seen that the transverse shear modulus G_{23} is overestimated. The transverse shear modulus can not be reliable determined from the measured frequencies since the plates were too thin ($h/a = 1/70$) for identification of this property. In this case thick plates should be used.

Verification of the results was performed by the finite element method (FEM) and through the independent experiments. For the finite element analysis the elastic constants $\mathbf{x}^*$ obtained by the identification procedure were used (see Table 3). Results are shown in Table 4. Residuals were calculated by the formulae

$$\Delta_i = \frac{f_i^{\text{FEM}}(\mathbf{x}^*) - f_i^{\text{exp}}}{f_i^{\text{exp}}} \times 100 \tag{15}$$

It is seen that differences between the experimental and numerical frequencies calculated by using elastic constants obtained by identification are very small. Exception is for the mode 15 since in this case the difference is 3.14%.

It is of interest to compare the elastic constants of the single layer of the cross-ply laminate with the

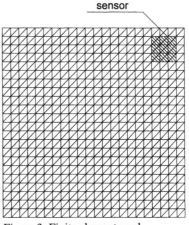

Figure 3: Finite element mesh

Table 3: Elastic properties of the single layer for cross-ply laminates

Property	PU10	PU11
E_1, GPa	38.89	38.20
E_2, GPa	12.78	11.95
G_{12}, GPa	5.06	4.77
G_{23}, GPa	11.70	9.55
ν_{12}	0.304	0.392

Table 4: Flexural frequencies f_i (Hz) and residuals Δ_i for plate PU10

Mode i	Exp.	FEM	Δ_i (%)
1	166	166.4	0.24
2	341	344.2	0.94
3	–	416.2	–
4	484	486.1	0.43
5	542	545.9	0.72
6	902	895.6	0.71
7	971	967.5	0.36
8	1090	1087	0.28
9	1155	1165	0.87
10	1273	1278	0.39
11	1523	1513	0.66
12	1643	1632	0.67
13	1898	1865	1.74
14	2003	2001	0.10
15	2290	2218	3.14
16	–	2275	–
17	2418	2379	1.61
18	–	2414	0.17
19	2733	2724	0.33
20	–	3010	–
21	3108	3157	1.58
22	3223	3226	0.22
23	–	3315	–
24	3605	3617	0.33

Table 5: Comparison of results for cross-ply and unidirectional laminates

Property	Cross-ply plate	UD plate
E_1, GPa	38.55	38.81
E_2, GPa	12.36	12.12
G_{12}, GPa	4.92	5.09
ν_{12}	0.347	0.255

properties obtained for the unidirectionally reinforced transversely isotropic plate made from the same material (Bledzki et al. 1999). Results are presented in Table 5, where the mean values of the elastic constants are presented. Good agreement of the results is observed for the constants E_1, E_2 and G_{12}. There is some difference for the Poisson's ratio. It can be explained since this property is less sensitive to fre-quencies as modulus of elasticity especially for the cross-ply laminate.

8 IDENTIFICATION OF DAMAGE ZONE

Damage and delamination detection in composite laminates is an active research area. Experimental results have showed (Penn et al. 1999) that due to delaminations and damage the natural frequencies have slightly changed. More sensitive to the presence of damage are curvature modes (Luo & Hanagud 1997) which give out more localized information.

Let us consider a numerical experiment in order to identify the size and location of a damage zone of the composite plate (see Fig. 4). In this case there are two parameters of identification: the location of the center of the damage zone x_1 and the width of the damage zone x_2. The damage of the transversely isotropic plate is introduced by reducing the stiffness of the damaged zone

$$E_1^* = x_3 E_1, \ E_2^* = x_3 E_2, G_{12}^* = x_3 G_{12} \qquad (16)$$

In the numerical experiment it was assumed that $x_3 = 0.5$ and the location of the damaged zone in the rectangular orthotropic plate ($a = b = 0.140$ m, h=2 mm) is given by $x_1^{\text{exp}} = 0.591a$ and $x_2^{\text{exp}} = 0.273a$. The finite element mesh was taken 22×22, i. e. the same as for the previous example (see Fig. 3). Employing these parameters the finite element model of the damaged plate was created. Further the 10 lowest eigenfrequencies of the damaged plate were calculated by solving the eigenvalue problem (5). These numerical frequencies were assumed as experimental frequencies of the damaged plate. Then the experiment design with 25 reference points was used. The domain of interest was assumed as follows

$$0.5a \le x_1 \le 0.7a, \ 0.15a \le x_2 \le 0.35a$$

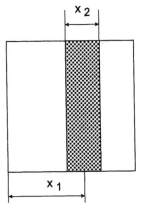

Figure 4: Location of damage zone of plate

In this domain of interest location of 25 reference points was calculated by eq. (11). In these reference points the 10 first eigenfrequencies were calculated by the finite element method. Employing these data the approximating functions were obtained for all 10 frequencies. These functions were used in the identification functional (3). Minimizing this functional the folowing parameters were obtained

$$x_1^* = 0.589a, \; x_2^* = 0.274a$$

It is seen that through the identification procedure about the same values of location and size of the damaged zone, which were introduced in the numerical experiment, have been obtained. It should be noted that for both examples - identification of elastic constants and location of the damage zone - the uniqueness of solution can be proved. Acctually, in both cases the stiffness matrix $\mathbf{K}$ of the eigenvalue problem (5) is a linear function of the parameters of identification $\mathbf{x}$. In this case the eigenvalues (natural frequencies) are convex functions of $\mathbf{x}$. This statement was proved in connection with solution of the optimum design problems of composite shells (Rikards 1980). Therefore, the functional (3) to be minimized is convex and the solution of the both identification examples outlined above is unique.

9 CONCLUSIONS

The elastic constants of the single layer of the crossply laminate have been determined by using the identification procedur based on the method of experiment design and the response surface approach. It was shown that elastic properties of the single layer are practically the same as constants obtained for the unidirectionally reiforced plate made from the same material. Performing the numerical experiment it was shown that the method of experiment design can also be applied for the identification of location of the damage zone of the transversely isotropic plate.

REFERENCES

Araújo, A. L., Mota Soares, C. M. & Moreira de Freitas, M. J. 1996. Characterization of material parametrs of composite plate specimens using optimization and experimental vibration data. *Composites. Part B.* 27B: 185-191.

Bathe, K.-J. 1982. *Finite Element Procedures in Engineering Analysis.* Englewood Cliffs: Prentice-Hall.

Bledzki, A. K., Kessler, A., Rikards, R. & Chate, A. 1999. Determination of elastic constants of glass/epoxy unidirectional laminates by vibration test of plates. *Compos. Sci. & Technol.* (in press).

Box, G. E. P. & Draper, N. R. 1987. *Empirical Model-Building and Response Surfaces.* New York: Wiley.

De Visscher, J., Sol, H., De Wilde, W. P., & Vantomme J. 1997. Identification of damping properties of orthotropic composite materials using a mixed numerical experimental method, *Appl. Compos. Mater.* 4(1): 13-33.

Frederiksen, P. S. 1992. Identification of temperature dependence for orthotropic material moduli, *Mechanics Mater.* 13: 79-90.

Frederiksen, P. S. 1997. Experimental procedure and results for the identification of elastic constants of thick orthotropic plates, *J. Compos. Mater.* 31(4): 360-382.

Luo, H. & Hanagud, S. 1997. An integral equation for changes in the structural dynamics characteristics of damaged structures. *Int. J. Solids Struct.* 34: 4557-4579.

Mota Soares, C. M., Moreira de Freitas, M., & Araújo A. L. 1993. Identification of material properties of composite plate specimens, *Composite Structures* 25: 277-285.

Pedersen, P. 1989. Optimization method applied to identification of material parameters. In H. A. Eschenauer & G. Thierauf (eds), *Discretization Methods and Structural Optimization - Procedures and applications*: 277-283, Berlin: Springer Verlag.

Penn, L. S., Jump, J. R. & Greenfield, M. J. 1999. Use of the free vibration spectrum to detect delamination in thick composites. *J. Compos. Mater.* 33(1): 54-72.

Rikards, R. 1980. Convexity of some classes of optimization problems for multilayer shells under conditions of stability and vibration. *Mechanics of Solids. Translation from Mekhanika Tverdogo Tela* 15(1): 130-137. Translated from Russian by Allerton Press Inc.

Rikards, R. 1993. Elaboration of optimal design models for objects from data of experiments. In P. Pedersen (ed.), *Optimal Design with Advanced Materials. Proceedings of the IUTAM Symposium, Lyngby, Denmark, 18 - 20 August, 1992*: 148-162, Amsterdam: Elsevier Science Publishers.

Rikards, R. & Chate, A. 1995. Optimal design of sandwich and laminated composite plates based on planning of experiments. *Structural Optimization.* 10(1): 46-53.

Rikards, R., Chate A. & Korjakin, A. 1995. Damping analysis of laminated composite plates by finite element method. *Engineering Computations.* 12: 61-74.

Rikards, R., Chate, A., Steinchen, W., Kessler, A. & Bledzki, A. K. 1999. Method for identification of elastic properties of laminates based on experiment design. *Composites. Part B.* 30: 279-289.

Roux, W. J., Stander, N. & Haftka, R. T. 1998. Response surface approximations for structural optimization. *Int. J. Numer. Meth. Engng.* 42: 517-534.

Sol, H. 1986. Identification of anisotropic plate rigidities using free vibration data. *PhD Thesis.* Free University of Brussels.

Vanderplaats, G. N. 1984. *Numerical Optimization Techniques for Engineering Design with Applications.* New YorkMc: Graw-Hill.

Yang, L. X., Steinchen, W., Schut, M. & Kupfer, G. 1995. Precision measurement and nondestructive testing by means of digital phase shifting speckle pattern shearing interferometry. *Measurement.* 16: 149-160.

Recent Developments in Durability Analysis of Composite Systems, Cardon, Fukuda, Reifsnider & Verchery (eds)
© *2000 Balkema, Rotterdam, ISBN 90 5809 103 1*

Damage modeling of composites structures-viscoelastic effects

M.Chafra
Institut Préparatoire aux Etudes d'Ingénieur, Mateur, Tunisia

Y.Chevalier
Institut Supérieur des Matériaux et de la Construction Mécanique, Paris, France

ABSTRACT : This approach is a tridimensional modeling used to study advanced fabric composite materials including damage and viscoelasticity. The model takes into account the features of loading (type and direction), and the mechanical behavior of the composite (linear or non linear). An increment procedure on stress and strain permits step by step computations for any type of loading which is also expressed as increments. Concerning the coupling between damage and viscoelasticity a new description of degradation in composite is proposed. The type of loading and the structural anisotropy of the material are also taken into account and we bring in a damage oriented parameter which depends on the strain velocity.

1 INTRODUCTION

Fiber's orientation creates directional effects in composite materials. This anisotropic behaviour generates difficulties towards mechanical behaviour modelling : some orientations are more sensitive to strain fracture and failure. This sensitivity depends on the kind of solicitation (Chevalier 1998). We adopt the isothermal and static loading, and study the composite in short time with small strains. Elastic, viscoelastic, plastic and non-linear properties are modelled with anisotropy and influence of loading orientation (off axes tests). Experimental studies on composites are widely developed but often with preferential proportional loading applied on samples which are cut at various angles with respect to fiber orientation (Vinh 1981, Baste 1991).

The matrix wich insures the link between fibers is generally organic mode of big molecules that represents a viscous character. The viscosity of the composite can equally have material heterogeneity origins. In fact, if the fiber-matrix bonding is middling ,micro-defect between different constituents causes an energy loss that can be globally interpreted by viscosity, and the relation between strains and deformations will depend on time (Lemaitre 1985). Concerning damage and viscoelasticity, a new description of degradation in composites is proposed, taking into account the features of loading and the strain velocity with the framework of linear behavior.

We suggest a model able to identify mechanical properties in various directions by using isostrain quadratic surfaces (Chafra 1995).

For non linear behaviour the loading process is spliced up into increments and each loading increment is analysed as a linear case with tangent viscoelastic properties. An identification procedure has been performed on carbon/ PMR15 fabric composites (Aussedat-Yahia 1997). The coupling between viscoelastic behaviour and damage is dominating in off axes loads: The effect of strain velocity is important.

2 ELASTIC BEHAVIOUR

To take into consideration all main effects in a composite material we introduce two 6-dimension vectorial spaces.

(i)   stress space $\sum$: presented with element of the base of vectors $\{\Sigma_\alpha\} \in \sum$ defined by :

$$\{\Sigma_\alpha\} = \left\{ \begin{array}{l} \Sigma_1 = \sigma_{11}, \Sigma_2 = \sigma_{22}, \Sigma_3 = \sigma_{12}, \\ \Sigma_4 = \sigma_{33}, \Sigma_5 = \sqrt{2}\sigma_{13}, \Sigma_6 = \sqrt{2}\sigma_{23} \end{array} \right\}^T \quad (1)$$

where σ_{ij} (i , j = 1, 2, 3) are the components of the stress tensor.

(ii) strain space $\mathcal{E}$: presented with element of the base of vectors $\{E_\alpha\} \in \mathcal{E}$ defined by:

$$\{\mathcal{E}_\alpha\} = \begin{cases} \mathcal{E}_1 = \varepsilon_{11}, \ \mathcal{E}_2 = \varepsilon_{22}, \ \mathcal{E}_3 = \sqrt{2\varepsilon_{12}}, \\ \mathcal{E}_4 = \varepsilon_{33}, \ \mathcal{E}_5 = \sqrt{2\varepsilon_{13}}, \ \mathcal{E}_6 \ \sqrt{2\varepsilon_{12}} \end{cases}^T \quad (2)$$

where ε_{ij} (i , $j = 1, 2, .3$). are the components of the strain tensor.

We define two tensorial invariants, the stress intensity and the strain intensity $\sum_0 = \sqrt{\sum_\alpha \sum_\alpha}$ and $\mathcal{E}_0 = \sqrt{\mathcal{E}_\alpha \mathcal{E}_\alpha}$, deduced from the second order strain (or stress) tensor.

To explain the unilateral effect we propose to characterise a loading in direction and type (Chafra 1996) :

-*Loading direction* : defined by a unit vector $\{\sigma_\alpha\} \in \sum$,with: $\sqrt{\sigma_\alpha \ \sigma_\alpha} = 1$ and $\sum_\alpha = \sum_0 \sigma_\alpha$

-*Type of loading:* is defined as a function with various possibilities associated with the sign of the stress tensor, hence the stress space is decomposed into zones characterised by the sign attributed to each component of stress. We define a parameter (χ) which corresponds to the number of lines of the matrix $[\ I_{\chi\alpha}\ ]$ and is equal to the line :

$\{$ sign (σ_1), sign (σ_2), ...,sign $(\ \sigma_6)\ \}$.

$$\text{sign}(\sigma_\alpha) = \begin{cases} 1 \text{ if } \sigma_\alpha \geq 0 \\ -1 \text{ if } \sigma_\alpha \langle 0 \end{cases}$$

During a proportional loading $\{\sigma_\alpha\}$ = cte, we define a generalised modulus which relates a stress intensity to a strain intensity

$$\sum_0 = \mathbf{E}^{(\chi)}\mathcal{E}_0 \text{ or } \mathcal{E}_0 = \mathbf{H}^{(\chi)} \sum_0$$
$$(3)$$
with $\mathbf{E}^{(\chi)} = (\mathbf{H}^{(\chi)})^{-1}$

and in the case of non-proportional loading, the constituve law is defined as an incremental elastic linear law.

$$\Delta \sum_\alpha (i) = \mathcal{E}_{\alpha\beta}^{(\chi)} \Delta \mathcal{E}_\beta (i) \text{ or}$$
$$\Delta \mathcal{E}_\alpha (i) = \mathbf{H}_{\alpha\beta}^{(\chi)} \ \Delta \sum_\beta (i) \quad (4)$$

To describe the variation of elastic compliance in linear domain which depends on the type and the direction of loading; we construct an isostrain quadratic surface in the space of the stress $\sum$:

$$\phi_E^{(\chi)} \equiv \mathfrak{R}_{\alpha\beta}^{(\chi)} \sum_\alpha \sum_\beta - \mathcal{E}_0^2 = 0 \quad (5)$$

$\mathcal{E}_0$ = constant ($\alpha, \beta = 1, ... 6$)

$\mathfrak{R}_{\alpha\beta}^{(\chi)}$ is related to the components of compliance matrix $\mathfrak{R}_{\alpha\beta}^{(\chi)} = \mathbf{H}_{\alpha\gamma}^{(\chi)}\mathbf{H}_{\gamma\beta}^{(\chi)}$ $\quad (6)$

and is related to the generalised modulus:

$$\mathbf{E}^{(\chi)} = \frac{1}{\sqrt{\mathfrak{R}_{\alpha\beta}^{(\chi)} \sigma_\alpha \sigma_\beta}} \quad (\{\sigma_\alpha\} = \text{constant}) \quad (7)$$

3 VISCOELASTIC BEHAVIOUR

For certain type of composite materials, the behaviour is strongly non-linear. Hence we note a viscoelastic behaviour within the matrix.
Consequently, the matrix is sensitive to the speed and mode of loading.

For each $\{ \sigma_\alpha \}$ = constant $\in \sum$ We decompose the process of loading into increments, successively evaluating tangent modulus

$$\mathbf{E}^{T(\chi)}(i) = \frac{\Delta \sum_0 (i)}{\Delta \mathcal{E}_0 (i)} \quad (8)$$

and secant modulus $\quad \mathbf{E}^{S(\chi)}(i) = \dfrac{\sum_0 (i)}{\Delta \mathcal{E}_0 (i)} \quad (9)$

with $\quad \begin{cases} \Delta \sum_0 (i) = \sum_0 (i+1) - \sum_0 (i) \\ \\ \Delta \mathcal{E}_0 (i) = \mathcal{E}_0 (i+1) - \mathcal{E}_0 (i) \end{cases} \quad (10)$

$$\mathbf{E}^{T(\chi)}(i) = \frac{\mathcal{E}_0 (i+1)}{\Delta \mathcal{E}_0 (i)} \mathbf{E}^{S(\chi)}(i+1) - \frac{\mathcal{E}_0 (i)}{\Delta \mathcal{E}_0 (i)} \mathbf{E}^{S(\chi)}(i) \quad (11)$$

For each level (i) we construct two quadratic isostrain surfaces :

a) *secant approach of* $\phi^{s(\chi)}$

$$\phi^{S(\chi)} \equiv \mathfrak{R}^{S(\chi)}_{\alpha\beta}(i) \sum_\alpha \sum_\beta - (\mathcal{E}_0(i))^2 = 0 \quad (12)$$

$\mathcal{E}_0(i)$ = constant ($i = 1, ... N$)

Identification process :

$$\Re^{S(\chi)}_{\alpha\beta}(i) = \mathbf{H}^{S(\chi)}_{\alpha\gamma}(i)\,\mathbf{H}^{S(\chi)}_{\gamma\beta}(i) \tag{13}$$

$$(\alpha,\beta,\gamma = 1,...6 \quad , \quad i = 1.....N)$$

is related to secant Young modulus by the following process :

$$\mathbf{E}^{S(\chi)}(i) = \frac{1}{\sqrt{\Re^{S(\chi)}_{\alpha\beta}\,\sigma_\alpha(i)\sigma_\beta(i)}} \tag{14}$$

$$(\alpha,\beta = 1,...6) \quad , \quad (i = 1,...N)$$

b) *Tangent approach of* $\phi^{T(\chi)}$

$$\phi^{T(\chi)} \equiv \Re^{T(\chi)}_{\alpha\beta}(i)\Delta\Sigma_\alpha\,\Delta\Sigma_\beta - (\Delta\mathcal{E}_0(i))^2 = 0 \tag{15}$$

$$\text{with } \Delta\mathcal{E}_0 = cte, i = 1,....N$$

Identification process :

$$\Re^{T(\chi)}_{\alpha\beta} = \mathbf{H}^{T(\chi)}_{\alpha\gamma}(i)\mathbf{H}^{T(\chi)}_{\gamma\beta}(i)$$
$$\left[\mathbf{E}^{T(\chi)}_{\alpha\beta}\right] = \left[\mathbf{H}^{T(\chi)}_{\alpha\beta}\right]^{-1} \tag{16}$$
$$\alpha,\beta,\gamma = 1,....6, i = 1,....N$$

$$\mathbf{E}^{T(\chi)}(i) = \frac{1}{\sqrt{\Re^{T(\chi)}_{\alpha\beta}(i)\sigma_\alpha(i)\sigma_\beta(i)}} \tag{17}$$

Relation between secant and tangent parameters

$$\sqrt{\Re^{T(\chi)}_{\alpha\beta}(i)\sigma_\alpha\sigma_\beta} = \frac{\Sigma_0(i+1)}{\Delta\Sigma_0(i)}\sqrt{\Re^{S(\chi)}_{\alpha\beta}(i+1)\sigma_\alpha\sigma_\beta}$$
$$-\frac{\Sigma_0(i)}{\Delta\Sigma_0(i)}\sqrt{\Re^{S(\chi)}_{\alpha\beta}(i)\sigma_\alpha\sigma_\beta} \tag{18}$$
$$(\alpha,\beta = 1,...6 \quad , i = 1,.......N)$$

Tangent moduli are used in incremental constituve equations

$$\Delta\mathcal{E}_\alpha(i) = \mathbf{H}^{T(\chi)}_{\alpha\beta}(i)\Delta\Sigma_\beta(i) \quad \text{or}$$
$$\Delta\Sigma_\alpha(i) = \mathbf{E}^{T(\chi)}_{\alpha\beta}(i)\Delta\mathcal{E}_\beta(i) \tag{19}$$
$$\alpha,\beta = 1,...6, i = 1,....N$$

In the case of biaxial loading (Σ_1 , Σ_2) : The constitutive law are written as follow :

$$\Delta\mathcal{E}_1 = \mathbf{H}^{T(\chi)}_{11}(i)\Delta\Sigma_1 + \mathbf{H}^{T(\chi)}_{12}\Delta\Sigma_2$$
$$\Delta\mathcal{E}_2 = \mathbf{H}^{T(\chi)}_{21}(i)\Delta\Sigma_1 + \mathbf{H}^{T(\chi)}_{22}\Delta\Sigma_2 \tag{20}$$

The isostrain quadratic surfaces are defined by the following equations :

$$\phi^{S(\chi)} \equiv \Re^{s(\chi)}_{11}\Sigma_1^2 + 2\Re^{s(\chi)}_{12}\Sigma_1\Sigma_2$$
$$+ \Re^{s(\chi)}_{22}\Sigma_2^2 - (E_0(i))^2 = 0 \tag{21}$$
$$\chi = 1,2,3,4$$

To determine three parameter, we should consider three stiffnesses in various direction 0°, 45° and 90°. The identification process requires the following equations :

$$\begin{cases}
\Re^{s(\chi)}_{11}(i) = (\mathbf{H}^s(0°))^2 \\
\Re^{s(\chi)}_{22}(i) = (\mathbf{H}^s(90°))^2 \\
\Re^{s(\chi)}_{12}(i) = (\mathbf{H}^s(45°))^2 - \\
\frac{1}{2}\left[(\mathbf{H}^s(0°))^2 + (\mathbf{H}^s(90°))^2\right]
\end{cases} \tag{22}$$

4 USE OF EXPERIMENTAL RESULTS

We often use an experimental method using samples out off axes for various directions (figure 1). The results provided by a test is the longitudinal modulus, while the results used in the models is the generalised modulus. By means of the transformation method, we establish the following relation between Σ_1, Σ_2 and σ_L .

$$\begin{cases}
\Sigma_1 = \sigma_L \cos^2\theta \\
\Sigma_2 = \sigma_L \sin^2\theta \\
\frac{1}{\sqrt{2}}\Sigma_3 = -\sigma_L \sin^2\theta\cos^2\theta
\end{cases}$$
$$\Rightarrow \quad tg\alpha = tg^2\theta$$

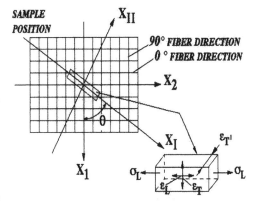

Figure 1 : Off axes tests

We define the longitudinal modulus and Poisson ratio and we calculate Σ_0 and E_0

$$\Sigma_0 = \sqrt{\Sigma_1^2 + \Sigma_2^2 + \Sigma_3^2} = \sigma_L$$

and $E_L^{(\chi)}(\theta) = \dfrac{\sigma_L}{\varepsilon_L}$

$$\mathcal{E}_0 = \sqrt{\varepsilon_L^2 + \varepsilon_T^2} = \varepsilon_L \sqrt{1 + \upsilon_{LT}^2}$$

$$\upsilon_{LT}^{(\chi)} = -\frac{\varepsilon_T}{\varepsilon_L}$$

then

$$\Sigma_0 = \overline{\Sigma}_0 \sqrt{\cos^4 \theta + \sin^4 \theta}$$

with $\overline{\Sigma}_0 = \sqrt{\Sigma_1^2 + \Sigma_2^2}$

and $\mathbf{E}^{(\chi)} = \dfrac{\Sigma_0}{\mathcal{E}_0} = \dfrac{\sigma_L}{\varepsilon_L \sqrt{1 + \upsilon_{LT}^2(\theta)}} = \dfrac{E_L^{(\chi)}(\theta)}{\sqrt{1 + \upsilon_{LT}^2(\theta)}}$

We establish the elastic moduli

$$\mathbf{E}'^{(\chi)}(\alpha) = \frac{\overline{\Sigma}_0}{\mathcal{E}_0} = \frac{E_L^{(\chi)}(\theta)\sqrt{\cos^4 \theta + \sin^4 \theta}}{\sqrt{1 + \upsilon_{LT}^2(\theta)}} \qquad (23)$$

with $\mathbf{H}'^{(\chi)}(\alpha) = (\mathbf{E}'^{(\chi)}(\alpha))^{-1}$

With three values of $\mathbf{H}'^{(\chi)}(\alpha)$, for $\alpha = 0°$, $45°$ and $90°$, we identify the parameters of surface $\mathfrak{R}_{11}^{(\chi)}$, $\mathfrak{R}_{22}^{(\chi)}$ et $\mathfrak{R}_{12}^{(\chi)}$ by the relation (22) and following for these parameters we express for each $\dot{\varepsilon}$ the mechanical characteristic for the constituve law (20)

$$\mathbf{H}_{11}^{(\chi)} = \sqrt{\frac{\mathfrak{R}_{11}^{(\chi)}}{1 + (\upsilon_{LT}(90°))^2}}$$

$$\mathbf{H}_{22}^{(\chi)} = \sqrt{\frac{\mathfrak{R}_{22}^{(\chi)}}{1 + (\upsilon_{LT}(0°))^2}}$$

$$\mathbf{H}_{12}^{(\chi)} = \sqrt{\frac{\mathfrak{R}_{11}^{(\chi)} + 2\mathfrak{R}_{12}^{(\chi)} + \mathfrak{R}_{22}^{(\chi)}}{1 + (\upsilon_{LT}(45°))^2}} \qquad (24)$$

$$- \sqrt{\frac{\mathfrak{R}_{11}^{(\chi)}}{1 + (\upsilon_{LT}(90°))^2}}$$

Experimental results concerning this formulation with different strain velocity are given in figure 2 (linear range) and figure 3 (non linear range) with carbon/PMR15 studied by Aussedat 1997.

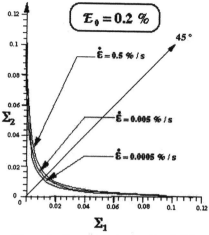

Figure 2: Isostrain criterion E_0 =0.2% Carbon/PMR15 fabric composites ($\chi = 1$, linear range)

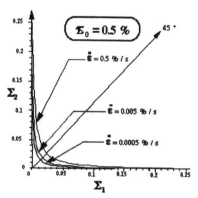

Figure 3: Isostrain criterion E_0 =0.5% Carbon/PMR15 fabric composites ($\chi = 1$, non-linear range)

The curves have shown that the viscoelasticity effects are very small on 90° or 0° direction and in strong coupling with the off axes loads. The effect of strain velocity is important.

5 DAMAGE MODELING

Concerning the coupling between damage and viscoelasticity a new description of degradation in composite is proposed. We introduce a damage oriented parameter which is dependent on the level of residual elastic deformation and the strain velocity: For each $\dot{\varepsilon}$:

$$D^{(\chi)}(E_0^P) = \frac{(E^{(\chi)} - E^{*(\chi)}(E_0^P))}{E^{(\chi)}} \qquad (25)$$

where $E^{(\chi)}$ is an initial elastic modulus, $E^{*(\chi)}$ is a modified elastic modulus, and $\mathcal{E}_0^P$ is a residual elastic deformation.

In the same manner of the linear or non linear case we construct an isostrain criterion $\phi_E^{*(\chi)}$:

$$\phi_E^{*(\chi)} \equiv \mathcal{R}_{\alpha\beta}^{*(\chi)} \sum_\alpha \sum_\beta -(\mathcal{E}_0')^2 = 0$$
$$\mathcal{E}_0' = \sqrt{\mathcal{E}' \mathcal{E}_\alpha'} = \text{constant} \qquad (\alpha, \beta = 1,...6) \qquad (26)$$

$$\mathcal{E}_0' = \mathcal{E}_0 - \mathcal{E}_{OM}^P, \ \mathcal{E}_0' \in \left[0, (\mathcal{E}_{OM} - \mathcal{E}_{OM}^P) \right]$$

The identification process is the following :

$$\mathcal{R}_{\alpha\beta}^{*(\chi)}(\mathcal{E}_{OM}^P) = H_{\alpha\gamma}^{*(\chi)}(\mathcal{E}_{OM}^P) H_{\gamma\beta}^{*(\chi)}(\mathcal{E}_{OM}^P) \qquad (27)$$

$$E^{*(\chi)} = \frac{1}{\sqrt{\mathcal{R}_{\alpha\beta}^{*(\chi)} \sigma_\alpha \sigma_\beta}}, (E_{\alpha\beta}^{*(\chi)})^{-1} = H_{\alpha\beta}^{*(\chi)} \qquad (28)$$

and the constituve law are written as follow :

$$\Delta \mathcal{E}_\alpha'(i) = H_{\alpha\beta}^{*(\chi)}(\mathcal{E}_{OM}^P) \Delta \sum_\beta(i)$$
$$\text{or} \ \ \Delta \sum_\alpha(i) = E_{\alpha\beta}^{*(\chi)}(\mathcal{E}_{OM}^P) \Delta \mathcal{E}_\beta'(i)$$
$$\alpha, \beta = 1,...6 \qquad (29)$$
$$\text{with} \ \ H_{\alpha\beta}^{*(\chi)} = \left[E_{\alpha\beta}^{*(\chi)} \right]^{-1}$$

We present an application with carbon/PMR15 fabric composites in table 1

Table 1 : Carbone/PMR15 fabric composites α

E^P_{OM}	$E^{*(1)}$ (45°) (Gpa)	$D^{(1)}$ (45°)
0.	21	0.
0.0007	14	0.33
0.0018	11.6	0.44
0.007	7	0.66

6 CONCLUSION

In this work, a tridimensional modeling is used to study advanced fiber and woven composite materials taking into account unilateral aspects. In some kind of composite materials it's additionally necessary to take into account the viscoelasticity effects. A non linear behaviour is detected and the apparent behaviour strengthens if the solicitation speed increases. The caricature of this phenomenon enhences its dependance versus time. The viscoelastic effects are dominating in off axes loads and the effect of strain velocity is important. An isostrain criterion is used to describe the variation of the viscoelastic properties in composite materials. We define a damage oriented parameter which depends on the strain velocity with the framework of linear behaviour. For non-linear behaviour the loading process is spliced up into increments and each loading increment is analysed as a linear case with tangent viscoelastic properties.

It would be interesting to complete the study by some tests to check the validity of model. The dynamic approach of damage must be taken into account in organic composite material.

ACKNOWLEDGEMENTS

The authors would like to acknowledge the work of their colleagues A.BALTOV (Mechan.Inst.Bulgaria) and T.VINH (ISMCM) who contributed to the development of basic concepts which are presented here.

REFERENCES

AUSSEDAT-YAHIA E. 1997. Comportement et endommagement du composite tissé Carbone/PMR15 soumis à des chargements mécaniques et thermiques. *Doctoral thesis in Mechanics.* . Ecole Nationale Supérieure des Mines de Paris .

BASTE S. 1991. Comportement non linéaire des composites à matrice fragile, théorie et mesure de leur endommagement. *Doctoral thesis in Mechanics.* Univ. Bordeaux I.

CHAFRA M., BALTOV A., VINH T. 1995 Tridimensional modelling of composite materials damage and failure. *Actes EUROMECH 334*, Lyon, pp154-163

CHAFRA M., BALTOV A., VINH T. - 1996 Modélisation tridimensionnelle des matériaux composites-Endommagement et rupture. *Revue des composites et matériaux avancés.* 56(1) : 49-72.

CHAFRA M., VINH T., CHEVALIER Y. - 1996 Damage and failure of composite materials:

Application to woven materials. Progress in durability analysis of composite analysis of *composite systems*. A.Cardon, H.Fukuda, K.Reifsnider ed. A.Balkema : Rotterdam-Brookfield.

CHEVALIER Y., LOUZAR M., CHAFRA M., MAUGIN G.A. - 1998.Damage modeling of composite structures : Static and dynamic approach; *Progress in durability analysis of composite systems*. K.Reifsnider, D.Dillard and A.Cardon (ed) A.Balkema : Rotterdam-Brookfield

LEMAITRE J., CHABOCHE J.L. 1985 . -*Mécanique des matériaux solides* - Dunod, Paris.

VINH T. 1981. Mesures ultrasonores des constantes élastiques des matériaux composites, *Sci, et techniques de l'armement*, vol.54, p.265-289.

Evolution of matrix cracking in cross-ply CFRP laminates: Differences between mechanical and thermal loadings

C. Henaff-Gardin, I. Goupillaud & M.C. Lafarie-Frenot
Laboratoire de Mécanique et de Physique des Matériaux, UMR 6617, ENSMA, Futuroscope-Chasseneuil, France

ABSTRACT : The aim of this study is to characterize the damage development in cross-ply composite laminates (T300/914) submitted to mechanical and thermal fatigue. Cyclic temperature variations induce cyclic biaxial in-plane stresses in each layer of the cross-ply laminate. The loading levels (temperature and applied stress amplitudes) have been chosen in order to induce similar transverse stress amplitudes in the 90° layers. A comparison of the matrix crack development throughout both types of tests has been undertaken : they have been found analogous, but with very different kinetics. A fracture mechanics approach, developed earlier and validated for mechanical fatigue loading, has been applied without modification to the case of thermal fatigue : it can predict initiation and crack density values at saturation but is inadequate to predict the damage development rates.

1 INTRODUCTION

In aeronautical applications, structural parts can be submitted to large cyclic temperature variations (Paillous & Pailler 1994). Large thermal stresses may develop in composite CFRP laminates due to the mismatch in the coefficients of thermal expansion of the fibers and the matrix, at a microscopic level, and consequently of adjacent plies stacked with different orientations in the laminate (Aswendt & Hofling 1993). So, in a cross-ply laminate for example, cyclic temperature variations induce cyclic biaxial in-plane strains in every layer (Datoo 1991). In that case, cyclic thermal exposure is likely to enhance damage similar to that observed under mechanical fatigue (Forsyth et al. 1994, Henaff-Gardin et al. 1995).

Under uniaxial mechanical tensile loading (static or fatigue), a characteristic pattern of matrix cracks develops very early in the specimen life, these cracks running parallel to the fiber direction in the most disoriented plies with respect to the loading axis. Subsequently, other damage modes usually occur (such as longitudinal cracking or delamination), the damage history depending on the material constituents, on the loading history and on the laminate stacking sequence.

Under thermal loading, crack patterns in cross-ply laminates consist of cracks in both 0° and 90° layers (Jennings et al. 1989, Boniface et al. 1992). In the case of cross-ply asymmetric $(0_m/90_n)$ laminates, residual thermal stresses may be sufficient enough to create a regular array of cracks in the 0° and 90° lay-

ers, and some out-of-plane curvature of the plate.

Damage mechanisms induced in CFRP laminates by mechanical fatigue have been extensively studied during the last twenty years (e.g. Charewicz & Daniel 1986, Tsai et al. 1987, Nairn & Hu 1994). Comparatively, only few experimental works have been concerned by the damage development during thermal fatigue of composite laminates (Herakovitch et al. 1980, Adams et al. 1986, Favre et al. 1996, McManus et al. 1996). Tests were generally undertaken for one temperature amplitude (Adams & Herakovich 1984, Jennings et al. 1989), and authors were focusing their attention on material influence (Forsyth et al. 1994) or on the influence of damage on thermal expansion coefficients (Boniface et al. 1991).

The aim of the present paper is to compare the matrix cracking development in cross-ply composite laminates subjected to 'equivalent' mechanical or thermal cyclic loadings. Moreover, this first experimental program has for objective to put in light the limitations of a purely mechanical approach (Henaff-Gardin et al. 1996 a, b) for the prediction of matrix cracking due to cyclic thermal exposure.

2 MATERIAL AND EXPERIMENTAL CONDITIONS

The material chosen for the present work (carbon/epoxy T300/914) was intended solely as a model material and not as a candidate for an elevated temperature usage. It was chosen essentially because it was previously well characterized in mechanical fatigue (Henaff-Gardin et al. 1997). The elastic constants and coefficients of thermal expansion (CTE) of this T300/914 material are the followings at ambient temperature :

longitudinal modulus : $E_1 = 140.4$ GPa
transverse modulus : $E_2 = 8.87$ GPa
Poisson ratio : $\nu_{12} = 0.337$
in-plane shear modulus : $G_{12} = 5.7$ GPa
longitudinal CTE : $\alpha_1 = 0.9 \ 10^{-7} \ °C^{-1}$
transverse CTE : $\alpha_2 = 28.7 \ 10^{-6} \ °C^{-1}$

Two cross-ply laminates $(0_3/90_3)_s$ and $(90_3/0_3)_s$ have been tested in both uniaxial tension fatigue and thermal cycling. In these stacking sequences, there are the same number of 0° and 90° plies, with two external layers of three plies and one internal layer of six plies. The position of the 90° layer (internal or external) will be particularly investigated under mechanical fatigue.

The thermal cycles are triangular, between +150°C and -50°C, with ±4°C/mn cooling and heating rates, corresponding to a $1.7 \ 10^{-4}$ Hz frequency. These tests have been carried out up to 500 cycles. The coupons used for thermal tests are parallelepipedic : 50 x 60 mm² and approximately 2 mm thick.

Fatigue tests are carried out under load controlled sinusoidal mode, at a frequency of 10 Hz and with a load ratio equal to 0.1, up to the failure of the specimens, after a few million cycles. The fatigue specimens are 250 mm long and 30 mm wide, with glass-epoxy end tabs 50 mm long for gripping in an Instron servohydraulic machine.

The values of the ply stresses during mechanical and thermal tests are given in table 1 ; they have been estimated using a thermoelastic analysis without edge effects (Datoo 1991).

For mechanical fatigue tests, the maximum stress value applied to the laminate was determined in order to have, in the 90° plies, about the same cyclic amplitude of σ_{22} (transverse ply stress) than that induced by the thermal variation in the specimens exposed to temperature. In our case, this value was found close to 50% of the static failure stress value $(0.5 \ \sigma_R = 365$ MPa). Nevertheless, because of the presence of large thermal residual stresses at ambient temperaure, it appears that the minimum transverse ply stress value in the 90° layer is really higher under fatigue loading, as illustrated in figure 1 (mechanical : 41 MPa ; thermal : 6.9 MPa). This implies that the maximum transverse stress value is approximately twice higher under mechanical fatigue.

Under thermal cycling, the transverse ply stresses σ_{22} are equal in all the layers, whereas under uniaxial mechanical fatigue σ_{22} is about three times higher in

Table 1. Stresses in the 0° and 90° layers of $(0_3/90_3)_s$ and $(90_3/0_3)_s$ laminates under mechanical fatigue and thermal cyclic loading.

		$\sigma_{22}^{90°}$ (MPa)	$\sigma_{11}^{90°}$ (MPa)	$\sigma_{22}^{0°}$ (MPa)	$\sigma_{11}^{0°}$ (MPa)
Fatigue	thermal residual stress (20°C)	36.7	-36.7	36.7	-36.7
	minimum mechanical stress	4.3	-1.3	1.3	685.9
	mechanical cyclic stress amplitude	38.5	-11.5	11.5	617.3
Thermal cycling	minimum thermal stress (-50°C)	6.9	-6.9	6.9	-6.9
	thermal cyclic stress amplitude	45.9	-45.9	45.9	-45.9

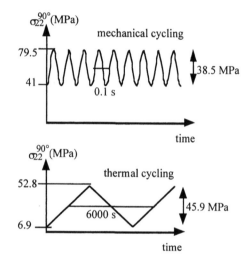

Figure 1. Characteristics of the applied mechanical and thermal cycles. Corresponding transverse ply stresses in the 90° layer.

the 90° than in the 0° one (table 1).

The damage development, particularly the 0° and 90° ply cracking, was investigated throughout tests by means of X-ray radiography, using a Zinc Iodide solution applied to the edges of the specimen, in order to enhance the picture contrast.

3 RESULTS

3.1 Cracking development under thermal cycling

X-radiographs illustrating damage development throughout thermal cycling are given in figure 2.

As far as thermal cycling is concerned, the two studied laminates $(0_3/90_3)_s$ and $(90_3/0_3)_s$ are identical, as temperature leads to biaxial loading. So, the different layers will be only refereed as 'external' or

crack
internal layer

crack
external layer

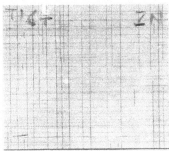

a. 50 cycles

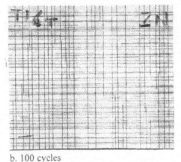

b. 100 cycles

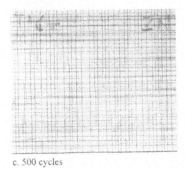

c. 500 cycles

Figure 2. X-ray patterns during thermal cycling.

'internal' layers of the laminate.

After 50 thermal cycles (Fig. 2.a.), cracks are present in both internal and external layers. These cracks initiated in the very first cycles of the tests, almost simultaneously in all the layers. In the external layers, the cracks initiate from the specimen free edges, and they are essentially crossing the entire coupon. Most of the cracks in the internal layer initiate at the free edges, and half of them are crossing the specimen width. The other half of these cracks propagate as thermal cycle number increases, but they never reach the opposite side because they are facing each other.

With increasing cycle number, the cracks in all the layers increase in number and length (Fig. 2.b.).

At the end of tests, after 500 thermal cycles, accurate observations show that the crack distribution in the internal layer is still heterogeneous, with a lower crack density in the earth of the coupon than at the free edges (Fig. 2.c.).

3.2 Cracking development under mechanical fatigue

The two figures 3 and 4 present examples of X-radiographs obtained under mechanical fatigue tensile loading, for the two studied laminates, respectively for $(0_3/90_3)_s$ and $(90_3/0_3)_s$ laminates.

Just half of the coupon is presented, the vertical dimension being the total width of the specimens (30 mm). One glass/epoxy end-tab is visible on the left side of each radiograph.

Transverse ply cracking always appears first, and is already developed after 50 cycles (Figs 3.a., 4.a.), with cracks spanning the entire specimen width. Although the loading axis is parallel to 0° fibers, some longitudinal cracks are present in $(90_3/0_3)_s$ laminate after 50 cycles (Fig. 4.a.). In $(0_3/90_3)_s$ laminate, such longitudinal cracks initiate after approximately $5 \cdot 10^4$ cycles (Fig. 3.b.).

The transverse and longitudinal cracks multiply and propagate when the cycle number increases, resulting in a regular array of cracks of both directions 0° and 90° that corresponds with a quasi saturation of fatigue damage (Figs 3.c., 4.c.).

For $(0_3/90_3)_s$ laminate, such a regular array of cracks is obtained far much later in the fatigue life of the coupons.

When matrix cracks are fully developed in both 0° and 90° layers, delaminations initiate near the edges and inside the coupon. They preferentially propagate along the cracks of the external layers.

3.3 Cracked surface area and edge crack density

Because of the complexity of the damage distribution, we have chosen to represent the cracking development through a global damage parameter : the cracked surface area per unit volume of the specimen, S_u, in every layer (external or internal) of the specimen. The figure 5 shows the development of cracked area values against the logarithm of the cycle number.

In figure 5.a., the cracked surface area values obtained in the external layers of the specimens tested

a. 50 fatigue cycles

b. 5 10^4 fatigue cycles

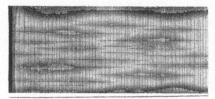

c. 2 10$^-$ fatigue cycles

Figure 3. Cracking development during mechanical fatigue of $(0_3/90_3)_s$ laminate (the loading axis is horizontal)

a. 50 fatigue cycles

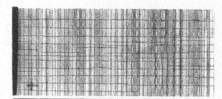

b. 5 10^4 fatigue cycles

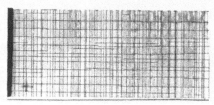

c. 2 10$^-$ fatigue cycles

Figure 4. Cracking development during mechanical fatigue of $(90_3/0_3)_s$ laminate (the loading axis is horizontal)

under thermal cycling of $(0_3/90_3)_s$ laminate are compared to those measured in the external 90° layers of the $(90_3/0_3)_s$ laminates subjected to mechanical fatigue loading. In the same way, in figure 5.b., the cracking development in the internal layer of the specimen exposed to temperature variations is compared to that of the inner ply of the $(0_3/90_3)_s$ laminate submitted to fatigue.

On this figure 5, it can be seen that the cracking development is far much faster in the specimens exposed to cyclic temperature variations than in those tested under mechanical fatigue.

In the external layers (Fig. 5.a.), the obtained ultimate values of the cracked surface area are analogous under thermal (0.24 mm^{-1}) and mechanical (0.22 mm^{-1}) cyclic loadings. In the internal layers, the ultimate S_u value remains lower under thermal cycling (0.37 mm^{-1}) than under mechanical fatigue (0.48 mm^{-1}). This last remark must be related to the preceding experimental observation that under thermal fatigue, after 500 cycles, the obtained cracking in the internal layer in still heterogeneous in the specimen width : an homogeneous distribution of damage in that case would have led to a cracked surface area higher, and closer to that measured under mechanical fatigue. Moreover, it must be noted that the ultimate cracked surface areas are higher in the internal layers.

In the external layers, the cycle number necessary to reach a quasi saturation of the cracked surface area is equal to 500 under temperature loading where as, under fatigue, 5 10^4 cycles are necessary to reach a real saturation in the external 90° layers of $(90_3/0_3)_s$ laminate. In the internal layers, a saturation stage of the cracked surface is reached after 1 10^6 cycles in $(0_3/90_3)_s$ laminate ; under thermal cycling, tests should be continued in order to have more precise information concerning the effective ultimate value of the cracked surface.

Because of the heterogeneity of damage in the internal layer under thermal loading,), as the density is different at the edge and in the heart of the specimen, the cracked surface area is not proportional to the edge crack density value. So, in figure 6, the development of the edge crack density measured on the edge of the coupons has been plotted versus the cycle number. On this figure, the edge crack density development is plotted for a single specimen in each case. When taking into account the observed experimental scattering between several specimens (± 0.05 /mm), the ultimate values of the edge crack densities can be considered as similar under thermal and mechanical fatigue. The corresponding ultimate mean values are about twice lower in the external layers : 0.44 mm^{-1} in the external layers, and 0.90 mm^{-1} in the internal ones.

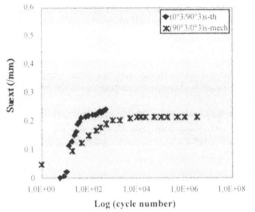

a. in the external layers

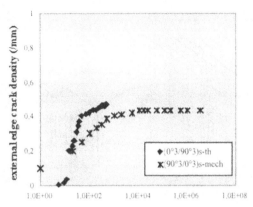

a. in the external layers

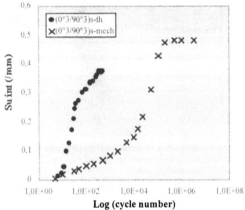

b. in the internal layers

Figure 5. Development of cracked surface area per unit volume of the specimen, versus loading cycle number : comparison between mechanical and thermal loadings

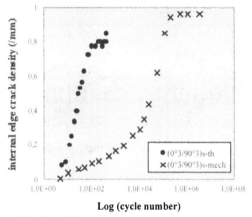

b. in the internal layers

Figure 6. Development of the edge crack density, versus loading cycle number : comparison between mechanical and thermal loadings

4 ANALYSIS

All these experimental results have been analyzed using a fracture mechanics approach developed earlier and validated for fatigue matrix cracking of various cross-ply laminates (Henaff-Gardin et al. 1996 a, b, Henaff-Gardin et al. 1997, Henaff-Gardin et al. 1999).

4.1 Presentation of the modeling

In the analysis, the studied cross-ply laminates are assumed to present a symmetric and periodic stacking sequence such as $(0_m, 90_n)_s$. The laminate is subjected to biaxial loading in both x and y directions, and to temperature variations ΔT. The z direction is that of the ply thickness (normal to the ply mean plane). The total 0° and 90° ply thicknesses are respectively noted $2t_0$ and $2t_{90}$. In order to represent in a simple analytical way the crack distribution in the specimen, we have chosen two damage parameters : the crack density d_{90} in inner 90° layer, and d_0 in each outer 0° layer. The cell geometry used for modeling the specimen is given in Fig. 6. The cracks are assumed to be plane, normal to the x axis in the 90° layer (and to the y axis in the 0° plies). They are uniformly spaced along the traction directions x and y (the crack spacings $1/d_0$ and $1/d_{90}$ are independent one from the other). Then, we can consider that the laminate, which presents a double periodicity along both x and y axes, is constituted by a double stacking of "repeating unit cells" limited by two pairs of consecutive cracks in 0° and 90° plies (see Fig. 7). The displacements are assumed to be parabolic in the cracked layer, and constant in the uncracked one (see Fig. 8).

73

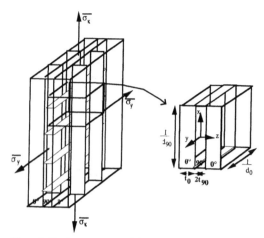

Figure 7. Crack geometry used for modeling the matrix cracking in cross-ply laminates

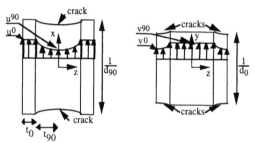

Figure 8. Assumptions on the displacement field

A critical value of the strain energy release rate, G_{Ic}, necessary for the initiation of a crack has been determined by performing quasi-static tests. It is equal to 83.5 J/m^2 for the T300/914 material used in this study. Moreover, a crack propagation law under fatigue, type of a power law has been introduced. The cracked surface propagation rate dS/dN are plotted versus the ratio G_I/G_{ini}, where G_I is the strain energy release rate value for the current crack density, and G_{ini} its initial value (the density tending towards zero). The obtained curve under mechanical fatigue is given below, in figure 9 (Henaff-Gardin et al. 1997).

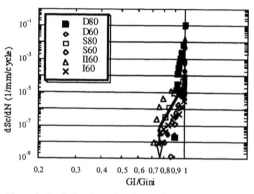

Figure 9. Evolution law of cracked surface propagation rate in an unit volume of cross-ply laminate, for four $(0_m/90_n)_s$ stacking sequences and two loading levels.

A general governing equation set is then derived for $(0_m, 90_n)_s$ laminates that involves only in-plane displacements in both layers, based on equilibrium, continuity and boundary conditions. Stress distributions, elastic constants and strain energy release rate under mode I opening (as the derivative of the stored elastic energy with respect to the cracked surface area per unit volume of the specimen) have been obtained as functions of the crack densities d_0 and d_{90} in 0° and 90° layers of the laminate (Henaff-Gardin et al. 1996 a, b).

The proposed analysis has then been used for previous experimental results. In $(0_m/90_n)_s$ laminates, cracks are created under predominant mode I opening.

During cyclic tests, once the first cracks are initiated, an increase of crack density is experimentally observed. The calculated corresponding value of the strain energy release rate is then decreasing throughout tests : this leads to a generation of new cracks more and more difficult and slow. The number of cracks must then tend towards a saturation value, in accordance with experimental observations.

Three stages appear in this propagation law :
- the initiation of cracks, corresponding to values of G_I/G_{ini} which are very close to 1, leading to a very quick evolution,
- the arrest of cracking, occurs for a threshold value of the ratio G_I/G_{ini} (approximately 0.75 for the studied material). This step corresponds with the experimental observation of the saturation in crack density and length.
- the propagation in itself, is observed between these two strain energy release rate limit values. The cracked surface propagation rate can be expressed as a power function of the strain energy release rate.

More recently, tests have allowed us to validate the preceding laws for damage evolution in the internal layer of $(0_m/90_n)_s$ laminates (Henaff-Gardin et al. 1998). The analysis has also been improved in order to predict damage development in the external layers of $(90_m/0_n)_s$ laminates (Henaff-Gardin et al. 1999). In every case, the 0° layers have been considered as uncracked.

4.2 Comparison of the experimental results with the prediction of the analysis

4.2.1 Crack initiation

In the case of cyclic tests, we note G_{ini} the value of the strain energy release rate when the first matrix crack initiates. G_{ini} is calculated for the maximum applied strain, and with a crack density approaching zero.

It has been showed previously, under mechanical fatigue, that when G_{ini} is lower than G_{Ic}, the initiation of the first matrix crack requires a micro-damage accumulation during the first fatigue cycles : no crack is observed before a N_{fpf} cycle number.

On the other side, when G_{ini} is higher than G_{Ic}, the first matrix cracks initiate at the first cycle.

Table 2 gives the values of G_{ini} for all the performed tests, as well as the corresponding predicted and experimental first ply failure cycle numbers.

For T300/914 laminates, G_{Ic} has been found equal to 83.5 J/m^2. As all the G_{ini} values are higher than G_{Ic}, the analysis predicts instantaneous cracking in all the tests, which is almost verified by the experiments.

Table 2. Strain energy release rate values at crack initiation ; predicted and experimental values of first ply failure cycle number

		mechanical fatigue		thermal cycling	
		$(0_3/90_3)_s$	$(90_3/0_3)_s$	$(0_3/90_3)_s$	
		internal layer	external layer	internal layer	external layer
G_{ini} (J/m^2)		141.5	141.5	110	110
N_{fpf}	predict.	1	1	1	1
	tests	10	1	5	10

4.2.2 Crack density at saturation and kinetics of cracking development

The propagation law under fatigue has allowed us to determine the predicted saturation crack densities for all the performed tests. These predicted values are compared to the experimental ones in the table 3. In the particular case of the internal layer under thermal cycling, where, at the end of tests, the damage is still heterogeneous in the specimen, the given experimental density corresponds to the edge one (indicated by *).

It can be observed in this table that there is a good agreement between the predicted and the observed edge crack density values, for every test conditions.

It has been shown that the analysis can predict the kinetics of damage development with cycle number, under mechanical fatigue loading. Unfortunately, under thermal cycling, the current modeling is unable to estimate the cycle number necessary to reach the saturation of crack density, as kinetics are observed far much faster under cyclic temperature. As an ex-

Table 3. Values of saturation crack densities : comparison of predictions and experiments for the two stacking sequences and the two types of loadings.
* : edge density values

			$(0_3/90_3)_s$		$(90_3/0_3)_s$	
		d_{sat} (mm^{-1})	intern. layer	extern. layer	intern. layer	extern. layer
mechanical fatigue	predict.	0.86	//	//	0.44
	tests	0.94	//	//	0.43
thermal cycling	predict.	0.86	0.44	0.86	0.44
	tests	0.84*	0.46	0.84*	0.46

ample, in the case of the internal layer of $(0_3/90_3)_s$ laminate, the analysis predicts a damage saturation after $2.1\ 10^6$ cycles ; experimentally, the saturation is obtained after $1\ 10^6$ cycles under mechanical fatigue, but only after 400 cycles under thermal cycling !

It must be noted that, in the present analysis, the variations of elastic constants, of coefficients of thermal expansion and of critical strain energy release rate with temperature have not been introduced. Moreover, the frequency of tests must have a strong influence on the damage development, where as this parameter doesn't appear in the model. On a physical and chemical point of vue, under thermal cycling, ageing of the material, and particularly of the matrix must play an important role as specimen almost reaches the glass transition temperature.

5 CONCLUSIONS

This study is concerned with the damage development under both mechanical fatigue and thermal cycling of $(0_3/90_3)_s$ and $(90_3/0_3)_s$ T300/914 laminates.

Under mechanical fatigue loading, cracks initiate in a first time in the 90° layers, followed by longitudinal cracks. Under thermal biaxial cyclic loading, cracks initiate almost in the same time in both internal and external layers. The cracked surface and edge crack density values at the end of tests are similar under thermal cycling and fatigue, but the kinetics of crack development are far much faster under thermal cycling.

A fracture mechanics analysis, previously developed in the case of mechanical fatigue loading in $(0_m/90_n)_s$ stacking sequences, has been applied without modification to the case of thermal cycling : it presently allows to predict crack initiation and crack density values at saturation. Nevertheless, this model still need to be modified to be able to predict the kinetics of damage development under thermal variations.

REFERENCES

Adams, D. S., D. E. Bowles and C. T. Herakovich 1986. "Thermally induced transverse cracking in graphite-epoxy cross-ply laminates." *Journal of reinforced Plastics and composites* 5: 152-169.

Adams, D. S. and C. H. Herakovich 1984. "Influence of damage on the thermal response of graphite-epoxy laminates." *Journal of Thermal Stresses* 7: 91-103.

Aswendt, P. and R. Hofling 1993. "Speckle interferometry for analysing anisotropic thermal expansion - application to specimens and components." *Composites* 24(8): 611-617.

Boniface, L., S. L. Ogin and P. A. Smith 1991. Fracture mechanics approaches to transverse ply cracking in composite laminates. *Composite materials : fatigue and fracture. Third volume.* K. O'Brien, American Society for Testing of Materials. ASTM STP 110: 9-29.

Boniface, L., S.L. Ogin and P.A. Smith 1992. The effect of temperature on matrix crack development in crossply polymer composite laminates. *Fifth european conference on composite materials*, Bordeaux, EACM.

Charewicz, A. and I. M. Daniel 1986. Damage mechanisms and accumulation in graphite/epoxy laminates. *Composite materials : fatigue and fracture.* H. T. Hahn, American Society for Testing of Materials. ASTM STP 907: 274-297.

Datoo, M. H. 1991. Residual stresses. *Mechanics of fibrous composites*, Elsevier Applied Science: 367-420.

Favre, J. P., H. Levadoux, T. Ochin and J. Cinquin 1996. Vieillissement des composites à matrice organique aux températures moyennes. Un premier bilan. *10èmes Journées Nationales sur les Composites JNC10*, Paris, France, AMAC.

Forsyth, D. S., S. O. Kasap, I. Wacker and S. Yannacopoulos 1994. "Thermal fatigue of composites - ultrasonic and SEM evaluations." *Journal of Engineering Materials and Technology - Transactions of the ASME* 116(1): 113-120.

Henaff-Gardin, C., J. L. Desmeuzes and D. Gaillot 1995. "Damage development due to cyclic thermal loading in cross-ply carbon/epoxy laminates." *Fatigue under thermal and mechanical loading, Ed : Bressers, J. ; Rémy, L., Kluwer Academic Publisher*: 285-293.

Henaff-Gardin, C., I. Goupillaud and M. C. Lafarie-Frenot 1998. Evolution de l'endommagement matriciel sous cyclage thermique de stratifiés à fibres longues :influence du matériau et du drapage. *11èmes Journées Nationales sur les Composites JNC11*, Arcachon, France, AMAC.

Henaff-Gardin, C., I. Goupillaud, M. C. Lafarie-frenot and S. Buhr 1999. Modelling of transverse cracking under uniaxial fatigue loading in cross-ply composite laminates : experimental validation. *12th International Conference on Composite Materials*, Paris, France, 5-9 July 1999.

Henaff-Gardin, C., M. C. Lafarie-Frenot and D. Gamby 1996 a. "Doubly periodic matrix cracking in composite laminates. Part 1 : general in-plane loading." *Composite Structures* 36(1-2): 113-130.

Henaff-Gardin, C., M. C. Lafarie-Frenot and D. Gamby 1996 b. "Doubly periodic matrix cracking in composite laminates. Part 2 : thermal biaxial loading." *Composite Structures* 36(1-2): 131-140.

Henaff-Gardin, C., M. C. Lafarie-Frenot and I. Goupillaud 1997. Prediction of cracking evolution under uniaxial fatigue loading in cross-ply composite laminates. *proc. of the 1rst International conference on fatigue on composites*, Paris, 3-5 Juin 1997.

Herakovitch, C. T., J. G. Davis and J. S. Mills 1980. Thermal microcracking in celion 6000/PMR-15 graphite/polyimide. *Thermal stress in severe environments.* H.

R. A. Kasselman D.P.H., Plenum Publishing Corporation: 649-663.

Jennings, M. T., D. Elmes and D. Hull 1989. Thermal fatigue of carbon fibre / bismaleimide matrix composites. *Third international conference on composite materials*, Bordeaux, France, Elsevier Applied Science.

McManus, H. L., D. E. Bowles and S. S. Tompkins 1996. "Prediction of thermal cycling induced matrix cracking." *J Reinf Plast Composite* 15(2): 124-140.

Nairn, J. A. and S. Hu 1994. Matrix microcracking (chapter 6). *Damage mechanics of composite materials.* R. Talreja, Elsevier. 9: 187-243.

Paillous, A. and C. Pailler 1994. "Degradation of multiply Polymer-Matrix composites induced by space environment." *Composites* 25(4): 287-295.

Tsai, G. C., J. F. Doyle and C. T. Sun 1987. "Frequency effects on the fatigue life and damage of graphite/epoxy composites." *Journal of Composite Materials* 21: 2-13.

Recent Developments in Durability Analysis of Composite Systems, Cardon, Fukuda, Reifsnider & Verchery (eds)
© *2000 Balkema, Rotterdam, ISBN 90 5809 103 1*

Effects of microstructure on damage evolution, strain inhomogeneity, and fracture in a particulate composite

C.T.Liu
Air Force Research Laboratory, Ol-AC AFRL/PRSM, Edwards, Calif., USA

ABSTRACT: In this study, the effects of microstructure of a particulate composite material on the local behavior near the crack tip was investigated. A large deformation digital correlation technique was used to determine the local strain fields within 2 mm of the crack tip and a real-time x-ray technique was used to determine the damage field near the crack tip. The experimental results were analyzed and are discussed.

INTRODUCTION

It is well known that a highly filled particulate composite material, on the microscopic scale, can be considered a nonhomogeneous material. Depending upon the degree of crosslink of the matrix material, filler particle size and distribution, and the bond strength at the interface of the particle and the matrix, the local stress and strength will vary in a random fashion. Therefore, when the material is strained, damage may develop in the material. The damage developed in the material may be in the form of microcracks or microvoids in the matrix or in the form of dewetting between the particle and the matrix. The damage will not be confined to a specific location; rather, it will diffuse into a relatively large area or zone. The growth of damage in the material may take place by tearing of the material or by successive nucleation and coalescence of the microvoids. Therefore, to gain an advanced understanding of the failure process in these materials requires a detailed knowledge of deformation process, damage initiation and evolution mechanisms, and crack-damage interaction.

In past years, a considerable amount of work has been done in studying local damage and crack growth behavior in highly filled polymeric materials (Liu 1991,1992,Liu & Ravi-Chandar 1996 ,Liu & Smith 1996, Smith et al.1992). In this study, the strain fields within a 2-mm region near the crack tip in a particulate composite, containing hard particles embedded in a rubbery matrix, were determined using a digital image correlation technique. The specimen (Fig.1) was subjected to a constant strain rate loading condition at room temperature. During the test, digital images were obtained at a given time interval, and they were analyzed, using a digital image correlation program, to determine the strain fields near the crack tip. In addition, Lockheed-Martin Research Laboratory's high energy real-time x-ray system (HERTS) was used to determine the damage field near the crack tip. The experimental data were analyzed and the results are discussed.

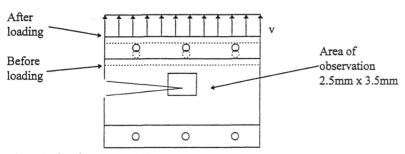

Figure 1 Specimen geometry.

THE EXPERIMENTS

In this study, a single-edge notched specimen was used to determine the local behavior near the crack tip under a constant strain rate condition at room temperature. Prior to testing, the specimen was conditioned at the test temperature for 1 hour, and loaded in a straining stage. A stepping motor was used to power the straining stage and to control the rate of deformation. To position the crack tip, a positioning stage was used to move the specimen in the x and y directions. The positioning stage is controlled by operating a joystick device. The fracture processes were a Nikon Metallurgical microscope, a CCD camera, and a personal computer with a frame grabber unit Figure 2 is a schematic representation of the test setup.

In this study, the damage field near the crack tip in a cracked sheet specimen subjected to a constant strain rate of 1.0 min^{-1} was investigated using the real-time x-ray technique. The specimen was made of the same particulate composite material as used in the crack propagation test. Prior to testing, a 23-mm crack was cut at the center of the specimen with a razor blade. During the test, Lockheed-Martin Research Laboratory's high-energy real-time x-ray system (HERTS) was used to investigate the characteristics of the damage field near the crack tip. The specimen was placed between the x-ray radiation source and the x-ray camera. The x-ray image exits from the specimen and strikes the screen that is in front of the x-ray camera. The screen converts the x-ray image into a light image. This image is reflected into a low-light-level television camera by a mirror placed at 45° to the beam in the back of the camera. This isocon TV camera then converts the light image into an electronic signal that can be routed into the main monitor and into the video tape recorder. A schematic representation of the x-ray testing setup is shown in Figure 3, and a detailed description of the HERTS can be founded in reference (Sklensky & Buchanan 1985).

DATA ANALYSIS

To determine the strain fields near the crack tip, a Large Deformation Image Correlation (LDDIC) program, developed by Vendroux & Knauss (1994) and Gonzalez (1997) was used. The LDDIC program was developed by modifying a Digital Image Correlation (DIC) program developed by Sutton al.et. (1986) for small deformations. The problem in applying DIC to compute strain fields in

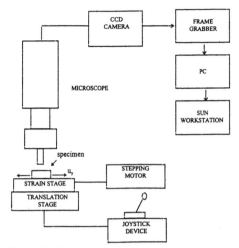

Figure 2   Experimental setup.

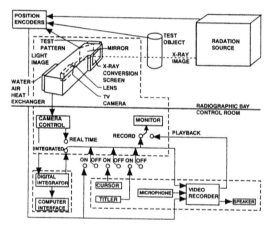

Figure 3   Block diagram of a real-time radiographic system.

a large deformation process is the failure of convergence of the DIC algorithm if the strain is larger than 10%. To circumvent this problem, the LDDIC takes intermediate images (or steps) of the deformation between the undeformed and the deformed states of the deformation , and, then, computes the displacements and the displacement gradients for every step of the deformation. The intermediate results are combined to produce displacements and displacement gradients for the global deformation. To determine the accuracy of the LDDIC program, a test was conducted on a and coated with microscopic speckles. The specimen was stretched to 80% strain. A sequence of 15 images was taken during the deformation process and the strains were calculated by the LDDIC program. A comparison of the prescribed and calculated strains shows that a maximum deviation of 1% strain occurs at the 40% strain level. This experimental result validate the LDDIC method (Gonzalez 1997).

The recorded x-ray data were processed to create a visual indication of the energy absorbed in the material. A region of high absorption (i.e., a low damage area) will be shown as a dark area, whereas a region of low absorption will produce a light or white area, with 254 shades of gray in between. Also, the x-ray image at a given applied strain level can be plotted in the form of iso-intensity contours of the transmitted x-ray energy to enhance the resolution of the damaged field.

RESULTS AND DISCUSSION

The effect of microstructure on the strain fields near the crack tip, plotted the contours of the iso-intensity principal strain , is shown in Figure 4 . The data is presented as maximum principal strain (MPS) rather than the usual components of strain because the crack opening and void formation depend on the local principal strains. In Figure 4, the crack is shown as the horizontal line with the crack tip at a location of x=0.4 mm and y = 1.25 mm (0.4mm or 1.25mm).The gray area surrounding the crack is the noise region in which the data can not be analyzed. This is because the digital image correlation program doesn't converge in places where new geometrical features appear. The cases of crack propagation and void formation are examples in which new features appear in the image. The digital image correlation program doesn't recognize the relationship between the features in the undeformed and the deformed images. Referring back to Figure 4, it is seen that, starting at the crack tip, there is a broken thin line that goes toward the right of the figure, representing the future crack propagation path. Note that the strain fields are highly nonhomogeneous and high strain regions are localized in the neighborhood of the crack tip. For example, the largest MPS, greater than 20%, is reached about 0.2 mm from the crack tip. There are several other high strain regions located near the crack tip. The average of these high strain regions has 400 microns in diameter. It is interesting to noting that the high strain region where a void will develop as the applied far field strain (FFS) is increased.

A plot of the MPS concentration factor, defined as a ratio of MPS to FFS, along the crack plane as a

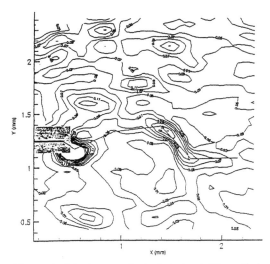

Figure 4   Maximum principal strain at 6% applied strain.

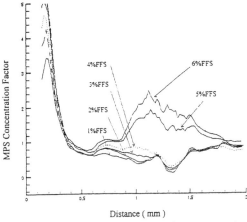

Figure 5   Maximum principal strain concentration factor along the crack plane.

function of the FFS is shown in Figure 5. From Figure 5, the curves representing FFS of 1% to 4% fall very close to each other indicating that the distribution of MPS is fairly linear with the FFS. However, when the FFS level increases to 5%, a significant increase in MPS concentration factor occurs in a region between x = 0.7 mm and x = 1.8 mm. This phenomenon is probably due to the separation of the particle and the binder, resulting in a significant increase in MPS. When the FFS is continuously increased, eventually a void is formed in this high strain region and the crack advances by the coalescence of the crack tip with the void.

In the above paragraphs, we discussed the effect of microstructure on the strain distributions near the crack tip. In the following paragraphs, the damage mechanisms and the local fracture behavior in the immediate neighborhood of the crack tip are discussed.

When crack occurs, the high stress at the crack tip will induce high damage near the crack tip region. The high damage zone at the crack tip is defined as the failure process zone, which is a key parameter in viscoelastic fracture mechanics. Experimental data reveal that when the local strain reaches a critical value, small voids are generated in the failure process zone. Due to the random nature of the microstructure, the first void is not necessarily formed in the immediate neighborhood of the crack tip. The formation of the voids is not restricted to the surface of the specimen where the maximum normal strain occurs. Since the tendency of the filler particle to separate from the binder under a triaxial loading condition is high, it is expected that voids or damage zones will be generated in the specimen's interior. Consequently, there are a large number of strands, which separate the voids and are essentially made of the binder material, that form inside the failure process zone. These damage processes are time-dependent and are the main factor responsible for the time-sensitivity of strength degradation as well as fracture behavior of the material.

Figure 6 is typical set of photographs showing the crack surface profile during opening and growth of a crack in the composite material specimen. Figure 6 shows that crack tip blunting occurs both before and after crack growth. Due to the heterogeneous nature of the composite material, the degree of blunting varies with the position of the advancing near the crack tip plays a significant role in the blunting phenomenon. During the blunting stage, voids developed in the failure process zone. The failure of the material between the void and the crack tip leads the crack to grow a short distance. In other words, the coalescence of the void and the crack tip leads the crack to grow into the failure process zone. This kind of crack growth mechanism continues until the main crack tip reaches the failure process zone tip. When this occurs, the crack tip resharpens temporarily. Thus, the process consists of a blunt-growth-blunt phenomenon which is highly nonlinear. Referring back to 6 Fig. a close look at the crack tip region reveals that the failure process zone has a cusp shape which is consistent with that predicted by Schapery (1975) in his study of fracture of viscoelastic material. In the failure process zone, the material can be highly nonlinear

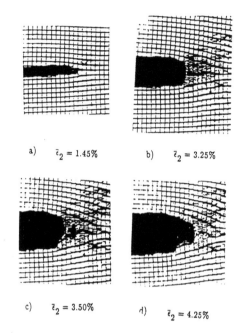

a) $\bar{\varepsilon}_2 = 1.45\%$ b) $\bar{\varepsilon}_2 = 3.25\%$

c) $\bar{\varepsilon}_2 = 3.50\%$ d) $\bar{\varepsilon}_2 = 4.25\%$

a) Early stage of crack opening
b) Development of stretch zone ahead of crack with tip blunting
c) Increased blunting with void formation ahead of crack
d) Crack extension by void coalescence (resharpening of crack follows)

Figure 6   Crack opening and growth in the specimen.

and suffer extensive damage. Experimental data indicate that the direction of the failure process zone with respect to the crack plane varies from

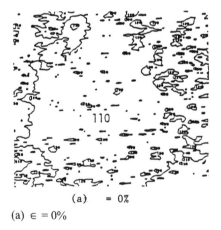

(a)   = 0%

(a) $\in = 0\%$

Figure 7   Iso-intensity contour plots of I_t near the crack tip.

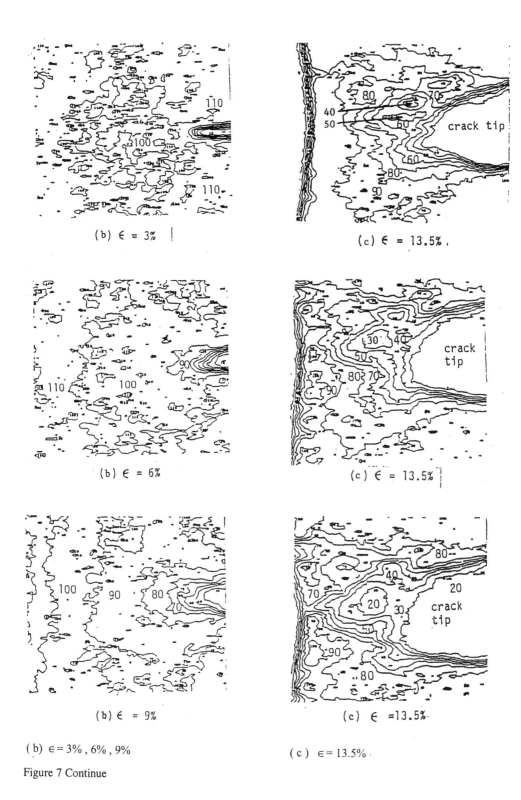

(b) $\epsilon = 3\%$

(c) $\epsilon = 13.5\%$.

(b) $\epsilon = 6\%$

(c) $\epsilon = 13.5\%$

(b) $\epsilon = 9\%$

(c) $\epsilon = 13.5\%$

(b) $\epsilon = 3\%$, 6%, 9%

(c) $\epsilon = 13.5\%$

Figure 7 Continue

specimen to specimen. This is believed to be related to the size of the highly strain region as well as the local microstructure of the material in that region. For a large magnitude of tip blunting, the size of the highly strained region is also large. Therefore, depending on the local microstructure, the direction of the failure process zone shows a relatively large variation. Experimental results reveal that before crack growth, the failure process zone develops either above, below, or along the crack plane. After crack growth, the successively developed failure zones at the tip of the propagation crack undulate about the crack plane, resulting in a zig-zag shape of crack growth. It is interesting and important to note that the crack has a tendency to grow in the average direction perpendicular to the applied load direction. The change of the stress concentration location as the result of crack tip blunting also contributes to the variation in failure process zone direction. When the crack tip is extensively blunted, the stress concentration location changes from the tip of the sharp crack to the upper and lower corners of the blunted crack. Therefore, the probability of developing a failure process zone near the corners of the blunted crack is greater.

To determine the size and damage intensity in the damage zone, iso-intensity contours of the transmitted x-ray energy, I_t, were plotted and are shown in Figure 7. In Figure 7 the number between the two contour lines is the range of I_t between the prior and the next intensity level. A small number indicates that the intensity of the transmitted x-ray energy is high or that the damage is high. These of the damage zone as well as damage intensity inside the damage zone.

Prior to conducting the test, the virgin specimen was imaged and the iso-intensity contours of I_t were plotted and are shown in Figure 7a. The nonuniform distribution of the values of I_t is an indication of the nonhomogeneity of the material. When the specimen is stretched to 3% strain, relatively high damage is developed near the crack tip region as shown in Figure 7b. When the specimen is stretched to the next strain level (6%), the damage intensity and the damage zone size increase. This type of damage evolution process continues as the magnitude of the applied strain is successively increased, as shown in Figure 7b. This phenomenon, under a monotonically-increasing loading condition, is expected because the magnitude of the stress and strain are increased due to the increase in the applied strain level. Consequently, the severity of the damage and the damage zone size increases.

Figure 7c shows the damage characteristics near the tip of the propagating crack when the specimen was held at 13.5% strain level. From Figure 7 c, we see that the crack tip is relatively sharp and two small damage zones with high damage intensities, I_t equal to 40 and 50, developed. As time elapses, the crack tip become blunted, the damage intensity near the crack tip increases, and the damage intensities I_t in the two highly damaged regions increase from 50 and 40 to 30. As the crack grows a short distance, the damage intensity near the crack tip increases further, and, a void is formed in the region where I_t equal to 20. Finally, the main crack tip and the void join, leading to fracture of the specimen.

CONCLUSIONS

In conclusion, the heterogeneity of the microstructure plays a key role for local damage and strain distributions near the crack tip. Experimental results indicate that the high strain field is localized within 1 mm of the crack tip. Also, the real-time x-ray technique is a promising technique to monitor damage initiation and evolution processes.

REFERENCES

Gonzalez, J. 1997. Full Field Study of Strain Distribution near the Crack Tip in Fracture of Solid Propellant Via Large Deformation Digital Aeronautical Engineer Thesis, California Institute of Technology, Pasadena, CA

Liu, C.T., 1991. Evaluation of damage Fields near Crack Tip in a Composite Solid propellant. Journal of Spacecracfts and Rockets, Vol.28, No. 1: 64-70.

Liu, C.T.1992. Acoustic Evaluation of Damage Characteristics in a Composite Solid Propellant," Journal of Spacecraft and Rockets, Vol.29, No. 5: 709-712.

Liu, C.T. & Ravi-Chandar 1996. Local Fracture and Crack growth in a Particulate Composite Material. Journal of Reinforced Plastic and Composites, Vol.15: 196-207.

Liu, C. T. & Smith, C.W. 1996. Temperature and Rate Effects on Stable Crack Growth in a Particulate Composite Material. Experimental Mechanics, Vol.36, No.3, 290-295.

Schapery, R.A. 1975. A Theory of Crack Initiation and Growth in Viscoelastic Media. Int. Journal of Fracture Mechanics, 11 ,PP.141-159.

Sklensky, A.F.& Buchanan 1985. Sensitivity of High-Energy Real-Time Radiography with

Digital Integration. ASTM SPT 716 ; .315-329.

Smith, C.W.,1 Wang,L. Mouille,H.,& Liu, C.T.,"
1992. Near Tip Behavior of Particulate
Composite Material Containing Cracks at
Ambient and Elevated Temperatures.Fracture
Mechanics, ASTM-STP 1189,23 ;775-787.

Sutton, M.A., Cheng,M.,Peters,W.H., Chao, Y.J.
& McNeill, S.R..1986 .Application of an
Optimized Digital Correlation Method to Planar
Deformation Analysis," Experimental
Mechanics. Vol. 4, No. 3:. 143-150.

Vendroux, G. & Knauss, W.G. 1994 Deformation
Measurements at the Sub-Micron Size Scale:
II. Reinforcements in the Algorithm for Digital
Image Correlation", GALCIT SM Report 94--5,
California Institute of Technology, Pasadena,
Ca..

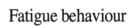
Fatigue behaviour

Recent Developments in Durability Analysis of Composite Systems, Cardon, Fukuda, Reifsnider & Verchery (eds)
© 2000 Balkema, Rotterdam, ISBN 90 5809 103 1

Fatigue in composite laminates – A qualitative link from micromechanisms to fatigue life performance

E. K. Gamstedt
Risø National Laboratory, Roskilde, Denmark

The fatigue life behaviour and damage mechanisms have been investigated in a variety of polymer-matrix composite materials. The progressive mechanism was either fibre-bridged cracking or interfacial debonding. A common denominator for all materials was that interfacial cracks growing in the fibre direction led to a rapid degradation and failure. Since the fibres have a distribution in strength, damage in the form of fibre breaks initiated at the first application of load. From these damage sites, interfacial cracks grew in the fibre direction in the composites with weak interfaces, leading to a continual redistribution of stress concentration in the neighbouring fibre segments. During fatigue, some of these fibres also failed and gave rise to new interfacial cracks, etc. Eventually damage coalesced, and final failure occurred. Just imminent to failure, localisation took place. Since interfacial cracking has shown to deteriorate the material, measures to suppress this mechanism are most likely to lead to more fatigue resistant materials.

1 INTRODUCTION

Fatigue is an increasingly important issue for load-carrying applications made of polymer-matrix composites. One way to deal with the fatigue problem is from a macroscopic and structural viewpoint, in which fatigue design criteria are sought. This would allow for more slender and lighter designs, resulting in e.g. lower fuel consumption for aerospace and automotive applications. Another aspect of fatigue concerns the microlevel and the composite material itself. By e.g. choosing a suitable type of fibre, interfacial treatment and matrix, one could improve the fatigue resistance of the material. Between the microstructural level of the composite material and the macroscopic level of the composite structure lies a spectrum of interacting complex mechanisms. To establish a direct quantitative link between these levels is a formidable task, which would require computer power and experimental techniques that are yet to be invented. A first step would be to find a qualitative link between the microstructure and global fatigue life performance by an experimental investigation of the underlying mechanisms. In this manner, the influence of e.g. a weak interface and a concomitant high debond rate on the fatigue life properties could be found. The variability of fibre strength is another important parameter, which is associated with the mechanism of fibre rupture. The fibre breaks can then act

as initiation sites for further damage growth. In the present study, micromechanisms of these kinds have been investigated in a number of carbon and glass fibre reinforced plastics with long continuous fibres, and analysed together with the resulting macroscopic fatigue behaviour. The aim is to shed light on the relation between the micromechanisms and fatigue performance for these polymer matrix materials, and thereby indicate how the material can be improved to suppress the identified detrimental mechanisms.

Due to the complexity that arise from the influence of applied loading mode and laminate lay-up on the mechanisms, some simplifications are required in order to narrow down the problem and to identify the dominating and critical microstructural properties. If all possible parameters are taken into consideration, e.g. a general laminate lay-up subjected to variable multiaxial tension-compression loading in service environment, the micromechanical scenario becomes too complicated to be analysed with a more exact and physical approach. We would then be forced and compelled to a macroscopic procedure, which does not allow for any extrapolative predictions or indicate how the material can be improved. The simplifications should be made judiciously to maintain relevance in view of applications. If we consider the stacking sequence of a general multidirectional laminate, it is the longitudinal ply that is the critical element since it is generally the main load-carrying constituent and the last ply to fail. Previous investigations (Dickson et al. 1989; Talreja 1993; Andersen et al. 1996) have also shown that the longitudinal plies control the fatigue life of multidirectional laminates when interpreted in terms of global applied strain. Focus is therefore placed on unidirectional composites with the fibres in the $0°$ direction. Concerning the loading mode, only tensile fatigue with a stress ratio of $R = \sigma_{min}/\sigma_{max} = 0.1$ is used at this stage. The simple loading modes must be clarified first before moving on to more complex loading involving e.g. compressive load excursions.

2 EXPERIMENTAL PROCEDURES

2.1 Materials

The tested materials were carbon-fibre reinforced epoxy (AS4/8552 from Hexcel) and carbon-fibre reinforced PEEK (polyetheretherketone; APC-2 from ICI). Both materials had the same kind of AS4 fibres, and were processed according to the manufacturer's recommendation. The chosen stacking sequence was $[0_4]_T$ with dimensions in accordance with recognised standards. Specimens of the same dimensions made of glass-fibre reinforced plastics were also analysed. These were made of polypropylene with and without a maleic anhydride modification, and reinforced by E-glass fibres. The purpose of the matrix modification was to improve the interfacial adhesion. The details of the manufacture of the specimens are reported elsewhere for the carbon-fibre reinforced plastic (Gamstedt & Talreja 1999) and glass-fibre reinforced polypropylene (Gamstedt et al. 1999).

2.2 Testing

The fatigue tests were carried out in an Instron 1272 tensile servo-hydraulic tensile machine at 10 Hz for the carbon-fibre reinforced plastic, and with a triangular waveform with an absolute strain rate of 10^{-2} s^{-1} for the glass-fibre reinforced plastic. The stress ratio was constantly $R = 0.1$. During the course of fatigue testing, the tensile machine was intermittently stopped, and replications were made on the surface of the composite specimens. These were later studied under optical microscope in chronological order to form a sequence of micrographs that show how the fatigue damage propagates. The replicas were made by applying pressure with a polymer film (cellulose acetate) onto the specimen surface. The film was made malleable with acetone and let harden on the specimen to give a negative of the surface topology.

3 RESULTS

Despite the large difference in constituent properties between the carbon-fibre reinforced plastic materials and the glass-fibre reinforced plastic materials,

they showed a striking similitude in fatigue damage mechanisms and in the influence of the mechanisms on the macroscopic fatigue performance. The major difference between the fatigue-sensitive and fatigue-resistant material was that progressive far-reaching debonding was present in the fatigue-sensitive material. The results presented in this section will therefore be partitioned according to the observed mechanism. Either the debonds grew in the longitudinal direction (along the fibre and load directions), or the progressive mechanism was matrix crack growth in the transverse direction (perpendicular to the fibre and load directions).

3.1 Transverse damage growth

In Figure 1, a fibre-bridged crack in the carbon fibre/epoxy material with superficial fibres is shown. This material showed strong interfaces, and matrix cracks were initiated from individual fibre breaks, and grew in the transverse direction. As the crack grew around and past adjacent fibres, the crack became bridged by the fibres. The progressive mechanism was matrix crack growth in the transverse direction. Final failure occurs when the bridging fibres are broken and the crack propagates catastrophically (Begley & McMeeking 1995).

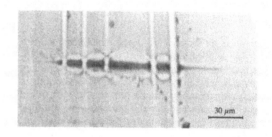

Figure 1: Fibre-bridged crack in fatigue of carbon fibre/epoxy at $\varepsilon_{max} = 1.1\%$ and 10^3 cycles

After some amount of cyclic loading, the crack opening profiles acquired a shape with squeezed tips (see inset in Figure 2). Such a shape is also indicative of cohesive tractions on the crack surfaces close the crack tips. In this way, the fibre bridging is shielding the crack tip, and reducing the effective stress intensity factor. The bridging is thus acting as a toughening mechanism which moderates the crack propagation rate. For monolithic materials showing Paris-type crack growth, the crack growth rate increases during fatigue. It can be seen in Figure 2 that the carbon fibre/epoxy material instead shows receding growth rates. In the subsequent section, the contribution of fibre bridging to the observed fatigue crack-growth toughness will be investigated.

In the glass-fibre reinforced plastic material with strong interfaces (maleic-anhydride modification of

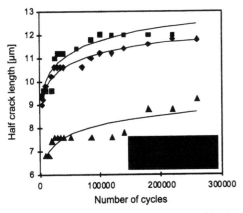

Figure 2: Crack growth curves for carbon fibre/epoxy at $\varepsilon_{max} = 0.98\%$ (bridged crack with squeezed tips in inset)

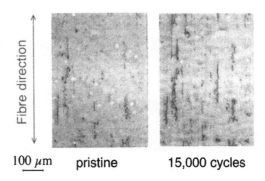

100 μm   pristine   15,000 cycles

Figure 3: Longitudinal crack growth for carbon fibre/PEEK at $\varepsilon_{max} = 0.96\%$

the polypropylene matrix), there was also matrix cracking in the transverse direction, although no signs of fibre bridging was observed (Gamstedt et al. 1999). A common feature with the carbon fibre/epoxy material, was that the transverse cracks were scarce and propagated independently at a relatively slow rate. This lack of coalescence and interaction with other damage sites warrants the use of fracture mechanics analysis of the most serve flaw for predictive purposes. Another common characteristic was that the fatigue damage originated from fibre breaks. Due to the variability in fibre strength, there will always be a number of broken fibres from which fatigue damage can develop even at low load levels.

The transverse damage growth was not prevalent at all strain amplitudes for the above materials. In carbon fibre/epoxy at low strain amplitudes close or below the fatigue limit, the transverse cracks were effectively arrested by limited fibre-matrix debonding (the so-called Cook-Gordon mechanism). This debonding acted as an arrest mechanism and not a progressive mechanism since it terminated the growth of the fibre-bridged cracks and did not propagate beyond a certain length itself. In the following, experimental results of debonding as a *progressive* mechanism will be presented.

3.2 Longitudinal damage growth

The fatigue damage in carbon fibre/PEEK developed in an entirely different manner. In Figure 3, it is evident that longitudinal cracks have grown. Also for this material, the progressive damage originated from distributed fibre breaks. At higher magnification, the longitidinal cracks were observed to form from progressive debonding.

In the glass-fibre reinforced plastic material with

weak interfaces (unmodified polypropylene matrix), there was also longitudinal crack growth at the fibre-matrix interface (Gamstedt et al. 1999). Here, the common feature with carbon fibre/PEEK was that progressive debonding originated from distributed fibre breaks, and in turn led to new fibre breaks, further debonding etc. and eventually to failure.

Furthermore, debonding as a progressive fatigue mechanism does not only have notable implications on the fatigue performance of the 0° ply. The formation of transverse cracks in 90° plies has also been shown to be a result of debond propagation, in particular for cyclic loading including compressive load excursions (Gamstedt & Sjögren 1999).

3.3 Damage localisation

For materials showing pronounced debonding, the accumulation of damage is relatively homogeneously distributed in the material during the largest part of the total fatigue life. This has been observed in unidirectional carbon/glass fibre hybrids by means of temperature field measurements (Gamstedt & Brøndsted 1999). The surface temperature of the specimen remained virtually constant and distributed during the first 99% of the expended lifetime (see Figure 4). Just imminent to failure, localisation took place and an overheated region appeared, which grew and caused final failure. The local heat dissipation can serve as a measure of the damage activities, since the heat is generated by irreversible effects in the materials such as hysteretic losses in the polymer matrix, and in particular the frictional sliding at crack surfaces. It was observed that an increase in damage (debonding, splits) was accompanied by an increase in local temperature during cyclic loading (Jacobsen et al. 1998; Gamstedt & Brøndsted 1999). If the transition from the distributed damage state to the localised damage state can be predicted, the fatigue life of the 0° can also be predicted. The transition is from a continuum damage state where the damage sites grow independently, to a localised damage state where the damage sites link up and progress catastrophically. The onset of damage coalescence is therefore of fundamental importance.

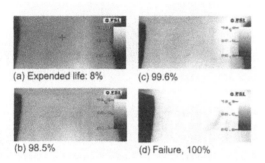

(a) Expended life: 8%   (c) 99.6%

(b) 98.5%   (d) Failure, 100%

Figure 4: Temperature field measurements of a unidirectional carbon-glass hybrid composite specimen: Heat localisation imminent to final failure

3.4 *Consequential fatigue performance*

Whether or not debonding is active as a progressive mechanism leads to a substantial difference in the growth rate and amount of damage in fatigue. This difference has a direct implication on the macroscopic fatigue lives of the specimens.

The fatigue-life diagrams of carbon fibre/epoxy and carbon fibre/PEEK are shown in Figure 5. Firstly, the scatter is larger for the carbon fibre/epoxy case. Secondly, the fatigue sensitivity, i.e. the slope of the sloping part in the fatigue-life diagram, is greater in carbon fibre/PEEK.

For the glass-fibre reinforced polypropylene, the matrix modification resulted in a stronger fibre-matrix interface (affirmed by post-mortem fractography), which suppressed debonding, and in turn led to a prolongation of the fatigue lives by approximately a decade (Gamstedt et al. 1999). Also, the axial stiffness remained virtually constant throughout the fatigue lives, displaying a 'sudden-death' type behaviour. In contrast, the unmodified glass fibre/polypropylene, with a relatively weak interface, showed a gradual decay of the axial stiffness, which indicates widely distributed fibre breaks caused by a progressive debond growth.

Despite the relative difference between the investigated carbon-fibre reinforced plastics and glass-fibre reinforced plastics, both groups showed the same influence of the micromechanisms on the fatigue performance. Progressive debonding led to more fatigue sensitive behaviour, whereas a strong interface led to localised transverse damage growth and a more fatigue resistant material.

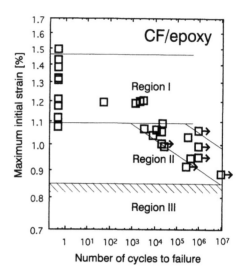

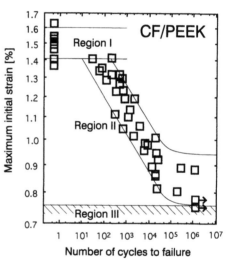

Figure 5: Fatigue life diagrams for carbon fibre/epoxy and carbon fibre/PEEK

4 QUANTITATIVE ATTEMPTS

4.1 *Fibre-bridged cracking*

Transverse matrix crack growth has been observed to be a progressive mechanism in the strong interface composites (carbon fibre/epoxy and glass fibre/maleic-anhydride modified polypropylene). In particular, fibre-bridged cracks arose as the matrix cracks grew around and past surrounding fibres in the carbon fibre/epoxy material (see Figure 1). In this way, the bridging fibres come to exert a cohesive action. Effectively, the stress intensity factor at the crack tip will be shielded by the cohesive tractions from the briding fibres. The fibre-bridging mechanism act as a toughening mechanism, and contributes to the observed crack retardation in Figure 2. Since the crack-opening displacements have been measured, and the da/dN-ΔK curves for various epoxies are given in handbooks, the relative contribution of the fibre-bridging can be estimated by fracture mechanics.

The bridged cracks are initiated from one or a couple of adjacent fibre breaks close to the specimen surface. When the transverse matrix cracks that propagates from these fibre breaks have crossed surrounding fibres, the cracks will become partially bridged. This can be illustrated by a the crack-opening profile in Figure 6b with traction-free crack surfaces in the centre and cohesive zones closer to the crack front, resulting in a crack with squeezed tips, such as the replica in the inset of Figure 2. Since the observed cracks emanated from the surface, a steady-state semi-elliptical crack front is expected after some time of propagation. However, experimental results on both metals and polymer composites have shown that the cracks acquire a semi-circular front when the geometry of the specimen is large compared to the crack dimensions (Newman & Raju 1981; Chermahini et al. 1993; Casado et al. 1997). Therefore, it is assumed that the crack front is semi-circular in the present investigation, and that the opening profile of the interior of the crack is given by the surface crack opening profile by symmetry of revolution (cf. Figure 6a).

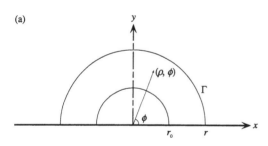

(a)

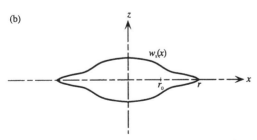

(b)

Figure 6: Rotational symmetric semi-circular surface crack viewed (a) in cross-section, and (b) from surface

The far-field uniform applied stress is denoted σ_0. If the cohesive traction is assumed to be proportional to the crack-opening displacement, $w(\rho, \phi) = w_s(\rho)$, the total crack surface traction (in the absence of the crack) would be $\sigma(\rho, \phi) = \sigma_0 - k\,w(\rho, \phi)$ in the bridged region and $\sigma(\rho, \phi) = \sigma_0$ in the unbridged central region. The crack-opening profile for an unbridged semi-elliptical crack subjected to uniform tension has been solved by Fett (1988). This solution is here used as a reference solution and denoted (σ_r, w_r). A relation between the reference solution and the experimental case can formulated with the Betti-Rayleigh reciprocal theorem,

$$\iint_S \sigma_r w \, dS = \iint_S \sigma w_r \, dS, \tag{1}$$

where S is the crack surface ($0 \leq \rho \leq r$, $0 \leq \phi \leq \frac{\pi}{2}$). Only a quarter circle is considered here due to symmetry. An explicit expression of the spring constant k can then be obtained as

$$k = \sigma_0 \frac{\iint_S (w_r - w) \, dS}{\iint_{S'} w w_r \, dS}, \tag{2}$$

where S' is the bridged region of the crack surface ($r_0 \leq \rho \leq r$, $0 \leq \phi \leq \frac{\pi}{2}$). By use of Equation 2, the evolution of the cohesive spring constant k during fatigue has been calculated from the experimental crack-opening profiles and the reference solution. The results are presented in Figure 7. Obviously, there is a cycle-dependent degradation of k, which results in an increasingly inefficient closure of the growing bridged zone. The reason for the decrease could be irreversible damage occurring at the interfaces and in the matrix material in the vicinity of the bridging fibres. These phenomena are not taken into account in the present elastic description.

The next step is to reduce the experimental data to a cycle-dependent nominal and effective stress intensity factor, $K_r(r, \phi)$ and $K(r, \phi)$ respectively. For the nominal unbridged case, the solution is (Newman & Raju 1981)

$$K_r(r, \phi) = \sigma_0 Y_r \omega(\phi) \sqrt{r}, \tag{3}$$

where $Y_r = 1.17$ when the crack is small compared to the specimen dimensions. The variation along the crack front is given by

$$\omega(\phi) = 1 + 0.1(1 - \sin\phi)^2. \tag{4}$$

The stress-intensity factor range at the surface crack tip (neglecting the non-square-root singularity at the vertex) can written as

$$\Delta K_{\text{nom}} = (1 - R) K_r(a, 0) \tag{5}$$

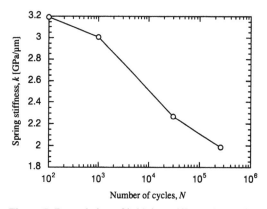

Figure 7: Degradation of bridging stiffness during fatigue

where R is the stress ratio. To compare with the growth rate for the neat matrix material, we can scale the stress-intensity factor range as if it were in the composite by energetic consideration (McCartney 1987)

$$\Delta K_{\mathrm{m}} = \sqrt{\frac{V_{\mathrm{m}} E}{E_{\mathrm{m}}}} \Delta K'_{\mathrm{m}}, \qquad (6)$$

where K_{m} and K'_{m} are the stress intensity factors in the composite and in the neat matrix material, respectively, V_{m} is the volume fraction of matrix, E_{m} is the Young's modulus of the matrix, and E is the effective Young's modulus of the composite defined below in Equation 10.

The $\mathrm{d}a/\mathrm{d}N$-ΔK curves for the neat epoxy matrix (Equation 6, data from Hertzberg & Manson, 1990) and for the carbon fibre/epoxy data (Equation 5) are displayed in Figure 8. Firstly, there is a large difference between the growth rates of the matrix material and the composite. Secondly, the growth rate of

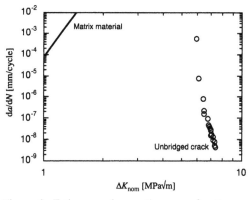

Figure 8: Fatigue crack growth curves for the unbridged crack and the matrix material

the composite has a downward trend, which indicates crack retardation (as seen in Figure 2). To investigate if the crack retardation and the difference in crack growth rates can be attributed to the bridging mechanism, the bridged case should be analysed by fracture mechanics in terms of the effective stress-intensity factor range.

For the effective (bridged) case, the stored energy can be explicitly calculated from

$$U(r) = \sigma_0 \iint_S w \, \mathrm{d}S - k \iint_{S'} w^2 \, \mathrm{d}S. \qquad (7)$$

The stored energy $U(r)$ is also related to the effective stress intensity factor by

$$U(r) = \int_{r_0}^{r} \left\{ \int_{\Gamma} \frac{K^2(r', \phi)}{\beta(\phi) E} \, \mathrm{d}\Gamma \right\} \mathrm{d}r', \qquad (8)$$

where $\beta(\phi)$ is a function that represents the change in stress state from plane stress to plane strain along the crack front (Wang 1995)

$$\beta(\phi) = \frac{1 - \nu^2 (1 - \sin \phi)^4}{1 - \nu^2}, \qquad (9)$$

where ν is the major Poisson's ratio, and E is the effective Young's modulus

$$E = 2E_{\mathrm{L}} \left\{ 2 \left(\sqrt{\frac{E_{\mathrm{L}}}{E_{\mathrm{T}}}} - \nu \right) + \frac{E_{\mathrm{L}}}{G_{\mathrm{T}}} \right\}^{-\frac{1}{2}}, \qquad (10)$$

expressed by the tensile and shear moduli of the orthotropic composite in the longitudinal (L) and transverse (T) directions.

Equations 7 and 8 can now be simplified to

$$\int_0^{\pi/2} \frac{K^2(r, \phi) r}{\beta(\phi)} \, \mathrm{d}\phi = E \frac{\partial U}{\partial r}, \qquad (11)$$

where $U(r)$ is calculated from cubic spline interpolation of the discrete experimental results. If it is assumed that $K(r, \phi)$ varies along the crack front in the same way as the nominal stress-intensity factor K_{r} (cf. Equation 3), such that

$$K(r, \phi) = \sigma_0 Y \omega(\phi) \sqrt{r}, \qquad (12)$$

a substitution into Equation 11 results in

$$Y = \frac{1}{\sigma_0 r \Omega} \sqrt{\frac{2E}{\pi} \frac{\partial U}{\partial r}}, \qquad (13)$$

where

$$\Omega = \sqrt{\frac{2}{\pi} \int_0^{\pi/2} \frac{\omega^2(\phi)}{\beta(\phi)} \, \mathrm{d}\phi}. \qquad (14)$$

When Y has been calculated from experimental data, the effective stress-intensity factor $K(r, \phi)$ is readily obtained from Equation 12. Similar to the unbridged nominal case, the effective stress-intensity factor range at the surface crack tip,

$$\Delta K_{\text{eff}} = (1 - R) K(a, 0), \qquad (15)$$

is calculated and displayed in a $da/dN\text{-}\Delta K$ plot in Figure 9. When the cohesive action from the bridging fibres is accounted for, there is a shift leftwards in the $da/dN\text{-}\Delta K$ plot, but the slope is still negative (indicative of crack retardation) and there is still a large gap to the curve of the matrix material in Figure 8.

The analysis of the fibre-bridged crack growth shows that the fibre bridging itself contributes to the observed fatigue toughness, but cannot account for the crack retardation. There has to be other toughening mechanisms that promote the crack retardation and fatigue crack growth resistance. One of them is crack front bowing around obstructive fibres (see Figure 10) which was not taken into consideration here. With crack front bowing the front becomes longer with an increasing rate compared with a semi-circular crack front as assumed in the fracture mechanics model.

4.2 Debond propagation

For the debond prone materials, a typical damage site was one or a set of planar fibre breaks from which debonds grew in the longitudinal direction. An example is shown in Figure 11, where debonds grew from a plane set of fibre breaks in a carbon-fibre reinforced plastic subjected to long-term fatigue.

The growth of debonds from a fibre break or a set of fibre breaks will surely change the stress profile in the neighbouring fibres. Since generally it is the neighbouring fibres that are the next to fail, and to take part

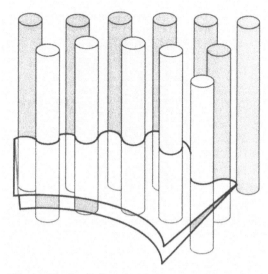

Figure 10: Crack front bowing as a propagation retarding mechanism

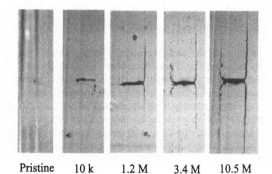

| Pristine | 10 k | 1.2 M | 3.4 M | 10.5 M |

Figure 11: Debond propagation from a set of plane fibre breaks at $\varepsilon_{\text{max}} = 0.89\%$ (——— 30 μm)

in the subsequent evolution of damage with interacting debonding and fibre breakage, it would be useful to parametrically analyse the simple scenario of one fibre break with growing debonds. Thus, the general trends of debond propagation and fibre strength variability on further damage accumulation (fibre breakage) could be assessed. A simple and useful shear-lag model developed by Beyerlein & Phoenix (1996) is employed here. The model is two-dimensional and infinite. In the present case, there is only one single fibre break from which four debonds are growing (see Figure 12).

If the fibres are assumed only to take axial stresses and the matrix only supports shear stresses, their constitutive relations together with the condition of static

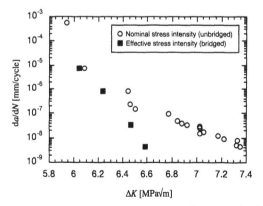

Figure 9: Fatigue crack growth curves for the bridged (nominal) and unbridged (effective) case

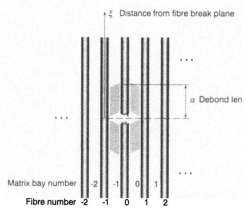

Figure 12: Single fibre break with debonds in an infinite two-dimensional composite

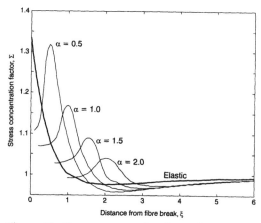

Figure 13: Stress concentration factor for different debond lengths

equilibrium result in

$$E_f A \frac{d^2 u_n(x)}{dx^2} =$$

$$-\frac{G_m h}{w}\left[u_{n+1}(x) - 2u_n(x) + u_{n-1}(x)\right], \quad (16)$$

where E_f is the Young's modulus of the fibres, A is the cross-sectional area of a fibre, $u_n(x)$ is the displacement of fibre number n at position x, G_m is the matrix shear modulus, h is the ply thickness, and w is the fibre spacing. With a normalisation, this differential equation would be independent of the material properties and the geometric parameters which would facilitate a systematic parametric investigation. If necessary, the normalised non-dimensional quantities can be converted back to physical dimensional values. The fibre displacement is asigned to be

$$U_n(\xi) = \frac{u_n(x)}{p^\star \sqrt{\dfrac{w}{E_f A G_m h}}}, \quad (17)$$

where the dimensionless axial coordinate is

$$\xi = \frac{x}{\sqrt{\dfrac{E_f A w}{G_m h}}}, \quad (18)$$

and the scaling factor is

$$p^\star = \sigma^\star \sqrt{\frac{w E_f A h}{G_m}}, \quad (19)$$

where $\sigma^\star$ can be arbitrarily chosen to keep the order of magnitude of the stresses close to unity. The fibre stress becomes

$$P_n(\xi) = \frac{p_n(x)}{p^\star}, \quad (20)$$

where $p_n(x)$ is stress in fibre number n at position x.

With this normalisation, the differential equation 16 is re-expressed in a more convenient form as

$$\frac{d^2 U_n(\xi)}{d\xi^2} + U_{n+1}(\xi) - 2U_n(\xi) + U_{n-1}(\xi) = 0, \quad (21)$$

from which solution, the fibre stress can be calculated,

$$P_n(\xi) = \frac{dU_n(\xi)}{d\xi}. \quad (22)$$

The elastic solution of Equation 21 for a single fibre break has been obtained by Hedgepeth & Van Dyke (1967). The influence of debonds can be accounted for by shear load couples, whose solution has been determined by Beyerlein & Phoenix (1996). For simplicity, the debonds were accounted for with an even distribution of shear couples along the debonds that counteract the elastic response. The resulting profile of the stress concentration factor along a fibre adjacent to the broken with various debond lengths, $\Sigma = P_1(\xi)$ or $\Sigma = P_{-1}(\xi)$, is shown in Figure 13. As the debond grow, the maximum stress concentration factor moves along the tip of the debond, but decreases in magnitude depending on the interfacial friction along the debond. This behaviour with a moving stress peak with growing debonds is something that has been observed experimentally in microcomposites with fibre monolayers by means of micro-Raman spectroscopy (Bennett & Young 1997; van den Heuvel et al. 1998).

During debond growth, the stress profile along the adjacent fibre changes. Some overloaded parts are relaxed, while the stress concentration increases in other parts of the fibre ahead of debond crack tip. Such a non-monotonic continuous change in stress profile may eventually lead to rupture of the fibre at a weak point subjected to an unprecedented stress. As successive fibre breakage sets out, further debonding and fibre breakage will continue in a self-escalating manner. Since the envelope of the stress concentration profile in the adjacent fibre increases monotonically at each point, so will the cumulative probability of failure of the fibre.

The probability of failure of the adjacent fibre can be analysed by Monte-Carlo simulations. The strength of the fibres are assumed to obey the Weibull distribution with the cumulative distribution function

$$P_f(\Sigma_f) = 1 - \exp\left\{-L\left(\frac{\Sigma_f}{\Sigma_0}\right)^\beta\right\}, \qquad (23)$$

where β is the shape parameter, Σ_0 is the scale parameter for unit fibre length (i.e. $L = 1$). To determine the effect of strength variability the mean value of the strength, $\bar{\Sigma}_f$, is kept constant, and the shape parameter is varied. The scale parameter is then determined as $\Sigma_0(\beta) = \bar{\Sigma}_f/\Gamma(1 + 1/\beta)$.

By inversion of Equation 23, the strength of a fibre segment i of length δ is

$$\Sigma_i = \Sigma_0(\beta)\left(-\frac{\ln Z_i}{\delta}\right)^{1/\beta}, \qquad (24)$$

where Z_i is a generated random variable from the uniform distribution between 0 and 1. Each of the 100 assigned fibre elements was then compared to the local stress, and the position of the weakest element with lowest positive strength to stress ratio was registered for 100,000 iterations for each stress profile and shape parameter. The mean axial distance between the the original fibre break and the secondary break in the adjacent fibre, $\bar{\xi}_f$, was thus calculated for each debond length and fibre strength distribution. The stress profile was the envelop or maximum stress profile from the point where there was no deboinding up to a debond of length α, i.e. $\Sigma_e(\alpha, \xi) = \max_{0 \le a \le \alpha} \Sigma(a, \xi)$. The results of the simulation are found in Figure 14.

With increasing debond lengths, the mean distance to the second fibre break increases. With debonding, the stress profiles in the neighbouring fibres become more spatially distributed, which gives rise to new fibre breaks further away from the orginal one. This means that the crack takes are more tortuous path, which macroscopically results in a more jagged fracture surface with sprawling fibres. This has been observed experimentally for glass fibre-reinforced polypropylene (van den Oever & Peijs 1998; Gamstedt et al. 1999) and for carbon fibre-reinforced plastics (Charewicz & Daniel 1986; Gamstedt & Talreja 1999) with varying degree of debond propensity. Figure 14 also shows a similar effect with respect to the fibre strength variability. For smaller values of the shape factor β, the expected inter-fibre break distance becomes larger. For the deterministic case ($\beta \to \infty$)

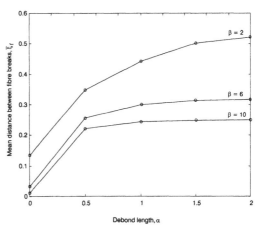

Figure 14: Mean distance to secondary fibre break with respect to debond length

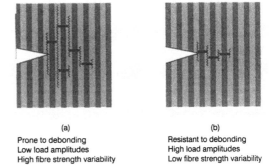

(a)
Prone to debonding
Low load amplitudes
High fibre strength variability

(b)
Resistant to debonding
High load amplitudes
Low fibre strength variability

Figure 15: Illustration of damage development depending on scatter in fibre strength, debonding and load level

with constant fibre strength, the fibres would fail in the same plane if they would fail at all, irrespective of the debonding process.

A schematic illustration of the influence of fibre strength variability and debonding (controlled by the load amplitude) is found in Figure 15. Both fibre strength variability and debonding strive for a distributed damage evolution. A wider statistic distribution of fibre strength will enhance the spatial distribution of fibre breaks, which in its turn will incite further debond growth to make existing fibre breaks link up. A scanning electron microscopic study on carbon-fibre-reinforced plastics by Theocaris & Stassinakis (1981) shows that a more winding and distributed crack propagation is caused by growing debonds and a larger scatter in fibre strength. Since debonding is a cycle depending mechanism, it will be more active at lower load amplitudes where it is given time to propagate. The progressiveness of this case with wide strength variability, debond prone material and low amplitudes is found in Figure 15a. The cycle-dependency will result in a fatigue sensitive material with a noticeable slope in the fatigue life curve.

On the other hand, if the material is resistant to debonding, or alternatively has a narrow distribution in fibre strength, the damage accumulation will be fairly localised as depicted in Figure 15b. This scenario is also prevalent at high load amplitudes, where the short lifetimes does not allow the debonds do grow noticeably. A material which is relatively resistant to debonding, such as carbon fibre/epoxy and glass fibre/maleic anhydride modified polypropylene, will show scarce local damage sites, and be less sensitive to fatigue. Since debonding and damage delocalisation is suppressed by a narrow distribution in fibre strength, similar trends will result in that case. Either the subsequent fibre close to the previous fibre break will fail or it will not, thus exhibiting a brittle behaviour. A trade-off between static notch sensitivity and fatigue sensitivity depending on the intended application is necessary for materials showing the above types of behaviour.

The effect of fibre-matrix debonding has been investigated here. However, other longitudinal damage mechanisms such as matrix cracking and matrix yielding would essentially have the same effects. In these cases, the shear stresses in the matrix close to fibre breaks would also relax, and have a distributing effect on the stress profiles in neighbouring fibres. The ensuing damage accumulation would in a general sense be similar to that of debonding.

5 DISCUSSIONS

5.1 *Qualitative link*

The fatigue behaviour of composites can be interpreted with fatigue-life diagrams (Talreja 1981), such as the one depicted in Figure 16. In a fatigue life diagram, the initial peak strain is plotted with respect to lifetime. Three different scatter bands are partitioned off based on distinct operative fatigue mechanisms. Failure in Region I derives from rupture of fibres, such as under quasi-static loading conditions. In the sloping scatter band, Region II, progressive mechanisms have to be active. The present study has identified fibre-bridged cracking or long-range debonding in this region. For the two types of composites reinforced by either carbon or glass fibres, there was a shift leftwards in the fatigue-life diagram with pronounced debonding, towards short fatigue lives. For the materials with fatigue resistant interfaces, the localised damage growth in the transverse direction resulted in relatively longer fatigue lives. At the fatigue limit in Region III, the driving force for damage propagation is below its threshold, and no substantial damage will form under reasonable testing times. Alternatively, the propagation of damage is efficiently

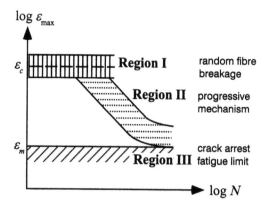

Figure 16: Fatigue-life diagram for unidirectional composites under loading parallel to the fibres

stopped by an arrest mechanism such as local debonding. In practice, fatigue-life diagrams with experimental data could for instance look like those in Figure 5.

The experiments have shown that the fatigue degradation of the composite, and ultimately also the fatigue life, are controlled by the rate at which the load-carrying constituent, i.e. the longitudinal fibres are being broken. The rate of fibre failure is controlled progressive mechanisms associated with the more fatigue sensitive matrix or fibre-matrix interface. It has been observed that debonding results in a more distributed damage state with a higher rate of fibre failure. On the other hand, transverse matrix crack growth results in localised and sparse damage sites, where few or no fibres are broken during the lion's share of the total fatigue life. The detrimental debond mechanism can be suppressed by increasing the fatigue crack growth resistance of the interface. This has been shown for glass-fibre reinforced polypropylene (Gamstedt et al. 1999). The same link presents itself for carbon-fibre reinforced composites. Carbon fibre/PEEK has shown lower maximum interfacial shear stresses compared with carbon fibre/epoxy in single-fibre composite tests (Vautey & Favre 1990). The ensuing mechanisms showed more longitudinal cracks and debonding in the carbon fibre/PEEK case (see Figure 3), which led to inferior fatigue performance (Curtis 1987; Gamstedt & Talreja 1999). An illustration of the qualitative link from interfacial efficiency, through micromechanisms, up to coupon or component fatigue behaviour is presented in Figure 17.

The debond mechanism observed in the present study showed to be progressive, i.e. it led to further damage accumulation by fibre breakage and eventually to failure. However, the results from Region III for the carbon fibre/epoxy material show that when debonding is not progressive, i.e. when it act as a crack arrest mechanism and terminates other pro-

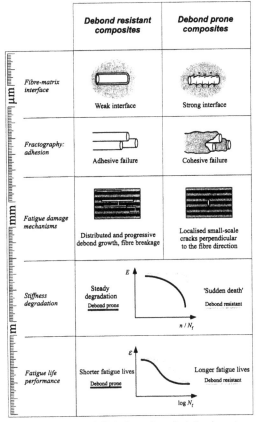

	Debond resistant composites	Debond prone composites
Fibre-matrix interface	Weak interface	Strong interface
Fractography: adhesion	Adhesive failure	Cohesive failure
Fatigue damage mechanisms	Distributed and progressive debond growth, fibre breakage	Localised small-scale cracks perpendicular to the fibre direction
Stiffness degradation	Steady degradation Debond prone	'Sudden death' Debond resistant
Fatigue life performance	Shorter fatigue lives Debond prone	Longer fatigue lives Debond resistant

Figure 17: Integral link from interfacial microstructure, micromechanisms to fatigue life performance

gressive damage modes, it has a beneficial effect on the fatigue performance. Schürmann & Knickrehm (1997) observed better fatigue properties for unidirectional glass fibre/polypropylene compared with glass fibre/epoxy, where the polypropylene based composite showed more debonding. In this case, debonding has a beneficial role with respect to the fatigue behaviour. This is in contrast to other investigations which has showed that debonding acts as a detrimental mechanism (Curtis 1991; Gamstedt et al. 1999; Gamstedt & Talreja 1999). The duality and mutually exclusive character of debonding must rely on its instigation of further fibre failures. The detrimental or beneficial role of debonding should be governed by whether its propagation results in further breakage of the neighbouring fibres or not. A growing debond gives rise to a continual non-monotonic change of the stress profile in the adjacent fibres, and thus increases the probability of failure of these fibres. Debonding that does not stop and grows beyond a certain short length will act as a detrimental and progressive fa-

tigue mechanism leading to inferior fatigue performance.

It is worthwhile to point out that an improvement of the static fracture toughness of the composite by use a tougher matrix may not increase in a corresponding increase in fatigue fracture toughness. This has been observed in fatigue delamination tests (Hojo et al. 1994). In fact, the use of a *statically* tougher PEEK matrix instead of e.g. epoxy may even result in a decrease in *fatigue* fracture toughness (Simonds et al. 1989). The point is that a change in static properties does not necessarily imply a concomitant change in fatigue properties, since the damage and failure mechanisms may be very different under static and fatigue loading conditions. For instance, the above experimental results for carbon fibre/PEEK show that fibre breakage dominates under static loading, whereas longitudinal crack growth such as debonding accompanied by further fibre breakage dominate under fatigue loading at intermediate amplitudes. This disparity in mechanisms should be the reason for the reversed order of toughness under static and fatigue loading respectively.

The fatigue damage mechanisms differ at different strain amplitudes. This also manifests itself in a discontinuity in slopes for distinct scatter bands of the fatigue-life diagram in Figure 16. Each mechanism will influence the activity of the other mechanisms in the case of variable-amplitude or block-amplitude loading. With the mechanisms in mind, we can try to make a mechanistic, albeit qualitative, analysis of observed influences of sequential block loading of a composite.

5.2 Variable-amplitude loading

The disparate nature of mechanisms at different strain amplitudes can also explain the influence of sequential variable amplitude loading in qualitative terms. A common behaviour for composite materials is that a high-low sequence of block amplitude loading results in relatively shorter lifetimes compared with a corresponding low-high order (e.g. Adam et al. 1994). It has shown that at high amplitudes the initiatory mechanism of fibre breakage prevails (Dharan 1975; Lorenzo & Hahn 1986). At lower amplitudes progressive debonding predominates, where it is given sufficient time to propagate to larger extent. These debonds tend to originate from fibre breaks. In a high-low sequence, an abundance of fibre breaks will form during the high amplitude interval, from which a multitude of debonds will propagate and cause new fibre breaks during the low amplitude interval. On the other hand, if the low amplitude interval precedes the one with a high amplitude, the debonds will have fewer fibre breaks from which they could propagate, and hence the accumulation of damage will be more mod-

High-low block sequence	Low-high block sequence
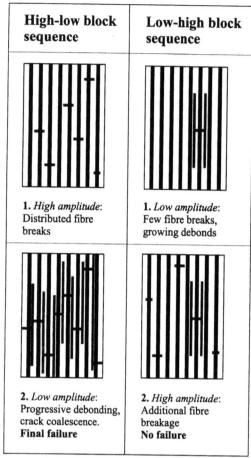	
1. *High amplitude*: Distributed fibre breaks	**1.** *Low amplitude*: Few fibre breaks, growing debonds
2. *Low amplitude*: Progressive debonding, crack coalescence. **Final failure**	**2.** *High amplitude*: Additional fibre breakage **No failure**

Figure 18: Mechanistic explanation of the influence of sequence order on block amplitude loading

erate. A schematic illustration of this behaviour is presented in Figure 18. This type of synergistic interaction of different fatigue mechanisms can explain the observed loading history dependencies and shortcomings of Palmgren-Miner's rule for composite materials. Experimental work on the mechanisms of the sequence effect is in progress.

5.3 Multidirectional laminates

As mentioned previously, the longitudinal 0° ply is the main load carrying member in a multidirectional laminate. There is experimental evidence that the 0° ply controls the fatigue life of multidirectional laminates when the fatigue behaviour is interpreted in terms of maximum strain (Dickson et al. 1989; Talreja 1993; Andersen et al. 1996). However, the strain in the 0° ply may change during fatigue as damage develop in the off-axis plies. This damage will lead to stiffness reduction in the off-axis plies, which in turn leads to increasing strain levels in the 0° ply when subjected to load controlled fatigue. This means that the critical element of the laminate, i.e. the 0° ply, effectively undergoes variable amplitude loading. In that case, constant amplitude fatigue-life data can therefore not be directly used to predict the fatigue life of the laminate. The cycle-dependent local strain must be considered instead of the global values of the applied stress or initial peak strain. Predictions can be made if the variation in strain of the critical element (longitudinal ply) is analysed (Reifsnider & Jamison 1986), which depends on the rate of stiffness reduction of the adjacent non-critical elements (off-axis plies). The stiffness reduction or strength degradation in the off-axis plies can either be modelled empirically with a power law (Halverson et al. 1996), or physically by micromechanical models which takes transverse cracking and delamination growth into concern (Akshantala & Talreja 1998).

For multiaxial tensile loading of laminates, the stress state becomes more complex, but the above reasoning still applies if we consider the axial strain level in the ply whose fibres are aligned with the direction of the largest principal stress component of the laminate. Since the fibres, whose direction is the same as that of largest principal stress component, constitute our critical element, these should be the prime object of analysis for improved design of fatigue resistant composites in multiaxial loading .

6 CONCLUSIONS

In view of the observed micromechanisms and accompanying fatigue life behaviour, it can be summarised that pronounced debond growth has a detrimental effect on the global fatigue properties. If this mechanism can be suppressed by a stronger and more resistant interface, longer fatigue lives of the material are expected. However, limited debonding can have beneficial effect if it acts as a crack-arresting mechanism, which prevents further damage propagation in the transverse direction. In this respect, an optimum in fatigue resistance is expected in terms of propensity to debond propagation.

The sequence effect in block amplitude loading of composite materials can qualitatively be explained by the synergism of differing damage mechanisms operative at the different load amplitudes. Fibre breakage is the dominating mechanism at high load levels, whereas progressive debonding prevails at lower load levels where it is given sufficient time to develop. Since the debonds initiate from fibre breaks, a high-low sequence leads to a more severe state of damage than a corresponding low-high sequence, and hence to a shorter fatigue life.

For improvement of the fatigue resistance, it can been recommended to monitor the development of fatigue damage microscopically, and then take measures to modify the composite composition to repress the deleterious mechanisms responsible for fatigue degradation and eventual failure. Micromechanism monitoring by surface replication can be recommended since it has proved to be straightforward and easy to use. Much further experimental characterisation and modelling work is needed to conceive a quantitative predictive model from the micromechanisms to fatigue life performance.

ACKNOWLEDGMENTS

The author wishes to thank Professors L. A. Berglund and R. Talreja for helpful discussions.

REFERENCES

Adam, T., Gathercole, N., Reiter, H., & Harris, B. (1994). Life prediction for fatigue of T800/5245 carbon-fibre composites: II. Variable-amplitude loading. *International Journal of Fatigue 16*(8), 533–547.

Akshantala, N. V. & Talreja, R. (1998). A mechanistic model for fatigue damage evolution in composite laminates. *Mechanics of Materials 29*(2), 123–140.

Andersen, S. I., Lilholt, H., & Lystrup, A. (1996). Properties of composites with long fibres. In R. M. Meyer (Ed.), *Design of Composite Structures against Fatigue*, Bury St. Edmunds, pp. 15–31. Mechanical Engineering Publications.

Begley, M. R. & McMeeking, R. M. (1995). Numerical analysis of fibre bridging and fatigue crack growth in metal matrix composites. *Materials Science and Engineering A200*(1–2), 12–20.

Bennett, J. A. & Young, R. J. (1997). Micromechanical aspects of fibre/crack interaction in an aramid/epoxy composite. *Composites Science and Technology 57*(8), 945–956.

Beyerlein, I. J. & Phoenix, S. L. (1996). Stress concentrations around multiple fiber breaks in an elastic material with local yielding or debonding using quadratic influence superposition. *Journal of the Physics and Mechanics of Solids 44*(12), 1997–2039.

Casado, J. A., Gutiérrez-Solana, F., Polanco, J. A., & Carrascal, I. (1997). Effect of stress level and waiting time on fracture criteria established for reinforced polyamides tested under fatigue. In S. Degallaix, C. Bathias, & R. Fougères (Eds.), *Proceedings of the First International Conference on Fatigue of Composites*, Paris, pp. 234–241. SF2M.

Charewicz, A. & Daniel, I. M. (1986). Damage mechanisms and accumulation in graphite/epoxy laminates. In *Composite Materials: Fatigue and Fracture, STP907*, Philadelphia. ASTM.

Chermahini, R. G., Palmberg, B., & Blom, A. F. (1993). Fatigue crack growth and closure behaviour of semicircular and semielliptical surface flaws. *International Journal of Fatigue 15*(4), 259–263.

Curtis, P. T. (1987). In investigation of the tensile fatigue behaviour of improved carbon fibre composite materials. In *Proceedings of the Sixth International Conference on Composite Materials*, London, pp. 54–64.

Curtis, P. T. (1991). Tensile fatigue mechanisms in unidirectional polymer matrix composite materials. *International Journal of Fatigue 13*(5), 377–382.

Dharan, C. H. K. (1975). Fatigue failure mechanisms in a unidirectionally reinforced composite material. In *Fatigue in Composite Materials, STP569*, Philadelphia, pp. 171–188. ASTM.

Dickson, R. F., Fernando, G., Adam, T., Reiter, H., & Harris, B. (1989). Fatigue behaviour of hybrid composites, Part 2 Carbon-glass hybrids. *Journal of Materials Science 24*(1), 227–233.

Fett, T. (1988). The crack opening displacement field of semi-elliptical surface cracks in tension for weight function applications. *International Journal of Fracture 36*, 55–69.

Gamstedt, E. K., Berglund, L. A., & Peijs, T. (1999). Fatigue mechanisms in unidirectional glass-fibre-reinforced polypropylene. *Composites Science and Technology 59*(5), 759–768.

Gamstedt, E. K. & Brøndsted, P. (1999). Damage dissipation and localization during fatigue in unidirectional glass/carbon fibre hybrid composites. In X. R. Wu & Z. G. Wang (Eds.), *Proceedings of the Seventh International Fatigue Congress*, Volume 3, pp. 1731–1736.

Gamstedt, E. K. & Sjögren, B. A. (1999). Micromechanisms in tension-compression fatigue of composite laminates containing transverse plies. *Composites Science and Technology 59*(2), 167–168.

Gamstedt, E. K. & Talreja, R. (1999). Fatigue damage mechanisms in unidirectional carbon-fibre-reinforced plastics. *Journal of Materials Science 34*(11), 2535–2546.

Halverson, H. G., Curtin, W. A., & Reifsnider, K. L. (1996). Fatigue life of individual composite specimens based on intrinsic fatigue behavior. *International Journal of Fatigue 19*(5), 369–377.

Hedgepeth, J. M. & Van Dyke, P. (1967). Local stress concentrations in imperfect filamentary composite materials. *Journal of Composite Materials 1*, 294–309.

Hertzberg, R. W. & Manson, J. A. (1980). *Fatigue of Engineering Plastics*. New York: Academic Press.

Hojo, M., Ochiai, S., Gustafson, C.-G., & Tanaka, K. (1994). Effect of matrix resin on delamination fatigue crack growth in CFRP laminates. *Engineering Fracture Mechanics 49*(1), 35–47.

Jacobsen, T. K., Sørensen, B. F., & Brøndsted, P. (1998). Measurement of uniform and localized heat dissipation induced by cyclic loading. *Experimental Mechanics 38*(4), 289–294.

Lorenzo, L. & Hahn, H. T. (1986). Fatigue failure mechanisms in unidirectional composites. In H. T. Hahn (Ed.), *Composite Materials: Fatigue and Fracture, STP907*, Philadelphia, pp. 210–232. ASTM.

McCartney, L. N. (1987). Mechanics of matrix cracking in brittle-matrix fibre-reinforced composites. *Proceedings of the Royal Society A409*, 329–350.

Newman, Jr., J. C. & Raju, I. S. (1981). An empirical stress intensity factor equation for the surface crack. *Engineer-

ing Fracture Mechanics 15(1–2), 185–192.

Reifsnider, K. L. & Jamison, R. (1986). A critical-element model of the residual strength and life of fatigue-loaded composite coupons. In H. T. Hahn (Ed.), *Composite Materials: Fatigue and Fracture, STP907*, Philadelphia, pp. 298–313. ASTM.

Schürmann, H. & Knickrehm, K. (1997). Zur Verbesserung der Biegeschwellfestigkeit hochbeanspruchter unidirektional endlosfaserverstärkter Faser-Kunststoff-Verbunde. *Vortrag 28. AVK-Tagung Baden-Baden Oktober.*

Simonds, R. A., Bakis, C. E., & Stinchcomb, W. W. (1989). Effects of matrix toughness on fatigue response of graphite fiber composite laminates. In P. A. Lagace (Ed.), *Composite Materials: Fatigue and Fracture, STP1012*, Philadelphia, pp. 5–18. ASTM.

Talreja, R. (1981). Fatigue of composite materials: Damage mechanisms and fatigue life diagrams. *Proceedings of the Royal Society A378*, 461–475.

Talreja, R. (1993). Fatigue of fiber composites. In T. W. Chou (Ed.), *Structure and Properties of Composites*, Weinheim, pp. 583–606. VCH.

Theocaris, P. S. & Stassinakis, C. A. (1981). Crack propagation in fibrous composite materials studied by SEM. *Journal of Composite Materials 15*, 133–141.

van den Heuvel, P. W. J., Peijs, T., & Young, R. J. (1998). Failure phenomena in two-dimensional multi-fibre microcomposites – 3. A Raman spectroscopy study of the influence of interfacial debonding on stress concentrations. *Composites Science and Technology 58*(6), 933–944.

van den Oever, M. & Peijs, T. (1998). Continuous-glass-fibre-reinforced polypropylene composites: Ii. Influence of maleic-anhydride modified polypropylene on fatigue behaviour. *Composites, Part A 29*(3), 227–239.

Vautey, P. & Favre, J. P. (1990). Fibre/matrix load transfer in thermoset and thermoplastic composites – single fibre models and hole sensitivity of laminates. *Composites Science and Technology 38*(3), 271–288.

Wang, G. S. (1995). A numerical procedure for the approximate WF solution of the mode I SIFs in 3D geometries under arbitrary load. *Computational Mechanics 15*(5), 177–208.

Recent Developments in Durability Analysis of Composite Systems, Cardon, Fukuda, Reifsnider & Verchery (eds)
© 2000 Balkema, Rotterdam, ISBN 90 5809 103 1

Long term prediction of fatigue life for FRP joint systems

Masayuki Nakada & Yasushi Miyano
Materials System Research Laboratory, Kanazawa Institute of Technology, Ishikawa, Japan

Stephen W. Tsai
Department of Aeronautics and Astronautics, Stanford University, Calif., USA

ABSTRACT: The tensile constant elongation-rate (CER) and fatigue tests for GFRP/metal conical shaped joint and GFRP/metal adhesive joints using a brittle epoxy and a ductile PMMA adhesives were carried out under various loading rates and temperatures. The fatigue failure loads as well as CER failure loads for these three types of FRP joint depend clearly on loading rate and temperature. The time-temperature superposition principle holds for fatigue failure loads as well as CER failure loads for these FRP joints, therefore the master curves for fatigue failure load can be obtained from these results. The dependence of these fatigue failure loads upon number of cycles to failure as well as time to failure and temperature can be characterized from master curves for these FRP joints.

1 INTRODUCTION

The mechanical behavior of polymer resins exhibits time and temperature dependence, called viscoelastic behavior, not only above the glass transition temperature T_g but also below T_g. Thus, it can be presumed that the mechanical behavior of polymer composites also significantly depends on time and temperature even below T_g which is within the normal operating temperature range. It has been confirmed that the viscoelastic behavior of polymer resins as matrices is a major influence on the time and temperature dependence of the mechanical behavior of fiber reinforced plastics (FRP) [1-8].

The time-temperature dependence of the tensile failure load under constant elongation-rate (CER) and fatigue loadings for GFRP/metal conical shaped joint and GFRP/metal adhesive joint using a brittle epoxy adhesives has been studied in our previous papers [9, 10]. It was observed that the fracture modes are almost identical under two types of loading over a wide range of time and temperature, and the same time-temperature superposition principle holds for CER and fatigue failure loads. Therefore, the master curve of fatigue failure load can be obtained from results of the CER tests under various CERs and temperatures and the fatigue tests at a single frequency under various temperatures.

In this paper, the tensile fatigue tests as well as tensile CER tests for GFRP/metal adhesive joint using a ductile PMMA adhesives are carried out under various loading rates and temperatures. The time and temperature dependence of tensile fatigue behavior for

this FRP joint is characterized by using the master curve of fatigue failure load and is compared with those for other joint systems mentioned above.

2 EXPERIMENTAL PROCEDURE

2.1 Preparation of FRP joints

The GFRP/metal conical shaped joint (Conical shaped joint) is made from a unidirectional GFRP rod, metal end fittings, and bonding resin as shown in Fig.1. The unidirectional GFRP rod consists of E-glass fibers and epoxy resin, and is produced by pultrusion molding. The weight content of fibers is approximately 83%. The bonding resin consists of epoxy resin and silica. The weight content of silica is approximately 66%. The metal end fitting is made from SCM 440 steel. The bonding resin taper and the bonding length of this conical shaped joint are respectively 6° and 25mm, as shown in Fig.1. The glass transition temperature T_g for the matrix resin of FRP rod and the bonding resin is 95°C.

The GFRP/metal adhesive joints (Adhesive joint) are made from a GFRP pipe, ductile cast iron rod, and adhesive resin as shown in Fig.2. Two types of adhesive resin are used. One is the brittle epoxy resin, Mavidon MA1790-1 A/B(Mavidon Corp.) with glass fillers. The weight content of glass fillers is approximately 41%. The other is PMMA resin, PLEXUS AO425 (ITW Adhesives). The GFRP pipe consists of glass cloth and epoxy resin. Ductile cast iron rod is made from ductile iron castings Grade 80-

55-06 (ASTM A 536-84). The adhesive resin thickness and length of adhesive joint are respectively 4mm and 28mm.

2.2 Test procedure

The tensile tests for three types of FRP joint under CER and fatigue loadings were carried out. The tensile CER tests were carried out under various loading rates and temperatures using an Instron type testing machine. The tensile load was applied at both end screws of the FRP joint. The loading rates (crosshead speeds) were 0.01, 1 and 100mm/min. The tensile fatigue tests were carried out under various temperatures at two frequencies f=5 and 0.05Hz using an electro-hydraulic servo testing machine. Load ratio R (minimum load/maximum load) was 0.05.

3 RESULTS AND DISCUSSION

3.1 Load-elongation curves

Typical load-elongation curves for three types of FRP joint; conical shaped joint, adhesive joint using a brittle epoxy, and adhesive joint using a ductile PMMA; are shown in Fig.3. The load-elongation curves for conical shaped joint are almost linear until the maximum load. The load-elongation curves for adhesive joint using a brittle epoxy resin show non-linear behavior, however the elongation at maximum load is very small. The load-elongation curves for adhesive joint using a ductile PMMA resin show remarkable non-linear behavior and the elongation at maximum load is very large.

3.2 Fracture appearances

The fracture appearances for three types of FRP joint under CER loading are shown in Fig.4 (a). The fracture of conical shaped joint occurs at the FRP rod within the contact area of the bonding resin. The failure mode changed with temperature. In the region of low temperature the conical shaped joint fails coaxially at the surface of FRP rod, and then the shearing crack propagates transversely in the FRP rod. In the region of high temperature the conical shaped joint also fails coaxially at the surface of FRP rod, and then the shearing crack propagates longitudinally in the FRP rod.

The fracture mode of adhesive joint using a brittle epoxy resin also changed with temperature. In the region of low temperature the adhesive joint mainly fails at the interface between the cast iron rod and adhesive resin. In the region of high temperature the adhesive joint mainly fails in the adhesive resin.

The fracture of adhesive joint using a ductile PMMA resin occurred in the adhesive resin nearby the interface between cast iron rod and adhesive resin in all region of temperature tested.

Fracture appearances for three types of FRP joint under fatigue loading are shown in Fig.4 (b). These fracture are similar to those under CER loading. In other words, all failed specimens are similar regardless of loading pattern. We consider, therefore, that the failure mechanisms are the same for CER and fatigue loadings for three types of FRP joint.

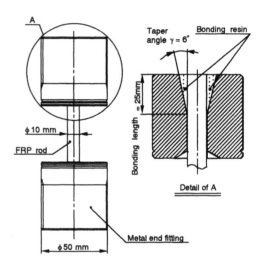

Figure 1. GFRP/metal conical shaped joint

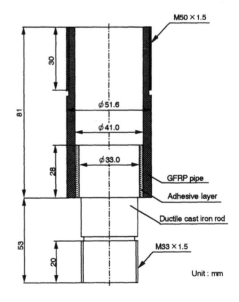

Figure 2. GFRP/metal adhesive joint

3.3 Master curve of CER failure load

Failure load curves for CER and for fatigue shown in Figs.5-10 will be presented below for three cases: (a) the tensile failure load for conical shaped joint, (b) the tensile failure load for adhesive joint using a brittle epoxy, and (c) the tensile failure load for adhesive joint using a ductile PMMA. The left sides of Fig.5 show the CER failure load P_s versus time to failure t_s, the time period from initial loading to maximum load.

The master curves for each P_s were constructed by shifting P_s at constant temperatures other than reference temperature T_0 along the log scale of t_s so that they overlap on P_s at the reference temperature or on each other to form a single smooth curve as shown in the right side of Fig.5. Since the smooth master curves for each P_s can be obtained, the time-temperature superposition principle is applicable for each P_s. The fracture modes A and B for conical shaped joint and adhesive joint using a brittle epoxy resin can be classified clearly on each master curve.

The time-temperature shift factor $a_{T_0}(T)$ is defined by

$$a_{T_0}(T) = \frac{t_s}{t_s{}'} \qquad (1)$$

where $t_s{}'$ is the reduced time to failure. The shift factors for each P_s obtained experimentally in Fig.5 are plotted

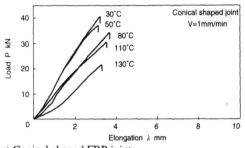

(a) Conical shaped FRP joint

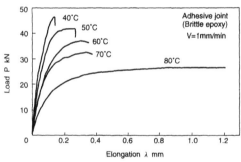

(b) Adhesive FRP joint (Brittle epoxy)

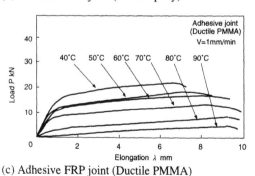

(c) Adhesive FRP joint (Ductile PMMA)

Figure 3. Load-elongation curves for three types of FRP joint

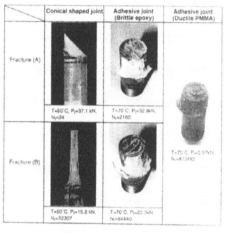

Figure 4(a). Fracture appearances for three types of FRP joint under CER loading

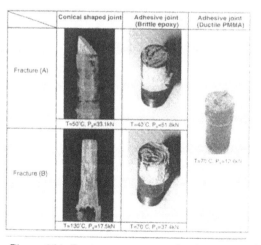

Figure 4(b). Fracture appearances for three types of FRP joint under fatigue loading

respectively in Fig.6. The dotted lines in these figures show the shift factors obtained experimentally for the creep compliance of the matrix resin of FRP rod for conical shaped joint and the adhesive resin for adhesive joint. The shift factors for tensile failure load of FRP joint agree with those for creep compliance of these resins, which are described by two Arrhenius' equations shown in Eq.(2) with different activation energies,

$$\log a_{T_0}(T) = \frac{\Delta H}{2.303G}\left(\frac{1}{T} - \frac{1}{T_0}\right) \quad (2)$$

where ΔH is activation energy [kJ/mol], G is gas constant 8.314×10^{-3} [kJ/(Kmol)].

From these results, the time-temperature dependence of CER failure load for conical shaped joint is controlled by the viscoelastic behavior of matrix

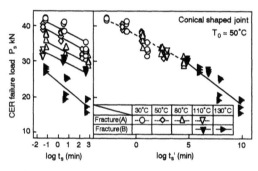

(a) Conical shaped FRP joint

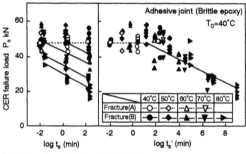

(b) Adhesive FRP joint (Brittle epoxy)

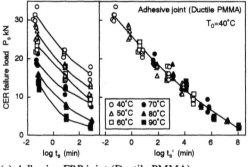

(c) Adhesive FRP joint (Ductile PMMA)

Figure 5. Mater curve of CER failure load for three tyeps of FRP joint

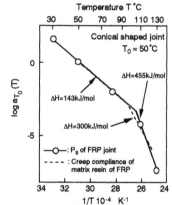

(a) Conical shaped FRP joint

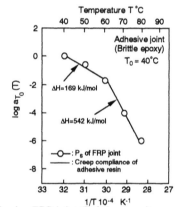

(b) Adhesive FRP joint (Brittle epoxy)

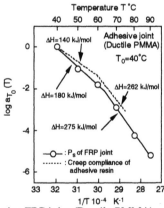

(c) Adhesive FRP joint (Ductile PMMA)

Figure 6. Time-temperature shift factors of CER failure load for three tyeps of FRP joint

resin of FRP rod, and those for adhesive joints are controlled by the viscoelastic behavior of adhesive resin.

3.4 Master curve of fatigue failure load

We turn now the fatigue failure load P_f and regard it either as a function of the number of cycles to failure N_f or of the time to failure $t_f=N_f/f$ for a combination of frequency f, temperature T and denote them by $P_f(N_f;$ f, T) or $P_f(t_f; f, T)$. Further, we consider the CER failure load $P_s(t_f; T)$ the fatigue failure load at $N_f=1/2$, R=0, and $t_f=1/(2f)$; this is motivated by closeness of the line connecting the origin and $(\pi, 1)$ and the curve $[1+\sin(t-\pi/2)]/2$ for $0<t<\pi$.

To describe the master curve of P_f, we need the reduced frequency f' in addition to the reduced time t_f', each defined by

$$f' = f \cdot a_{T_0}(T), \quad t_f' = \frac{t_f}{a_{T_0}(T)} = \frac{N_f}{f'} \quad (3)$$

We introduce two alternative expressions for the master curve: $P_f(t_f'; f', T_0)$ and $P_f(t_f'; N_f, T_0)$. In the latter expression, the explicit reference to frequency is suppressed in favor of N_f. Note that the master curve of fatigue failure load at $N_f=1/2$ is regarded as the master curve of CER failure load. Equation (3) enables one to construct the master curve for an arbitrary frequency from the tests at a single frequency under various temperatures.

Figure 7 displays the fatigue failure load P_f versus the number of cycles to failure N_f (P_f-N_f curve) at a frequency f=5Hz together with CER failure load which is regarded as the fatigue failure load at $N_f=1/2$. The fracture modes of conical shaped joint and adhesive joint using a brittle epoxy are also classified into two modes on each P_f-N_f curve. In the region of small N_f the fracture mode is A, where the P_f decreases scarcely as N_f increases. In the region of large N_f the fracture mode is B, where the P_f decreases clearly as N_f increases.

In Fig.8, fatigue failure load versus the reduced time to failure at the reference temperature are depicted using the shift factor for CER failure load; the master curve for CER failure load is included in the figure. The master curves of P_f for fixed N_f are constructed by connecting the points of the same N_f on the curves of each frequency as shown in Fig.9.

To predict the P_f-N_f curve at frequency f* and T*, we note from Eq.(3) that

$$t_f' = \frac{N_f}{f^* \cdot a_{T_0}(T^*)} \quad (4)$$

and read a pair (P_f, t_f') from Fig.9 which gives a pair

(P_f, N_f). The P_f-N_f curves at f=0.05Hz predicted in this manner are displayed in Fig.10 together with test data. Since the P_f-N_f curves predicted on the basis of the superposition principle capture test data satisfactorily, the time-temperature superposition principle for CER failure load also holds for fatigue failure load. Therefore, the validity for the construction of master curves of fatigue failure load by using the

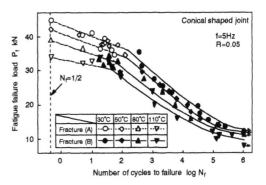

(a) Conical shaped FRP joint

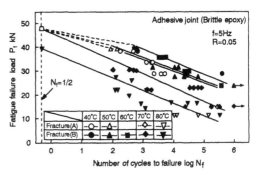

(b) Adhesive FRP joint (Brittle epoxy)

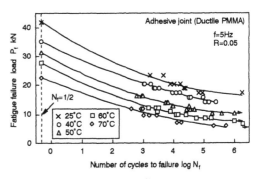

(c) Adhesive FRP joint (Ductile PMMA)

Figure 7. P_f-N_f curves for three types of FRP joint at frequency f=5Hz time-temperature shift factor for CER failure load is confirmed.

3.5 Comparison of the master curves of fatigue failure load

From the master curves of tensile fatigue failure load for three types of FRP joint as shown in Fig.9, the time-temperature dependent fatigue behavior for these three types of FRP joint can be characterized. The tensile fatigue failure load for conical shaped joint depends slightly on time to failure and temperature, however the fatigue failure load decreases clearly with increasing number of cycles to failure N_f. The tensile fatigue failure load for adhesive joint using a brittle epoxy adhesives depends on time to failure, temperature, and N_f. The fatigue failure load for adhesive joint using a ductile PMMA adhesives depends clearly on time to failure and temperature, however the fatigue failure load decreases scarcely with increasing N_f.

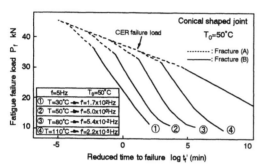

(a) Conical shaped FRP joint

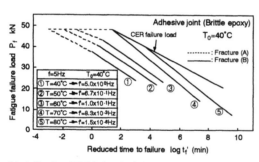

(b) Adhesive FRP joint (Brittle epoxy)

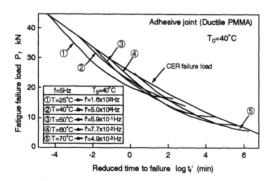

(c) Adhesive FRP joint (Ductile PMMA)

Figure 8. Fatigue failure load at several fixed reduced frequencies for three types of FRP joint

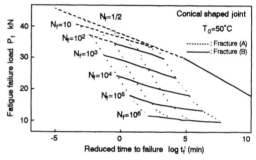

(a) Conical shaped FRP joint

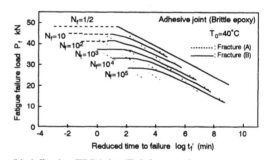

(b) Adhesive FRP joint (Brittle epoxy)

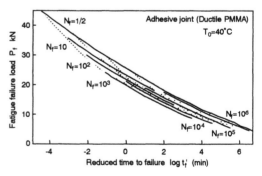

(c) Adhesive FRP joint (Ductile PMMA)

Figure 9. Master curves of fatigue failure load for three types of FRP joint

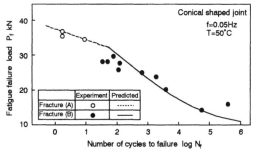

(a) Conical shaped FRP joint

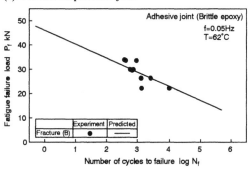

(b) Adhesive FRP joint (Brittle epoxy)

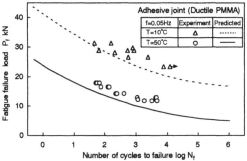

(c) Adhesive FRP joint (Ductile PMMA)

Figure 10. P_f-N_f curves for three types of FRP joint at frequency f=0.05Hz

4 CONCLUSION

The time and temperature dependence of tensile fatigue behavior for GFRP/metal adhesive joint using a ductile PMMA adhesives is determined experimentally and is compared with those for other joint systems, GFRP/metal conical shaped joint and GFRP/metal adhesive joint using a brittle epoxy adhesives. The time-temperature superposition principle holds for tensile fatigue failure loads for all of these FRP joints, therefore, the master curves of fatigue failure load for these joints can be obtained. The master curves of fatigue failure load show very characteristic behavior due to the structure and the combination of materials of FRP joints.

REFERENCES

1. Aboudi, J. and G. Cederbaum, Composite Structures, 12 (1989), p.243.
2. Ha, S.K. and G. S. Springer, J. Composite Materials, 23 (1989), p.1159.
3. Sullivan, J.L., Composite Science and Technology, 39 (1990), p.207.
4. Miyano, Y., M. Kanemitsu, T. Kunio, and H. Kuhn, J. Composite Materials, 20 (1986), p.520.
5. Miyano, Y., M. K. McMurray, J. Enyama, and M. Nakada, J. Composite Materials, 28 (1994),p.1250.
6. Miyano, Y., M. K. McMurray, N. Kitade, M. Nakada, and M. Mohri, Advanced Composite Materials, 4 (1994), p.87.
7. Miyano, Y., M. Nakada, and M. K. McMurray, J. Composite Materials, 29 (1995), p.1808.
8. Miyano, Y., M. Nakada, M. K. McMurray, and R. Muki, J. Composite Materials, 31 (1997), p.619.
9. Miyano, Y., M. Nakada, and R. Muki, Mechanics of Time-Dependent Materials, 1 (1997), p.143.
10. Miyano, Y., S. W. Tsai, M. Nakada, S. Sihn, and T. Imai, Proc. ICCM/11,(1997), VI, p.26.

Recent Developments in Durability Analysis of Composite Systems, Cardon, Fukuda, Reifsnider & Verchery (eds)
© 2000 Balkema, Rotterdam, ISBN 90 5809 103 1

Influence of fiber-matrix adhesion on the fatigue behavior of cross-ply glass-fiber epoxy composites

J.Gassan

Institut für Werkstofftechnik, University of Kassel, Germany (Presently; Fraunhofer Institut für Kurzzeitdynamik – EMI, Freiburg, Germany)

ABSTRACT: It was shown that the loss energy under tension-tension loadings in load increasing mode is a sensitive tool to characterize the nature of fiber-matrix adhesion in glass-fiber/epoxy cross-ply composites. Critical load for damage initiation/propagation and rate of progress of damage was found to be significantly affected by the interphasial properties.

1 INTRODUCTION

The fatigue behavior of composite materials has been a subject of active research in recent years. The damage process in composites subjected to fatigue loading is significantly different from that observed in conventional materials. Four main damage models have been observed in laminates composites under fatigue loadings:

- matrix cracking,
- fiber-matrix debondings,
- delaminations, and
- fiber fracture.

Typically, matrix cracking and delamination occur early in the life, while fiber-matrix debonds and fiber fractures initiate during the beginning of the life and accumulate rapidly towards the end, leading to final failure. It has been observed that the stiffness of the laminate reduces during the process of damage accumulation in laminated composites by using stiffness change as non-destructive fatigue damage parameter (Subramanian 1995). Crack propagation (crack bridging or debonding) plays an essential role in fatigue behavior of composites. When interface bonding is relatively weak, debonding and frictional sliding occur readily upon crack extension, allowing fibers to remain intact and bridge the crack. Upon cyclic loading of the composite, the frictional sliding at the interface changes interfacial properties such as roughness. When the fibers are frictionally bonded to the matrix, the interfacial sliding can be fully characterized by the interfacial sliding shear stress 'τ'. This interfacial sliding shear stress has found to be not constant, the changes in 'τ' have been attributed to fiber surface abrasion, asperity wear,

and matrix plasticity. A strong interface would inhibit interface sliding and lead to fiber fracture instead of crack bridging by intact fibers (Cox 1991, Ramakrishnan 1993, Bao 1993).

More practically oriented papers (Subramanian 1995, Keusch 1998) summarized the influence of interface on the fatigue behavior of carbon fiber/epoxy cross-ply composites. It was found that the fatigue performance is improved by increasing interface strength. Further, it has also been reported that fatigue performance is reduced by increasing interface strength in the case of brittle epoxies because these matrices have been shown to initiate matrix cracks.

Limited information is available regarding the interface effects on the fatigue behavior of glass fiber based composites. For instance Keusch et al. (1998) looked into the fatigue behavior of cross-ply epoxy based laminates [0°/90°/90°/0°] under tension-tension loadings with different fiber-matrix adhesion in the 0° and 90° plies. The fatigue strength of these laminates was measured to be in order: good adhesion in the 0° and 90° plies, good adhesion in the 0° plies and a reduced one in the 90° plies, reduced adhesion in the 0° plies and a good one in the 90° plies, and at least, reduced adhesion in both, the 0° and 90° plies. In an earlier paper of the author (Gassan in press) on the interfacial effects of cross-ply glass-fiber/epoxy composites to the S-N it was found that the S-N curve was shifted to approximately 30 % higher applied max. loads for the composites with well-bonded fibers. Furthermore, the damage as measured by stiffness reduction was more significant for the composites with poor bonded fibers as for the well-bonded ones. The loss energy at a given strain amplitude was found to be significantly higher for composites with poorly bonded fibers.

Van den Oever et al. (1998) showed for unidirectional glass-fiber polypropylene composites by 10° off-axis fatigue tests that the interface has a large influence on damage development at a normalized lifetime n/N > 10%, while below both types of composites showed the same stiffness.

2 MATERIALS AND TEST PROCEDURE

2.1 Materials

In this study continuous E-glass fibers with a specially developed epoxy compatible sizing (EP-sizing) for a strong fiber-matrix adhesion and with a polyethylene sizing (PE-sizing) to generate a weak fiber-matrix interaction were used and embedded in an epoxy resin (LY 556 / HY 917 / DY 070 from Ciba Geigy GmbH). The EP-sizing based on an uncured bisphenol A epoxy binder with γ-aminopropyltriethoxy silane, while the PE-sizing is pure high molecular weight polyethylene (Hordamer PE 03 from Hoechst AG).

The laminates were manufactured by using a two step process: unidirectional prepreg tapes were produced by filament winding technology and hot pressed under vacuum and 80°C to cross-ply composites. Finally, the composites were additionally cured by 1 h at 100°C and 8 h at 140°C. The fiber volume fraction in each of the cured panels was determined using DIN EN 60. The average fiber volume fraction in both types of panels was 0.4. The 150 mm long, 16 mm wide, and 2 mm thick specimens were cut from cross-ply laminates. Cross-ply end tapes were bonded to the coupons.

Previously published data (Wacker 1998) under quasi-static loadings for unidirectional composites containing these both types of fibers with a similar epoxy resin are listed in table 1.

2.2 Test Procedure

All of the fatigue tests were performed on a servo-hydraulic MTS test machine under load controlled mode. A 25 mm extensometer was used to monitor strain continuously during the

Table 1. Overall mechanical properties of unidirectional EP and PE-sized composites (Wacker 1998)

Property	EP-sizing	PE-Sizing
0° - Tensile Strength	796 MPa	783 MPa
Fiber content	46.5 vol.%	49.7 vol.%
90° - Tensile Strength	58.1 MPa	18.4 MPa
Fiber content	50.7 vol.%	49.2 vol.%
ILSS	72.2 MPa	44.8 MPa
Fiber content	50.3 vol.%	50.6 vol.%

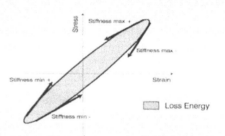

Figure 1. Dynamic stress-strain cure and definition of the characteristic values used (Lazan 1968, Walls 1996)

fatigue test. Tension-tension fatigue test in load increasing mode with different stress-ratios 'R' (defined as minimum to maximum applied load) and frequencies 'f' were done for the different material systems. Damage was monitored by recording the dynamic stress-strain curves during the test. Further different types of modulus characteristics and the loss energy as the area of the hyseresis loop were calculated. The definition of the characteristic values used are illustrated in Fig. 1.

3. RESULTS AND DISCUSSION

It is obvious that in fibrous composite materials energy is dissipated, loss energy, during crack initiation and propagation by a multiplicity of microfracture events occurring at the crack tip including fiber fracture, matrix cracking, interfacial breakdown, fiber 'relaxation', and fiber pull-out (Beaumont 1974). Because of the fact that each of these different microfracture events is related to a defined amount of consumed energy, loss energy seem to be an effective and sensitive tool for measuring damages directly in materials as was shown and previously discussed (Gassan in press, Bledzki 1997 and 1998).

Changes in the tangent modulus, i.e. 'stiffness min +' at the start of the loading and 'stiffness max +' at the end of the loading, as measure for the non-linearity of the loop was found to be a second very sensitive tool to characterize interfacial properties of composites (Walls 1996, Gassan in press). It was discussed that this non-linearity could be a result of fiber-matrix sliding (dependent on interfacial adhesion and applied load) in the 0° plies as well as the possibility of crack closure arising from frictional effects, consequently these cracks will not open unless a certain stress is applied (Pryce 1992). In consequence, crack density affect non-linearity of the loop.

Because of this, loss energy and tangent modulus was used throughout this study to discuss the influence of fiber-matrix adhesion on the fatigue behavior of glass-fiber/epoxy cross-ply composites.

3.1 Loss energy

Fig. 2 show the loss energy vs. applied max. load for cross-ply composites containing EP and PE sized glass-fibers tested at a frequency of 10 Hz and a stress ratio of 0.1. One can see that the critical load for damage initiation/propagation is lower for the PE-sized as for the EP-sized composites, followed by a more significant period of continuous damage propagation for the EP-sized composites. For these composites the rate of damage progress for a given applied max. load was found to be lower.

As a result of both, the load at failure was measured to be 230 MPa and 310 MPa for the composites containing PE-sized and EP-sized glass-fibers, respectively.

It has long been recognized that the stress-ratio has a strong influence on fatigue response (El Kadi 1994) as was also found in our tests (Fig. 3). One can see for EP-sized composites that rate of damage propagation increases with a decrease in stress-ratio, while critical load for damage initiation seem to be constant for all ratios used.

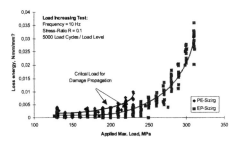

Figure 2. Influence of fiber-matrix adhesion on the loss energy vs. applied max. load

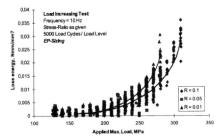

Figure 3. Influence of stress-ratio on the loss energy vs. applied max. load for EP-sized composites

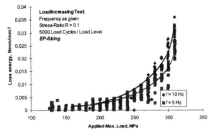

Figure 4. Influence of frequency on the loss energy vs. applied max. load for EP-sized composites

In contrast and known from investigation on the fatigue behavior (S-N-curves) of glass-fiber/phenolic composites by Richardson et al. (1997), the authors concluded that frequencies between 1.5 and 25 Hz have only a minor but increasing influence on the fatigue strength. The results illustrated in Fig. 4 on EP-sized composites show that an increase in test frequency lead to a more rapid increase in loss-energy. This is perhaps not only because of differences in damage mechanisms but could be

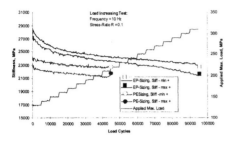

Figure 5. Influence of fiber-matrix adhesion on the stiffness of glass-fiber epoxy composites

superposed by the viscoelastic mechanisms of the material. Load at failure was measured to be more or less independent from frequency.

3.2 Dynamic modulus

Dynamic modulus vs. load cycles and applied max. load, respectively, for both types of composites is shown in Fig. 5. One can see for PE-sized composites, that the dynamic modulus is in general lower and characteristic damage state (CDS) is reached at lower applied loads. Further because of damages, hysteresis stress-strain loops are characterized by changing area (chapter 3.1) and decreasing main slope and tangent modulus as mentioned already. Both composites tested possess a strong non-linearity in the loading part of the stress-strain loop (ratio of 'stiffness max +' to 'stiffness min +') which is more significant for the PE-sized composites. Reasons for this non-linearity could be fiber-matrix sliding dependent on interfacial adhesion in the 0° plies as well as crack closure arising from frictional effects. A similar discussion is published by Walls et al. (1996) for metal matrix composites, where the reduction in modulus is connected with matrix cracking, while the non-linearity in the stress-strain response was more a result of interfacial sliding. Both of these structural mechanisms led to a more significant non-linearity for the PE-sized composites during the whole test and independent from applied load.

4. CONCLUSION

The effect of the interfacial properties of glass-fiber/epoxy cross-ply composites on the fatigue behavior in load increasing mode was studied by using differently sized glass-fibers. The composites tested only differed in the interface, all other conditions being kept constant. To generate a poor fiber-matrix adhesion a PE-sizing was used, while the EP-sizing led to a strong adhesion.

It was shown that the loss energy is a sensitive tool to characterize the nature of the fiber-matrix adhesion. The critical load for damage initiation/propagation was significant lower (~190 MPa) for the composites with the poor adhesion than for those composites containing well-bonded fibers (~230 MPa). Further, rate of damage propagation was reduced for the second case. Furthermore, the damage as measured by stiffness reduction was more significant for the PE-sized composites as was found for the EP-sized ones. The degree of non-linearity of the stress-strain loop, beside others, as measure for the interfacial strength was much lower for the EP-sized composites as for the PE-sized ones.

Recent Developments in Durability Analysis of Composite Systems, Cardon, Fukuda, Reifsnider & Verchery (eds)
© 2000 Balkema, Rotterdam, ISBN 90 5809 103 1

Influence of quasi lifetime treatments on the static and fatigue durability of continuous glass fiber-reinforced polypropylene

J.F. Neft & P. Schwarzer
Volkswagen AG, Wolfsburg, Germany

K. Schulte
Technical University Hamburg-Harburg, Germany

ABSTRACT: This study investigated the mechanical properties, including static, and fatigue behaviour, of a continuous GFRT, with the trademark name Plytron© from Borealis (Norway). The material consisted of prepregs with E-glass fibre as reinforcement and isotactic polypropylene as matrix. Both unidirectional and cross-ply stacking sequences were used. The material was investigated in three different conditions: $I_)$ as delivered, $II_)$ aged (90°C, 100 days) and $III_)$ aged under a special 125 day climate cycle. It was tested under tension, compression and bending stresses. Fatigue testing was performed at two different stress ratios, R=0.1 and R=-1. The treatments showed no pronounced effects on the mechanical properties which is thought to be due to the similar degree of cristallinity found in the as delivered material and the aged conditions. The transverse tensile strength was the property that was effected the most by the material ageing and resulted in a degradation of the tensile strength

1. INTRODUCTION

Fibre reinforced polymers (FRP) have been of great interest to aerospace industry due to their high specific moduli and strength, their excellent corrosion and impact behaviour and their very good fatigue durability. These properties are also very attractive for applications in the automobile industry, where engineers try to increase the fuel efficiency of cars by an overall weight reduction. However low material and processing costs are of equal importance for a successful substitution of currently used steels with new materials. A fairly new group of FRP, the glass fibre-reinforced thermoplastics (GFRT), seem to fulfil these requirements and thus have been studied intensively in the last years.

GFRT, especially glass/polypropylene laminates (glass/PP), have been investigated with regard to different fabrication and processing methods. Prepregs and textile preforms as semifinished products turned out to be most efficient [1, 2, 3]. Davies investigated the influences of processing on the mechanical properties of glass/PP [4]. He showed the importance of cooling rate on the matrix morphology and the laminate mechanical properties: the slower the cooling rate the higher the matrix cristallinity and the higher the numbers of

defects at interspherulitic regions. This simplified crack propagation and reduced mechanical properties of the laminate. The fibre/matrix interface strength [5, 6, 7, 8] has a big impact on the laminate properties. The interface strength can be improved by coupling agents, e. a. silane. Pomies and Carlsson investigated the influence of water absorption on the transverse tensile properties and the shear fracture toughness of glass/PP [9, 10, 11]. They showed that water had no pronounced effect on these properties due to the high cristallinity and inherent character of the polypropylene matrix. The results of temperature ageing of composite materials has been summarised by Hancox [12, 13]. However, up to date no studies on the effect of thermal and climate treatments on the mechanical behaviour, such as tensile creep and fatigue, of GFRT have been performed.

Paramount to the introduction of GFRT into automotive applications is a wide database of their durability behaviour. A comprehensive understanding of the mechanical behaviour and the damage mechanisms under service conditions is necessary for a successful application. Therefore this study investigated the influence of different ageing treatments on the mechanical durability of glass/PP laminates. Tensile, transverse tensile, compression, bending tests as well as fatigue tests under tension-tension and tension-compression loading were performed.

2. EXPERIMENTAL

2.1 Material and specimen preparation

In this study unidirectional glass fibre-reinforced (E-glass) isotactic polypropylene with the trademark name Plytron© from Borealis (Norway) was investigated for two different lay ups: unidirectional (UD) and cross-ply stacking (CP) sequences. The fibre volume fraction of both lay ups was about 35%. Test specimen geometries were chosen according to DIN 29971 [14]. For fatigue testing, static tension specimens were used. Test plates of 2 and 3 mm thickness were manufactured. Lay-up sequences and specimen thicknesses are shown in table 1.

Table 1: Stacking sequences and thickness of test specimens.

thickness [mm]	laminate UD	laminate CP
2	$[0°]_{10}$	$[0°,90°,0°,90°,0°]_s$
3	$[0°]_{14}$	$[0°,90°,0°,90°,0°,90°,0°]_s$

Test plates were consolidated in an autoclave under vacuum (1 bar) at 240°C for 90 minutes. The test specimens were then cut with a water-cooled diamond saw. For tension and compression testing glass/polyester tabs were glued onto the primered (Loctite 770) specimen ends with Loctite 406.

The material was investigated in its as delivered condition and in two differently aged conditions. Within the first ageing treatment (ageing I), which lasted for a total of 125 days, the material was exposed to 100 so called climate cycles. Each cycle consisted of 8 steps of different temperature, humidity and exposure time, as can be seen in table 2. After every 25th cycle an ageing treatment at 90°C for 600 hrs was added. In the second ageing treatment (ageing II) the material was aged at 90°C for 100 days.

2.2 Static and fatigue tests

For static testing a Zwick 1485 universal test machine with a load capacity of 100 kN was used. Loading conditions were chosen according to DIN 29971 [14]. The strain measurements during tension and bending tests were performed with a Zwick extensometer with a base length of 100 mm. For compression strain measurement a strain-gauge exten-

Table 2: Sequences of one climate cycle.

sequence	program	length [minutes]
1	45°C/95% r. h.	120
2	heating to 90°C/dry	50
3	90°C/dry	240
4	cooling to 15°C/dry	80
5	15°C/dry	60
6	cooling to −25°C/dry	40
7	−25°C/dry	60
8	heating to 45°C/95% r. h.	70

someter (DLR) with a base length of 45 mm was mounted symmetrically onto the centre on the wide edge surfaces of the specimens. For compression and tension-compression fatigue testing a special antibuckling device, described by Schulte [15], with a free unsupported length of 12.5 mm, was used. Fatigue testing was performed on a Schenk servohydraulic test machine with a load capacity of 10 kN. Strains were detected online with the same straingauge extensometer used for compression testing. The sinusoidal fatigue load was applied at stress ratios R (minimum stress/maximum stress) equal to 0.1 (tension-tension) and −1 (tension-compression) and frequencies of 10 Hz and 5 Hz, respectively. Tests were performed under load control until complete fracture of the specimens. All tests were performed at 23°C and 50% relative humidity.

3. RESULTS AND DISCUSSION

3.1 Static tests

Figures 1 and 2 show the strength of both laminates and all three conditions for tensile, transverse tensile, compression and three-point bending tests. The strength values are also listed in Table 3 along with their standard deviation. It is evident, that the CP laminates had up to 45% lower strength values than the equivalent UD laminates. This is thought to be due to the lower fibre fraction along the stress axis of the CP laminates. The comparably high compression strength of CP laminates are thought to depend on the antibuckling like behaviour of the 90°-fibres in the laminate [15].

The different treatments had no pronounced effects on any of the measured properties. Only a slight improvement of bending strength and a slight decrease in compression and tension strength were observed.

However these differences might be explained by scatter of the data. The only exception were the transverse tensile strengths. They decreased clearly for the aged laminates. The transverse tensile tests

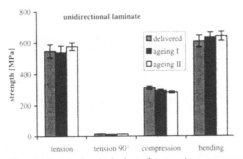

Figure 1: Mechanical data for tension, transverse tension, compression and bending for unidirectional laminates.

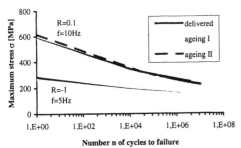

Figure 2: Mechanical data for tension, compression and bending for cross-ply laminates.

Table 3: Strength and standard deviation values for both laminates for static loading conditions.

	delivered σ [MPa]	ageing I σ [MPa]	ageing II σ [MPa]
tension UD, 0°	548 ±42	538 ±43	577 ±24
tension UD, 90°	15,2 ±1,3	11,6 ±1,6	13,8 ±0,8
tension CP		329 ±15	345 ±34
bending UD	608 ±38	633 ±33	641 ±29
bending CP	396 ±35	403 ±33	416 ±18
compression UD	311 ±10	294 ±8	284 ±6
compression CP	230 ±11	215 ±35	193 ±26

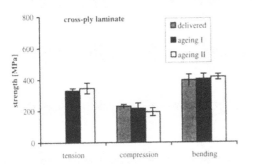

Figure 3: S-N curves for unidirectional laminates and all treatments, R=0.1 and R=-1.

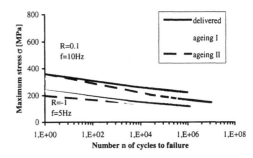

Figure 4: S-N curves for cross-linked laminates and all treatments, R=0.1 and R=-1.

are very sensitive to influences on the fibre/matrix interface and the matrix properties. Therefore it is obvious that either both ore one of these properties were affected by the treatments. Because glass fibres are not affected by water or a long term temperature of 90°C ageing [16], which was the ageing maximum temperature, the strength degrading processes must had happen in the matrix ore the fibre/matrix interface which was verified by the test results.

3.2 Fatigue Testing

Temperature rise during fatigue loading
During the fatigue tests a thermocouple which was fixed to the centre of the specimen was used to measure the temperature change due to thermal heating in the specimen. A maximum temperature rise of 4°C was measured for the tension-tension fatigue tests and of 6°C for the tension-compression tests with the highest stress amplitude. Because of these negligible temperature rises no influence on the damage behaviour was expected and this test frequencies were kept constant. The higher temperature rise for tension-compression tests was due to friction processes between the antibuckling guide and the specimens.

S-N curves
The S-N curves of all fatigue tests are shown in figures 3 and 4. For a better overview only the approximation curves obtained by a last square fit are plotted. The tension-tension tests resulted in a superior fatigue behaviour than the tension-compression tests, for both CP and UD laminates. This can be explained by the better tensile behaviour of FRP compared to their compression behaviour.

Figure 5: Fracture surface morphology of delivered specimen for transverse tension load.

Figure 6: Fracture surface morphology of delivered specimen for transverse tension load.

Similar trends of mechanical properties measured in static tests were also found in fatigue tests up to a number of 10^5 cycles. From here the effect of ageing influences were too small to show an effect on the fatigue behaviour of the laminates. Further testing is necessary to verify these results.

Fracture surface morphology
The as delivered laminate fracture surfaces show a high crystalline character (figures 5 and 6). The degree of matrix cristallinity was determined by Differential Scanning Calorimetry (DSC). It was found that the as delivered laminates have a degree of cristallinity of 55%. This high value was thought to be due to the long consolidation time and was further increased up to 57% for both ageing treatments. The rise of 2% is suggested to lead to larger spherulites and therefore to the presence of more defects at interspherulitic regions.

This observations have a good correspondence with Davies [4]. However analysis of thin polished sections and wide-angle X-ray (WAXS) investigation are necessary to confirm this hypotheses. No difference in fibre/matrix delamination between the as delivered and the aged conditions was observed.

4. CONCLUSION

The influence of different treatments on the static and fatigue properties of a glass/PP laminates was investigated. However treatments showed no pronounced effects on the mechanical properties which is thought to be due to the similar degree of cristallinity found in the as delivered material and the aged conditions. The transverse tensile strength was the property that was effected the most by the treatments ageing of the material resulted in a degradation of the tensile strength. Fractography of transverse tensile specimens revealed a higher matrix cristallinity of the aged specimens, larger spherulites and thus a higher number of defects in interspherulitic regions. This resulted in higher crack propagation rates. To verify these results further matrix microstructure characterisation has do be done.

REFERENCES

[1]   K. F. Friedrich, R. Reinicke in VDI-Berichte (Nr.: 1420), **1998**, 97-113
[2]   R. Funck, M. Neitzel, O. Christen in Proceedings of ICCM-11, **1997**, I-419 – I-425
[3]   H. Stumpf, Study on the manufacture of thermoplastic composites from new textile preforms, PhD-D Thesis, TU Hamburg-Harburg, **1998**
[4]   P. Davies, W. J. Cantwell, Composites, **1994**, 25, 869-877
[5]   H. A. Rijsdijk, M. Contant, T. Peijs, Composites Science and Technology, **1993**, 48, 161-172
[6]   J. L. Thomason, G. E. Schoolenberg, Composites, **1994**, 3, 197-203
[7]   J. Karger-Kocsis, E. Moos, T. Czigany, Advanced Composite Letter, **1997**, 2, 31-36
[8]   E. K. Gamstedt, L. A. Berglund, T. Peijs, Composite Science and Technology, **1999**, 59, 759-768
[9]   F. Pomies, L. A. Carlsson, J. W. Gillespie in Composite Materials, ASTM STP 1230 (Ed. R. H. Martin), **1995**, 283-303

[10] F. Pomies, L. A. Carlsson, Journal of
      Composite Materials, **1994**, 28, 22-35
[11] P. Davies, F. Pomies, L. A. Carlsson, Journal
      of Composite Materials, **1996**, 30, 1004-1019
[12] N. L. Hancox, Materials and Design, **1998**,
      19, 85-91
[13] N. L. Hancox, Materials and Design, **1998**,
      19, 93-97
[14] N.N., DIN 29971, **1986**
[15] K. Schulte, in ASTM STP 1185 (Ed. S. E.
      Groves, A. L. Highsmith), **1994**, 278-305
[16] G. S. Springer, Environmental Effects on
      Composite Materials, **1981**, Technomic

Recent Developments in Durability Analysis of Composite Systems, Cardon, Fukuda, Reifsnider & Verchery (eds)
© *2000 Balkema, Rotterdam, ISBN 90 5809 103 1*

Long term prediction of fatigue life of unidirectional CFRP

Yasushi Miyano & Masayuki Nakada
Materials System Research Laboratory, Kanazawa Institute of Technology, Ishikawa, Japan

Rokuro Muki
Civil and Environmental Engineering Department, University of California, Los Angeles, Calif., USA

ABSTRACT: The tensile test for resin impregnated carbon fiber strand (CF/Ep strand) and the flexural tests for longitudinal and transverse directions of unidirectional CFRP laminates were carried out. For both the CF/Ep strand and CFRP laminates, constant strain-rate (CSR) tests at three loading rates and fatigue tests at two frequencies were performed under various temperatures; all fatigue strengths as well as CSR strengths depend clearly on loading-rate and temperature. The time-temperature superposition principle holds for all fatigue strengths as well as CSR strengths, therefore the master curves for all fatigue strengths can be obtained from these results. The dependence of these fatigue strengths upon number of cycles to failure as well as time to failure and temperature can be characterized from these master curves.

1 INTRODUCTION

The mechanical behavior of polymer resins exhibits time and temperature dependence, called viscoelastic behavior, not only above the glass-transition temperature T_g but also below T_g. Thus, it can be presumed that the mechanical behavior of FRP using polymer resins as matrices also depends on time and temperature even below T_g which is within the normal operating-temperature range. These examples are shown by Aboudi et al. [1], Sullivan [8], Gates [3], and Miyano et al. [5].

The time-temperature dependence of the flexural strengths for satin-woven CFRP laminates under constant strain-rate (CSR) and fatigue loadings has been studied by Miyano et al. [2,4,6,7]. It was observed that the fracture modes are almost identical under two types of loading over a wide range of time and temperature, and the same time-temperature superposition principle holds for CSR and fatigue strengths. Therefore, the master curve of fatigue strength can be obtained from the CSR strengths under various CSRs and temperatures and the fatigue strengths at a single frequency under various temperatures.

In this paper, the tensile test for resin impregnated carbon fiber strand (CF/Ep strand) used for unidirectional CFRP and the flexural tests for longitudinal and transverse directions of unidirectional CFRP laminates were carried out. For both the CF/Ep strand and CFRP laminates, CSR tests at three loading rates and fatigue tests at two frequencies were performed under various temperatures. For all three cases, the master curves for CSR strength and the master curves of fatigue strength for fixed time to failure and temperature as well as for fixed number of cycles to failure are constructed from these test results.

2 EXPERIMENTAL PROCEDURE

2.1 Preparation of specimen

Epoxy resin impregnated carbon fiber strand (CF/Ep strand) was employed as the tensile test specimen for the longitudinal direction of unidirectional CFRP, which consists of high strength carbon fibers TORAYCA® T400-3K (TORAY) and a general purpose epoxy resin EPIKOTE® 828 (YUKA SHELL EPOXY). These specimens were produced by filament winding method. The glass-transition temperature T_g of epoxy resin is 112°C. The diameter of CF/Ep strand is approximately 1 mm.

Unidirectional CFRP laminates was employed as the flexural test specimen for the longitudinal and transverse directions, which consists of high strength carbon fiber TORAYCA® T300-3K (TORAY) and a general purpose epoxy resin #2500 (TORAY). These specimens were produced by hot pressing of the prepreg sheets made from these fiber and resin. The glass transition temperature T_g of 2500 is 130°C. The fiber volume fraction of CFRP laminates was approximately 55%.

2.2 Test procedures

The tensile CSR and fatigue test specimens of CF/Ep strand are prepared as shown in Fig.1. Pulling-out of strand from the grip ends during loading occurred initially in both CSR and fatigue testing. This pulling-out was suppressed by two improvements: use of specimen with taper-shaped configuration at both ends fixed by adhesive resin in the grips as shown in Fig.1, and use of the small temperature chamber as shown in Fig.2 to keep the grip at room temperature. After the improvements, fracture of all specimens tested for both CSR and fatigue loadings occurred within the central 70mm region of the specimen.

The tensile CSR tests were carried out under various constant temperatures by using an Instron type testing machine with a small constant temperature chamber as shown in Fig.2. Loading rates (crosshead speeds) were 0.01, 1, and 100 mm/min. The fatigue tests were carried out under several constant temperatures at 2 frequencies f=2 and 0.02Hz by using an electro-hydraulic servo testing machine with a small constant temperature chamber. Stress ratio R (minimum stress/maximum stress) was 0.1. The tensile CSR and fatigue strengths σ of CF/Ep strand are defined by

$$\sigma = P_{max} \frac{\rho}{t_e} \qquad (1)$$

where P_{max}, ρ, t_e are the maximum load of CF/Ep strand, the density of fiber strand, the tex of fiber strand, respectively, and their dimensions in this order are [N], [g/cm^3], [g/10^{6}m].

Three point bending CSR tests for longitudinal and transverse directions of unidirectional CFRP laminates (to be abbreviated as longitudinal tests and transverse tests below) were carried out by using an Instron type testing machine with a constant temperature chamber. The nominal dimensions of the test specimens were 80, 10, and 3 mm (length, width, thickness) for the longitudinal tests and were 65, 15, and 3 mm for the transverse tests. The span of the test fixture was 60 mm for the longitudinal tests and was 50 mm for the transverse tests. The tests were carried out under various constant temperatures at three loading-rates V=0.02, 2, 200 mm/min.

Specimens with the same dimensions were used for three point bending fatigue tests. The fatigue tests were carried out under several constant temperatures at f=2 and 0.02Hz by using an electro-hydraulic servo testing machine with a constant temperature chamber. Stress ratio R was 0.05. The flexural CSR and fatigue strengths σ of CFRP laminate are defined by

$$\sigma = \frac{3 \cdot P_{max} \cdot L}{2 \cdot b \cdot h^2} \qquad (2)$$

where P_{max}, L, b and h are the maximum load, the span, width and thickness, respectively, and the dimension of the maximum load is [N] while that of the rests are [mm].

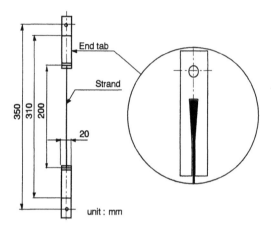

Figure 1. Tensile test specimen for CF/Ep strand

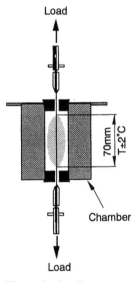

Figure 2. Small constant temperature chamber for CF/Ep strand

3 RESULTS AND DISCUSSION

3.1 Master curve of CSR strength

Strength curves for CSR and for fatigue will be presented below for three cases: (a) tensile strength for CF/Ep strand, (b) the flexural strength in the longitudinal direction of unidirectional CFRP laminates, and (c) the flexural strength in the transverse direction of unidirectional CFRP laminates. The left sides of Fig.3 show the CSR strength σ_s versus time to failure t_s, the time period from initial loading to maximum load.

The master curves for each σ_s were constructed by shifting σ_s at constant temperatures other than reference temperature T_0 along the log scale of t_s so that they overlap on σ_s at the reference temperature or on each other to form a single smooth curve as shown in the right side of Fig.3. Since the smooth master curves for each σ_s can be obtained, the time-temperature superposition principle is applicable for each σ_s.

The time-temperature shift factor $a_{T_0}(T)$ is defined by

$$a_{T_0}(T) = \frac{t_s}{t_s{}'} \qquad (3)$$

where $t_s{}'$ is the reduced time to failure. The shift factors for each σ_s obtained experimentally in Fig.3 are plotted respectively as three kinds of circle in Fig.4. The solid and dotted lines in this figure show the shift factors for the creep compliance of matrix epoxy resin, Epikote 828 and #2500, respectively. The shift factors for each σ_s agree well with that for the creep compliance of corresponding matrix resin, which are described by two Arrhenius' equations shown in Eq.(4) with different activation energies ΔH. Therefore, the time-temperature dependence of three kinds of CSR strength for unidirectional CFRP is controlled by the viscoelastic behavior of matrix resin.

$$\log a_{T_0}(T) = \frac{\Delta H}{2.303 \cdot G}(\frac{1}{T} - \frac{1}{T_0}) \qquad (4)$$

where G is the gas constant, 8.314×10^{-3} [kJ/(K · mol)].

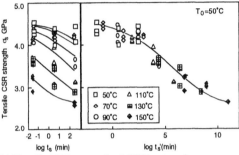

(a) Tensile CSR strength of CF/Ep strand

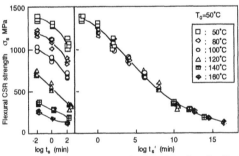

(b) Flexural CSR strength in the longitudinal direction of unidirectional CFRP laminates

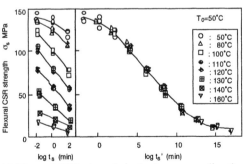

(c) Flexural CSR strength in the transverse direction of unidirectional CFRP laminates

Figure 3. Master curves of CSR strength for three kinds of loading direction

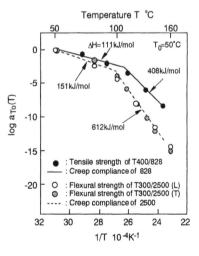

Figure 4. Time-temperature shift factors for CSR strength

3.2 Master curve of fatigue strength [7]

We turn now the fatigue strength σ_f and regard it either as a function of the number of cycles to failure N_f or of the time to failure $t_f = N_f/f$ for a combination of frequency f, temperature T and denote them by $\sigma_f(N_f; f, T)$ or $\sigma_f(t_f; f, T)$. Further, we consider the CSR strength $\sigma_s(t_f; T)$ the fatigue strength at $N_f = 1/2$, R=0, and $t_f = 1/(2f)$; this is motivated by closeness of the line connecting the origin and $(\pi, 1)$ and the curve $[1+\sin(t-\pi/2)]/2$ for $0 < t < \pi$.

To describe the master curve of σ_f, we need the reduced frequency f' in addition to the reduced time t_f', each defined by

$$f' = f \cdot a_{T_0}(T) , \quad t_f' = \frac{t_f}{a_{T_0}(T)} = \frac{N_f}{f'} \qquad (5)$$

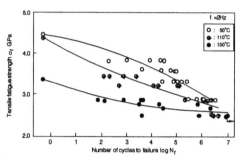

(a) Tensile fatigue strength of CF/Ep strand

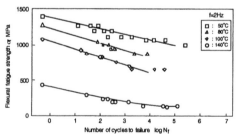

(b) Flexural fatigue strength in the longitudinal direction of unidirectional CFRP laminates

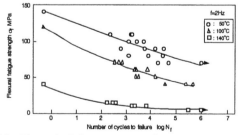

(c) Flexural fatigue strength in the transverse direction of unidirectional CFRP laminates

Figure 5. S-N curves for three kinds of loading direction

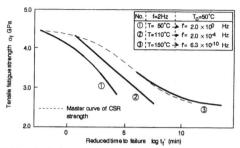

(a) Tensile fatigue strength of CF/Ep strand

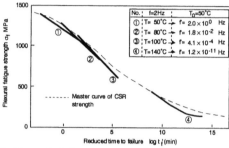

(b) Flexural fatigue strength in the longitudinal direction of unidirectional CFRP laminates

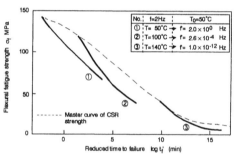

(c) Flexural fatigue strength in the transverse direction of unidirectional CFRP laminates

Figure 6. Fatigue strength versus reduced time to failure for three kinds of loading direction

We introduce two alternative expressions for the master curve: $\sigma_f(t_f'; f', T_0)$ and $\sigma_f(t_f'; N_f, T_0)$. In the latter expression, the explicit reference to frequency is suppressed in favor of N_f. Note that the master curve of fatigue strength at $N_f = 1/2$ is regarded as the master curve of CSR strength. Equation (5) enables one to construct the master curve for an arbitrary frequency from the tests at a single frequency under various temperatures.

Figure 5 displays the fatigue strength σ_f versus the number of cycles to failure N_f (S-N curve) at a frequency f=2Hz together with CSR strength which is regarded as the fatigue strength at $N_f = 1/2$. In Fig. 6, fatigue strength versus the reduced time to failure at the reference temperature $T_0=50°C$ are depicted in solid curves using the shift factor for CSR strength; the master curve for CSR strength is included in the figure in dashed curve. The master curves of σ_f for fixed N_f are constructed by connecting the points of the same N_f on the curves of each frequency as shown in Fig.7.

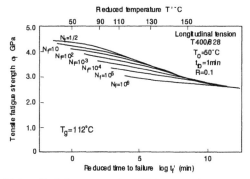

(a) Tensile fatigue strength of CF/Ep strand

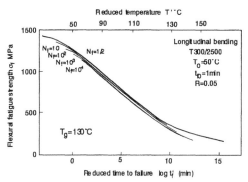

(b) Flexural fatigue strength in the longitudinal direction of unidirectional CFRP laminates

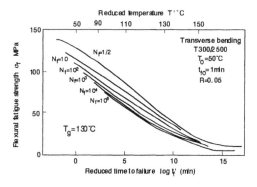

(c) Flexural fatigue strength in the transverse direction of unidirectional CFRP laminates

Figure 7. Master curves of fatigue strength at constant number of cycles to failure for three kinds of loading direction

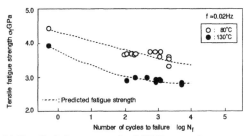

(a) Tensile fatigue strength of CF/Ep strand

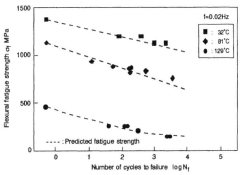

(b) Flexural fatigue strength in the longitudinal direction of unidirectional CFRP laminates

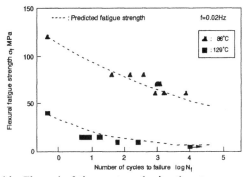

(c) Flexural fatigue strength in the transverse direction of unidirectional CFRP laminates

Figure 8. Experimental and predicted S-N curves at frequency f=0.02Hz for three kinds of loading direction

To predict the S-N curve at frequency f* and T*, we note from Eq.(5) that

$$t_f' = \frac{N_f}{f^* \cdot a_{T_0}(T^*)} \qquad (6)$$

and read a pair (σ_f, t_f') from Fig.7 which gives a pair (σ_f, N_f). The S-N curves at f=0.02Hz predicted in this manner are displayed in Fig.8 together with test data. Since the S-N curves predicted on the basis of the superposition principle capture test data satisfactorily, the time-temperature superposition principle for CSR strength also holds for fatigue strength. Therefore, the validity for the construction of master curves of fatigue strength by using the time-temperature shift factor for CSR strength is confirmed.

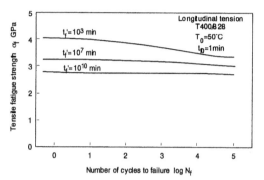

(a) Tensile fatigue strength of CF/Ep strand

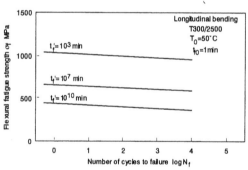

(b) Flexural fatigue strength in the longitudinal direction of unidirectional CFRP laminates

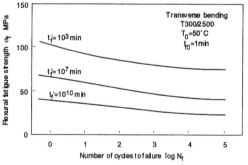

(c) Flexural fatigue strength in the transverse direction of unidirectional CFRP laminates

Figure 9. Master S–N curves at constant reduced time to failure for three kinds of loading direction

3.3 Characterization of three kinds of fatigue strength

The fatigue fracture mode for the tensile test of CF/Ep strand is tensile in all ranges of time to failure, temperature, and number of cycles to failure N_f. The fatigue fracture mode for the flexural test in the longitudinal direction of unidirectional CFRP laminates is the compressive fracture on the compression side of specimen triggered by the microbuckling of fiber in all ranges of time to failure, temperature, and N_f. On the other hand, the fatigue fracture mode for the flexural test in the transverse direction of unidirectional CFRP laminates is the tensile fracture on the tension side of specimen in all ranges of time to failure, temperature, and N_f.

The viscoelastic behavior of matrix resin directly controls the flexural strength in the transverse direction of CFRP laminates and, in almost same degree, the flexural strength in the longitudinal direction as shown in Fig. 7. It is remarkable that the time-temperature dependence is observed in tensile strength for CF/Ep strand even though much weaker than those for flexural strength.

The master S-N curves for several reduced time to failure t_f' are constructed from Fig. 7 and shown in Fig. 9. The tensile fatigue strength of CF/Ep strand decreases clearly with N_f in the range of short time to failure and low temperature. On the other hand, the tensile fatigue strength decreases scarcely with N_f in the range of long time to failure and high temperature in the vicinity of glass transition temperature. The flexural fatigue strength in the longitudinal direction of unidirectional CFRP laminates decreases clearly with time to failure and temperature, however this strength depends scarcely on N_f in all ranges of time to failure and temperature. The flexural fatigue strength in the transverse direction of unidirectional CFRP laminates decreases clearly with time to failure, temperature, and N_f in all ranges of time to failure and temperature.

4 CONCLUSION

The tensile test for resin impregnated carbon fiber strand (CF/Ep strand) and the flexural tests for longitudinal and transverse directions of unidirectional CFRP laminates were carried out. For both the CF/Ep strand and CFRP laminates, constant strain-rate (CSR) tests at three loading rates and fatigue tests at two frequencies were performed under various temperatures. The time-temperature superposition principle holds for all fatigue strengths and CSR strengths of which shift factors for CF/Ep strand and for CFRP laminates agrees well with that of creep compliance for each constituent matrix resin. Based on the test results, the master curves of fatigue strength for several number of cycles to failure were constructed for all three cases described above; from the master curves so constructed and the time-temperature superposition principle, we can determine a master S-N curve for arbitrary pair of frequency and temperature.

Further, we observed substantial time-temperature dependence of the tensile strength for CF/Ep strand though weaker than the dependence of flexural strength. It is quite remarkable that viscoelastic behavior of the matrix resin seems to affect the tensile strength of uniaxial CF/Ep strand. In contrast, the flexural strength for the transverse direction is directly influenced by the viscoelastic behavior of matrix resin.

REFERENCES

1. Aboudi, J. and G. Cederbaum, "Analysis of Viscoelastic Laminated Composite Plates", Composite Structures, 12 (1989), 243-256.
2. Enyama, J., M. K. McMurray, M. Nakada and Y. Miyano, "Effects of Stress Ratio on Flexural Fatigue Behavior of a Satin Woven CFRP Laminate", Proceedings of 3rd Japan SAMPE, Vol. 2: (1993), 2418-2421.
3. Gates, T., "Experimental Characterization of Nonlinear, Rate Dependent Behavior in Advanced Polymer Matrix Composites", Experimental Mechanics, 32 (1992), 68-73.
4. McMurray, M. K., J. Enyama, M. Nakada and Y. Miyano, "Loading Rate and Temperature Dependence on Flexural Fatigue Behavior of a Satin Woven CFRP Laminate", Proceedings of 38th SAMPE, No. 2 (1993), 1944-1956.
5. Miyano, Y., M. Kanemitsu, T. Kunio and H. Kuhn, "Role of Matrix Resin on Fracture Strengths of Unidirectional CFRP", Journal of Composite Materials, 20 (1986), 520-538.
6. Miyano, Y., M. K. McMurray, J. Enyama and M. Nakada, "Loading Rate and Temperature Dependence on Flexural Fatigue Behavior of a Satin Woven CFRP laminate", Journal of Composite Materials, 28 (1994), 1250-1260.
7. Miyano, Y., M. Nakada, M. K. McMurray and R. Muki, "Prediction of Flexural Fatigue Strength of CFRP Composites under Arbitrary Frequency, Stress Ratio and Temperature", Journal of Composite Materials, 31 (1997), 619-638.
8. Sullivan, J., "Creep and Physical Aging of Composites", Composite Science and Technology, 39 (1990), 207-232.

Recent Developments in Durability Analysis of Composite Systems, Cardon, Fukuda, Reifsnider & Verchery (eds)
© *2000 Balkema, Rotterdam, ISBN 90 5809 103 1*

Fatigue behavior of single lap adhesive composite joints

Ian Saunders & Yuris A. Dzenis
Department of Engineering Mechanics, University of Nebraska-Lincoln, Nebr., USA

Adhesive bonding has a high potential for aerospace, automotive, and other structural applications. Adhesively bonded composite patches can be used to repair aerospace parts with minimum deterioration of aerodynamic contours. Although adhesive joints have been extensively studied during the past decades, the mechanisms of fatigue failure and life of joints are not yet sufficiently understood. The objective of this paper was to study mechanical behavior and mechanisms of damage and failure in single-lap composite joints under fatigue.

Single-lap joint specimens were manufactured by the secondary curing from a high temperature, Boeing certified graphite-epoxy composite and an adhesive film used in the Air Force. Unidirectional and cross-ply adherends were utilized. Filets of epoxy resin, formed at the ends of the overlap, were removed before testing. Instrumented fatigue testing of joints was performed on a digitally controlled servohydraulic testing machine (Fig. 1). Sinusoidal tension-tension load function with the load ratio 0.1 was applied. The load amplitude varied from 0.4 to 0.8 of the quasistatic failure load. The resulting number of cycles to failure varied between 74 and 63,166. Damage and fracture evolution was simultaneously studied on-line by acoustic emission and optical video microscopy. AE location calibration was performed by a pencil lead break test (Fig. 2).

Substantial acoustic emission was acquired from the specimens throughout the fatigue tests (Fig. 3). Acoustic emission signal analysis was performed after removal of the frictional noise created by fretting crack faces. Higher concentration of the AE events was observed near the ends of the overlap zone indicating damage development in these areas. However, failure of the joints with both unidirectional and cross-ply adherends occurred through the bond line. Bond cracks initiated within 40-60% of fatigue life. Increase in the rate of AE accumulation marked the crack initiation. Gradual crack growth in the bond line was then observed to failure by the on-line video microscopy. The S-N curves for joints with unidirectional and cross-ply adherends overlapped in the life range from 1,000 to 70,000 cycles to failure (Fig. 4) but might have different slopes. The fracture surfaces of the failed specimens were fractographically examined and the final fracture areas under fatigue were identified and measured. Plotted against the fatigue load amplitude (Fig. 5), these fracture areas exhibited a linear variation. This effect can be used for the development of a final failure criterion for fatigue. A nonlinear FEM model of a single-lap adhesive composite joint with a bond crack is being developed. The model will be used for evaluation of strain energy release rates in cracked joints and formulation of the relevant fatigue crack growth criterion.

This work was funded in part by the Air Force Office of Scientific Research and the National Science Foundation.

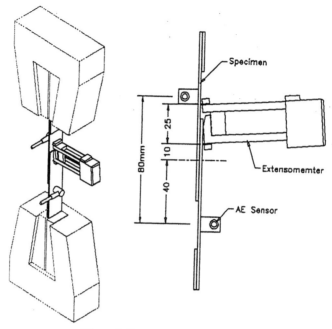

Figure 1. Experimental setup

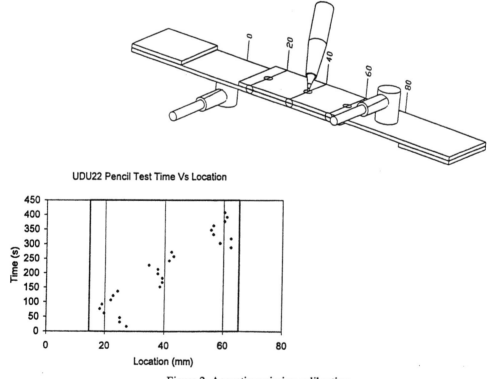

Figure 2. Acoustic emission calibration

Unidirectional Specimens

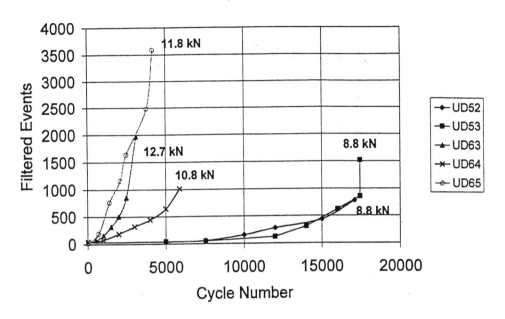

Cross-ply Specimens

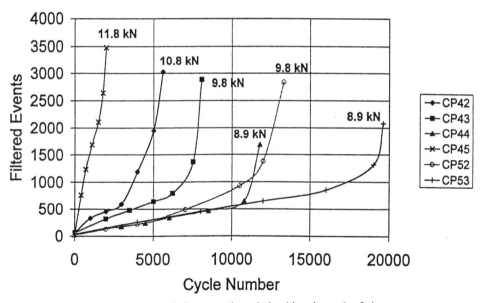

Figure 3. Cumulative acoustic emission histories under fatigue

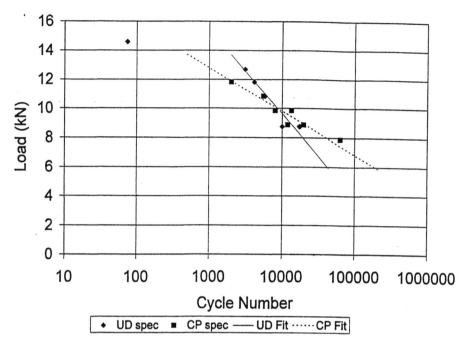

Figure 4. S-N curves for joints with unidirectional and cross-ply adherends

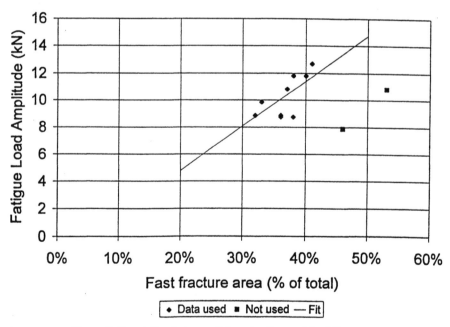

Figure 5. Correlation between fatigue loading and final fracture area

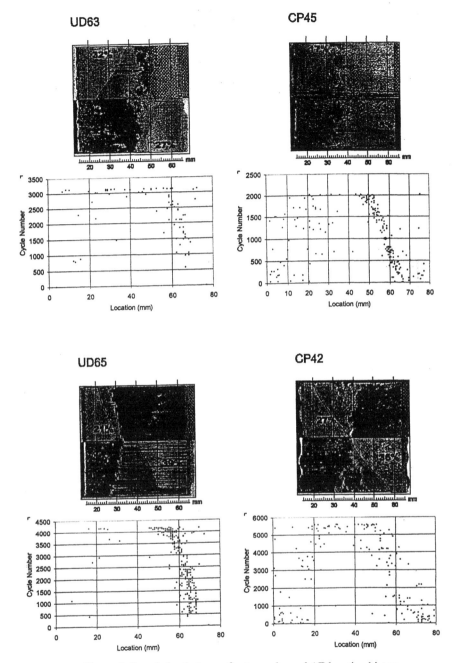

Figure 6. Correlation between fractography and AE location history

Recent Developments in Durability Analysis of Composite Systems, Cardon, Fukuda, Reifsnider & Verchery (eds)
© 2000 Balkema, Rotterdam, ISBN 90 5809 103 1

Fatigue damage in woven ceramic matrix composites

A. Haque, M. Rahman & S. Jeelani
Center for Advanced Materials, Tuskegee University, Ala., USA

ABSTRACT: This paper investigates the damage development in SiC/SiNC woven ceramic matrix composites (CMCs) under tensile and cyclic loading both at room and elevated temperatures. The ultimate strength, failure strain, proportional limit and modulus data at a temperature range of 23°C-1380°C are generated. The stress/strain plot shows a pseudo-yield point at 25% of the failure strain (ε_f) which indicates damage initiation in the form of matrix cracking. The evolution of damage beyond 0.25 ε_f both at room and elevated temperature comprises of multiple matrix cracking, interfacial debonding and fiber pullout. Although the nature of the stress/strain plot shows damage-tolerant behavior under static loading both at room and elevated temperature, the life expectancy of SiC/SiNC composites degrades significantly under cyclic loading at elevated temperature. This is mostly due to the interactions of fatigue damage caused by the mechanically induced plastic strain and the damage developed by the creep strain. The in situ damage evolutions are monitored by acoustic event parameters, ultrasonic C-scan and stiffness degradation.

1 INTRODUCTION

The ability of ceramic matrix composites to sustain load degrades overtimes as the damage initiates and develops due to external loading conditions and presence of various environments. The important information that need to be identified in the overall damage development process are the mechanisms that initiates the damage, then the damage growth rate and finally the failure state. All these factors eventually determine the life expectancy of the CMCs. Although some works are performed in the past relating to damage mechanics and durability of whisker, particulate and continuous fiber reinforced ceramic matrix composites (Yen & Jones 1997, Talreja 1990, Pastor et al. 1998, Burr et al.), the demand for textile or woven ceramic matrix composites are increasing significantly in recent years due to their enhanced multidirectional and out-of-plane properties. The type that has been most commonly used in producing ceramic composites is the "satin weave". In most cases, the satin weaves are expected to provide superior properties than the plain weave because of larger section of unbent fiber. As a result this paper focuses the damage development and material responses of SiC/SiNC woven composites under monotonic tensile and tension-tension cyclic loading at room and elevated temperatures. The damage initiation and growth rate are monitored by acoustic emission, ultrasonic C-scans and modulus degradation measurements and both the tensile strength and fatigue life are correlated with the developed damage parameters.

2 EXPERIMENTAL WORK

2.1 Materials Specification

The ceramic matrix composite was composed of Nicalon silicon carbide woven fabric and Silicon Nitrogen Carbon (SiNC) matrix. The Nicalon[TM] ceramic fiber was basically beta-SiC crystallites and the SiNC matrix was an amorphous mixture of silicon, carbon and nitrogen. The diameter of the fiber was approximately 15 microns with an average density of 2.55 gm/cm[3]. The CMC was manufactured through polymer impregnation and pyrolysis process. The matrix precursor was first impregnated into the eight-harness satin weave quasi-isotropic [0/±45/90]$_s$ fabric lay-up, then cured by low temperature processing. The composite was then pyrolized to convert the pre-ceramic matrix polymer into a ceramic.

2.2 Specimen Design

An edge-loaded specimen geometry shown in Figure 1 was used for all the tests at room and elevated temperatures. This specimen geometry eliminates the mounting of external tabbing materials and/or holes to accommodate the specimens in the grip of the test machine. The design of the specimen was initially proposed by Holmes and has been used to generate reliable tensile and fatigue test results without any slippage(Holmes).

2.3 Test Environment

Tensile and fatigue testing of the CMC was performed using Instron-8502 servohydraulic frame equipped with a self-aligning supergrip. A two zone short furnace equipped with Kanthal super 33 heating elements (molybdenum disilicate elements) was used to perform the elevated temperature tests. The length of the furnace hot zone was 101.6 mm. The edge-loaded specimen geometry and the self-aligning grips provided accurate and reproducible alignment.

3 RESULTS AND DISCUSSIONS

3.1 Tensile Responses

The stress-strain plot for SiC/SiNC composites under monotonic tensile loading at room, 700°C and 1380°C temperatures are shown in Figure 2. The results show very negligible degradation in failure strength and rather slightly improved failure strain at the elevated temperatures. This minimal change in both failure stresses and modulus are due to sustained interfacial bonding between the SiC fiber and SiNC matrix at elevated temperature. This is due to the fact that the oxidation resistance of the SiNC matrix is shown to be very good up to 1400°C. The oxidation resistance of SiNC matrix depends on the formation of SiO_2 – protection layer. Usually porous non- oxide materials show mass changes greater than 2.5 % due to the oxidation of the pore walls. The oxidation behavior of pyrolyzed SiNC-ceramic is different and the complete mass gain is shown to be less than 0.7% at 1400°C after 24 hours, with higher oxidation temperature yielding increasing mass gains (Weibelzahl et al. 1999). A proprietary coating in Nicalon fiber by Dow Corning protects the oxidation of the fiber in the SiC/SiNC composite at elevated temperatures in the range 1200 – 1250°C.

The stress-strain curves both at room and elevated temperature showed extreme linearity at the initial

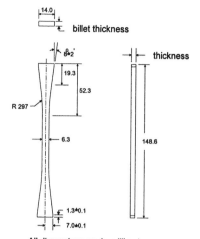

All dimensions are in millimeters
Figure 1. An edge-loaded tensile specimen.

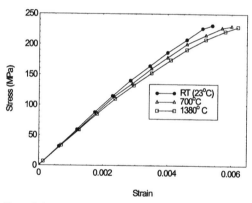

Figure 2. Stress vs. strain plot at room temperature(23°C), 700°C and 1380°C.

stage confirming that the load is carried out by both the matrix and the fiber. This stage continued up to 0.12% strain. Beyond 0.12% strain the stress-strain plots continued to deviate from the linearity due to crack initiation and growth until the final failure occurred at approximately 0.55% strain. No major activity was also observed in the energy, time and amplitude plots of the acoustic emission data shown in Figure 3 during the period when load was carried by both the matrix and the fiber. This data in Figure 3 was acquired during room temperature tensile test. The time period at which the nonlinearity begins in the stress/strain plot coincides with the time span in the acoustic emission plot shown in Figure 3. Figure 3 shows no major activity up to 25% of the total time period and consequently, Figure 3 shows extreme linearity up to approximately 25% of the total failure strain. Since the load was applied at constant displacement rate, 25% of the failure strain obviously occurred at 25% of the total loading period. Some of the microcracks are even shown to occur at a very earlier stage before the transition point. These microcracks are identified by low amplitude acoustic events in the range of 55-70 dB shown in Figure 3. The crack initiation mostly proceeds with matrix crack in the transverse fiber bundles of 0 and 90° weave and then eventually proceeds to + 45 and –45° weaves. Crack growth and propagation basically started at the beginning of the transition point and this stage is also identified with the acoustic event data in the amplitude range of 70-95 dB. During the crack propagation stage, the fibers in the fill direction of +45° and -45° fabrics are fractured which are shown by some high energy and amplitude level peaks in Figure 3. Low amplitude fracture (55-70 dB) with matrix crack also continued during crack growth until the final failure occurred. The acoustic events with high amplitude (95-100 dB) and high energy contents (above 2970 Joules) indicate excessive fiber breakage (0° orientation) prior to final failure.

Figures 4-5 show the micrographs of the fractured surface of SiC/SiNC composites under tensile loading at room and elevated temperatures. The failure modes in the form of matrix microcrack, interfacial debonding, fiber pull out and fiber fracture were observed in Figures. 4-5. The interfacial debonding and fiber pull out occurred almost at the same rate both at room and elevated temperatures. Figure 5 shows cracked matrix particles on the fiber surface indicating strong interfacial bonding between the fiber and matrix at elevated temperature. The fractured surface morphology of the fiber at room temperature is shown to be comparatively rough and coarse than the morphology at elevated temperature. This possibly allowed little decrease in strength at elevated temperatures.

3.2 Fatigue Responses

The fatigue tests were performed at room and elevated temperature levels (23°C and 700°C) under load control at a frequency of 5 Hz and stress ratio (R = $\sigma_{min}/\sigma_{max}$) of 0.1. A comparison of S-N plots at room temperature and 700°C are shown in Figure 6. The results show that elevated temperature (700°C) has a tremendous effect on the fatigue life of SiC/SiNC composites. At room temperature, the endurance limit is considerably high (approx. 80% of σ_{ult}). But at 700°C, the endurance limit is expected to be much lower (45% of σ_{ult}) which is

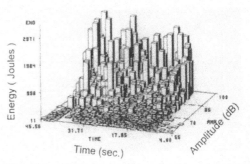

Figure 3. Amplitude, energy and time plot of acoustic emission data under tensile loading.

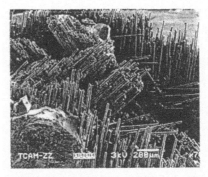

Figure 4. SEM micrograph of fractured surface under tensile loading at room temperature showing matrix crack, fiber pull out, interfacial debonding and fiber fracture (×70).

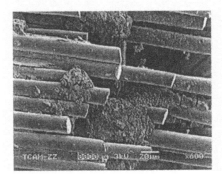

Figure 5. SEM micrograph of fractured surface under tensile loading at 1380°C showing matrix crack, fiber pull out and fiber breakage (×600).

still under investigation. The data that has been generated at 700°C shows that the specimen fails after 0.15 x 10^6 cycles at 0.60 σ_{ult}. For most polymer matrix as well as ceramic matrix composites the fatigue life diagram for a laminate can be categorized into three domains depending on

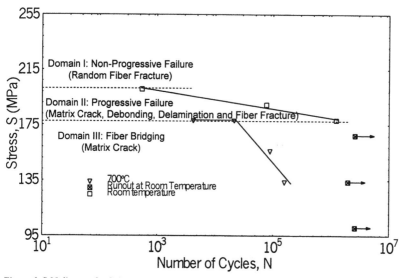

Figure 6. S-N diagram for fatigue test at room and elevated temperature.

the operative damage mechanisms inside the laminate. The mechanism operating in Domain I is fiber failure of catastrophic type. In this domain, the matrix cracking is expected to have been essentially completed in the first cycle and the fibers thus exposed to carry the applied load. Failure of fibers occur at random by a stochastic process. Domain II primarily indicates progressive damage consisting of fiber-bridged matrix cracking, debonding, delamination and fiber breakage. The lower limit to this domain is Domain III, where the applied stress is insufficient to cause the fiber breakage. Domain III is the region in which the crack growth is either eventually arrested or continues without significant fiber failures. The three domains shown in Figure 6 are defined for the fatigue life diagram at room temperature. Similar domains can also be defined for the fatigue life diagram at 700°C. Figure 7 shows the micrograph of fractured surface under fatigue loading at room temperature. A comparison of Figures 4, 7 indicates an extended fractured surface with more matrix cracks and delamination for fatigue specimens than the tensile specimens. This shows that the extent of fatigue damage is more severe than the static damage. It is also observed that the extent of fractured surface increases as the stress level decreases providing a slow crack growth rate with higher fatigue life. Figure 8 shows the micrograph of the fractured specimen subjected to fatigue loading at 700°C. The micrograph of the fiber surface at room temperature is shown to be different than that at 700°C. At room temperature a coarse fiber texture with more matrix bonding are

Figure 7. SEM micrograph of the fractured specimen subjected to fatigue loading at room temperature and at 90% of σ_{ult} showing matrix crack, fiber pull out and fiber breakage (×600).

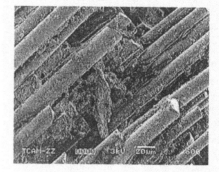

Figure 8. SEM micrograph of the fractured surface under fatigue loading at 700°C and at 80% of σ_{ult} showing matrix crack, fiber pull out and fiber breakage (×600).

Mid surface at RT,
90% of σ_{ult} and N=543

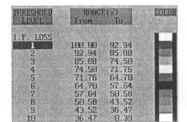

Top surface at RT,
90% of σ_{ult} and N=543

Mid surface at RT,
80% of σ_{ult} and N=1225297

Top surface at RT,
80% of σ_{ult} and
N=1225297

Mid surface at 700°C, 80% of σ_{ult} and N=21410

Top surface at 700°C 80% of σ_{ult} and
N=1225297

Mid surface at 700°C, 60% of σ_{ult} and N=158571

Top surface at 700°C, 60% of σ_{ult} and
N=158571

Figure 9. C-scan images of the fractured specimens subjected to fatigue loading at RT and 700°C.

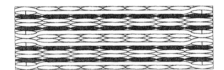

Crack initiates in the transverse fiber at the
interlaced region (stage-1)

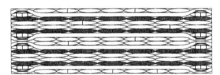

Delamination initiates at the interlaced region
(stage-4)

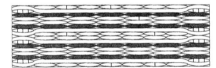

Crack density increases (stage-2)

Longitudinal fiber fails at the interlaced region
(stage-5)

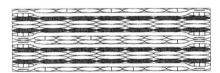

Crack develops in the longitudinal fiber at the
interlaced region (stage-3)

Final failure takes place(stage-6)

Figure 10. Schematic of damage development under fatigue loading

observed which indicates higher interfacial bond strength. But at 700°C, the fiber surfaces are shown to be more smooth and clear with some localized matrix accumulation. This means that the bonding between fiber and matrix became weaker at this temperature and the fiber pulled out took place under less force allowing poor fatigue strength at700°C. The morphology of the accumulated matrix also shows some chemical change of SiNC matrix with more pores at 700°C.

Figure 9 shows the C-scan images of fractured specimens tested under fatigue loading at room and

137

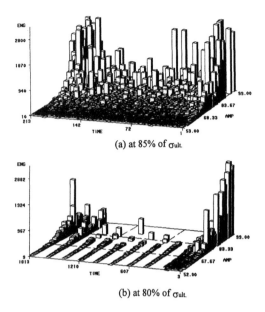

(a) at 85% of $\sigma_{ult.}$

(b) at 80% of $\sigma_{ult.}$

Figure 11. Energy, amplitude and time plot of acoustic emission data under fatigue loading

elevated temperatures. The C-scan images of top and middle surface basically represent different section across the thickness of the specimen, which are obtained by setting multiple gates across the thickness during the scanning process. The scanned images show comparatively extended fatigue

damage at lower stress levels which occurred in prolonged period providing higher fatigue life cycles. Moreover the extent of damage in the top section are shown to be comparatively more than the middle section.

Figure 10 shows the development of damage under fatigue loading. At the beginning, cracks initiate in the warp (transverse to loading) direction along the fibers. On continued loading longitudinal cracks appeared between fibers in the loading direction in the undulation regions where the fill fibers cross over the warp fibers. The crack densities increased with load cycling and a uniformly distributed pattern of orthogonal cracks appeared to develop. On further cycling delaminations take place which are primarily confined to the undulation regions. Thus a uniform distribution of delaminations are developed in the interlaminar planes. Towards the end of fatigue life, fiber bundles fail in the undulation regions. A progressive weakening of the laminate strength then led to final failure.

Figure 11 shows the plots of acoustic emission events under fatigue loading at room temperature. At 85% of $\sigma_{ult.}$, the specimen failed at 1308 cycle. Acoustic emission data was taken up to 1000 cycles. The low amplitude and low energy level events are the matrix cracks, the medium amplitude and medium energy level events are crack growth and delaminations and the high amplitude and high Figure 11a shows some high-amplitude and high-

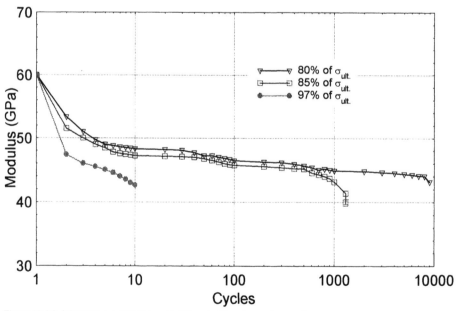

Figure 12. Modulus versus cycles plot under fatigue loading at room temperature.

138

energy activities indicating fewer fiber fracture at the beginning. This is the reason of sharp fall in modulus after the first cycle as shown in Figure 12. Some low-amplitude and low-energy activities in the form of matrix cracks also occur along with the fiber fracture at the beginning. During the intermediate stage, some events corresponding to matrix cracks and crack growth occur. These events are responsible for the gradual decrease in modulus. At or around 1000 cycles, there are some medium and high energy events which corresponds to delaminations and fiber fracture prior to final failure. Again this is the reason of sharp fall in modulus prior to failure. The above scenario is more evident from Figure 11b. The specimen was tested at 80% of $\sigma_{ult.}$ and it failed at 9440 cycle. Acoustic emission data was taken continuously up to 1000 cycles and then at a regular interval of 1000 cycles up to 9000 cycles.

Figure 12 shows the plot of degraded modulus against the number of fatigue cycles. At all loading level the modulus falls sharply after the first cycle as well as at the end of fatigue life. In between there is a gradual decrease in modulus. The sharp fall in modulus after the first cycle is attributed to matrix crack and fewer fiber failure whereas the sharp fall before the final failure is attributed to delaminations and fiber fracture. This phenomenon is supported by the acoustic emission plot shown in Figure 11.

4 CONCLUSSIONS

The tensile strength and modulus of SiC/SiNC woven composites have been observed to decrease negligibly with increased temperature beyond the linear portion of the stress/strain plot. Such linearity in the stress/strain plot is continued up to 25% of the failure strain (ε_f). The failure process initiates and propagates beyond $0.25\ \varepsilon_f$ both at room and elevated temperature with matrix cracking, debonding and fiber pullout. The failure strain, ε_f is observed to increase with elevated temperature. The acoustic emission data is shown to monitor the activity of various failure modes during tensile and fatigue loading at room temperature. Fatigue strength at room temperature is found to be considerably high (80% of σ_{ult}). But elevated temperature has a remarkable effect on the fatigue strength. At 700°C, the fatigue strength is considered to be approximately 45% of σ_{ult} which is still under investigation. During fatigue loading, initiation of cracks take place at the interlaced region, which eventually leads to delaminations and final failure. Stiffness of the material decreases with increase number of cycles. There is a sharp decrease in modulus immediately after first cycle as well as before final failure. At the intermediate stage, the degradation of modulus occurs at a very slow rate. This phenomena is also shown to be supported by the acoustic emission data.

5 REFERENCES

Burr, A., Hild, F., & Leckie, F.A. Damage, Fatigue, and Failure of Ceramic-Matrix Composites, Applications of Continuum Damage Mechanics to Fatigue and Fracture, ASTM STP 1315, D. L. McDowell, Ed., American Society for Testing and Materials: 83-96.

Holmes, J. W. J. of Materials Science. Vol. 26: 1808-1814.

Pastor, M.S., Case, S.W., & Reifsnider, K.L. 1998. Durability of Ceramic Matrix Composites. AD-Vol.56, Recent Advances in Mechanics and Aerospace Structures and Materials – 1998, ASME 1998: 61-66.

Talreja, R. 1990 Fatigue of Fiber-Reinforced Ceramics. Structural Ceramics- Processing, Microstructure and Properties, Proceedings of the 11th Rise International Symposium on Metallurgy and Materials Science: 145-159.

Weibelzahl, W., Mutz, G., Suttor, D., & Ziegler, G. 1999. Corrosion Stability and Mechanical Properties of Polysilazane – Derived SiCN – Ceramics. Key Engineering Materials. Vols. 161 – 163: 111-114

Yen, C.F. & Jones, M.L. 1997 Material Modeling for Cross-Ply Ceramic Matrix Laminates with Progressive Damages and Environmental Degradation. MD-Vol.80, Composites and Functionally graded Materials, ASME 1997: 189-202

Influence of environmental conditions (moisture-temperature)

Recent Developments in Durability Analysis of Composite Systems, Cardon, Fukuda, Reifsnider & Verchery (eds)
© 2000 Balkema, Rotterdam, ISBN 90 5809 103 1

Aging and long term behaviour of composite tubes

P. Davies, R. Baizeau & D. Choqueuse
Marine Materials Laboratory, IFREMER Centre de Brest, Plouzané, France

L. Salmon & F. Nagot
EDF, Centre de Recherche des Renardières, Moret-sur-Loing, France

ABSTRACT: This paper will discuss results from studies undertaken by IFREMER and EDF over the last ten years. The aim of these studies has been to examine the aging and long term behaviour of glass fibre reinforced epoxy composite tubes for cooling water system applications. First, aging of resins and composites in water is discussed. Results from tests to establish the kinetics of resin hydrolysis are used in a simple model to predict composite degradation with time, and correlated with results from panels immersed for 8 years at 20, 40 and 60°C. Then results from creep tests on tubes under internal pressure with closed ends, lasting up to 18 months are given. Creep strains are shown to be lower than those measured in similar tubes with free ends reported previously. Damaged and assembled tubes have been tested. Finally current research activities are discussed.

1 INTRODUCTION

Thin wall filament wound glass fibre reinforced composite tubes are being used extensively today, but these structures have been available for over 40 years. Early applications were mostly military, such as missiles and rocket casings (e.g. Kies 1962), but they can now be found in chemical engineering plants (Mallinson 1988), fishing boats (Croquette 1992), offshore platforms (Gibson 1993), and many other industrial cooling systems. In oil and gas production the main use is small diameter low pressure pipe for water flood systems. Williams (1999) reports that Shell now has over 2250 kilometres of FRP (fibre reinforced plastic) piping materials in service round the world. Much of the development work for these applications was performed in the 1960's and 1970's (e.g. Bax 1970, Spencer 1978).

The main incentive for using FRP pipe systems to replace steel is their good resistance to corrosion, and failure rates in sea water in service have been shown to be significantly lower for FRP (de Bruijn 1996). It was mainly for this reason that in 1991 EDF (Electricité de France) decided to introduce composite circuits in their new nuclear power station at Civaux near Poitiers. The circuits involved bring river water to cooling and fire systems and the stringent safety requirements for such applications required extensive full scale testing at EDF and the CEA (French Atomic Energy Authority). Several collaborative research projects were also run. The present paper presents an overview of results from one such project, with IFREMER, but results from other studies on glass/epoxy tubes with the Applied Mechanics Laboratory, (LMARC) in Besançon (Maire 1992, LeMoal 1993, Thiebaud 1994, Perreux 1995, Suri 1995), Ecole Centrale in Paris (Bai 1996), and on glass/polyester tubes at ENS Cachan (Ghorbel 1996) are also available.

The collaboration between IFREMER and EDF focused on long term behaviour and aging. At the IFREMER Brest Centre glass/epoxy pipework has been used successfully for nearly 30 years in a sea water distribution system, but for the power station application the safety authorities required guarantees concerning the long term behaviour. A programme of aging and creep tests was therefore initiated and results from the first phase were presented at DURA-COSYS in 1995 (Baizeau 1995).

In the first part of the present paper the aging of resin and composite samples will be discussed. In 1995 results from 3 year immersion aging of composites were described. Results after 8 years will now be presented, and the basis for a simple lifetime prediction method based on resin behaviour will be shown.

The second part of the paper concentrates on the behaviour of tubes under internal pressure, including creep behaviour of undamaged and damaged tubes. Tubes assembled by mechanical and adhesive systems have also been tested. Finally current research areas and future requirements are discussed.

2 AGING OF EPOXY RESIN AND GLASS/EPOXY COMPOSITES IN WATER.

The degradation of composites in water may result from many different mechanisms, involving the fibre, fibre-matrix interface and matrix resin. In order to treat aging practically simplifying assumptions are therefore necessary. One approach is to determine, for each mechanism, a characteristic time before which the mechanism can be ignored but beyond which it plays a significant role in the durability of the material. This approach will be illustrated below for the case of wet aging of glass reinforced anhydride cured epoxy, by considering the hydrolysis of the matrix.

2.1 Resin properties

Anhydride cured epoxies contain esters which are susceptible to hydrolysis. The chemical reaction is: RCOOR' + H_2O → RCOOH +R'OH. The hydrolysis of the polymer leads to molecular chain breakage and water molecules can fix to the material. This leads to an increase in sample weight of 18 g/mole of broken chains. The resin weight change will therefore be a monotone function of the number of breakages. A simple weight measurement should therefore enable quantitative information on the kinetics of hydrolysis to be obtained. This assumes that weight gain due to water diffusion into the polymer can be separated from the overall weight change, (and that other damage does not occur simultaneously) which can be ensured by using very thin specimens. Figure 1 shows this schematically.

Samples of epoxy resin (diglicidyl ether of bisphenol A, DGEBA) with methyl tetra hydrophthalic anhydride (MTHPA) were immersed at three temperatures, 60, 70 and 80°C. A kinetic model was used used based on the following assumptions: the reaction is homogeneous throughout the material, the kinetics are of second order (one with respect to the ester functions concentration, one with respect to water concentration), and solubility of water varies little with temperature. The time to reach a given degree of advancement depends on the water content, which can be calculated for all temperatures using an Arrhenius model and an activation energy calculated from the resin weight gains at different temperatures. The determination of a critical level of hydrolysis leading to mechanical property changes is then required.

By measuring the failure stress and strain as a function of water absorbed on samples of resin aged in water at 60°C, a critical value of percentage weight gain (and hence of hydrolysis) corresponding to a drop in mechanical properties was determined, Figure 2.

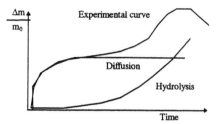

Figure 1 Resin weight change during immersion

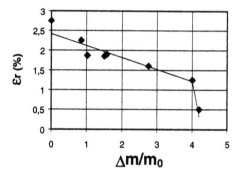

Figure 2. Failure strain versus % water absorbed.

For the epoxy/anhydride system studied here that value is around 4%. By introducing this critical value (4% water absorbed in the resin) in the kinetic model a lifetime can be determined for the matrix hydrolysis mechanism. This varies from several centuries at 20°C to less than one year at 80°C. These characteristic times show that matrix hydolysis is not a problem under normal operating conditions but may become critical at elevated temperatures.

2.2 Composite panel weight gain

Figure 3 shows the results from 8 year immersion in distilled water at different temperatures (20, 40 and 60°C) of 3 mm thick composite panels, filament wound at ±55°. The curve for immersion at 60°C shows a similar form to those in Figure 1, with an increase in weight gain after about 3 years. However it should be noted that of the five panels tested at 60°C two showed a more rapid increase in weight gain than the other three. This will be discussed further below. Tg measurements were made periodically by DSC, then after 7 years specimens at each temperature were removed and dynamic mechanical analysis was performed in 3 point flexure (temperature increased at 2°C/minute). Results are shown below, Table 1, indicating that the glass transition temperature of 60°C aged samples (which showed the large increase in weight) has dropped significantly. Similar

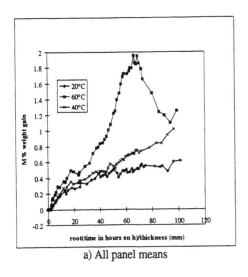

a) All panel means

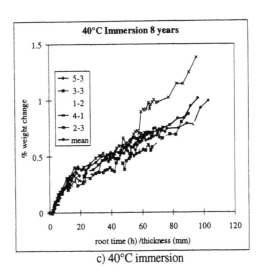

c) 40°C immersion

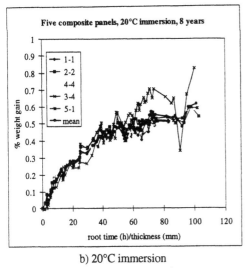

b) 20°C immersion

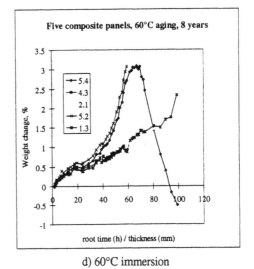

d) 60°C immersion

Figure 3. Weight gains, 20, 40 and 60°C, 8 years.

Table 1. Glass transition temperatures, °C (DSC & DMA), aged composite panels

Aging temperature	Initial state DSC	10 months DSC	22 months DSC	35 months DSC	7 years DMA
20°C	129	130	129	131	116°C
40°C	129	130	129	132	119°C
60°C	124	123	119	115	86°C

drops in Tg were measured by DMA on aged resin samples, Figure 4.

Scanning electron microscopy was performed on polished sections and fracture surfaces of regions on the exterior and centre of specimens at all three temperatures. Figure 5a shows the specimen thickness and it is apparent that some voids are present. Specimens aged at 20°C showed no evidence of interfacial debonding after 7 years. Those aged at 40°C show some signs of interface debonding on the surface but not in the centre, while those aged at 60°C show very close to the surface (in a 0.2 mm thick skin layer) clear interfacial degradation, Figure 5b, but again in the majority of the thickness of the specimens there is no evidence of debonding, Figure 5c.

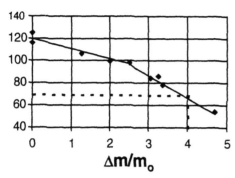

Figure 4. Change in resin Tg (°C), 60°C aging.

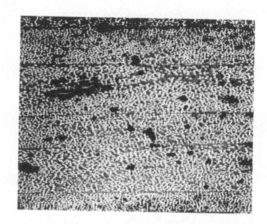

Given these results it may be justified to assume that the bulk of the specimen is degrading in a homogeneous manner and to apply the kinetic model of resin hydrolysis to predict an order of magnitude of the state of hydrolysis in the composite panels. This would suggest that when the composite has reached a weight gain around 1.6 to 2% (when the 40% by mass of resin matrix in the composite is at 4% weight gain plus the uncertain contribution of voids which will fill up with water) then there will be a drop in mechanical properties. This weight gain is achieved after a few years at 60°C but not after 8 years at 40°C (Figure 3). The predicted time for a 4mm thick resin panel to reach 4% water content is 55 years. The influence of this weight gain on composite mechanical properties will be discussed in section 2.4 below.

2.3 Correlation with behaviour of tubes

Samples taken from tubes fully immersed for 5.8 years at 60°C were also examined and the same degradation mechanism was observed, a little interfacial debonding near the surface but none in the centre of the tube wall. This suggests that the simplified approach based on resin hydrolysis kinetics, combined with regular checks on glass transition temperature, may provide a useful tool in determination of the service life of these tubes. As tubes in service are only exposed to water on the inside wall there may need to be an adjustment to the weight gain kinetics however, as previous results (Baizeau 1995) indicated that weight increases in this case were much lower than expected based on plate or fully immersed tube results. Some additional tests involving machining of inner and outer layers were performed (Suri 1995) to try to explain this phenomenon but did not clarify the reasons. As far as the degradation mechanisms are concerned this effect should be positive in reducing the degradation rate. Another aspect which can affect the correlation is the presence of voids, high levels of which (>5%) can be introduced during

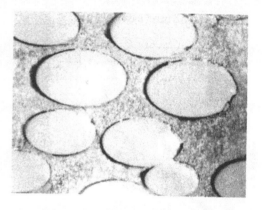

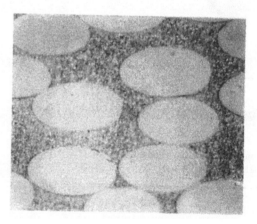

Figure 5. SEM photos 60°C aged panels after 7 years' immersion,
a) all thickness
b) outer surface (0.2mm skin layer)
c) centre

filament winding. Chiou (1996) has also reported anomalies in long term behaviour of tubes, which were attributed to the presence of voids.

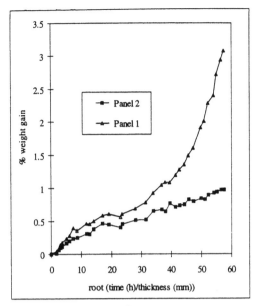

Figure 6. Weight gains of two panels removed after 3 years aging at 60°C, for mechanical tests.

2.4 Mechanical property changes after aging

A small number of flexure and short beam shear tests were performed on samples taken from the panels aged at 60°C after 3 years. Two panels were tested, one which had shown a rapid increase in weight gain (to 3%), while the other had only gained 1% weight, Figure 6. These should therefore be above and below the critical weight gain corresponding to loss in resin properties (Figure 2).

The reasons for the difference are not yet clear and are being studied, but may be caused by formulation or curing differences (the two panels which gained weight the fastest come from the same panel). It should be noted however that after 8 years the remaining panels at 60°C had also gained 2% weight or more (Figure 3). In flexure there is a clear drop in strength for both, Table 2, but this property is governed by the surface degradation (interface debonds). In interlaminar shear however, governed by shear of the resin in the centre of the sample, the loss in strength is only significant for the sample with high weight gain.

Table 2. Results from tests on composite panels after 3 years at 60°C

Panel	Flex strength, MPa		ILSS, MPa	
	35°	55°	35°	55°
Reference (unaged)	410	127	44	19.5
Panel 1 (3%)	231	87	30	13
Panel 2 (1%)	302	120	43	18.5

Further projects are now underway to develop this model further, including tests in which tube and resin samples of the same epoxy with three hardeners, (anhydride, aliphatic diamine and aromatic cycloamines) are being aged at different temperatures. Studies of the degradation of fibre-resin interfaces are also being performed. This may have a short characteristic time and be the critical degradation mechanism in some cases.

2.5 Accelerating tests with pressure

While raising temperature is often a convenient way to accelerate aging tests, increasing hydrostatic pressure may also be used. This may be more appropriate for applications such as underwater pipelines, and some tests have been run to look at how pressure affects diffusion kinetics in these materials. Examples of results for the same epoxy resin with three different fibre types, for specimens cut from filament wound tubes wound at ±55°, aged at 60°C are shown below, Figure 7.

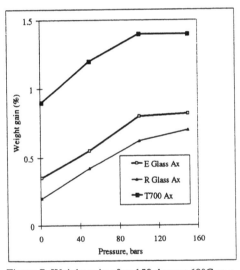

Figure 7. Weight gain after 150 days at 60°C as a function of pressure.

This figure shows that there is a significant effect of pressure for all three materials, with an increase in weight gain as pressure is increased up to 100 bars. Further increase to 150 bars does not appear to have a significant effect. This effect of pressure is related to the porosity common to filament wound materials.

3 CREEP OF TUBES AND TUBE ASSEMBLIES UNDER INTERNAL PRESSURE LOADING

Two series of tests have been performed to investigate the long term behaviour under internal pres-

Table 3. Summary of creep tests performed
A: Straight tube, B: Mechanical joint at centre, C: Tulip
adhesive joint at centre, D: Damaged
Hoop stresses applied to damage:
D1 35 MPa, D2: 67 MPa, D3: 130 MPa

Series	Pressure MPa (σ_q)	Temperature, °C (Number)	Duration Creep/ Recovery	Details
1.	2.7 (38)	25, 40, 55	12m/1m	All type
No liner	4.0 (56)	25, 40, 55	9m/1m	A
Free ends	2.7 (38)	25, 40 (2), 55	6m/2.5m	
$\sigma_z=0$	5.4 (76)	55 (2)	6m/2.5m	
2.	0.8 (11)	20 (3)	17m/1m	A,B,C
With liner	1.7 (24)	40 (5)	17m/1m	A, B, D1-3
Fixed ends,	3.8 (53)	40	10m/1m	B
$\sigma_z=\sigma_q/2$	3.8 (53)	60 (3)	10m/1m	A (2), C

sure. A total of 24 tubes, each 1 to 1.5 metres long, has been tested, Table 3. The internal diameter is 150 mm, wall thickness is nominally 5 mm. The first phase of the project, a series of 12 creep tests on tubes without internal liner and loaded with free ends ($\sigma_z = 0$) at different temperatures and pressures, was described by Baizeau et al. at DURACOSYS 95 (1995). A description of the second phase, which involved tests on tubes with internal liners, will now be presented. The aims of these tests were to establish the influence of creep on residual stiffness, to compare the free end and fixed end loading conditions, to examine how damage affects subsequent creep behaviour and to see how bonded and mechanical joints behaved.

Long term testing of tubes requires a number of precautions to be taken. First, the temperature and pressure must be continuously monitored. An extract from the temperature log is shown in Figure 8.

The 20°C chamber was installed in an existing test laboratory with temperature and humidity regulation and posed no particular problems. The 40°C and 60°C chambers used heated air blowers with three thermocouples per chamber to regulate temperature and maximum (exceptional) variations of 8°C were recorded at 60°C. The pressure was also recorded throughout the tests, and could be regulated to

within ±2 bars at the highest pressure. In addition to following the pressures and temperatures it is also necessary to record the strain response by strain gauges throughout the creep and recovery cycle. This requires a reliable data logger which can handle over 100 strain, temperature and pressure channels. A *Scorpio* logger was used here, which allows low current impulsion strain gauge excitation, minimising local heating problems. In addition a dummy gauge was bonded on an identical tube section in each temperature chamber to allow strain gauge drift to be measured. Photo 1 shows one of the creep chambers.

Pressure tests were also performed on the tubes before and after this cycle. These involved subjecting tubes to low pressure cycles with two loading conditions, fixed and free ends (i.e. with and without axial stress), and allowed the influence of the creep cycle on residual stiffness to be assessed.

3.1 *Creep results*

Given the large amount of data collected, there is only space here to show some examples of the creep behaviour. Figure 9 shows strains measured at the start and end of creep and recovery.

At first sight these creep strains appear low, lower than creep under free end loading conditions reported for Phase 1 of the project, Figure 10. The

Photo 1. 20°C creep chamber with tubes of type A, B and C.

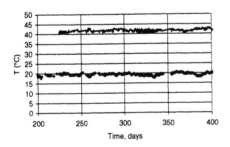

Figure 8. Extract from temperature log.

148

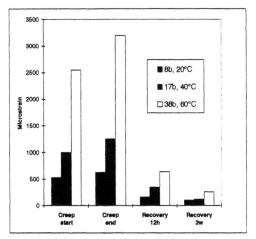

Figure 9. Creep hoop strains (10 months 38 bars, 17 months 8 & 17 bars) and recovery (1 month)

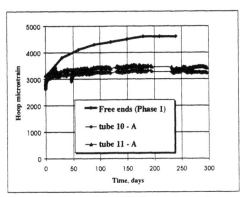

Figure 10. Creep of two fixed end tubes and one with free ends, 60°C 38-40 bar.

biaxial stress state apparently reduces the hoop deformation, and a ply stress analysis suggests shear stresses are higher in the ply for free end conditions (LeBras 1995), which may explain the difference. From a practical viewpoint free end loading conditions are subject to friction due to the need for a joint between the sleeve and the tube so fixed end load conditions are easier to control experimentally.

3.2 Influence of damage

Different levels of damage were introduced into three tubes by loading under internal pressure. The procedure for these tests was the following:
First, the tubes were pressurised at EDF. The pressures (hoop stresses) chosen were:
D1 19 bars (35 MPa), just above the first acoustic emission recording, at the elastic limit.
D2: 38 bars (76 MPa), an intermediate damage state

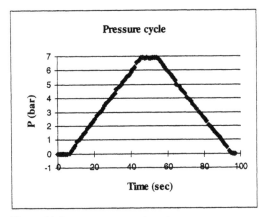

Figure 11. Low pressure cycle before and after creep.

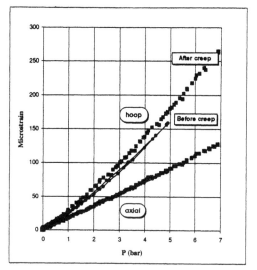

Figure 12. Strain versus pressure before (solid line) and after creep cycle, tube D3.

D3: 76 bars (130 MPa), corresponding to extensive damage.
These values should be considered in relation to the pressure at failure by leakage (weeping) of these cylinders at 197 MPa (LeBras 1995).
They were then shipped to IFREMER and low pressure cycles (up to 7 bars) were performed to establish their apparent initial stiffness in the damaged state, Figure 11. (Note that the initial 'damaging' pressure cycle resulted in some small permanent hoop strains, which are not taken into account in the subsequent stiffness determination as new strain gauges were bonded at IFREMER so it appeared that the D2 and D3 damaged tubes were stiffer than the undamaged tubes). The three tubes were then loaded in creep at 40°C and 17 bars for an 18 month cycle, including 17 months of creep and 1 month's

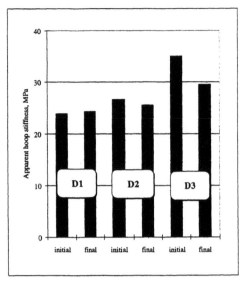

Figure 13. Apparent hoop rigidities (hoop stress/hoop strain), damaged tubes, before and after creep & recovery cycle.

recovery. The presence of the damage affected the creep behaviour, increasing the creep strains slightly and causing more scatter in strain readings. At the end of this period a second low pressure cycle was applied to determine apparent residual stiffness.

The comparison of the strain gauge response at low pressure before and after creep showed a loss in apparent stiffness for the D2 and D3 tubes, Figures 12 and 13.

The tube is less stiff in the hoop direction after creep, while the axial response is unchanged (curves superposed).

3.3 Tube assemblies

The bonded and mechanically-joined assemblies behaved in a satisfactory way throughout the tests, and no leaks were noted. The presence of the mid-section joints did not affect the strain gauge recordings. Displacement transducers measuring relative axial displacements of the two half tubes showed no movement.

4 CURRENT AND FUTURE RESEARCH

Research work at IFREMER and EDF is continuing in this area. Current projects are concentrated in three areas.

First, more work is being carried out on the resin behaviour in order to quantify resin degradation kinetics and the influence of hardeners on long term behaviour. Interface studies are also underway.

Second, studies of tolerance of tubes and bonded joints to defects, delaminations and impact damage are being performed. An example is the application of fracture mechanics tests to curved, multi-layer specimens (Ozdil 1999).

The third area is NDT, particularly applied to bonded joints. The use of acoustic emission in-situ is being studied.

Finally, the long term behaviour of composite cylinders under external pressure is also being examined with laboratory tests and deep sea measurements.

5 CONCLUSIONS

This paper gives a brief overview of some of the results available from aging and creep tests performed within a collaborative project between IFREMER and EDF over the last 10 years. The development of a resin degradation model has enabled predictions to be made of the useful lifetime of composite panels and then applied to tubes. Creep tests have shown that creep strains in the hoop direction are lower when fixed end conditions (biaxial stress) are applied than for the free end conditions used in previous studies. Damage introduced by a pressure overload is shown to increase creep strains slightly and apparent stiffness measured before and after tests is shown to decrease. The two assembly systems studied behaved satisfactorily.

REFERENCES

Bai J et al., 'Mechanical behaviour of ±55° filament wound glass fibre/epoxy tubes, Parts I-IV, Comp. Sci & Tech., 1996.

Baizeau R, Davies P, Choqueuse D, LeBras J, 'Evaluation of the integrity of composite tubes', DURACOSYS 95, ed Cardon, Fukuda & Reifsnider, Balkena 1996, p225.

Bax J, 'Deformation behaviour and failure of glass fibre reinforced resin material', Plastics & Polymers, 1970, Feb. p27.

Chiou P-L, Bradley WL 'Moisture-induced degradation of glass/epoxy filament wound composite tubes', J. Thermopl. Comp., 9, April 1996, p118.

Croquette J,'Parquic JC, Forestier JM, Dufour X, Proc. IFREMER conf., Composite materials for Marine Applications, 1992, p412.

de Bruijn JCM, van den Ende CAM, 'GRP pipes are safer than steel ones', Reinforced plastics Feb. 1996 , p40.

Ghorbel I, Spiteri P, 'Durability of closed end pressurized GRP pipes under hygrothermal conditions', Parts I & II, J. Comp. Mats., 30, 14, 1996, p1562 & p1581.

Gibson AG, Chapter 11 in 'Composite materials

in marine structures', ed Shenoi & Wellicome, Cambridge Ocean Technology, 1993.

Spencer B, Hull D, 'Effect of winding angle on the failure of filament wound pipe', Composites 9, 1978, p263.

Kies JA, Bernstein H, 'Recent advances in glass fiber reinforced plastic rocket motors', Proc 17th Annu conf. Reinf. Plastics, 1962 6-B.

LeBras J, 1995, Proc. AMAC-JST meeting, 24 January Besançon.

LeMoal P, 1993, PhD Université de Franche Comté

Maire J-F, 1992, PhD Université de Franche Comté

Mallinson JH, 'Corrosion resistant plastic composites in chemical plant design', Dekker 1988.

Ozdil F, Carlsson LA, Davies P, 1999, 'Characterization of delamination toughness of angle-ply glass/epoxy cylinders', Proc ICCM12, Paris, paper 242.

Perreux D, Varchon D, LeBras J, 'The mechanical and hygrothermal behaviour of composite pipes', Proc. ENERCOMP 95, Montréal, p819.

Suri C, 1995, PhD Université de Franche Comté

Thiebaud F, 1994, PhD Université de Franche Comté

Williams JG, Silverman SA, 'Composites Technology used onshore with synergy to offshore applications', OTC Paper 11062, 1999.

Recent Developments in Durability Analysis of Composite Systems, Cardon, Fukuda, Reifsnider & Verchery (eds)
© 2000 Balkema, Rotterdam, ISBN 90 5809 103 1

The influence of moisture on the physical aging response of epoxy: Experimental results and modeling considerations

W. H. Han & G. B. McKenna
Polymers Division, NIST, Gaithersburg, Md., USA

ABSTRACT: We have investigated the hygrothermal effects in thin epoxy films (c.a. 50 μm thick) using mass uptake, swelling, and uniaxial creep compliance measurements inside a dew formation-free transparent chamber capable of arbitrary relative humidity and temperature controls. We show that the physical aging behavior in humidity down-jumps is equivalent to temperature down-jumps such that the time-aging time superposition principle holds for both cases. Furthermore, we demonstrate that moisture-jump experiments exhibit memory and asymmetry of approach similar to those found in temperature-jump experiments. A modified form of the TNM-KAHR model of structural recovery is proposed to account for the observations.

1. INTRODUCTION

In previous work [1] we presented evidence that a relative humidity jump experiment results in mechanical responses that are similar to those obtained in temperature jumps. Hence, we were able to make the argument that the models of structural recovery or physical aging that govern a glass forming material's viscoelastic responses should be able to extend to moisture effects as well. In the following paragraphs we briefly review that work and present new experimental results that further substantiate those observations. We also summarize the TNM-KAHR [2-5] models and present an extension of the models that we propose to account for the humidity induced structural recovery or physical aging response. We present preliminary results needed to obtain the TNM-KAHR material parameters for moisture induced structural recovery.

2. HUMIDITY-TEMPERATURE "EQUIVALENCE"

Conceptually it is possible to think that changes in the moisture content of a glassy polymer have much in common with changes in temperature because of the well known effect of moisture as a plasticizer [6]. Because water depresses the glass transition temperature, one could imagine that an isothermal change in moisture content could be treated as a change in temperature, i.e., a change in the distance from the glass transition temperature T_g.

Previously [1], we showed that a moisture jump experiment in an epoxy material produced results similar to a temperature jump experiment to the same final conditions. Figure 1 shows the comparison of the aging time shift factors a_{te} vs aging time t_e after a humidty or temperature jump to the same final temperature and humidity conditions. Clearly, the responses are similar. The interpretation of such behavior is given schematically in Figure 2 where we depict a 3-dimensional Volume-Temperature-Humidity surface. If the mobility of the material depends on the volume, then, performing a RH-jump from above the RH-induced glass transition H_g to below it is equivalent to performing a temperature-jump from above to below T_g.

3. EXTENDED TNM-KAHR MODEL

The temperature history dependence of the relaxation response in glassy materials is often analyzed in terms of the so-called TNM-KAHR [2-5] model of structural recovery. The time evolution of the departure δ from the equilibrium volume is written in the form of a convolution integral:

$$\delta(z) = -\Delta\alpha \int_0^z R(z-z') \frac{dT}{dz'} dz'$$

where (1)

$$z = \int_0^t \frac{d\xi}{a_T a_\delta}$$

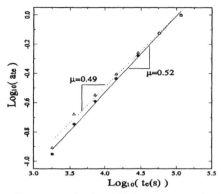

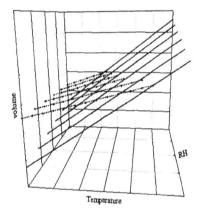

Figure 1. Aging time shift factors vs aging time for an epoxy glass subjected to RH-jump or T-jump conditions. Filled circles are for a RH-jump from RH=0.75 to RH=0.50 at T=68 °C. Open triangles are for T-jump from 85 °C to 68 °C at RH=0.50.

Figure 2. Schematic of Volume-Temperature-RH surfaces for equilibrium and glassy states of a moisture sensitive polymer glass. (See text for discussion).

and a_T and a_δ are shift factors for temperature **T** and departure from equilibrium δ. The departure from equilibrium is defined as $\delta=(v-v_\infty)/v_\infty$ where **v** and v_∞ (or v_{liq}) are the volume at time **t** and in equilibrium, respectively, for a given temperature **T**. $\Delta\alpha$ is the change in the coefficient of thermal expansion upon traversing the glass transition temperature. **R(t)** is a viscoelastic response function for the structural recovery and **z** is the reduced time. Similar equations can be written for the enthalpy as well, but we limit ourselves here to the volume formulation. We add that the concept of physical aging [7] results in the viscoelastic response (creep or relaxation) times shifting with the same a_T and a_δ shift factors as the structure. The shift factors have been represented by [2]:

$$\frac{\tau_i(T,\delta)}{\tau_{i,r}} = a_T a_\delta = e^{-\theta(T-T_r)} e^{-\frac{(1-x)\delta\delta}{\Delta\alpha}} \tag{2}$$

where the first exponential term is a_T and the second is a_δ. $\tau_i(T,\delta)$ is the relaxation time at the relevant values of temperature and structure and the $\tau_{i,r}$ refers to the relaxation time at the reference state, generally taken for $T_r=T_g$ and δ=0. The parameter **x** is a partition parameter $(0 \leq x \leq 1)$ that determines the relative importance of temperature and structure on the relaxation times. The parameter θ is a material constant that characterizes the temperature dependence of the relaxation times in equilibrium. KAHR used $\theta \approx E_a/RT_g^2$ where E_a is an activation energy and **R** is the gas constant.

In the case of the temperature-jump experiment, the parameters to be determined are **x**, θ (or E_a), **R(t)** and $\Delta\alpha$. This can be a non-trivial task, though procedures for parameter estimation have been developed [8,9]. To extend this sort of formalism to the case of the humidity-jump experiment, we assume that the effects of volume and moisture are additive and, then, equation 1 can be modified as follows:

$$\delta(z) = \int_0^z dz' \left[\Delta\alpha_T R_T(z-z') \frac{dT}{dz'} + \Delta\alpha_h R_h(z-z') \frac{dh}{dz'} \right] \tag{3}$$

$$where \quad z = \int_0^t \frac{d\xi}{a_T a_\delta a_h}$$

and the aging problem now requires not only the temperature factors to be obtained but also appropriate parameters for the humidity terms. The reduced time now includes a shift factor a_h due to the changing moisture in the material.

To first approximation, one would anticipate that $R_T(t)=R_h(t)$ and the mechanical test data leading to Figure 1 support this contention [1]. The shift factors a_T and a_h can be obtained by performing isothermal and constant humidity experiments, as appropriate. The determination of a_δ requires implementation of procedures similar to those used in temperature experiments [8,9]. The value of the change of volume with humidity $\Delta\alpha_h$ at H_g is more problematic, as will be seen in the results section. However, the formalism offers the possibility of describing the change of the relaxation response of a material when both temperature and humidity are changed. In the following we present data typical of that required to determine the function $\Delta\alpha_h$. We also present further results from RH-jump experiments that justify the continued examination of equation 3 to describe moisture induced aging effects.

4. EXPERIMENTAL [10]

4.1 Materials

Diglycidyl ether of bisphenol A(DGEBA, DER 332, Dow Chemical, USA) epoxide monomer was mixed with a curing agent, amine terminated poly(propylene oxide) (Jeffamine, T403, designation of Texaco Chemical Company) in a stoichiometric ratio. The mixture was degassed for 20 min at 23 °C, then cast into smooth TEFLON molds with dimensions 6 mm X 0.06 mm X 210 mm. To prevent bubble formation, thin samples were cured at high pressure (2MPa) and at temperature of 100 °C for 16 hours. The samples were then slowly cooled overnight. The glass transition temperature of this material measured by standard DSC is 72.3 °C [11].

4.2 Moisture Uptake Measurements

The in-situ moisture uptake measurements were done in a double layer transparent hygrothermal chamber built in this laboratory [1]. The temperature of the mixture of air and moisture inside the hygrothermal chamber is controlled within ± 0.1 °C. The humidity was controlled to within ± 0.5 % relative humidity (RH). In-situ measurements of weight changes of the epoxy films were carried out using cathetometers and quartz springs which allow us to detect ± 1 µg changes. Volume change was monitored by using a thin epoxy strip which was hung freely. To this was attached a tiny magnetic core for a hermetically sealed LVDT, the voltages from which were read via a PC. The expanded uncertainty in the measurements of the relative volume change is estimated to be 1.25×10^{-5} based on sample geometry and the response characteristics of the LVDT [1].

4.3 Creep Measurements

Subsequent to a humidity or temperature change, aging experiments were performed following Struik's protocol [7]. The thin films were loaded in a miniature tensile machine maintained at the same temperature for creep compliance tests at different aging times. The apparatus uses linear differential transformers (LVDTs) (Schaevitz Inc.) connected to a microprocessor based signal conditioner (MP/2000, Schaevitz Inc.) and sets of weights which can be loaded and unloaded by a geared lift powered by a high humidity temperature resistant D.C. motor (Motor Technology Inc). An AD/DA converter card installed in an IBM PC receives the creep compliance voltages from the LVDT conditioners, and send out signals to a motor controller for the motor operation. Repeat experiments provide an estimated expanded

uncertainty in the measurements of approximately 5 % of the measured load. The uncertainty of a single set of aging experiments is considerably less than this.

5. RESULTS

5.1 Coefficients of Thermal and Moisture Expansion

At a constant cooling rate of 0.1 °C/min, the length changes of the thin epoxy strips were measured in constant humidity conditions. Similarly,

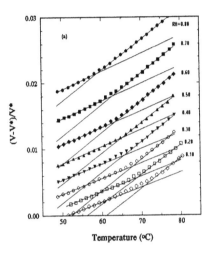

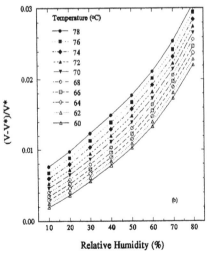

Figure 3. Epoxy-Moisture Effects: **(a)** Relative volume vs Temperature at constant RH. **(b)** Relative volume vs RH at constant temperature. V* is the volume at 55 °C and RH=0.20. (See text for discussion).

isothermal measurements were performed by reducing the RH at a rate of 0.02/min. From the sample strip length data, assuming isotropic dimensional change, we directly calculate the corresponding volume change. In independent experiments of mass uptake, the diffusivity of the epoxy sample was experimentally measured to be approximately $1.6 \times 10^{-8} cm^2/s$ using the Fickian diffusion equation [12] for the lowest temperature (T=45 °C). For this value of diffusivity and the sample thickness (60 μm), the diffusion half time is less than 2 min, which is significantly shorter than the experimental time scale. Figures 3a and 3b depict the volume-temperature response of the epoxy at different relative humidities and the volume-RH response at different temperatures, respectively. Clearly the T_g decreases with decreasing RH, however, the behavior of the volume as a function of RH does not lead to a clear definition of a humidity induced glass transition H_g. Hence, while it is possible to determine $\Delta\alpha_T$ for input into equation 3, the value of $\Delta\alpha_{\prime\prime}$ is less clearly ascertained from the data. Figure 4 indicates this to be the case. In 4a we plot the value of the coefficient of thermal expansion from the constant humidity experiments both above and below T_g. It can be seen that the difference $\Delta\alpha_T$ between the glassy and liquid values is constant within the experimental error. Fitting polynomials to the low humidity and high humidity data of Figure 3b provides a means to estimate $\Delta\alpha_{\prime\prime}$. The result of such a treatment for the estimated behavior of $\Delta\alpha_{\prime\prime}$ is shown in Figure 4b where it is clear that the isothermal values of $\Delta\alpha_{\prime\prime}$ are functions of both temperature and humidity.

5.2 Asymmetry of Approach Experiments

One other point to be made here arises from volumetric measurements subsequent to RH-jumps. Previously we had shown that the retardation times in RH-jumps followed a similar behavior to temperature-jumps (see Introduction). One of the classic T-jump experiments that show the non-linearity of structural recovery is the so-called asymmetry of approach. Here we illustrate the first asymmetry of approach results from experiments in RH-jump conditions.

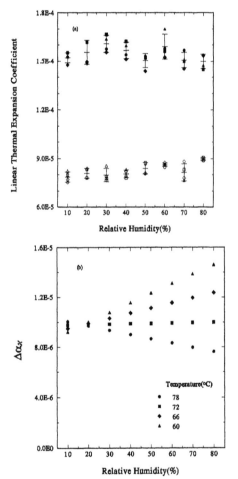

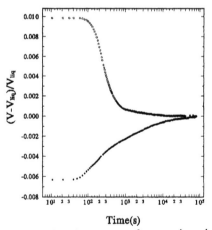

Figure 4. Expansion coefficients in epoxy: **(a)** Thermal expansion in the glassy (open symbols) and rubbery (filled symbols) states vs RH. Points represent individual experiments. Error bars represent single standard deviation from the repeat experiments. **(b)** Estimates of the parameter $\Delta\alpha_{\prime\prime}$ in equation 3 from data of Figure 3b at different temperatures. (See text for discussion.)

Figure 5. Asymmetry of approach: volume departure from equilibrium vs log time following relative humidity jumps from RH=0.75 to RH=0.50 (upper curve) and from RH=0.28 to RH=0.50 (lower curve) for an epoxy glass at T=65.5 °C. (See text for discussion).

In the T-jump experiments of Kovacs [12], the sample was tested in up- and down-jumps to a final temperature T_0. The magnitude of the up and down-jumps was the same and non-linear response was deduced from the observation that the up-and down-jump δ vs log (time) curves were not symmetric about the $\delta=0$ line. For the RH-jump experiments, we take a similar approach, expecting the responses from nearly similar values of ΔRH in magnitude would be mirror images in the up- and down-jump conditions if the response were linear. Figure 5 shows first results from such an experiment in which a sample was equilibrated at RH=0.75, T=65.5 °C and jumped to RH=0.50, T=65.5 °C and the volume recovery followed. Also plotted in the figure is the result for a sample equilibrated at RH=0.28, T=65.5 °C and jumped to RH=0.50, T=65.5 °C. The figure shows clearly that the response in the up-jump experiment is much slower than in the down-jump. This behavior is very similar to what was observed by Kovacs [12] for T-jumps and rationalized by equations 1 and 2. The similar result shown in Figure 5 provides further rationalization for an equation of the sort of equation 3 in which the humidity history has a similar impact on glassy behavior as does the thermal history. Further work will establish the level of reproducibility and uncertainty in such measurements.

6. DISCUSSION

Clearly, the behavior seen in humidity-jump and temperature-jump experiments is similar in nature. The plasticizing effect of the moisture in the sample leads to a depression of the glass transition. Then, in an isothermal experiment, a change in moisture content (induced by a change in RH) is the equivalent of a change in temperature because the 'distance' from the T_g has been changed. Equation 3 partially quantifies this observation.

At the same time, the formalisms that have been explored for determining the material parameters for temperature-jump experiments, have not been developed for humidity-jump experiments. Models for changing moisture content and its effect on T_g [6,14,15] do exist. To our knowledge, however, equation 3 has not been previously proposed. That being said, the determination of the moisture or RH parameters in equation 3 that are equivalent to the temperature parameters for equation 1 (given in equation 2) are not so straight-forward. In this work, values of $\Delta\alpha_{,_{IJ}}$ have been determined. Further work is required to determine how well the other parameters can be obtained from the RH-jump experiments themselves.

7. CONCLUSIONS

An equivalence between humidity-jump and temperature-jump experiments has been demonstrated. Mechanical experiments show that physical aging occurs after RH-jump experiments in a fashion similar to classical T-jump experiments. An equation building on the TNM-KAHR formalism has been proposed to take into account the moisture (RH) history of the sample. Experiments were performed in isothermal conditions to obtain the values of the coefficient of moisture (RH) expansion and its change $\Delta\alpha_{,_{IJ}}$ at H_g. The current results suggest that $\Delta\alpha_{,_{IJ}}$ is not a constant and further work is required to implement the model. Finally, the first experiments that exhibit an asymmetry of approach in RH-jump conditions were presented, further justifying continued exploration of the value of extending the TNM-KAHR formalism to plasticizing environments.

8. ACKNOWLEDGEMENTS

The major portion of this work was carried out in the laboratories of the Polymers Division at NIST. W.H. Han is grateful to NIST for providing a guest scientist position.

9. REFERENCES

1. A) Han, W.H. and McKenna, G.B. (1997) SPE ANTEC, II, 1539. B). Han, W.H. and McKenna, G.B. (1997), Proc. NATAS, **25**, 382.

2. Kovacs, A.J., Aklonis, J.J.,Hutchinson, J.M., and Ramos, A.R. (1979) Journal of Polymer Science, Polymer Physics Edition, **17**, 1097.

3. Tool, A. Q. (1946) Journal of Research of the National Bureau of Standards (USA) , **37**, 73.

4. Narayanaswamy, O.S.(1971) Journal of the American Ceramic Society, **54**, 491.

5. Moynihan, C.T., Macedo, P.B. Montrose, C.J., Gupta, P.K. DeBolt, M.A., Dill, J.F., Dom, B.E., Drake, P.W., Esteal, A.J., Elterman, P.B., Moeller, R.P., Sasabe, H., Wilder, J.A. (1976) Annals of the New York Academy of Sciences, **279**, 15.

6. Knauss, W.G. and Kenner, V.H. (1980) J. Appl. Phys., <u>51</u>, 5131.

7. Struik LCE (1978) *Physical Aging in Amorphous Polymers and Other Materials*, Elsevier, Amsterdam.

8. Hodge, I.M. (1994) Journal of Non-Crystalline Solids, **169**, 211.

9. Scherer, G.W. (1986) *Relaxation in Glass and Composites*, Wiley, New York.

10. Certain commercial materials and equipment are identified in this paper to specify adequately the experimental procedure. In no case does such identification imply recommendation or endorsement by the National Institute of Standards and Technology, nor does it imply necessarily that the product is the best available for the purpose.

11. Lee,A. and McKenna,G.B.(1988) Polymer, **29**, 1812.

12. Vieth, W.R. (1991) *Diffusion In and Through Polymers: Principles and Applications*, Hanser, New York.

13. Kovacs, A.J. (1963) Fortschritte der Hochpolymeren-Forschung, **3**, 394.

14. McKenna GB (1989) in *Comprehensive Polymer Science: Volume 2 Polymer Properties,* ed. by C. Booth and C. Price, Pergamon, Press, Oxford, 311-362.

15. Gibbs, J.H. and DiMarzio, E.A. (1958) Journal of Chemical Physics, **28**, 373.

Interactions between moisture and flexural fatigue damage in unidirectional glass/epoxy composites

A. Chateauminois
Ecole Centrale de Lyon, UMR IFoS 5621, Ecully, France

ABSTRACT: The interactions between hygrothermal ageing and mechanical damage are reviewed in the context of the three-point bending fatigue of unidirectional glass/epoxy composites. The effects of the hygrothermal history of the composite on its endurance properties are examined in the light of the stress corrosion cracking processes occurring in the glass reinforcement during water ageing. From a detailed analysis of the damage micro-mechanisms during the early stages of the fatigue life, the decreases in the endurance properties of the composite have been rationalised within the framework of fatigue-life diagrams. Some of the implications of the interrelations between chemical degradation and fatigue damage on the development of accelerated ageing procedures are also considered. The relevance of the overall water uptake to quantify the extent of the hygrothermal damage is especially questioned.

1 INTRODUCTION

The durability of polymer matrix composites in hot/wet environments involves several physico-chemical and mechanical damage processes occurring at very different time and length scales. From matrix plasticization (or hydrolysis) up to crack propagation, the characteristic length scales of the phenomena can vary by several orders of magnitude. In a similar way, the kinetics of the chemical and mechanical degradation mechanisms can range from minutes to years. The prediction of the long-term behaviour of the composites is further complicated by the potential interactions between the mechanical and physico-chemical damage processes. In this context, the development of reliable durability models can only be envisaged if the main degradation mechanisms and their interrelations are identified through a multidisciplinary analysis. Such an approach is also a prerequisite for the much-needed development of relevant accelerated ageing procedures, which implies the knowledge of the activation energies of the individual mechanical and physico-chemical degradation processes.

In this paper, the development of such a multidisciplinary analysis is considered in the context of the flexural fatigue of unidirectional glass/epoxy composites subjected to water ageing. For that loading configuration, the fracture of fibres may be regarded as the primary event which controls the process of damage development and eventual accumulation to failure. The delayed failure of the fibres under fatigue loading is largely controlled by the sub-critical growth of surface defects. In the case of bulk glass, strong interactions between crack propagation and environment have been largely identified and described within the framework of stress corrosion cracking (SCC) concepts (Wiederhorn & Bolz 1970, Wiederhorn 1978). Accordingly, the sub-critical crack growth rate of the surface defects is a function of both the applied stress and the environment (temperature, humidity, pH). In addition to the activation of SCC mechanisms, water exposure may also induce some changes in the statistical population of surface defects, due to the creation of flaws by an ion exchange mechanism between the glass and the surrounding media (Charles 1958, Metcalfe & Schmitz 1972). These chemical reactions are also known to generate detrimental tensile stresses on the glass surface.

These various processes can be replaced in the context of glass/epoxy composites exposed to humidity. During water sorption into the material, the fibres embedded into the matrix may experience changes in their physico-chemical environment by virtue of the accumulation of moisture at the interface. Low molecular weight species resulting from the hydrolysis of the matrix can also be leached at the interface, where they will contribute to modify the local value of the pH and to weaken the load bearing capacities of the glass fibres.

The fatigue analysis of unidirectional composite therefore appears as a peculiarly relevant tool to investigate some of the features of the coupling

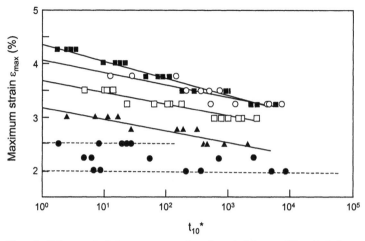

Figure 1. S-N curves of a S-glass epoxy composite under static fatigue conditions (R=1, three-point bending). (■) unaged; aged 100 days at (O) 30°C, (□) 50°C, (▲) 70°C and (●) 90°C. t_{10}^* is the time to 10% stiffness loss normalised to the loading time.

between mechanical damage and hygrothermal ageing in polymer matrix composites. On the basis of different investigations we have carried out over the past few years, these interactions are reviewed in this paper, with a special emphasis on their implications regarding the definition of accelerated ageing procedures.

2 CHARACTERISTIC TIME SCALES

The way hygrothermal ageing and mechanical damage can interact strongly depends upon their relative characteristic time scales.

The driving force for the physico-chemical ageing processes is the sorption of water within the macromolecular network or at the interface. Although complex interactions between the water molecules and the macromolecular network are involved, the associated diffusion times are known to be thermally activated and to increase approximately as the square of the thickness of the structure (Schen & Springer 1981).

The times associated to the nucleation and the propagation of a fatigue damage depends on the test frequency, the strain ratio R and the strain/stress level.

By varying these ageing and fatigue parameters, many practical situations may be encountered between the following two extremes:

(i) If the time scales for water diffusion are much larger than those for damage nucleation and growth, no significant decrease in the lifetimes can be expected from defects created hygrothermally in the bulk material. Only local interactions between water and the crack tip can be envisaged. Such processes involve the capillary flow of water along the cracks,

i.e. characteristic times which are much shorter than those involved in water sorption into the macromolecular network. This situation may be encountered for example if low temperatures are associated with high strain levels.

(ii) If water saturation occurs before the propagation of the mechanical damage, significant changes in the lifetimes can result from defects which are hygrothermally induced through the thickness of the material.

In this paper, only the latter situation will be examined. In service conditions, it may be encountered for high temperatures/low strains combinations. An experimental methodology to study this kind of interactions consists in performing fatigue experiments using preconditioned specimens. During the preliminary ageing stage, coupons are saturated with water for different relative humidity, temperature and exposure times. They are subsequently tested in fatigue during a shorter time, while preventing the specimens from desorption.

3 MACROSCOPIC FATIGUE BEHAVIOUR

3.1 *Effects of the hygrothermal history*

At the macroscopic scale, the usual route to investigate the effects of ageing on the fatigue behaviour is to determine the residual lifetimes of water saturated specimens. Accordingly, the relative moisture uptake, M_t, of a composite material is often considered as a relevant parameter to describe its ageing state before mechanical testing. From a chemical point of view, the extent of a hydrolysis reaction depends on the time and the temperature as well as on the water concentration itself. In a situation where the fatigue response is dependent upon chemically induced damage, the residual lifetimes will therefore depend upon the whole hygrothermal history of the specimens rather than upon their final moisture uptake before the fatigue experiments. This can be clearly demonstrated by considering separately the effects of the temperature, the immersion time and relative humidity on the endurance properties of aged composites.

The results reported in Figure 1 give an example of the residual lifetimes of an aged composite under a three-point bending static fatigue loading (i.e. relaxation; strain ratio R=1). The specimens were exposed to water for the same amount of time at four different temperatures. For this high-Tg DGEBA/DDM matrix system, a quasi-fickian sorption behaviour was observed and the water saturation levels reached before fatigue testing were roughly independent on the ageing temperature (Fig. 2). Despite these similar water uptakes, the fatigue properties were strongly dependent on the ageing temperature. These results clearly demonstrate that M_t cannot be considered as a relevant parameter to describe the damage state of the composite after water ageing.

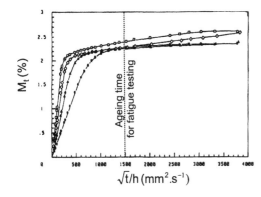

Figure 2. Sorption curves of a DGEBA/DDM composite showing a quasi-Fickian behaviour at short times. Immersion at ($*$) 30°C; ($\triangle$) 50°C; ($\diamond$) 70°C; ($\square$) 90°C. h is the specimen thickness.

In the specific case of this hydrolysis resistant system (Chateauminois et al. 1995), the degradation of the fibre/matrix interface can account for a significant part of the temperature dependence. Thermo-gravimetric analysis of the aged composite (Chateauminois et al. 1994) revealed that exposure to 70°C and 90°C resulted in enhanced desorption rates, which were attributed to the occurrence of capillary flows at debonded interfaces. Due to this interfacial degradation, more liquid water may be available on the fibre surface for chemical attack of the silica network. As a result, additional defects will be created on the fibre surface by thermally activated chemical processes. In addition, stress corrosion

mechanisms will also be enhanced during the mechanical loading of the fibres, due to the increased amount of water available at the crack tip.

A similar dependence upon hygrothermal history has also been observed in the case of dynamic fatigue (R=0.1). In Figure 3, the ageing strategy consisted in exposing the materials to liquid water at different temperatures and for various exposure times in order to obtain the same water content (about 1.5%). Once again, the strong differences in the residual fatigue properties indicated that the amount of hygrothermal damage was not properly quantified by the water uptake. In that situation, the increase in the exposure time was more detrimental than an increase in the ageing temperature. This suggests that the endurance properties of a saturated composite could evolve with time, even if there is no corresponding changes in M_t.

Similar fatigue experiments were also carried out after exposure to relative humidity rather than to liquid water. The resulting effects of moisture were found to be much less detrimental to the fatigue properties, although the water uptakes were not always significantly lower than in immersion (Vauthier 1996).

For this composite system, a large part of these effects can be explained by the strong interrelations between the chemical hydrolysis of the epoxy matrix and the degradation of the fibre properties. The origins and the implications of these mechanisms will be reviewed in the next section.

3.2 Coupling effects between mechanical and chemical degradation

The occurrence of a decrease in the cross-link density of the epoxy network due to hydrolysis can be conveniently investigated by D.M.T.A. If the analysis is carried out using wet specimens, the measured lowering of the thermo-mechanical properties integrate the effects of both reversible plasticization and irreversible chemical hydrolysis. The D.M.T.A. experiments can also be carried out using specimens that were aged and subsequently dried at room temperature. In that situation, most of the effects of water plasticization are eliminated and the observed drop in the thermo-mechanical properties can only be attributed to the decreases in the network cross-link density. The latter can be quantified from the depression in either the glass transition temperature or the rubbery modulus (Chateauminois et al. 1990). In that sense, D.M.T.A. can be considered a useful tool to discriminate between plasticization and hydrolysis in epoxy networks.

Figure 4 shows an example of such an approach in the case of the composite used for the dynamic fatigue experiments presented in Figure 3. This graph indicates that a significant part of the decrease in the glass transition temperature of the aged com-

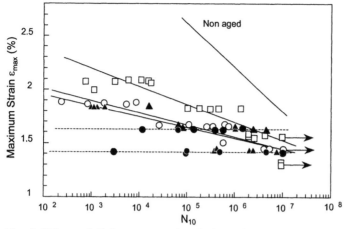

Figure 3. S-N curves of a E-glass epoxy composite under dynamic fatigue conditions (R=0.1, three-point bending).
aged (●) 180 days at 50°C; (○) 77 days at 60°C, (▲) 51 days at 70°C (□) 10 days at 90°C. N_{10} is the number of cycles to 10% stiffness loss.

posite is due to the chemical degradation of the macromolecular network. The epoxy matrix was a DGEBA/DICY system, where a substantial part of the DICY hardener particles remained unreacted after curing. As reported by other authors (Kasturiarachi & Pritchard 1984, Shah et al. 1985), these residual particles are soluble into water and can be leached in the ageing environment. Moreover, a substantial increase in pH of the ageing media (up to 10) can occur due to the formation of ammonia from the hydrolysis of the amine groups of the DICY (Vauthier et al. 1996). In addition to promoting the hydrolysis of the epoxy matrix, high values of the pH are known to induce an extensive hydrolysis of the silica network (Mazer & Walther 1994). Most of the reduction in the fatigue properties can therefore be attributed to the enhanced chemical degradation of the glass fibres.

These processes can account for the observed difference in lifetimes after exposure to either relative

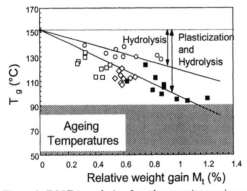

Figure 4. D.M.T.A analysis of aged composite specimens. Aged at (□) 50% R.H., (◇) 80% R.H., (■) immersion; (○) aged and redried specimens.

humidity or liquid water. In the case of the water immersion, the chemical changes in the ageing media resulting from the leaching of the hydrolysis products will promote further chemical degradation of the composite. In such a situation, it becomes necessary to pay some additional attention to the ageing procedure, in order to ensure a reproducible physico-chemical environment for the specimens. This means that the volume of water, the amount of material and the total specimen surface exposed to the ageing solution must be kept constant. The extrapolation to real service conditions of such confined ageing tests remains, however, questionable. It may be more realistic to envisage a water circulation in the ageing container throughout the conditioning time.

The chemical degradation of the epoxy network is generally associated to non-Fickian sorption behaviour with no apparent water saturation level. In such a situation, the overall weight change M_t can include several different contributions:

(i) the sorption of water in the macromolecular network. Contrary to the assumptions of Fickian diffusion, the kinetics of water ingress is then coupled with chemical reactions. Although Langmuir-type laws have sometimes been used to describe non-Fickian behaviour (Carter & Kibler 1978, Gurtin & Yatomi 1979, Suri & Perreux 1995), it should be pointed out that they are totally inadequate in that context, because they do not take into account the change in the diffusion characteristics of the network after chemical degradation. To our best knowledge, there is no model presently available to describe such effects in epoxies.

(ii) The filling of hygrothermally induced micro or macro-cracks by the liquid water.

(iii) The leaching of the low molecular weight compounds resulting from the hydrolysis of the

epoxy network or the dissolution of unreacted products.

Regarding the fatigue behaviour, one of the main implications of the chemical ageing processes is the development of an heterogeneous degradation state. By virtue of the long transient steps involved during water sorption, the hygrothermally-induced damage will be distributed throughout the specimen thickness, even if a saturated state is considered for mechanical testing. The resulting property gradients can be considered as the equivalent - at the microscopic scale - of the hygrothermal history effects reported in the previous section. This heterogeneous nature of the water-induced damage strongly questions the validity of the extrapolation of fatigue tests on aged specimens to various loading configurations and geometrical shapes, even if the damage mechanisms remain the same. These limitations mainly arise from the complex interactions between the stress and the ageing profiles throughout the material, which determine the average residual properties measured at the macroscopic scale.

3.3 Quantitative analysis of fatigue losses

A quantitative description of the fatigue properties of the composites after ageing can be provided within the framework of a classical Wöhler's relationship:

$$\varepsilon_{max} = A - B \, Log \, N_{10} \qquad (1)$$

Where A is the strain level leading to 10% stiffness loss at the end of the first cycle ($N_{10} = 1$) and B is the rate of decrease of the sustainable strain per decade of fatigue cycles. In the case of static fatigue (R=1), $Log \, t_{10}^*$ can be substituted to $Log \, N_{10}$ in equation (1), where t_{10}^* is the time to 10% stiffness loss normalised to the loading time.

Table 1. Residual quasi-static failure strain (ε_r) and coefficients of the Wöhler's relationship (A,B) before and after ageing in water. (same ageing times as for Figure 3)

Ageing Temp. (°C)	A (%)	B (%/decade)	ε_r (%)	A/ε_r
unaged	4.7	0.42	3.5	1.34
60	2.3	0.12	2.1	1.09
70	2.2	0.10	2.1	1.05
90	2.7	0.16	2.6	1.04

After ageing in water, a simultaneous decrease in both the values of A and B was systematically recorded. In the case of dynamic fatigue, it was also observed that the difference between A and the quasi-static failure strain decreased (see Table 1 as an example). For unaged unidirectional glass/epoxy composites, A may be significantly greater than ε_r. This difference has been previously attributed to

loading rate effects (Mandell 1982, Salvia et al. 1997): quasi-static tests were performed at 2 mm.min^{-1}, whereas the loading rate during a fatigue test at 25 Hz is close to 0.6 m.s^{-1}. In that situation, the loading rate effect can mainly be related to sub-critical crack growth processes in the glass fibres. As the loading rate is increased, the time available for crack growth is reduced. Due to the resulting delay in fibre breakage, the failure strain (or the value of A) is thus increased. The reduction in the ratio A/ε_r after ageing therefore denotes a reduced strain rate sensitivity of the composite. A possible explanation for this would be to consider that water ageing induced a significant increase in the density of defects having a critical size in the investigated strain range. As a result, the relative contribution of SCC mechanisms to the overall fibre failure process will be reduced, an increased part of the fibre being broken by catastrophic failure.

Although the Wöhler's approach was found to be suitable in most cases to quantify the endurance losses after ageing, some limitations have been encountered for the most drastic environmental exposures. As it can be seen in Figure 1 (for the composite aged at 90°C) and in Figure 3 (for the composite aged at 50°C), the fatigue results for these severe ageing conditions were enclosed in an horizontal scatter band, where the lifetimes can vary by several order of magnitude at a given strain level. Due to this large scatter in lifetimes, it was no longer possible to calculate a reliable estimate of the parameters A and B of equation (1). In the case of static fatigue, a detailed analysis of the stiffness loss curves revealed that this scatter was associated with a transition from progressive to non-progressive crack propagation.

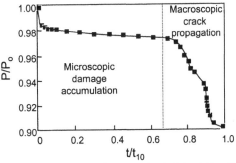

Figure 5. Typical stiffness loss curve (static fatigue). P and P_o are respectively the load at time t and at the end of the loading.

From a typical stiffness loss curves, two steps can be generally be identified in the fatigue life of the specimens (Fig. 5):

(i) a first stage where no significant loss in stiffness can be measured apart from the initial viscoelastic relaxation of the matrix. This period corresponds to the progressive accumulation of broken fibres at the microscopic level,

(ii) During the second stage of the fatigue life, the propagation of macroscopic matrix cracks is associated with a progressive drop in the stiffness. The matrix crack propagation involves successive fibre failures and interfacial debonding at the crack tip.

After ageing in the most severe condition (90°C), it was observed (Fig. 6) that most of the lifetime of the specimens was taken up by the first stage corresponding to microscopic damage accumulation. The crack propagation step occurred in a catastrophic way over a very limited part of the lifetime, sometimes after a long incubation time with no apparent macroscopic damage. These observations emphasise the necessity of a more detailed analysis of the damage micro-mechanisms before the macroscopic crack propagation, especially regarding the accumulation of fibre failures.

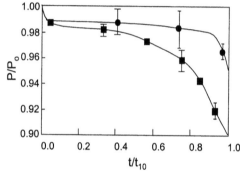

Figure 6. Average stiffness loss curves for a S-glass epoxy composite under static fatigue conditions (R=1, $\varepsilon_{max} = 0.7\ \varepsilon_R$). (■) unaged; (●) aged 100 days at 90°C.

4 DAMAGE MICRO-MECHANISMS

4.1 *In-situ analysis of fibre failures*

The interrelations between the chemical degradation of the matrix and the interface and the weakening of the fibre reinforcement during water ageing have been reviewed in the previous section. These effects are of peculiar importance during the first stages of the fatigue life, where the nucleation of macroscopic cracks is largely controlled by the tolerance of the composite to the progressive accumulation of the fibre failures at the macroscopic scale. These processes are basically dependent upon two main factors:

(i) the stress transfer processes occurring at the extremities of the broken fibres (Reifsnider 1994). When a fibre fails, it becomes ineffective in supporting the applied load over a certain distance from the fibre failure position. This 'ineffective length' depends on the stress transfer occurring at the interface. If the interface is strong, the stresses transfer back into the broken fibre very quickly and the inef-

fective length is small in size. In that situation, the local stress concentration near the broken fibre is high because of the rapid nature of the stress transfer in that region. The tendency for a 'brittle' fracture to propagate across the specimen in the region of the first fibre failure will be great. If the material surrounding the broken fibre is compliant, the ineffective length is increased and the stress concentrations are redistributed over an enlarged region; As a result, a progressive failure involving the interactions of different sites of fibre failure will be favoured. This latter situation may be promoted after ageing due to the interfacial degradation

(ii) As mentioned above, the statistical distribution of fibre defects and their sub-critical growth are the other controlling processes for damage accumulation after ageing.

In the case of a flexural loading, the first fibre breaks are localised over a narrow zone located on the tensile side of the specimens and beneath the loading span. Glass/epoxy composites being translucent, it is therefore possible to quantify the density of the broken fibres from the microscope observations of the specimen surface under loading (Vauthier et al. 1998). Using this technique, the first damage can be quantified in the real physico-chemical environment encountered by the glass fibres during ageing.

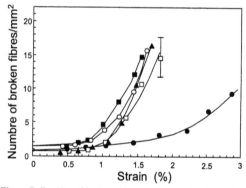

Figure 7. Density of broken fibres on the tensile side of flexural specimens *vs* applied strain. Aged at (○) 50°C, (▲) 60°C, (■) 70°C and (□) 90°C (quasi-static loading, R.T., same ageing times as for Fig. 3).

Figure 7 shows an example of the increase in the density of the broken fibres as a function of the strain during a quasi-static loading (2mm.min⁻¹). These measurements have been carried out using the same DGEBA/DICY composite system and the same conditioning procedure as for the dynamic fatigue experiments reported in Figure 3. The fibre failure processes have been monitored until the onset of the first macroscopic damage (cracking, bundle failure). Above this threshold, the stress distribution on the tensile side of the specimens is too greatly modified to make reliable measurements. The last data point reported on each of the curves plotted in Figure 7 therefore corresponds to the transition from the microscopic to the macroscopic damage.

After ageing, a much more rapid increase in the density of the broken fibres was observed as a function of strain. Changes in the statistical distribution of fibre defects during the ageing stage can account for this observation. Two main effects resulting from the water chemical attack can be envisaged:
- a broadening of the distribution of the defect sizes, which will lower the strain threshold for the first fibre breaks,
- an increase in the overall density of flaws.

During the mechanical loading, these changes are cumulated with the enhancement of the sub-critical crack growth rates to induce an increased density of broken fibres at a given strain level.

From Figure 7, it can also be noted that, after ageing, the transition from the microscopic to the macroscopic damage occurred for higher densities of broken fibres. This can at first be explained by the shift of the distribution of fibre failures in a low strain range: at low strain levels, the composite is more tolerant to the accumulation of fibre breaks. A contribution of the degraded interfaces may also be envisaged. Due to the increased debonding length at the ends of the broken fibres, the stress concentrations may be redistributed over larger volumes around the failure locations. As mentioned above, such processes are known to favour microscopic damage accumulation instead of localised crack propagation.

4.2 Fatigue-life diagrams

A comprehensive description of the changes in the damage mechanisms during ageing can be proposed within the conceptual framework of the so-called fatigue-life diagrams. Introduced by Talreja (Talreja 1987), these diagrams provide a description of the predominant damage mechanisms as a function of the applied strain and the number of cycles. In the case of the tensile fatigue of unidirectional composites, two mains domains were identified by Talreja (Fig. 8a):

(i) an horizontal scatter band (domain (I)) centred about the strain to failure, which corresponds to fibre breakage and interfacial debonding. In this domain, the fatigue failure results from the progressive accumulation of fibre breaks in the volume of the composite, until a specimen cross-section is stressed high enough to break the remaining fibres in it. These mechanisms represent the non-progressive part of the fatigue damage, in that sense that they do not involve the progressive growing of a damaged area from the early stages of the fatigue life to the final failure.

(ii) the second damage area (domain (II)) is a sloping band located between the lower bound of the fibre-breakage scatter band and the horizontal line representing the fatigue limit of the matrix, ε_m. This region corresponds to progressive matrix cracking and interfacial shear failure. This damage is cycle-dependent and its rate is strain-dependent.

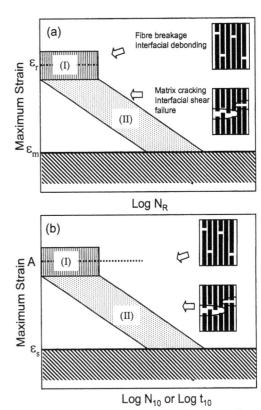

Figure 8. Fatigue-life diagram for (a) tensile fatigue (from (Talreja 1987)) and (b) flexural fatigue of unidirectional composites. (see comments in text for areas (I) and (II)).

Similar fatigue-life diagram can be established for flexural fatigue if the following modifications are taken into account:

(i) The upper fibre-breakage scatter band is centred about the value of the parameter A of the Wöhler's relationship (equation (1)), instead of about the failure strain ε_r. Nevertheless, the damage in this area remains controlled by the accumulation of fibre breaks at randomly scattered sites on the tensile side of the specimens.

(ii) the lower bound for the progressive propagation of matrix cracks cannot be directly set as the fatigue limit ε_m of the unreinforced matrix. It corresponds to the strain limit below which no cracks or only non-propagating cracks are initiated in the matrix. This limit, denoted as ε_s, depends upon the stress transfer processes occurring at the ends of the

broken fibres as well as on the fatigue properties of the unreinforced matrix.

After hygrothermal ageing, the boundaries of the different damage areas in the fatigue-life diagram can evolve due to the degradation of the fibre, matrix and interface properties.

(i) Domain I

In-situ microscopic observations indicated that the accumulation of broken fibres was enhanced at low strains due to the chemical attack of the glass fibres. The corresponding change in the fatigue-life diagram will be a shift to low strains of the fibre-breakage scatter band.

(ii) Domain II

The Wöhler's analysis showed that ageing results in a decrease in the slope, B, of the band corresponding to progressive crack propagation. This lowering corresponds to an enhanced strain sensitivity of the crack propagation rates. To some extent, the latter can be attributed to the alterations in the distribution of fibre strength. In the domain (II), the crack propagation rate is effectively partially controlled by the localised failure of fibres at the crack tip. In the previous section, it was indicated that, after ageing, the density of fibre breaks increased more rapidly as a function of strain. This enhanced strain-dependence of the fibre failure processes can in turn result in a more pronounced increase in the crack propagation rate as a function of strain.

Regarding the changes in ε_s during ageing, different aspects must be considered:

(i) The effects of water plasticization on the intrinsic resistance of the matrix to cracking. Basically, plasticization processes results in an increased mobility of the macromolecular chains. This can improve the relaxation of stress concentrations at the crack tip, due to the enhanced creep ability of the plasticized matrix. At the same time, plasticization can also reduce significantly the gap between the glass transition temperature and the ageing temperature. This is evident from Figure 4, which shows that the Tg of the aged specimens can be depressed up to the ageing temperature by virtue of the combined effects of plasticization and hydrolysis. In such conditions, physical ageing processes in the epoxy network are strongly activated, with a possible increase in the brittleness of the matrix. This effect of water plasticization on physical ageing may therefore have a detrimental effect on ε_s.

(ii) By redistributing the stress concentrations over a larger volume around the broken fibres, the degradation of the interface shear strength can also prevent localised crack propagation and thus increase the value of ε_s.

As an example, a fatigue-life diagram showing the changes in the damage areas has been plotted in Figure 9a. For the considered strain range, it was not possible to determine the strain limit ε_s. Some evidence of this limit may, however, be provided in the most severe ageing conditions when a transition from progressive to non progressive crack propagation was observed (cf section 3.3). If the fibre strength properties are sufficiently decreased, it may be envisaged that the fibre-breakage scatter band is lowered below the limit strain, ε_s, corresponding to the progressive cracking (Fig. 9b). Defects in the form of broken fibres can thus be initiated but they can no longer propagate because $\varepsilon_{max} < \varepsilon_s$. As a result, the progressive damage sloping band disappears from the fatigue life-diagram. The final failure of the specimen therefore involves the progressive accumulation of fibre breaks until a material cross-section is weakened enough to induce the final failure of the specimen. In that situation, the scattering in the lifetimes mainly reflects the statistical distribution of the fibre strength after ageing.

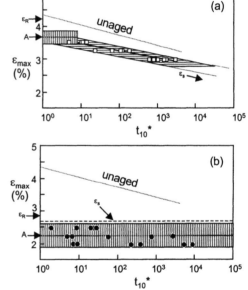

Figure 9. Fatigue life diagrams after afeing 100 days at (a) 50°C and (b) 90°C. (Static fatigue, R.T.).

5 CONCLUSION

The interactions between hygrothermal ageing and mechanical damage have been reviewed in the context of the bending fatigue of unidirectional glass/epoxy composites. It was demonstrated that the overall weight change M_t of the material cannot be considered as a relevant parameter to quantify the amount of damage induced by water sorption. The dependence of the residual mechanical properties upon the whole hygrothermal history implies that the various environmental parameters, namely the temperature, the humidity level and the exposure time should be considered separately in the definition of the ageing procedures and in the prediction of the residual lifetimes.

A detailed analysis of the mechanical damage processes also showed that the changes in the macroscopic fatigue behaviour could be qualitatively related, at the microscopic level, to changes in the statistical distribution of the fibre failures during the early stages of the fatigue life. The latter were related to the chemically induced defects on the fibre surface and to stress corrosion mechanisms, both being strongly dependent upon the chemical degradation of the matrix during ageing. The *in-situ* quantitative assessment of the residual load bearing capacity of the fibres therefore appears as a key point to be able to predict the durability of unidirectional glass/epoxy composites. This may be envisaged through the use of SCC concepts, which could allow to take into account separately the effects of the stress, the temperature, the time and the phyiscochemical environment within a kinetics model for the fibre degradation.

ACKNOWLEDGEMENTS

The author is very grateful to E. Vauthier, V. Pauchard and Prof. L. Vincent for helpful and stimulating discussions about this work. Many thanks are also due to André Barrier for his very efficient technical support.

REFERENCES

Carter, H. G. & Kibler, K. G. 1978. Langmuir-type model for anomalous moisture diffusion in composite resin. *Journal of Composite Materials* 12: 118.
Charles, R. J. 1958. Static fatigue of glass-I. *Journal of Applied Physics* 29(11): 1549.
Chateauminois, A., Chabert, B. & Soulier, J. P. 1990. Long time ageing of glass-epoxy composites. In P. Hamelin & G. Verchery (Eds.), *Proc. Int. Symp. Textile in Building Construction, Lyon, July 1990.* Paris: Pluralis.
Chateauminois, A., Chabert, B., Soulier, J. P. & Vincent, L. 1995. Dynamic mechanical analysis of epoxy composites plasticized by water: artifacts and reality. *Polymer Composites* 16(4): 288-296.
Chateauminois, A., Chabert, B., Souliert, J. P. & Vincent, L. 1994. Interfacial degradation during hygothermal ageing. Investigations by sorption/desorption experiments and viscoelastic analysis. *Polymer* 35(22): 4765-4774.
Gurtin, M. E. & Yatomi, C. 1979. On a model for two phase diffusion in composites material. *Journal of Composites Materials* 13: 126.
Kasturiarachi, K. A. & Pritchard, G. 1984. Free dicyandiamide in crosslinked epoxy resins. *Journal of Materials Science Letters* 3: 283-286.
Mandell, J. F. 1982. Fatigue behaviour of fibre-resin composites. In G. Pritchard (Ed.), *Developments in reinforced plastics*: 67-107. London: Appl. Sci. Publ.

Mazer, J. J. & Walther, J. V. 1994. Dissolution kinetics of silica Glass as a function of pH between 40°C and 85°C. *Journal of Non-Crystalline Solids* 170: 32-45.
Metcalfe, A. G. & Schmitz, G. K. 1972. Mechanism of stress corrosion in E-glass fibres. *Glass Technology* 13(1): 5.
Reifsnider, K. L. 1994. Modeling of the interphase in polymer matrix composite material systems. *Composites* 25(7): 461-496.
Salvia, M., Fiore, L. & Fournier, P. 1997. Flexural fatigue behaviour of UDGFRP. Experimental approach. *International Journal of Fatigue* 19(3): 253-262.
Schen, C. H. & Springer, G. S. 1981. Moisture absorption and desorption of composites material. In G. S. Springer (Ed.), *Environmental effects on composite materials*: 15. Technomic Publications.
Shah, M. A., Jones, F. R. & Bader, M. G. 1985. Residual Dicyandiamide (DICY) in Glass-Fibre Composites. *Journal of Materials Science Letters* 4: 1181.
Suri, C. & Perreux, D. 1995. The effects of mechanical damage in a glass fibre/epoxy composite on the absorption rate. *Composites Engineering* 5(4): 415-424.
Talreja, R. 1987. *Fatigue of composite materials.* Basel: Technomic Publ. Co. Inc.
Vauthier, E. (1996). PhD Dissertation. Lyon, Ecole Centrale de Lyon.
Vauthier, E., Abry, J. C., Bailliez, T. & Chateauminois, A. 1998. Interactions between hygrothermal ageing and fatigue damage in unidirectionnal glass/epoxy composites. *Composites Science and Technology* 58: 687-692.
Vauthier, E., Chateauminois, A. & Bailliez, T. 1996. Fatigue damage nucleation and growth in a unidirectional glass/epoxy composite subjected to hygrothermal ageing. *Polymer and Polymer Composites* 4(5): 343-351.
Wiederhorn, S. M. 1978. Mechanisms of Subcritical Crack Growth in Glass. In R. C. Bradt (Ed.), *Fracture Mechanics of Ceramics*: 549-580. Plenum Press.
Wiederhorn, S. M. & Bolz, L. H. 1970. Stress Corrosion and Static Fatigue of Glass. *Journal of the American Ceramic Society* 53: 543-548.

Recent Developments in Durability Analysis of Composite Systems, Cardon, Fukuda, Reifsnider & Verchery (eds)
© 2000 Balkema, Rotterdam, ISBN 90 5809 103 1

Contribution to the durability analysis of a thick composite structure under complex mechanical and hygroscopic loads

Y. Poirette, F. Pierron & A. Vautrin
Département Mécanique et Matériaux, Ecole des Mines de Saint-Etienne, France

ABSTRACT: This paper presents a methodology suitable for durability analysis of a glass/epoxy composite structure. In particular, the effects of hygrothermal ageing on specific mechanical properties have been investigated. These properties were regarded as most critical properties issued from the finite element analysis of the whole structure. Thus, a new curved specimen has been designed and tested together with Iosipescu specimens. Dynamic tension-compression tests have been also performed. Hygroscopic ageing of coupons led to 3D diffusion parameters determination. Changes of mechanical properties with moisture content have been monitored. Results can be used in a finite element model to simulate specimen behaviour subjected to moisture gradient.

1 INTRODUCTION

During the past five years, novel applications of glass fibre polymer reinforced composites have gradually emerged in transport system or civil and building engineering. When long term structural performances are required, in case of rail applications for instance, the present lack of design standards and data bases is an actual difficulty that highly restricts the application field of this type of materials. Design of composite structural components subjected to hygro-thermo-mechanical loadings is still uncertain and should rely on refined analyses. The structural performances can be affected in different ways (Weistman 1995) and local drops of stiffness and strength can occur finally. Furthermore, time dependant hygrothermal gradients can result in coupling effects with fatigue behaviour. This problem becomes more severe when thick components are concerned since 3D effects may arise. The design process that has been used in the present work is relying on a specific combination of experimental and numerical studies to aim at reliable values of safety factors. The present approach can be of interest to set up more confident guide routes to support the use of composites as structural materials.

2 PRESENTATION OF THE PROBLEM

The present work is part of an important industrial study dealing with the design and development of a novel power transmission for locomotives.

The existing transmission is a complex system of several tens of pieces weighing about 450 kg.

The new solution must be able to fulfil the same specifications than the current system, namely :

- to transmit a maximum torque at the start of the train,

- to tolerate a maximum vertical misalignment of several tens millimetres due to railway defects.

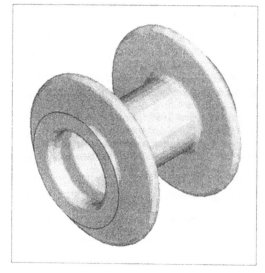

Figure 1. Geometry of the new component.

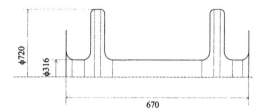

Figure 2. Dimensions in mm of the composite transmission

To achieve this goal, the transmission must combine a high torsion stiffness as well as a flexural stiffness as low as possible.

In the existing solution, it is obtained by a complex assembly of elastic knee joints which ensure the Cardan effect.

To fulfil the requirements, a 10 mm thick composite shaft with an evolutionary shape has been finally proposed (Fig. 1). The elastic design has been carried out in the framework of a feasibility study (Cerisier et al. 1996). Figure 2 shows dimensions of the composite transmission.

The flanks at the extremities of the shaft perform the same functions as that of cardan.

The allowed movements of the composite component are the same than that the existing transmission.

The material chosen for the component is a glass fibre balanced weave embedded in an epoxy resin with 45° orientation with respect to the shaft axis.

The main advantage of the composite solution is the important mass reduction. Indeed, its mass has been reduced to about 50 kg. Another advantage of this solution is the suppression of a great number of parts which facilitates assembling.

3 MATERIAL AND SPECIMENS

3.1 Material

The material used in this work is the E-glass fabric reinforced epoxy composite. Plates were obtained by compression moulding of 7781/XE85AI HEXCEL prepregs (Cerisier 1998) using a specific cure cycle of 1h30mn at 120°C and 5 bars.

Plates of 1 mm, 2 mm, 5 mm and 18 mm thickness were manufactured in order to characterize the mate-

rial behaviour. The average ply thickness was about 450 μm. The achieved fibre volume fraction of the plate was 49-51%. Material properties are listed in Table 1.

3.2 Iosipescu specimens

Iosipescu specimens were cut out with a diamond saw in 2 mm and 18 mm thick plates to measure the in-plane and through-thickness shear moduli. Figure 3 shows the geometry of the Iosipescu specimens.

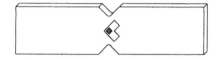

Figure 3. Iosipescu specimen geometry.

3.3 Curved specimens

Because of the cost of the manufacturing of the whole transmission, it is impossible to carry out tests on full-size prototype to optimize the structure and validate the solutions. Therefore, special samples were designed from finite element analysis of the structure. Figure 4 shows the shell model of the complete structure. This study, carried out with the ANSYS 5.2 package, showed that flexural load case is the worst for the transmission with respect to the material strengths. In particular, some areas of the transmission were determined as critical areas. These are the joining zones between the cylinder and the flanges. One important feature of this part of the transmission is that curvature is of the same order as the thickness. It means that three dimensional effects can be more neglected.

A 3D finite element sub-model (Fig. 5) was extracted from whole structure shell model to compute the full 3D stress state. This analysis showed the importance of the through-thickness stresses. The results have allowed to design a simple substructure in the shape of curved specimen (Fig. 6). The 3D stresses field in this specimen subjected to tensile load is similar to that of the transmission subjected to flexural load.

Table 1. Material properties.

	E_{11} Gpa	E_{22} Gpa	E_{33} Gpa	G_{12} Gpa	G_{13} Gpa	G_{23} Gpa	v_{12}	v_{13}	v_{23}	β_1	β_2	β_3	ρ kg/m³
Glass/epoxy	25.4[1]	25.4[1]	14[2]	4.9[2]	3.8[2]	3.8[2]	0.13[1]	0.3[3]	0.3[3]	0[3]	0[3]	0.6[3]	2000

(1) from (Cerisier 1998)
(2) from present work.
(3) from (Tsai 1986)

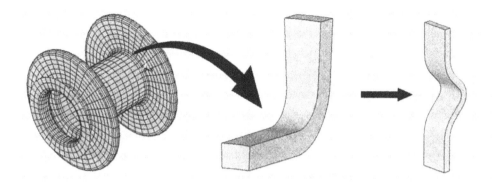

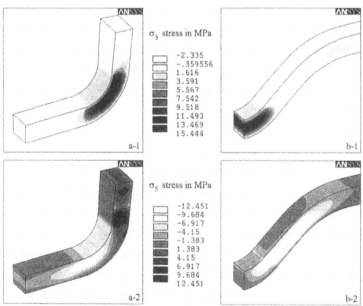

Figure 8. Comparison between stress fields of through-thickness tension (1) and shear (2) in the transmission (a) under flexural load and in the curved beam (b) in tension.

Three stress ratios have been computed, namely :

$$R_2 = \frac{\sigma_2}{\sigma_1} ; \quad R_3 = \frac{\sigma_3}{\sigma_1} ; \quad R_5 = \frac{\sigma_5}{\sigma_1} .$$

The indexes 1, 2, 3 and 5 refer to contraction notation (Tsai 1986). The influence of each parameter on the specimen stress ratios has been studied to approach the transmission stress ratios.
It appeared that :

- increasing the radius decreases R_3 and R_2 but increases R_5,

- increasing the thickness decreases R_2 but increases R_3 and R_5,

- height and width both influence R_5 and R_2 respectively.

This investigation led to the design of two geometries for the curved specimen, as shown in Table 2.
Thick specimens have been chosen to keep the same radius and thickness as the transmission.
Thin specimens keeping the same radius to thickness ratio have been chosen for conditioning purposes.
For every geometry, some specimens have been manufactured with 0° and 45° fabric orientation from axis 1. Later on, they will be noted 0° and 45° curved specimens respectively.

Table 2. Stress ratio issued from transmission and selected specimens numerical analysis.

Curved beam (R,H,T,W) * in mm	R_2	R_3	R_5
(20,32,10,40)	0.35	0.1	0.09
(6,8,3,20)	0.46	0.1	0.09
Transmission	0.56	0.1	0.07

* (R,H,T,W)=(Radius, Height, Thickness, Width).

4 EXPERIMENTAL PROCEDURE

4.1 Environmental exposure

Samples were conditioned in vapour atmosphere at different relative humidities and temperatures. Four different conditions have been obtained using saturated salt solutions, namely 60%, 80% and 96% relative humidity (R.H.) at 60°C and 96% R.H. at 40°C. A 60°C maximum temperature has been chosen as accelerating factor of moisture absorption to avoid any thermal damages which may occur at higher temperature.
Prior to humid exposure, all specimens have been dried in an oven until they reached constant weight. The moisture uptake has been monitored by weighing the samples to within 0.1 mg using a precision balance.
Absorption coupons have been cut out from 1 mm and 18 mm thick plates to determine the diffusion coefficients in the directions parallel and normal to the fabric plane. Others have been cut out from 5 mm plate for 3D diffusion.

4.2 Mechanical monotonic tests

Iosipescu in-plane and through-thickness tests were achieved under displacement control at a rate of 1mm/min to evaluate shear properties. Tests have been performed using the Iosipescu EMSE fixture designed by Pierron (1997). Six dried specimens have been tested for each of the measured properties. Furthermore, specimens have been conditioned to evaluate the influence of moisture content on the shear material response. Six instrumented specimens have been exposed in the 60°C enclosures. The weight of three non-instrumented ones was monitored during conditioning for reference.
Monotonic tensile tests were carried out on curved specimens under displacement control at a rate of 1 mm/min until failure occurred. First, dried 10 mm thick specimens have been tested to determine the failure mode under the complex mechanical service loads. Strain gauges were bonded to the convex surface of the specimen in the middle of the central curved section.
Then, to evaluate the influence of moisture content on curved specimen behaviour, 0° dried 3 mm thick specimens have been tested for reference. Furthermore, six samples have been exposed in each conditioning enclosures at 60°C. Non-instrumented specimens have been weighed during conditioning for reference. Conditioned specimens will be tested when the reference samples reach moisture saturation.
Lastly, since the transmission is a thick composite structure, it will be saturated after several years. Then, it will be submitted to hygroscopic gradients in most cases. In order to take this into account, 45° thin curved specimens are to be tested before moisture saturation. Six samples have been exposed to the humid atmospheres at 60°C. Again, dried specimens have been tested for reference.
All tests have been carried out at room temperature.

4.3 Mechanical dynamic tests

Since the transmission is submitted to dynamic loads, it is essential to perform fatigue tests on the curved specimens. Tension-compression fatigue behaviour has been investigated. The tests were performed under displacement control at a frequency of 10 Hz. Two load signal amplitudes were chosen : ± 0.85 mm and ± 1.7 mm which corresponds respectively to the nominal and maximum vertical shift of the transmission in service. Two 45° thick specimens were tested for each load case. This specimen has been chosen because of its configuration : plies number and orientation are equivalent to those transmission. It allows to simulate the transmission stress state at best.

Specimens stiffness was recorded until failure occurred.

5 RESULTS

5.1 Absorption behaviour

For all samples, absorption curves have not reached the saturation level according to Fick's law. Figure 9 shows an example of moisture absorption curve for specimens conditioned at 60°C/96% R.H. ; the moisture content is plotted versus the square root of the exposure time to thickness. Each point of the curve is an average of four data points.

Nevertheless, since the beginning of the curves seems linear, the main absorption phenomenon is assumed to be Fickian diffusion for the simulations. Then, an apparent maximum moisture content M_m and apparent diffusivities along and normal to the fibres D_x, D_y and D_z have been evaluated using the solution of 3D Fick equation :

$$M_t = M_m \left[1 - \left(\frac{8}{\pi^2} \right)^3 \Sigma_x \Sigma_y \Sigma_z \right] \quad (1)$$

in which

$$\begin{cases} \Sigma_x = \sum_{j=0}^{\infty} \frac{\exp\left[-(2j+1)^2 \pi^2 \left(D_x t / L^2\right)\right]}{(2j+1)^2} \\ \Sigma_y = \sum_{j=0}^{\infty} \frac{\exp\left[-(2j+1)^2 \pi^2 \left(D_y t / l^2\right)\right]}{(2j+1)^2} \\ \Sigma_z = \sum_{j=0}^{\infty} \frac{\exp\left[-(2j+1)^2 \pi^2 \left(D_z t / h^2\right)\right]}{(2j+1)^2} \end{cases} \quad (2)$$

where M_t is the moisture content at time t, M_m the moisture content at saturation, L, l and h the specimen dimensions and D_x, D_y, and D_z are the diffusion coefficients along the length, the width and the thickness of the specimen respectively.

Here, it has been assumed that D_x and D_y were equal since the composite is reinforced by balanced fabric. Furthermore, M_m was supposed to be the same for each coupon.

M_m, D_x and D_z have been computed by an optimization method. Using least square optimization, the solutions are obtained by minimising the function :

$$\sum_t \left[M(t) - M_{exp}(t) \right]^2 \quad (3)$$

where M(t) is the analytical solution of Fick problem defined as M_t in Equation (1) and $M_{exp}(t)$ is the moisture content measured at time t.

M_m, D_x and D_z values resulting from these calculations are listed in Table 3.

The results show that :

- the moisture content basically depends on the relative humidity, and
- the apparent diffusion coefficients seem to be mainly affected by temperature.

These results are in agreement with results from others studies (Loos & Springer 1981).

The solid line in Figure 9 represents the theoretical weight gain assuming Fickian diffusion with 60°C/96% R.H. parameters.

Table 3. Apparent maximum moisture content and diffusivities in Fickian diffusion assumption.

	M_m (%)	D_x (mm²/s)	D_z (mm²/s)
60°C/60%RH	0.4	7.1 10⁻⁷	1.2 10⁻⁷
60°C/80%RH	0.5	7.6 10⁻⁷	1.5 10⁻⁷
60°C/96%RH	0.6	6.3 10⁻⁷	1.4 10⁻⁷
40°C/96%RH	0.6	3.4 10⁻⁷	1.1 10⁻⁷

5.2 Change in shear properties with moisture content

The in-plane shear modulus of dried specimens was measured to be about 4.9 GPa, while the through-thickness shear modulus was 3.8 GPa.

Tests on conditioned specimens have been performed after 6 months of humid exposure. It appeared that 6 months was widely sufficient to reach the calculated value of pseudo Fickian saturation.

It can be observed in Figure 10 that both in-plane and through-thickness shear moduli are not significantly affected by the absorbed moisture.

Failure tests have been performed on dried and 60°C/96% R.H. conditioned specimens. Experiments revealed a decrease of about 14% on the ultimate in-plane shear stress and about 6% on the ultimate through-thickness shear stress (Fig. 10). Those results may affect the failure mode of the aged transmission.

5.3 Monotonic tension test on curved beam

First tests were performed on 0° thick specimens. Failure occurred in delamination in a brittle manner. Figure 11 shows a typical load-deflection curve of curved specimen in tension.

Furthermore, 3D finite element analysis was carried out using the ANSYS 5.4 package. The effects of the change in curvature due to large deflections were neglected. Assuming linear behaviour, comparison between experimental measurements (strains and global rigidity) and calculation results led to evalu-

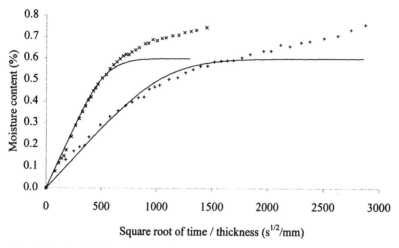

Figure 9. Experimental and theoretical absorption curves at 60°C/96% H.R. ; + 1mm specimen ; × 2mm specimen (from 18mm thickness plate) ___ Fick model

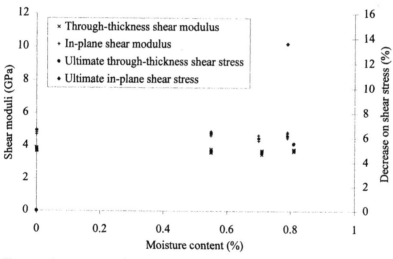

Figure 10. Shear properties as function of moisture content.

ate through-thickness modulus at 14 GPa. In the same way, strength has been evaluated at 40 MPa. These estimations of through-thickness modulus and strength are similar to values published on equivalent materials (Goetschel & Radford 1997).

Other tests performed on 45° specimens led to the same failure mode in through-thickness tension.

To evaluate the influence of moisture on curved beam behaviour, six dried thin specimens have been tested for reference.

In the same way that for Iosipescu samples, non-instrumented specimens have been weighed during conditioning. Tests will be performed when weight gain reaches pseudo saturation level calculated using

Fick solution. Nevertheless, for the simulation purpose we have made the assumption that the transverse tensile modulus remains constant and independent on the moisture, since no significant changes have been recorded for the shear behaviour.

5.4 Fatigue tension-compression test on curved beam

No significant stiffness loss was recorded for the lower signal amplitude even after 20 million cycles, but a characteristic whitening of shear damage was observed early in the test on the concave surface of the curved section.

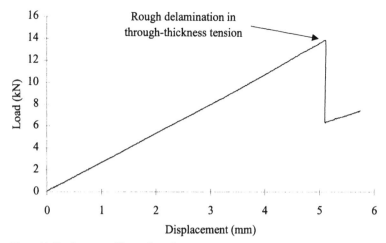

Figure 11. Tension test on 0° curved specimen.

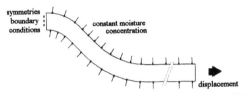

Figure 12. Hygroscopic and mechanical boundary conditions in finite element analysis.

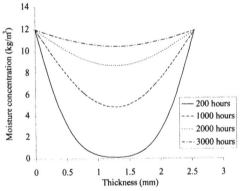

Figure 13. Moisture profiles in specimen after 200, 1000, 2000 and 3000 hours of conditioning.

Failure in delamination occurred during the first 6000 cycles for the two specimens tested with the higher signal amplitude. The reason for this critical response is not yet found because the load level reached in the beginning of the test is significantly lower than the static failure load.

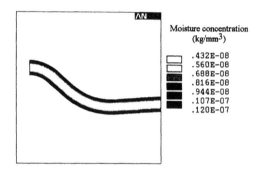

Figure 14. Moisture concentration field simulation after 100 hours of conditioning.

6 MODELLING

6.1 Moisture gradient

The diffusion parameters calculated above were use to introduce a moisture gradient in the finite eleme model of the curved specimen. This calculation w achieved using the thermal analysis of the ANSY programme since the heat conduction problem al Fick's model are of the same type. Thus, in place a thermal field, the programme can evaluate tl moisture concentration field in the sense of Fickia diffusion and take it into account in a structur analysis.

The first analysis has been performed on 45° th specimen using the following parameters :

$D_x = D_y = 6.3 \ 10^{-7} \ mm^2/s$,
$D_z = 1.4 \ 10^{-7} \ mm^2/s$,
$M_m = 0.6\%$.

Material is assumed to be initially dried, therefo the initial concentration field is uniform and equa

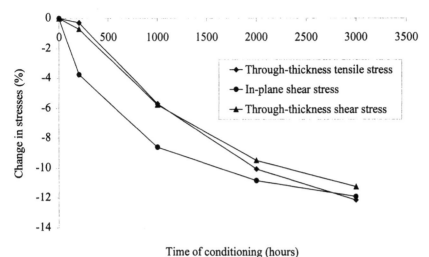

Figure 15. Change in maximum stresses with conditioning time.

to zero. A constant moisture concentration is applied as hygroscopic boundary conditions on the two faces of the specimen. This moisture concentration is calculated using Equation (4).

$$M_m = \frac{c_m}{\rho} * 100 \qquad (4)$$

where c_m is the moisture concentration applied as boundary conditions and ρ is the material density in the dry state.

Mechanical boundary conditions have been applied : conditions of symmetry on curved extremity and a longitudinal displacement at the free end.

Both the hygroscopic and mechanical boundary conditions are depicted in Figure 12.

Four calculations have been performed to simulate the moisture absorption after 200, 1000, 2000 and 3000 hours of conditioning. Figure 13 shows the moisture profiles through the thickness of the specimen and Figure 14 the moisture concentration field after 1000 hours of conditioning. It can be noticed that the calculated moisture gradient is similar to a moisture gradient through a 10 mm thick specimen after an exposure of approximately 16000 hours in the same conditions, assuming Fickian diffusion and identical maximum moisture content whatever the thickness.

The comparison between calculations perform both under dry and hygroscopic conditions do not reveal any strong dependence between the state of stress and moisture gradient.

However, the moisture gradient leads to a decrease of about 10% on the whole stresses after about 3000 hours of conditioning of thin specimen (Fig. 15).

7 CONCLUSION

The present approach is based on the concept of representative sub-structures. After the stiffness and strength static validation of the design within elasticity framework, critical areas have been determined by coarse finite element calculation and subsequently representative sub-structures defined to carry out refined numerical and experimental analyses.

This point is a key step in the future optimization process. The part sub-structuring has led to the definition of special specimen geometries that will allow the experimental and cost effective :

- characterization of relevant mechanical properties,

- identification of major coupling effects,

- validation of service hygro thermo mechanical states modelling.

This type of testing is complementary to the usual mechanical characterization based on standardized specimens. Indeed, it is absolutely necessary to investigate both numerically and experimentally complex internal states due to service loading. In particular, time-dependant gradients and fatigue performances have to be studied.

In the future, it is expected that more reliable safety factors, taking into account time dependant service loaded could result from such an approach.

ACKNOWLEDGEMENTS

The authors wish to thank the ALSTOM TRANSPORT company for the financial support of this study.

REFERENCES

Cerisier, F. 1998. Conception d'une structure travaillante en matériaux composites et étude de ses liaisons. *PhD thesis, Université Jean Monnet, Saint-Étienne*

Cerisier, F., Grédiac, M., Pierron, F. & Vautrin, A. 1996. Design of a locomotive transmission in composite materials. *3rd biennal joint conference on Engineering Systems Design and Analysis.*

Goetschel, D.B. & Radford D.W. 1997. Analytical Development of Through-Thickness Properties of Composite Laminates. *Journal of advanced materials*: 37-46.

Loos, A.C. & Springer G.S. 1981. Moisture Absorption of Graphite-Epoxy Composition Immersed in Liquids and Humid Air. *Environmental Effects on Composite Materials (1)*: 34-50.

Pierron, F., & Vautrin A. 1997. Measurement of the in-plane shear strengths of unidirectional composites with the Iosipescu test. *Composites Sciences and Technology,* 57, 1653-1660.

Tsai, W.S. 1986. Composites Design, 4th edition. *Ed. Think Composite.*

Weistman, Y.J. 1995. Fluids Effects on Polymeric Composites – A Review. *Technical Report N° ESM 95-3.0 CM.*

Recent Developments in Durability Analysis of Composite Systems, Cardon, Fukuda, Reifsnider & Verchery (eds)
© *2000 Balkema, Rotterdam, ISBN 90 5809 103 1*

The damping and dynamic moduli of fibre reinforced polymer composites after exposure to hot, wet conditioning

R.D.Adams
Department of Mechanical Engineering, University of Bristol, UK

M.M.Singh
Department of Engineering Materials, University of Southampton, UK

ABSTRACT: Several polymer matrix composites were exposed to steam at 100°C to provide accelerated ageing. Moisture uptake, shear modulus, damping and interlaminar shear strength were correlated to provide a means of nondestructively evaluating the moisture-induced degradation.

INTRODUCTION

Fibre-reinforced polymer composites are widely used in aerospace applications, and are increasingly finding a place in the automotive and sports industries. Among their attractions are their resistance to corrosion, and their inherently good vibration damping properties, which can be designed into a composite structure, avoiding the use of 'add-on' damping treatments. In wet of humid conditions, however, these materials absorb water which has a plasticising effect on the matrix and causes a loss in the mechanical performance. The extent of this loss depends on the nature of the polymer, as well as on the hygrothermal loading history. To assess the durability of polymer composite components, lengthy experiments are required; if accelerated testing is used, the validity of extrapolating the results to in-serve conditions is always questionable. A means of assessing the integrity of a composite component nondestructively, while in service, so that it can be replaced before failure, is highly desirable, and would allow designers to turn to composites with greater confidence. Since damage and degradation of fibre-reinforced polymers are often not clearly manifest on the surface, visual inspection is usually insufficient and more sophisticated methods are required.

Measurements of modulus and damping capacity derived from vibration tests have been used as a nondestructive evaluation tool. One of the objectives of this work was to investigate the usefulness of such measurements for the nondestructive assessment of moisture-induced degradation of the mechanical performance of fibre-reinforced polymers. Further, changes in the macroscopic mechanical properties reflect changes to the microstructure, and insight into the mechanisms of moisture uptake in composites can be gained from studying the effects of moisture and temperature on the overall dynamic response of relatively large specimens.

MATERIALS

A number of polymer composites was studied: these were three different carbon-fibre-reinforced epoxy resins, a glass-fibre-reinforced epoxy, and carbon-fibre-reinforced PEEK (poly(ether ether ketone)). The epoxy resins were thermosetting, brittle resins, such as are now common as matrix materials in composites used in the aircraft industry. PEEK, however, is an aromatic thermoplastic resin, which has recently gained popularity because of its stability in hot, wet environments, its relative ease of component manufacture, and the potential for recycling and reusing the material.

ENVIRONMENTAL CONDITIONING

Steam was chosen as a suitably aggressive environment for this work. At higher temperatures and high humidity, some fibre-reinforced polymers exhibit non-Fickian diffusion, which may be ascribed either to structural damage in the form of micro-cracks, or to hydrogen bonding of the water

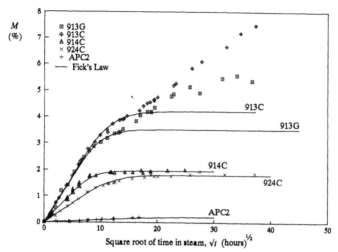

Fig. 1. Percentage mass increase for the five composite materials as a function of the square root of steam.

molecules to the polymer. The combination of temperature and relative humidity at which non-Fickian behaviour occurs depends on the resin system. Because one aim of this study was to explore the possibilities of using changes in the dynamic properties as the basis of nondestructive evaluation of mechanical degradation, it was of interest to attempt to damage some of the specimens. Although, because of their known response to such environments, fibre-reinforced composite components are unlikely, in practice, ever to be subjected to steam, 100% relative humidity at 100°C provides the worst possible conditions at atmospheric pressure. The high temperature and humidity also ensured a relatively short time to reach equilibrium moisture levels. Steam was chosen in preference to boiling water, to avoid uncertainties arising from leaching of constituents into the water.

EXPERIMENTAL METHODS

The dynamic shear properties of all the materials were determined at intervals during conditioning of the specimens in steam, in hot air and during recovery. To find the longitudinal shear modulus and the corresponding loss factor, a torsion pendulum was used.

RESULTS AND DISCUSSION

Moisture absorption

The percentage mass increase for the five composite

materials is shown as a function of the square time in steam in Fig. 1. The lines on Fig. 1 re the curves for Fickian diffusion that most ne the experimental data.

The diffusion parameters, D and M_α determined by an iterative method and are pr in Table 1.

Collings[1] derived empirical relationsh predict the values of D and M_∞ for the carbo reinforced composites 913C, 914C and 924C varying conditions of relative humidit temperature. The predictions for these mate 100°C and 100% relative humidity are also g Table 1.

It is clear that the diffusion parameters 914C and 924C materials are reasonabl predicted. From Fig. 1, it can be seen that th 924C and APC2 materials exhibited ess Fickian behaviour, even under these conditions. It is, perhaps, worth notir conditions of 70°C and 85% relative humi now chosen as the standard for cond aerospace materials.

Collings' empirical relationships for conditions would predict a saturation n content of 1.47% for 914C and 1.14% fol somewhat lower than observed at 100°C an relative humidity.

Effect on loss factor

To compare more directly the behaviour c different composites and the unreinforced r change in the longitudinal shear loss factor,

180

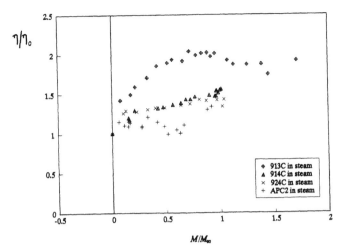

Fig. 2. Change in the longitudinal shear loss factor, η, as a function of M/M_∞

Table 1. Measured diffusion parameters of fibre-reinforced composites and unreinforced matrix resins exposed to steam. 1R = RAE predictions.

Material	M_∞ (%)	D_c (mm^2 s^{-1})
913G, 0°	3.45	1.90 x 10^{-6}
913C, 0°	4.20	1.16 x 10^{-6}
913C^{1R}	2.26	2.2 x 10^{-6}
914C, 0°	1.95	1.77 x 10^{-6}
914C^{1R}	1.95	1.5 x 10^{-6}
924C, 0°	1.80	0.80 x 10^{-6}
924C^{1R}	1.52	0.98 x 10^{-6}
APC, 0°	0.21	0.34 x 10^{-6}
913 unreinforced	8.54	2.15 x 10^{-6}

the materials has been plotted as function of M/M_∞, which is a measure of how close the moisture content is to the saturation level (Fig. 2). The change in loss factor is defined by η/η_0 where η_0 is the value of the dry material before moisture conditioning. Each point represents the average of the values obtained. Initially, three specimens were tested each time, then two, when the first specimen was removed for drying, and finally, only one specimen, in the last stages of conditioning when the second specimen had been removed for drying.

The extent of the influence of the matrix on the shear loss factor is illustrated by the difference between the loss factors of the reinforced epoxies and that of the carbon-fibre-reinforced thermoplastic, APC2. Whereas all the fibre-reinforced epoxies had an initial loss factor in the region of 0.013, that of the APC2 was only one third of this, about 0.004. Even

the APC2, however, showed a slight increase in l factor with absorbed moisture, although the quan of water was only about 10% of that absorbed by epoxies (see Table 1). The observed scatter in results for the APC2 is due to the fact that the factor is so small, so that small changes due experimental error cause relatively large fluctuati in the ratio η/η_0.

All the epoxy composites experienced an in sharp increase in loss factor, followed by a n gradual increase with rising moisture content shown in Fig. 2. The carbon-fibre-reinforced and 924 resins behaved in an almost iden manner, but the 924C demonstrated a sharper ir increase in loss factor than the 914C. The composites experienced a greater increase in factor than the other composites. A stage reached, however, when the loss factor bec almost constant as the moisture level approache saturation level, and then was reduced again a moisture content increased further, when it ca assumed that the moisture uptake was no lc predominantly by diffusion. The glass trans temperatures, T_g, of the 913, 914 and 924 resin given by Ciba-Geigy as 131, 180 and 1 $\mathcal{C}$ respectively. Absorbed moisture in polyme known to lower the glass transition temperatur and the larger increase in loss factor of the composites may be due to the reduction of the n T_g, even though all measurements were ma 20°C, well below the dry T_g of all the materials.

Effect on longitudinal shear modulus

The change in shear modulus G/G_0, of th materials during moisture uptake is shown

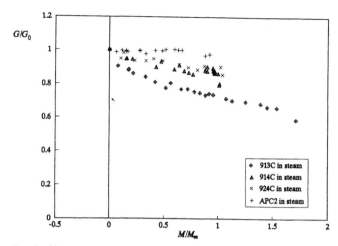

Fig. 3. *Change in shear modulus, G/G₀, during moisture uptake as a function of M/M∞*

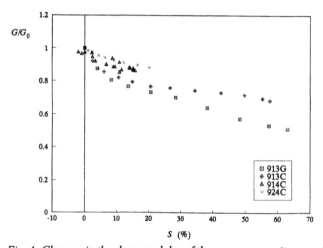

Fig. 4. *Changes in the shear modulus of the epoxy composites as a function of the loss in ILSS, S.*

function of M/M_∞ in Fig. 3. Once again the order of degradation was the same: the APC2 was least affected, the 914C and 924C materials (which had a similar response), the unreinforced, carbon-fibre-reinforced and glass-fibre-reinforced 913 followed in that order. The greater loss of modulus of the 913 composites than of the plain resin could be due to the gross damage in the form of delamination and due to changes in the interface or interphase between the fibres and the matrix that cannot occur in the unreinforced polymer.

Correlation of dynamic shear properties with the loss of ILSS

In Figs 4 and 5, the changes in the shear modulus and in the loss factor, respectively, of the epoxy composites are shown as a function of the loss in ILSS, S, expressed as a percentage of the initial preconditioning strength.

Thus S is given by:

$$S = \frac{ILSS_0 - ILSS_M}{ILSS_0} \times 100\% \quad (1)$$

where $ILSS_0$ and $ILSS_M$ are the interlaminar strength initially and at a moisture content M, respectively. Figure 4 shows that, for all these composites, there was a strong correlation between the dynamic shear modulus and the loss in interlaminar shear strength, the larger was the reduction in shear strength, the

182

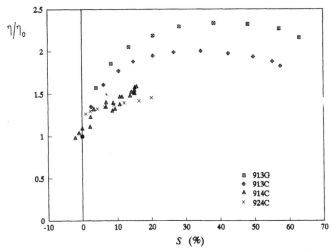

Fig. 5. Changes in the loss factor of the epoxy composites as a function of the loss in ILSS, S.

greater was the fall in shear modulus. The loss factor was more sensitive to the loss in strength (Fig. 5), but for the 913C and 913G materials, which continued to experience a fall in shear strength with prolonged exposure to steam, the loss factor levelled out at high moisture contents. This means that, for these materials, the loss factor was a less useful gauge of the strength of the material. For the 914C and 924C materials, both the loss factor and the longitudinal shear modulus were indicators of the material strength. Modulus measurements are, therefore, better as a nondestructive indication of the interlaminar shear strength of a hydrophilic composite material than are loss factor measurements.

CONCLUSIONS

Samples of several fibre-reinforced epoxies and fibre-reinforced PEEK were exposed to steam. The moisture absorption characteristics, longitudinal shear modulus and loss factor, and the interlaminar shear strength were monitored, with a view to assessing the feasibility of the use of dynamic measurements as a nondestructive tool for determining the structural integrity of polymer composites degraded by moisture. The degree of damage inflicted by steam depended on the type of polymer matrix. Epoxy resin reinforced with glass fibre was more degraded than when reinforced with carbon fibre, suggesting that the interphase plays a role in the response to moisture conditioning. The longitudinal shear modulus was found to be a reliable indicator of the interlaminar shear strength

of polymer composites. Changes in the she: factor were more sensitive to the loss of interl; shear strength, but were less useful for ext: hydrophilic materials.

REFERENCES

1. Collings, T.A., *Environmental Str Testing*, Royal Aerospace Establis Technical Report TR 880683 (Report), 1988.

2. Carter, H.G. & Kibler, K.G., Langmι model for anomalous moisture diffu: composite resins. *J. Comp. Mater.*, 118-131, 1978.

3. McKague, L., Environmental synergi: simulation in resin matrix com; *Advanced Composite Material Environmental Effects*, ASTM STP 6 J.R. Vinson, pp. 193-204, 1978.

Recent Developments in Durability Analysis of Composite Systems, Cardon, Fukuda, Reifsnider & Verchery (eds)
© *2000 Balkema, Rotterdam, ISBN 90 5809 103 1*

Modeling of composite interfaces subjected to thermal stresses and physical aging

D.-A. Mendels, Y. Leterrier, & J.-A. E. Månson
Laboratoire de Technologie des Composites et Polymères (LTC), Ecole Polytechnique Fédérale de Lausanne (EPFL), Switzerland

ABSTRACT: Coupling between internal stress and the aging process is of considerable importance to characterize the durability of a composite. The present work investigates how interfacial properties are modified upon physical aging, using single fiber model composites. Firstly, the micro-mechanics of the interface and their evolution upon aging are studied through the microbond test. Secondly, this theory is coupled to that of the energy release rate and their equivalence is assessed by experimental investigation. These methods were tested in the case of freshly cured and aged epoxy droplets on glass fibers, and were shown to yield similar results. Providing that the relaxation of internal stresses and the shift of relaxation times of the matrix when it is physically aged are accounted for, the interface is described by one single parameter. This allows the determination of the intrinsic shear strength of the interface and its energy release rate.

1 INTRODUCTION

The durability of composite materials is the result of a complex interaction between the reinforcing fiber, the polymer matrix and the fiber/matrix interface [1, 2]. In the composite, the homogeneous stress state usually found in the neat polymer is radically modified by the internal stress state of the matrix, as well as by additional stress resulting from loading of the composite structure [3].

Internal stresses build up as a consequence of different thermal and mechanical properties between matrix and fibers, and are particularly sensitive to processing conditions and thermal cycling. Simultaneously, when the polymer matrix is cooled below its glass transition temperature T_g, it departs from thermodynamic equilibrium. The subsequent slow evolution of the polymer towards equilibrium, termed structural recovery, has been shown to affect its mechanical properties significantly, by a process known as physical aging [4]. Changes in viscoelastic properties of the composite during physical aging have been studied for instance using classical laminate theory [5, 6] or finite element method (FEM) [7-10].

The influence of local stress concentrations at the fiber/matrix interface on the structural recovery and aging of the polymer has not been clearly established, although it has been shown that structural recovery profoundly changes the viscoelastic response of the polymer [11]. Such a phenomenon hampers the analysis of the failure of composites, controlled to a large extent by the interface strength.

The present work investigates how interfacial behavior evolves upon physical aging. Firstly, the micro-mechanics of the interface and their evolution upon aging are studied through the microbond test. Secondly, this theory is coupled to that of the energy release rate and their equivalence is assessed by experimental investigation.

2 THEORY

2.1 *Modeling of the interfacial shear strength for the microbond test*

The microbond test has been extensively used to assess the shear strength of polymer/fiber interfaces [12]. In this test, the interfacial shear strength is determined from the force required to debond a droplet of cured resin from an individual fiber. Figure 1 shows a typical force vs. displacement plot obtained for such a specimen.

The analysis of the microbond test requires determining precise dimensions of the droplet L and r_d (respectively length and radius), together with the fiber radius r_f. The apparent contact angle of the solid droplets, the volume and radius of a cylinder equivalent in volume and of the same length as the droplets - referred to as the equivalent cylinder in the

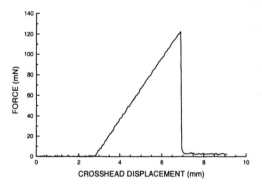

Figure 1: typical force vs. displacement plot of the microbond pullout test

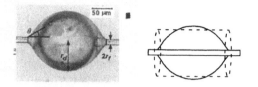

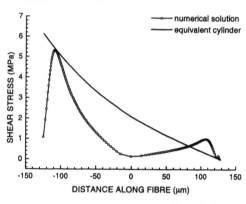

Figure 2: droplet and its simulated profile, and shear stress along droplet interface produced by loading of the droplet on the left side

following - are subsequently determined assuming that the Carroll equation [13] describing the droplet profile holds for the solid droplet, i.e. the unduloid shape of the initial droplet remains unduloid after processing.

Further, the test is conducted and its validity checked, according to the procedure described later in the experimental section. The recorded load at debonding F_{max} is then obtained, as well as the friction force after debonding.

The experimental shear strength is then determined from the force at debond F_{max} as:

$$\tau_{exp} = \frac{F_{max}}{2\pi r_f L} \tag{1}$$

In a recent work, the classical analysis of the test has been modified from a novel elastic stress transfer model [14], using two different approaches: by introducing the exact droplet profile into the analysis, or by using the equivalent cylinder method [15]. The former requires numerical integration, whereas the equivalent cylinder model results in an analytical formulation. These two approaches were shown to be equivalent and result in the determination of the same interfacial shear stress (Fig. 2). Interestingly, the two maxima observed on the numerical solution provide an insight into the commonly observed cone of debonding phenomenon.

The analytical expression of the experimental interfacial shear strength τ_{exp} is used in the present study to compare interfacial regions subjected to different levels of internal stresses, such as those resulting from physical aging [8]. It has been shown previously [15] that τ_{exp} results from two distinct contributions, namely that of the intrinsic shear strength τ_0 and that of the internal stresses, introduced through the factor K in the following equation:

$$\tau_{exp} = \tau_0 \frac{\tanh(\beta L)}{\beta L} - K \frac{\tanh(\beta L/2)\tanh(\beta L)}{\beta L} \tag{2}$$

The parameter β is determined from the elastic properties of the system (fibre and matrix moduli E_f and E_m, matrix Poisson's ratio v_m) and from its geometry [14]:

$$\beta = \left[\kappa\left(r_f E_f + \left(r_d^2 - r_f^2\right)E_m\right)\middle/ r_f E_f\left(1 + v_m\right)\right]^{1/2} \tag{3}$$

and κ is a structural parameter given by:

$$\kappa = 12.\left(2\left(r_d - r_f\right) + \xi\left(r_d^2 - r_f^2\right)\right)\middle/ r_f\left\{24r_f\right.$$
$$\times\left(r_d^2 - r_f^2\right) - 16\left(r_d^3 - r_f^3\right) - 3\xi\left(r_f^2 - r_d^2\right)^2 \tag{4}$$
$$\left.+ 6r_d\left(2 + \xi r_d\right)\left[2r_d^2 \ln\left(\frac{r_d}{r_f}\right) - \left(r_d^2 - r_f^2\right)\right]\right\}$$

where

$$\xi = -\frac{E_m}{r_f E_f} \tag{5}$$

The micro-mechanical study is completed by the determination of the internal stress factor K in the following section. Rather than using the simple formulation of a thermo-elastic stress, it accounts for time - temperature - aging time effects.

2.2 Internal stresses and their evolution upon aging

The present section establishes the necessary basis for modeling the relaxation of the internal stresses near the interface, i.e. the time-dependence of the so-called interphase. One of the most classical mechanical models presented to describe the relaxation behavior of polymers is based on a parallel combination of Maxwell elements, yielding the elementary relaxation response of an element i [16]:

$$E_i(t) = E_{0i} \exp(-t/\tau_i) \qquad (6)$$

where E_{0i} is the original modulus at t=0 associated with the ith element and τ_i is the characteristic relaxation time for this element. The relaxation modulus is then expressed for a distribution of N relaxation times:

$$E(t) = \sum_{i=1}^{N} E_{0i} \exp(-t/\tau_i) \qquad (7)$$

When the spectrum of relaxation times is expanded to a continuous distribution $\Psi(\tau)$, the above equation becomes:

$$E(t) = \int_0^\infty \Psi(\tau)\exp(-t/\tau)d\tau \qquad (8)$$

A useful expression of this integral was derived by Kohlrausch [17], which approximates equation (8) using a stretched exponential law:

$$E(t) = E_0 \exp\left(-(t/\lambda)^w\right) \qquad (9)$$

where E_0 is the unrelaxed modulus, λ is the global relaxation time and w is characteristic of the width of the relaxation time spectrum. This approach is used because of its great simplicity, the total viscoelastic response being described by means of three parameters only, and also because it has been established by several independent workers on microstructural bases [18, 19]. However, the most general form of the stretched exponential was shown to describe accurately the relaxation modulus of amorphous polymers below their T_g [20], expressed as:

$$\frac{E(t) - E_\infty}{E_0 - E_\infty} = \exp\left(-(t/\lambda)^w\right) \qquad (10)$$

where E_∞ is the relaxed modulus. In the simple case where no structural recovery occurs, the global relaxation time generally satisfies:

$$\tau = \tau_r a_T \qquad (11)$$

where a_T is a global temperature shift factor generally determined by the Williams-Landel-Ferry (WLF) law [16]. A simple linear viscoelastic law, i.e. an exponential retardation function, is not sufficient to describe the salient features of structural recovery [2], and the stress relaxation modulus $E(t)$ of a quenched and annealed amorphous glass changes not only with the load time t but also with the aging time t_e. In this sense, equation (10) together with

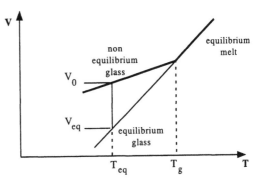

Figure 3: schematic representation of the volume-temperature behavior of a glass forming material

(11) is inadequate in describing the stress relaxation in the glassy state, since a structural recovery term δ is required to describe the aging behavior.

Indeed, the model is chosen in order to describe accurately the polymer behavior when it is aged below its glass transition temperature T_g. If one cools an amorphous polymer from above to below T_g as depicted in Figure 3, and then keeps the temperature fixed, the volume of the material evolves spontaneously towards equilibrium [21]. Such behavior is also obtained if the enthalpy instead of the volume is measured, so that the evolution of the glassy polymer's thermodynamic state has been termed structural recovery.

This change in internal state is materialized by some physical changes, such as the engineering properties of the polymer [22-25]. The ensemble of these changes has been termed physical aging, to distinguish it from chemical aging, since all the changes are completely reversible when the polymer is reheated to above its Tg.

A Narayanaswamy-Moynihan [26, 27] formalism was chosen together with a more classical Kohlrausch-Williams-Watts (KWW) [17, 28] stretched exponential function to describe the relaxation tests, and afterwards the internal stresses parameter K.

It was demonstrated in reference [29] that Equation (10) is consistent with the Narayanaswamy-Moynihan development of the KWW form. The Narayanaswamy-Moynihan development and the KAHR model of Kovacs, Aklonis, Hutchinson and Ramos [30] are equivalent in that the former involves a continuous distribution of relaxation times, while the latter deals with a discrete spectrum of relaxation times. The KAHR model formally uses a multiple ordering parameter model to define the viscoelastic response function for the glass, and to describe the volume departure from equilibrium.

Additionally, Equation (11) is replaced by:

$$\tau_i(T,\delta) = \tau_{i,r} a_T a_\delta \qquad (12)$$

187

to account for structural recovery. The shift factors are defined by:

$$a_T = \tau_i(T,0)/\tau_i(T_r,0) = \exp\{-\Theta(T-T_r)\} \quad (13)$$

$$a_\delta = \tau_i(T,\delta)/\tau_i(T,0) = \exp\{-(1-x)\Theta\delta/\Delta\alpha\} \quad (14)$$

where T_r is a reference temperature, and $\delta = 0$ denotes equilibrium. Θ and $0<x<1$ are material parameters, $\Delta\alpha = \alpha_l - \alpha_g$ with α_l and α_g being the liquid and glass coefficient of thermal expansion respectively. Importantly, one notices that the whole spectrum of retardation times is shifted along the time axis, i.e. its shape does not change. It thus supposes that w is a constant versus the aging time.

Equations (10), (12) to (14) provide a formalism for the description of the relaxation behavior of the glass once the functions for the shift factors a_T, a_δ and the material parameters are known.

Since all the experiments were performed at the same temperature, the temperature shift factor a_T was set to 1. Additionally, $\Delta\alpha$ was obtained from a quasi-linear plot obtained from the dilatometer experiments of Duran and McKenna [31].

The internal stress relaxation is assumed to follow a relaxation law similar to that of the relaxation stress of the bulk matrix:

$$\frac{K(t)-K_\infty}{K_0-K_\infty} = \exp\left(-(t/\lambda)^w\right) \quad (15)$$

where K_0 is the unrelaxed internal stress term, determined by the thermo-elastic formula [32]:

$$K_0 = a\alpha E_f\left(\alpha_m - \alpha_f\right)\Delta T/2 \quad (16)$$

and λ is obtained from Equations (12) to (14).

One should note that the interfacial region ages at a rate which may differ from that of the bulk, since the relaxation time λ accounts for the stress state calculated using our model.

2.3 Energy release rate formulation

The energy release rate at the interface is derived from Equations (2) & (4), combined with the analysis introduced by Liu & Nairn [33], who devised the limiting energy release rate G_∞ as a function of the droplet length L, taken as crack length, including friction stresses at the debonded interface and residual stresses. Observing that in microscopic specimens it is very difficult to observe crack growth, the above authors suggest interfacial toughness is estimated by assuming the peak debonding force corresponds to a debond length that is equal to the droplet length. It is thus assumed that the debond propagates stably along the interface with increasing force and

that a drop in force only occurs when the debond reaches the droplet length. The analysis is based on the exact thermo-elastic energy release rate for an n-phase composite derived in [34], for traction only boundary conditions, written as:

$$G = \frac{d}{aA}\left(\frac{1}{2}\int_S \vec{T}^0\cdot\vec{u}\,dS\right) + \frac{V\Delta T}{2}\sum_{i=1}^n v_i\alpha^{(i)}\cdot\frac{d\bar\sigma^i}{dA} \quad (17)$$

where $\vec{T}^0$ and $\vec{u}$ are the surface tractions and displacements (both mechanical and thermal), ΔT is the temperature difference between the specimen temperature and the stress-free temperature, V is the total volume, V_i is the volume fraction of phase i, $\alpha^{(i)}$ is the thermal expansion tensor of phase i, and:

$$\bar\sigma^i = \frac{1}{Vv_i}\int_{V_i}\sigma^i dV \quad (18)$$

is the phase-averaged stress in phase i, where V_i indicates integration over the volume occupied by phase i. Based on FEM results, it was further suggested that an acceptable energy release rate for a crack length of $a = L$ can be estimated by calculating $G(a)$ from the energy release curve for a droplet that has a much larger length than the actual droplet. In other words, the limiting energy release rate is obtained as

$$G_\infty(a) = \lim_{L\to\infty} G(a) \quad (19)$$

In the present case, the equivalent cylinder is therefore taken as infinite to avoid the singularity existing when the crack length equals the embedded length (Fig. 4).

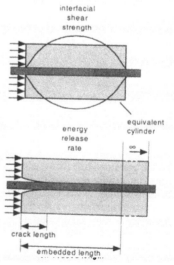

Figure 4: interfacial shear strength vs. energy release rate approach

One obtains:

$$G_\infty(L) = \frac{a}{2}\Big[C_{33s}(\sigma_d - kL)^2 + D_{3s}$$

$$\times\Big(2\sigma_d - k\Big(2L - \frac{1}{\beta}\Big)\Big)\Delta T \qquad (20)$$

$$+\Big(\frac{D_3^2}{C_{33}} + \frac{v_m(\alpha_T - \alpha_m)^2}{v_f A_0}\Big)\Delta T^2\Big]$$

where σ_d is the debond axial stress, obtained from the peak force in the force displacement plot, k is the frictional stress transfer rate parameter detailed later, α_T and α_m are the respective coefficients of thermal expansion (CTE) for fiber and matrix, and ΔT is the difference between glass transition (freshly cured samples) or aging (aged samples) temperature and room temperature.

The factor β in this analysis is taken from the previous model (Eq. (2) & (4)), and therefore corresponds to the equivalent cylinder stress transfer parameter; v_m and v_f are respectively the matrix and fiber volume fractions. The other coefficients are obtained using the following equations, in which subscripts A and T stand for axial and transverse directions:

$$A_0 = \frac{v_m(1 - v_T)}{v_f E_T} + \frac{1 - v_m}{E_m} + \frac{1 + v_m}{v_f E_f} \qquad (21)$$

$$A_3 = -\Big(\frac{v_A}{E_A} + \frac{v_f v_m}{v_m E_m}\Big) \qquad (22)$$

$$C_{33} = \frac{1}{2}\Big(\frac{1}{E_A} + \frac{v_f}{v_m E_m}\Big) - \frac{v_m A_3^2}{v_f A_0} \qquad (23)$$

$$C_{33s} = \frac{1}{2}\Big(\frac{1}{E_A} + \frac{v_f}{v_m E_m}\Big) \qquad (24)$$

$$D_3 = -\frac{v_m A_3}{v_f A_0}(\alpha_T - \alpha_m) + \frac{1}{2}(\alpha_A - \alpha_m) \qquad (25)$$

$$D_{3s} = \frac{1}{2}(\alpha_A - \alpha_m) \qquad (26)$$

By expressing the frictional stress as Coulombian we let the frictional parameter k account for lateral shrinkage during curing, cooling to the test temperature and aging. To do so, the radial shrinkage is introduced following the two concentric cylinders approach, and assuming that Poisson's effects are negligible (see for example ref. [35]).

3 EXPERIMENTAL PROCEDURE

The resin employed is a Diglycidyl Ether of Bisphenol A (DGEBA, Dow Chemicals DER332), cured with a diamine (Jeffamine D400) in a stoechiometric amount. The resin is preheated during 3 hours at 60°C to melt any residual crystals. The two constituents are then stirred manually until the solution is clear, and carefully degassed in a vacuum oven at 40°C for 10 min. Reproducible droplet sizes are obtained using a fiber that is dipped into the resin and removed at a relatively high speed. The droplets that form on the fiber are subsequently transferred to the experimental fiber by simple contact. The cure cycle is chosen as described in previous publications [22], i.e. 24 hours at 100°C and then quench to the room or aging temperature. A matrix Young's modulus of 1.85 GPa and Poisson's ratio of 0.35 are determined by Electronic Speckle pattern Interferometry (ESPI) [3, 36] in the strain range 10^{-4} to 10^{-3}.

E-type glass fibers of diameter of ca 11 μm, modulus $E_f = 72$ GPa, and a γ-APS (OSI Specialties aminosilane 1100) surface treatment, are used.

In order to determine the droplet dimensions, the fibers are mounted under zero pre-tension on lateral supports on a microscope slide. The fibers and droplets are then stuck on a support using a cyanocrylate glue and the assemblage mounted on a 4 N load cell. Extreme care is exercised in aligning the support and fiber perpendicular to the jaws of the microbond device.

The pullout set-up used in this work uses Teflon jaws (2.5mm thick) clamped by a pneumatic set-up. The device allows a negligible friction force on the fiber, while the radial clamp force on the fiber is controlled within 5 kPa. The jaws are mounted on a stable support, and pullout is carried out through vertical displacement of the load cell, mounted on the crosshead of a tensile testing machine at a speed set of 0.5mm/min.

Conventionally, when a large population of droplets are tested, a scatter in debonding force with a standard deviation as high as 20% is usual. In order to limit this scatter, the following criteria were defined from the force versus displacement curves up to debonding [15]. While the ideal force vs. displacement graph would be such as Figure 1, multiple breaks, anomalies at the initiation of the load, and trials where cohesive rather than adhesive failure has occured are discarded from the analysis. Obviously, the tests where the fiber breaks before the interface are also discarded.

The resin prepared in the conditions described above was cast into silicone moulds and cured into the oven at the same conditions. Dog-bone shaped samples were then milled in the cast coupons, and further tested in small-strain relaxation, as described in the following.

The resulting glass transition temperature T_g of 42.1°C was measured by Differential Scanning Calorimetry (heating rate 20°C/min). The aging procedure for both casted coupons and microbond specimen was: first, annealing 2 hours at 70°C, then quenching to the aging temperature (37.0°C), isothermal aging during the aging time t_e, and finally testing at room temperature of 22°C.

The dynamic viscoelastic data cover a two-decade frequency range from 1 to 100 rad/s and temperatures from 17°C to 67°C, it was recorded on a Dynamic Mechanical Thermal Analyzer RSA II (Rheometrics).

Next, small-strain relaxation tests were performed at different aging times on the same epoxy, according to the following procedure. The specimen was brought to the fixed initial strain $\varepsilon_0 = 0.01$ with a fast loading rate of 2.10^{-2} s^{-1}. The strain was then maintained at this value by means of a closed loop control system. The stress evolution $\sigma(t)$ was recorded as a function of time for a relaxation period of 1 hour. Following previous studies, the results presented in Figure 6 were expressed in terms of the relaxation modulus $E(t) = \sigma(t) / \varepsilon_0$. In order to increase the resolution of E_0 it was further measured using ESPI in the strain range 10^{-4} to 10^{-3}.

4 RESULTS AND DISCUSSION

4.1 Viscoelastic behavior of the matrix subject to physical aging

The exponent w characterizing the width of the spectrum of relaxation times was first determined by the dynamic viscoelastic properties of the epoxy through the glass transition region.

In this experiment, the complex tensile modulus E^* was studied. It can be separated into the real and imaginary parts E' and E":

$$E*(z) = E'(z) + iE''(z) \qquad (27)$$

where z is a non dimensional frequency.

Starting from KWW Equation (10), Chow has demonstrated using one Fourier transform followed by one Laplace transform that the storage and loss moduli can be efficiently calculated, respectively, from [37]:

$$\frac{E'-E_\infty}{E_0-E_\infty} = 1 + \sum_{m=1}^{\infty} \frac{(-1)^m \Gamma(m\lambda+1)}{m! z^{m\lambda}} \cos(m\lambda\pi/2) \qquad (28)$$

and

$$\frac{E''-E_\infty}{E_0-E_\infty} = \sum_{m=1}^{\infty} \frac{(-1)^{m+1} \Gamma(m\lambda+1)}{m! z^{m\lambda}} \sin(m\lambda\pi/2) \qquad (29)$$

A comparison between the theory and experiment for the epoxy resin is carried out in Figure 5. The experimental points in Figure 5 have been shifted horizontally along the time axis to form the 'master curve'. The reference temperature for the master curve is 41.2°C. The full curves present the theoretical calculation where $(E_0, E_\infty) = (1.85, 0.01)$ GPa and the exponent w equals 0.33.

The relaxation modulus of the epoxy network is shown in Figure 6 as a function of aging time, and described by a KWW type model [23]. This type of representation is used to obtain the time – aging time shift factor a_{te}, as well as the aging rate [4].

This experiment was completed by the determination of the Young's modulus variation with aging time by ESPI (Fig. 7).

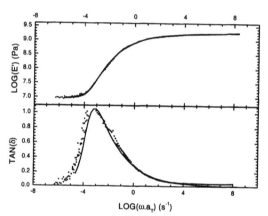

Figure 5: storage modulus and loss tangent of the DER332/D400 epoxy as measured by RSA

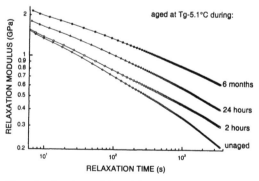

Figure 6: relaxation modulus of the epoxy network at 20°C after aging at Tg-5.1°C up to 6 months

The horizontal shifts, a_{te}, along the time axis were then obtained as:

$$a_{te} = \lambda(t_e) / \lambda(t_e(ref)) \qquad (30)$$

where $\lambda(t_e)$ is the value of λ at the relevant aging time and $\lambda(t_e(ref))$ is the λ at the reference aging time. The reference aging time was chosen as 7 min, since a perfectly unaged state of the material cannot be reached. Since the time to cool and mount the sample was set to 14 min, a $t_e/2$ correction was applied in the calculations, i.e. the aging time was increased by $t_e/2$ [38]. This correction appears to be negligible when the aging time is greater than two hours, otherwise it is quite important. In Table 1, we present the parameters determined for each aging experiment following Equation (10), and the resulting double logarithmic plot of aging time shift factor a_{te} versus aging time is shown in Figure 8.

The linearity obtained allows the determination of the double-logarithmic shift rate μ, defined as:

$$\mu = d\log(a_{te}) / d\log(t_e) \qquad (31)$$

From the present set of experiments the shift rate was determined as $\mu = (7.9\pm0.1)\times10^{-2}$ in Figure 8. It was further found that the exponent w equals 0.33 ± 0.04 from the relaxation tests depicted in Figure 6. It is thereby verified that within experimental error viscoelastic properties of the epoxy studied involve one unique distribution of relaxation times, and the shape of the relaxation time spectrum defined by w does not depend upon the aging time.

4.2 Effect of aging on the interfacial shear strength

Conventionally, the interfacial shear strength τ_{exp} is plotted versus the droplet length and is found to decrease with increasing specimen size (Fig. 9). A marked increase, close to 40%, in τ_{exp} is found upon aging as shown in Figure 9, which is attributed to the presence of internal stresses and to the densification of the matrix. It was verified that no chemical aging was present in a separate experiment by re-heating a series of aged microbond samples above T_g and performing the test, which resulted in similar values of τ_{exp}. To clarify this considerable increase, internal stresses were calculated according to Equation (15).

The resulting simulation shown in Figure 9 compares well with the experiment, and allows the determination of a unique τ_0 value, independent of aging time, and equal to 22.1 ± 0.2 MPa. This result justifies the hypothesis that aging properties of the matrix alone affect those of the interfacial region.

It should also be noticed that when internal stresses are neglected in the analysis, one finds values for τ_0 of 25.1 (unaged) and 37.6 MPa (aged).

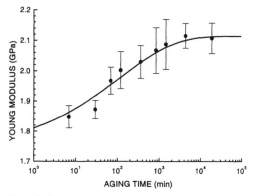

Figure 7: Young's modulus vs. aging time determined by ESPI

Table 1: Equation (10) parameters from relaxation experiments on epoxy aged at various times

aging time t_e (min)	E_0 (MPa)	E_∞ (MPa)	λ (s)
7	2210	201	46.8
127	2170	290	51.2
367	2330	302	66.7
727	2345	338	71.0
1447	2430	361	78.5
259200	2745	551	107.8

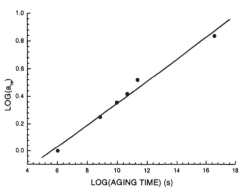

Figure 8: shift factor a_{te} vs. aging time

Since the chemistry of the interface was kept unchanged, it seems unlikely that an intrinsic parameter, describing the strength of the interface, would have changed with aging time.

This interesting finding, which indicates that physical aging does not change the intrinsic properties of the interfacial region, needs to be further investigated. To this end, the energy release rate approach is used as an alternative and is detailed in the following section.

191

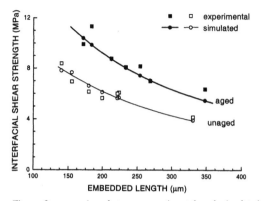

Figure 9: comparison between experimental and simulated IFSS vs. embedded length

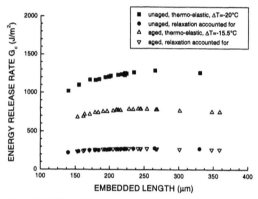

Figure 10: Effect of internal stress level on R-curves (Coulombian friction is accounted for)

4.3 Energy release rate formulation

In the present section, the previous experimental results analyzed through the interfacial shear strength criterion are used. It is found that the average friction estimated from the experimental data after debonding closely matches that determined using least squares with minimization of the standard deviation on G_∞.

The energy release rate is determined following several stress levels, as shown in Figure 10. These stress levels were calculated for the thermo-elastic case or including their relaxation upon aging, using an expression similar to Equation (15).

Due to the high CTE of the epoxy matrix, important internal stresses are developed. Neglecting the relaxation of these stresses would lead to large overestimation of G_∞. Similarly, one would also determine a considerable decrease in G_∞ upon aging, which would contradict the previous strength analysis. This paradox is resolved by introducing the relaxation of internal stresses into the analysis, leading to a close match, with $G_\infty = 262\pm14$ J/m² (unaged),

versus $G_\infty = 266\pm10$ J/m² (aged). This analysis therefore yields the same results as the previous one, which is that physical aging does not modify the intrinsic properties of the interface, rather, it modifies the relaxation rate of process-induced internal stresses.

It is essential to notice that in the analysis of the energy release rate, the internal stress contribution is of primary importance, since it represents the major part of the energy involved in the process. If internal stresses were ignored, one would find a very low G_∞ of about 10 J/m². The important contribution of the friction factor was also demonstrated in reference [33], its major effect being to flatten the R-curves.

Although this study only considered mode II failure of the interface, the results should be of a broad significance also for other failure modes, and therefore relevant for composite laminates.

5 CONCLUSIONS

Two methods for analyzing the microbond test were presented; namely the interfacial shear strength and the energy release rate analyses. These methods were tested in the case of freshly cured and aged epoxy droplets on glass fibers, and were shown to yield similar results. In the case of the model system chosen, no changes in the intrinsic mechanical properties of the interface are found upon structural recovery of the polymer. This result is obtained providing that the relaxation of internal stresses and the shift of relaxation times of the matrix when it is physically aged are accounted for. This coupling between structural recovery and stress relaxation appears to be critical in describing the behavior of the model composite system considered, when both analyses are carried out.

6 ACKNOWLEDGEMENTS

The authors are indebted to the Swiss National Science Foundation for financial support, and also wish to thank Prof. J. A. Nairn for fruitful discussions.

7 REFERENCES

1. Reifsnider, K.L., 1994. *Composites*, **25**(7): 461.
2. McKenna, G.B., 1994. *J. Res. NIST*, **99**(2): 169.
3. Kim, P., 1995. PhD Thesis. EPFL.
4. Struik, L.C.E., 1978. in *Physical Ageing in Amorphous Polymers and Other Materials*. Amsterdam: Elsevier Science Publishers.
5. Monaghan, M.R., L.C. Brinson, and R.D. Bradshaw, 1994. *Composites Engineering*, **4**: 1023.

6. Brinson, L.C. and T.S. Gates, 1995. *Int. J. Solids Structures*, **32**: 827.

7. Gurjar, A., D.G. Zollinger, and T. Tang, 1996. *Transp. Res. Rec.*, **1529**: 95.

8. Chambers, R.S., F.P. Gerstle, and S.L. Monroe, 1989. *J. Am. Ceram. Soc.*, **72**: 929.

9. Chiang, M.Y.M. and G.B. McKenna, 1994. *Polym. Eng. Sci.*, **34**: 1815.

10. Roy, S. and S. Denduluri, 1996. Proc. 1996 11th *Techn. Conf. of the Am. Soc. Compos.*, Atlanta: Technomic Publ Co Inc, 603.

11. McKenna, G.B., Y. Leterrier, and C.R. Schultheisz, 1995. *Polym. Eng. Sci.*, **35**: 403.

12. Zhandarov, S.F. and E.V. Pisanova, 1997. *Compos. Sci. Technol.*, **57**: 957.

13. Carroll, B.J., 1976. *J. Colloid Interf. Sci.*, **57**(3): p. 488-495.

14. Mendels, D.-A., Y. Leterrier, and J.-A.E. Månson, 1999. *in press in J. Compos. Mater.*

15. Mendels, D.-A., Y. Leterrier, and J.-A.E. Månson, 1999. *to be subm. to J. Compos. Mater.*

16. Ferry, J.D., 1970. in *Viscoelastic Properties of Polymers*. New York: John Wiley and Sons.

17. Kohlrausch, F., 1847. *Pogg. Ann. Phys.*, **12**: 393.

18. Chow, T.S., 1985. *J. Non-Crystal. Sol.*, **75**: 209.

19. Ouali, N., M.B.M. Mangion, and J. Perez, 1993. *Phil. Mag. A*, **67**: 827.

20. Chow, T.S., 1990. *J. Mater. Sci.*, **25**: 957.

21. Kovacs, A.J., 1963. *Fortschr. Hochpolym.-Forsch.*, **3**: 394.

22. Lee, A. and G.B. McKenna, 1988. *Polymer*, **29**: 1812.

23. Lee, A. and G.B. McKenna, 1989. *Polymer*, **31**: 423.

24. McKenna, G.B., *et al.*, 1991. *J. Non-Cryst. Sol.*, **131-133**: 497.

25. G'sell, C. and G.B. McKenna, 1992. *Polymer*, **33**: 2103.

26. Moynihan, C.T., *et al.*, 1976. *Ann. N.Y. Acad. Sci.*, **279**: 15.

27. Narayanaswamy, O.S., 1971. *J. Am. Ceram. Soc.*, **54**: 491.

28. Williams, G. and D.C. Watts, 1970. *Trans. Faraday Soc.*, **66**: 80.

29. Chow, T.S., 1984. *Macromol.*, **17**: 2336.

30. Kovacs, A.J., *et al.*, 1979. *J. Polym. Sci., Polym. Phys. Ed.*, **17**: 1097.

31. Duran, R.S. and G.B. McKenna, 1990. *J. Rheol.*, **34**(6): 813.

32. Gorbatkina, Y.A., 1992. in *Adhesive Strength in Fiber-Polymer Systems*. New York: Ellis Horwood.

33. Liu, C.-H. and J.A. Nairn, 1999. *Int. J. Adh. Adhes.*, **19**: 59.

34. Nairn, J.A., 1997. *J. Applied Mech.*, **64**: 804.

35. Timoshenko, G., 1970. in *Theory of Elasticity*: McGraw-Hill.

36. Leendertz, J.A., 1970. *J. Phys. E.: Scientific Instruments*, **3**: 214.

37. Chow, T.S., 1988. *Polymer*, **29**: 1447.

38. Crissman, J.M. and L.J. Zapas, 1989. *Polymer*, **30**: 447.

Recent Developments in Durability Analysis of Composite Systems, Cardon, Fukuda, Reifsnider & Verchery (eds)
© 2000 Balkema, Rotterdam, ISBN 90 5809 103 1

Accelerated evaluation of environmental ageing monitored by fatigue crack growth experiments

Volker Altstädt, Sven Keiter & Michael Renner
Polymer Engineering, Technical University Hamburg-Harburg, Germany

Alois Schlarb
Braun Medical AG, Escholzmatt, Switzerland

ABSTRACT: This article describes fatigue crack growth experiments to investigate the degradation of the durability of polymers due to fluid environments. The degradating effect of media causing stress cracking can be observed on the fracture surfaces of tested samples by scanning electron microscopy. Strategies to improve environmental stress cracking like changes in molecular weight, orientation, toughening with rubber particles of different sizes are discussed. Fatigue crack growth experiments can be employed as a very fast and effective screening method.

1 INTRODUCTION

Fatigue crack propagation (FCP) experiments can be employed as a fast and effective screening method for determining long-term mechanical properties of polymers. Advantages of this method, like the need for a very small quantity of test material (<10 g), the broad range of fatigue crack propagation rates (from 10^{-2} mm/cycle to 10^{-7} mm/cycle) measured within one specimen and a well defined stress state at the crack tip favor the use of FCP experiments instead of traditional S-N curves (Altstädt & Loth & Schlarb 1995). These experiments can also be conducted in the presence of a specific environment and are able to provide valuable information about environmental effects on crack propagation.

If the FCP experiment is carried out in the presence of a critical fluid environment, a more or less progressive degradation of mechanical properties (embrittlement) depending on the polymer can be observed. This degradation is caused by enhanced disentanglement and chain scission of the molecules affected by a solving liquid. The process is called environmental stress cracking (ESC). ESC is defined as the simultaneous action of stress and contact with specific fluids. Approximately 15% of all failures of polymer components are due to ESC. Since prediction in many cases is exceedingly difficult, assessment through suitable laboratory tests becomes important (Kambour 1973). Standard tests for environmental stress cracking of polymers are the *Ball or Impression Method* (ISO 4600), the *Bent Strip Method* (ISO 4599) and the *Constant Tensile Stress Method* (ISO 6252).

In these standards either a constant static load or a constant deformation is applied. In the case of the

bent strip method and the impression method the residual strength is quantified through a succeeding impact or tensile test. This, however, does not imply that the stress state at the crack tip in the moment of failure is well defined. Moreover, the common standardized test methods do not consider a well defined stress state within the test specimen. This could be achieved by applying the principles of linear elastic fracture mechanics to ESC test procedures in combination with a Compact Tension (CT) specimen. In contrast to tensile or impact tests, the stress state at the crack tip of a CT-specimen is well defined. A specific fluid is able to penetrate to the front of the crack tip. Under cyclic loading the propagation of a fatigue crack through the bulk material of the CT-specimen is affected by the presence of the fluid. From the dynamics of crack propagation an information about the interaction between the crack tip, designated as a microscopically small probe, the fluid environment and the ESC sensitivity of a specific polymer, polymer blend or polymer composite can be achieved.

1.1 Fatigue crack growth

All fatigue failures in polymers or polymer composites involve one phase in which a defect zone such as a craze or microcrack initiates, followed by a propagation phase to final fracture. Based on the assumption, that the fatigue lifetime is determined by the propagation phase, a preexisting flaw is assumed. The stress state at the tip of the crack is defined by the stress intensity factor range ΔK. For the case of fatigue, Paris (1964) showed, that a linear relationship predicted by a simple power law for a double logarithmic scale exists between the FCP rate da/dN and

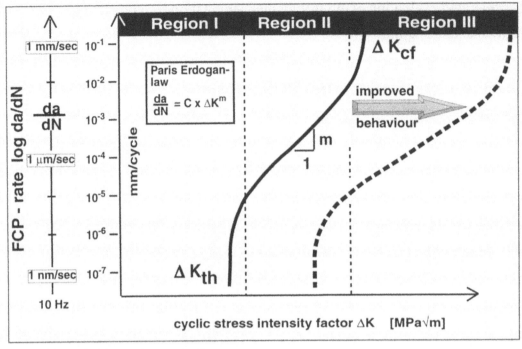

Figure 1. Scheme of a fatigue crack propagation diagram

the applied ∆K (Figure 1).

The linear dependence is frequently observed only over an intermediate range of growth rates. When investigating a wide range of da/dN, deviations from this linear behavior may be observed, as illustrated schematically in Figure 1. That is, FCP-rates decrease rapidly to vanishingly small values as ∆K approaches the threshold value ΔK_{th}. This ∆K level defines a design criterion that is analogous to the fatigue limit determined from traditional S-N curves (Hertzberg 1980, Janzen 1991). FCP rates increase markedly as ∆K approaches K_{cf}, at which unstable fracture occurs within one loading cycle. From the standpoint of evaluating a materials fatigue resistance, any decrease in FCP rates at a given value of ∆K or, alternately, any increase in ∆K to drive a crack at a given speed is, of course, beneficial.

As shown in Figure 2, the experiments can be conducted also under constant ∆K conditions. In this case, specific influences of a medium could be detected by a change in the crack propagation rate as a function of exposure time. For metallic materials fatigue crack growth experiments under environment are described in ASTM E 647, for polymers no standard procedure exists until now.

Environmental stress cracking (ESC) of polymers is a phenomenon which has been researched over a period of more than 40 years. The phenomenon involves so many influential variables that the behavior cannot be predicted with sufficient accuracy. The only alternative is testing. For this any advances in test methods are important to make research more efficient and effective.

2 EXPERIMENTAL

2.1 Materials

Amorphous thermoplastics are sensitive to ESC. Particular Polystyrene (PS) is well known to be sensitive against fluid permeation causing crazing (Kambour 1973). To study the effect of molecular weight and various types of rubber modification on the ESC behavior under fatigue crack growth conditions, commercial PS polymers were selected for this investigation. The materials were kindly supplied by BASF AG, Ludwigshafen, Germany and are listed in Table 1.

For the investigations regarding lipid resistance two different polycarbonates have been used. The specimens were kindly supplied by B. Braun Medical AG, Switzerland.

All polymers were processed by injection molding under standard conditions and tested considering the injection molding direction.

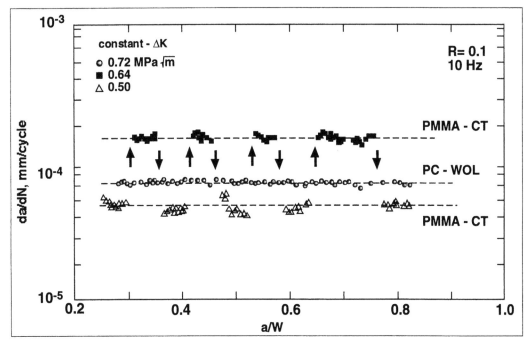

Figure 2. Experiments under constant ΔK (Lang 1984).

2.2 Test Methods

Compact-type (CT) specimens were cut from rectangular injection molded plates with 4 mm nominal thickness and precracked by a razor blade. All fatigue crack growth tests were run on a Schenck PSA® servohydraulic test system at a frequency of 10 Hz under sinusoidal loading in tension-tension with a minimum to maximum load ratio (R) of 0.1 at room temperature. The tests were run under ΔK control with a software designed by Fracture Technology Associates, Inc.. A ΔK-decreasing portion and a ΔK-increasing portion were measured separately with two different specimens of the same material and combined to one FCP-diagram. The crack length was monitored by a compliance technique, as published by Saxena & Hudak (1978).

The tests under environment were carried out by applying the critical fluid through a soaked sponge,

which was fixed on both sides of the specimen as an unlimited source. By this the crack was always covered with the soaked sponge. For the test with Polystyrene (PS) a commercial sunflower oil (tradename Livio®) was used. The tests with Polycarbonate (PC) were performed with a fat emulsion (Lipofundin® MTC 20%) for parenteral nutrition.

In this investigation all tests were run within load amplitude limits under ΔK control. The fatigue behavior would be expected to deteriorate because under the softening effect of a liquid medium, the samples would be strained more extensively for each loading cycle.

2.3 Fractography

Fracture surfaces were studied with a field emission electron microscope LEO 32 . The microscope was operated at an accelerating voltage of 0.5-1 kV. Because of the low accelerating voltage no gold coating of the specimen was necessary and the specimens could be investigated directly after fracture of the specimen in the fatigue experiment.

3 RESULTS AND DISCUSSION

3.1 Effect of molecular weight

Physical properties of polymers are strongly dependent on the average molecular weight M_w and se-

Table 1. Investigated materials.

®Trademark of BASF AG, Germany

material	property
PS 148H®	polystyrene - M_w 238,000 g/mol
PS 168N®	polystyrene - M_w 354,000 g/mol
PS 486M®	polystyrene impact modified by small PB particles
PS 2710®	polystyrene impact modified by large PB particles

quence distribution M_W/M_N, because the presence of molecular entanglements can significantly affect the mechanical behavior. As previously shown by Altstädt (1997) for different commercial PS systems prepared by free radical polymerization and by anionic polymerization the number average M_N of the molecular weight corresponds well to an increase of the fatigue crack growth behavior. Further experiments were conducted in this study to explore the effect environmental ageing.

The effect of molecular weight on the ESC behavior was investigated for two different PS systems. These were tested with and without oil environment. As shown in Figure 3 - 4 both PS systems exhibit a severe decrease in fatigue crack growth resistance when tested under oil environment. In both cases, ΔK for a given fatigue crack growth rate is significantly reduced and the slope of the curve is increased, simultaneously a decrease in ΔK_{th} is observed.

The comparison of both PS systems clearly shows an improved ESC resistance under fatigue loading of the PS with the higher molecular weight. This is reflected in a smaller slope of the linear portion of the FCP-diagram and a higher ΔK_C and ΔK_{th} for PS 168N.

3.2 Effect of orientations

To investigate possible orientations of the molecules due to the injection moulding process on the ESC resistance under fatigue loading conditions, samples of polystyrene were tested parallel and perpendicular to the injection molding direction. A specimen prepared from granules by compression molding, which should have a minimum of orientation, was included in the investigation as a reference. To minimize the effect of a plastic deformation possibly induced by precracking with the razor blade, the precrack was extended for a minimum 2 mm by fatigue loading before starting the experiment. In some cases it was difficult to carry out the test, because the crack grew parallel to the injection direction or an additional crack was initiated on the fixing holes of the samples.

As shown in Figure 5 the FCP-diagram of the specimens tested perpendicular to the injection moulding direction is shifted significantly to higher ΔK values compared to the specimens tested parallel. While the reference specimens processed by compression molding behave similar to those with parallel orientation. Obviously it is easier for the fatigue crack to propagate in the direction of the oriented entanglement network. This can be explained by the fact, that crack propagation in PS is accompanied by crazing. Molecules which are already stretched in one direction are losing the ability to fibrillate in the other direction. By this the probability for chain scission is increasing and the breakdown of the material drawn into the process zone occurs at lower ΔK values.

If polystyrene is tested perpendicular to the injec-

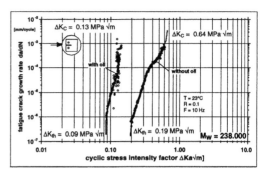

Figure 3. *FCP with and without a stress cracking media - PS 148H.*

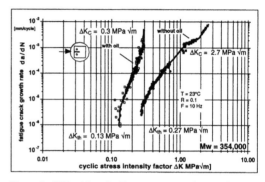

Figure 4. *FCP with and without a stress cracking media - PS 168N.*

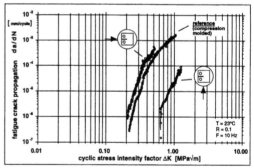

Figure 5. Molecule orientation effects on crack propagation - *PS 148H.*

tion molding direction and a stress cracking media is present at the same time, an additional embrittlement can be observed by the steep increase in the FCP-curve which makes it almost impossible to measure the dynamic of crack propagation in a broad range of crack propagation rates as usual (Figure 6).

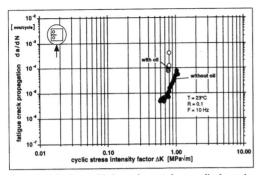

Figure 6. Injection molded samples tested perpendicular to the injection direction   - PS 168N.

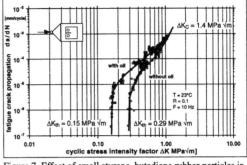

Figure 7. Effect of small styrene–butadiene rubber particles incorporated in PS   - PS 486M.

3.3 Toughness modification by rubber particles

A common strategy to improve the mechanical behavior of thermoplastics is to incorporate uniformly distributed rubber particles in the polystyrene matrix. In the case of HIPS (high impact polystyrene) styrene-butadiene block copolymers are used. This strategy is also applied to improve the environmental stress crack resistance of polystyrene.

For this investigation two commercial PS grades with different average diameter of the rubber particles were selected (PS 486M: Ø ≈ 2 μm, PS 2710: Ø ≈ 5 μm). The systems were tested with and without exposure to vegetable oil. The corresponding FCP-diagrams are shown in Figure 7 - 8.

The FCP behavior of the two rubber modified polystyrenes is comparable in the threshold region and in region III (Figure 1) at high propagation rates. A significant difference is visible in region II of intermediate crack propagation rates 10^{-5} to 10^{-4} mm/cycle. Particular PS with larger rubber particles is more effective in decreasing the slope of the crack growth curve.

Compared to the unmodified PS 148H and PS 168N (Figure 3 - 4) the observed ranking in terms of FCP behavior is PS 148H < PS 486M = PS 2710 < PS 168N. This ranking is changed completely if the tests are conducted in the presence of vegetable oil: PS 148H << PS 168N < PS 486M < PS 2710. Obviously rubber toughening is more efficient for the improvement of the ESC resistance under fatigue crack growth conditions as the increase in molecular weight corresponding to PS 148H and PS 168N. A remarkable difference between the two rubber modified systems is visible in the threshold region. The ΔK_{th} value of PS 2710 is not affected by the presence of the medium while ΔK_{th} of PS 486M is reduced by a factor of two.

The observation of the fracture surfaces in Figures 9 - 12 reveals the different size of the rubber particle of the two PS systems (Figure 9 - 11). Tested in air, in both cases the crack propagates through the rubber particles which is only possible in the case of good interfacial bonding. Tested in vegetable oil, the fracture surface appears to be very smooth. In the case of PS 486M some marks from the underlying morphology are visible.

It is generally accepted that rubber particles act as stress concentrators, initiate crazes as well as participate in their termination. The contribution of toughness depends on the concentration of the rubber, the particle size and the interfacial bonding. In the case of fatigue crack growth small particles are more efficient at low crack propagation rates because the plastic zone size is smaller, while larger particles are more efficient at higher propagation rates and the corresponding larger plastic zone diameter.

Under the influence of a stress cracking medium it seems that in the case of large rubber particles small crazes may easily develop. As a consequence the diffusion rate of oil is reduced, because of the increased fibril density. Small rubber particles show less intensive crazing and the crazes are easier overloaded because of a higher effective opening of the crack tip. In the latter case the oil can penetrate easier through the crazes. This effect is stronger in region I of the FCP curve as compared to region II (Figure 1). At high crack propagation rates, it seems to be possible, that the oil is not able to penetrate fast enough to the crack tip which explains the same ΔK_C values of the two rubber modified systems.

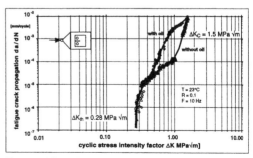

Figure 8. Effect of large styrene–butadiene rubber particles incorporated in PS   - PS 2710.

199

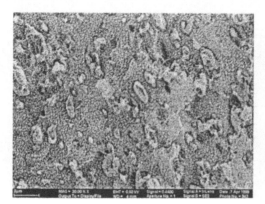

Figure 9. FCP fracture surface of PS 486M tested under air environment.

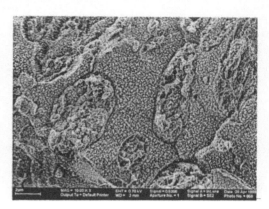

Figure 11. FCP fracture surface of PS 2710 tested under air environment.

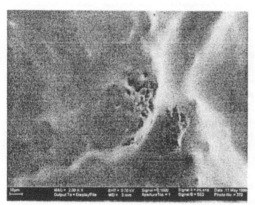

Figure 10. FCP fracture surface of PS 486M tested under oil environment.

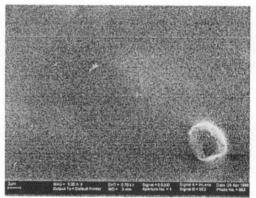

Figure 12. FCP fracture surface of PS 2710 tested under oil environment.

Figure 13. Three way stop cock as a component for medical infusion systems   - PC1.

It should be mentioned at this point, that our interest here is rather in the effect of critical fluids on fatigue crack propagation. The total fatigue lifetime

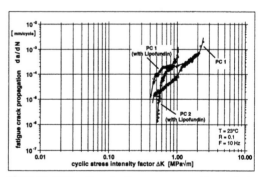

Figure 14. *FCP with and without a stress cracking media* in polycarbonate PC1 (reference) and PC2 (improved).

of unnotched specimens is determined by a initiation and a propagation phases. The initiation phase could be significantly effected by the presence of rubber particles (Sauer & Chen 1983). The effect of critical fluids on the initiation phase has to be studied separately.

3.4 Lipid resistance of medical devices

For medical devices, in particular for components of infusion systems, transparent polymers are indispensable. Because of the multitude of drugs possibly flowing through these systems, the need for disinfection of the devices and the possibility to connect varying components by force locking, the ESC resistance plays a decisive roll for the selection of a suitable material. Figure 13 shows exemplary a three way stop cock as a important component of such a system.

In principle the ESC resistance can be controlled within limitations, which are mostly given by the viscosity for processing, by the molecular weight and molecular weight distribution of the polymer. Because of this reason polycarbonate with a molecular weight of above 30,000 g/mol is used for this application.

Specially in the case of parenteral nutrition with lipid containing emulsions the occurrence of stress cracking by the intravenous infusion in three way stop cocks made by polycarbonate could not be completely excluded.

A significant improvement by the administration of a of lipid containing emulsion was achieved with a special additive to polycarbonate. As shown in Figure 14, the better behavior of the new polycarbonate PC2 proved into practice could also by verifyed by fracture mechanical fatigue crack growth experiments. In the presence of the fat emulsion the higher lipid resistant PC2 shows a higher fatigue threshold value ΔK_{th} as well as an improvement by a factor of two for ΔK_{cf} in comparison to PC1 used so far.

4 CONCLUSIONS

Fatigue crack propagation (FCP) experiments can be employed as a fast and effective screening method for the evaluation of environmental. As shown for PS, molecular weight, molecular orientation and rubber toughening are playing an important role on the ESC behavior.

ACKNOWLEDGMENT

We thank Dr. F. Ramsteiner and Dr. W. Loth from BASF AG, Polymer Research Division, Ludwigshafen, Germany for the helpful discussions.

REFERENCES

Altstädt V. & Loth W. & Schlarb A. 1995, *Comparison of Fatigue Test Methods for Research and Development of Polymers and Polymer Composites*; Proceeding of the International Conference on Progress in Durability of Composite Systems. Brussels, Belgium, 16-21.7.1995.

Altstädt V. 1997, *Fatigue Crack Propagation in Homopolymers and Blends with High and Low Interphase* Strength; Proceedings "European Conference on Macromolecular Physics - Surfaces and Interfaces in Polymers and Composites", Lausanne, June 1st-6th 1997.

Bubeck 1981, Polym. Eng. Sci. 21, p 624.

Hertzberg R.W. & Manson J.A. 1990, *Fatigue of Engineering Plastics* New York: Academic Press.

Janzen, W. & G.W. Ehrenstein 1991. *Bemessungsgrenzen von glasfaserverstärktem PBT bei schwingender Beanspruchung.* Kunststoffe 81-3, p. 231.

Kambour, R.P. & Gruner C.L. & Romagosa E.E. 1973, J. Polymer. Sci., Polym. Phys., Vol.11, p.1879.

Lang, R.W. 1984, *Applicability of Linear Elastic Fracture Mechanics to Fatigue in Polymer and Short - Fiber Composites;* Ph.D. Dissertation Lehigh University, Bethlehem, PA, USA.

Paris, P. C. 1964, Proceedings of the 10th. Sagamore Conference. Syracuse Univ. Press, NY, p. 107.

Sauer, J. A. & Chen C. C., *Crazing and Fatigue Behavior in One- and Two-Phase Glassy Polymers*; Advances in Polymer Science, 52/53, Editor: H. H. Kausch.

Saxena A. & Hudak S.J. & Donald J. K. & Schmidt D.W. 1978, J. Test Eval. 6 p 167.

Recent Developments in Durability Analysis of Composite Systems, Cardon, Fukuda, Reifsnider & Verchery (eds)
© *2000 Balkema, Rotterdam, ISBN 90 5809 103 1*

Interphasial stress and strain fields in multi-phase particulates: Influence of moisture content

N. K. Anifantis, P. A. Kakavas & G. C. Papanicolaou
Composite Materials Group, Department of Mechanical and Aeronautical Engineering, University of Patras, Greece

ABSTRACT: The degradation mechanisms in metal-filled polymeric matrix particulates due to water absorption and the role that play both the existence itself and the specific nature of the interphase to these mechanisms are investigated. It is assumed that in the area between the matrix and filler, both considered as homogeneous and isotropic, an inhomogeneous interphase develops, whose properties and volume fraction can be determined from the overall behavior of particulate. Next, the interphase properties can be used to describe the degree of adhesion between the two main phases of the composite. Based on theoretical considerations and water absorption experimental findings, the effect of the extent of the interphase layer, the change in interphasial properties and their law of variation on the interphasial stress and strain fields developed in the area surrounding the inclusion is investigated. The results give a better insight to the micro-mechanisms responsible for the overall macroscopic behavior observed in particulates under water absorption conditions.

1 INTRODUCTION

In polymeric matrix composites the environmental conditions may cause severe changes in their overall properties due to the physical and/or the chemical degradation of the polymeric matrix. These changes are associated to the temperature variations, moisture absorption and volumetric expansion, thus producing reduction in material moduli, loss of adhesion, debonding even corrosion cracking. The increased sensitivity of such materials to environmental degradation is closely related to the complex material structure based on the existence and cooperation of different materials within the same unit cell, and accordingly, to the presence of step inhomogeneities. These phenomena usually occur either within the bulk matrix and/or in the area close to the filler-matrix interface. Material quality and reliability reasons enforce thus the investigation of the influence of the detailed material structure on the environmental degradation.

Water absorption by the epoxy resin leads to a softening of the resin with a loss of stiffness and strength. The degradation increases as the conditions become more severe (Soutis & Turkmen, 1997). It is well known that composites having high filler contents absorb little moisture and show negligible change in modulus. On the other hand, composites having matrix-dominated behavior exhibit the most moisture pickup and the greatest reduction in modulus (Shen & Springer, 1976). The effects of temperature and moisture on the nonlinear stress-strain response of composite laminates has been investigated qualitatively (Surrel and Vautrin, 1989). The applied stress also has an effect on moisture absorption in polymers and polymer composites. Thus, in composites the residual stresses may cause increased water absorption (Marom & Broutman, 1981, Kakavas et al. 1995). However, a systematic study of the effect of water absorption on the elastic response of composites can hardly be found in literature.

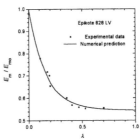

Figure 1. Degradation of the tensile modulus of elasticity of Epikote 828 LV polymer with moisture absorption.

In a composite material consisted from two main phases, due to mechanical or chemical interactions as well as to specific manufacturing conditions, a third phase is always developed laying along the matrix-reinforcement. This third phase is the so-called interphase and it is characterized by strong inhomogeneities resulted mainly from existing impurities and flaws and thus operating as low reliability and failure site region. Amongst the various models which are taking into account the existence of the interphase, the most reliable, accurate and simple are those considering exponential variation of properties in the radial direction (Kakavas et al. 1995). These models may account on for the adhesion conditions between the matrix and the filler, by setting proper discontinuities of material moduli along the filler surface (Anifnatis et al. 1997, Kakavas et al. 1998).

The effect of moisture absorption on the extent of the interphase in particulate composites was previously studied and the results are reported elsewhere (Theocharis et al. 1983). It was shown that during the process of moisture absorption, a change in the extent of the interphase develops, which is related to the degradation of the mechanical behavior of the composite and to the percentage amount of moisture absorbed. A theoretical model was developed for the evaluation of the elastic modulus in particulate composites taking into account the existence of the interphase between the main phases of the polymeric matrix composites (Sideridis et al. 1986). The degradation of composite strength or stiffness can be estimated by using a combination of empirical relations and micromechanics equations (Chamis, 1987). The objective of the present study is to characterize the effects of moisture content on the stress and strain fields developed within an heterogeneous interphase in particulate composites. The micromechanical model developed is based on the method of homogeneous subdomains. It is assumed that the filler and the matrix are homogeneous materials while the interphase is inhomogeneous with properties varying in the radial direction. Also, both the modulus of elasticity and the interphase thickness are known functions of the moisture content. Between filler and matrix proper adhesion conditions are further valid. In the present study,. It is assumed that the interphase consists from a number of homogeneous subdomains. Inside each subdomain material properties are defined in averaged sense to account for the variation of material properties within the interphase while proper equilibrium equations, and on interfacial appropriate compatibility and continuity conditions have been considered. Application of general solutions to those conditions leads to a linear algebraic system of equations for the parameters of the problem. Numerical solution of these equations yields the solution of the problem. The degree of accuracy of this method depends on the number of the subdomains selected to approach the material inhomogeneity. As the number of subdomains increases, the exponential variation of the modulus of elasticity is better approximated and consequently better accuracy is obtained. Considering the filler volume fraction, the moisture content and the adhesion efficiency as global design variables, general solutions may be derived for any virtually particulate composite. Thus, by assigning proper values to these variables particular cases may be studied.

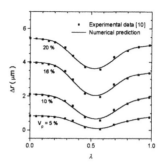

Figure 2. Interphase thickness with moisture absorption.

2 EXPERIMENTAL PROCEDURE

In the present work, without loss of generality, a particulate composite consisted from Epikote 828 LV polymeric matrix reinforced with iron particles is considered. Moisture absorption affects mainly material properties and thus, for the purposes of the present study, the drop of the tensile modulus with moisture content is experimentally investigated. Moisture content affects insignificantly Poisson's ratio and thus in the following analysis the degradation process of this material property is neglected. In order to measure the tensile modulus of the polymer, standard dog-bone specimens were prepared, which were cured at 100 °C for 24 hours and were cooled afterwards at room temperature. Diethylene-tetramine DETA, containing 8% epoxy was used as a curing agent. Voids produced during the mixing procedure, were removed by placing mixtures in vacuum containers. To investigate the moisture affected degradation process, specimens were immersed in distilled water bath with constant temperature 40 °C for various time periods. Then, specimens were removed from the bath, air dried and weighted for the determination of moisture content. Tensile tests were executed in a tensile Instron machine with crosshead speed 1 mm/min.

3 MODELLING

In Figure 1, the normalized experimental data concerning the modulus of elasticity of Epikote 828 LV polymer versus relative moisture absorption are shown . The modulus of elasticity in this Figure is normalized over the dry modulus of the polymer which was found to be E_{mo} = 3.26 GPa. In the same Figure, the moisture content is expressed in terms of the relative moisture absorption, defined as $\lambda = M / M_{max}$, where M is the percentage by weight current value of moisture absorption and M_{max} the corresponding value at the saturation point. Numerical interpolation over experimental data allows for writing the degraded modulus of elasticity of the polymer in the form:

$$E_m = E_{mo} F(\lambda) \tag{1}$$

where

$$F(\lambda) = 1 - (1 - H) \lambda e^{(1-\lambda)\gamma} \tag{2}$$

In the above equation, $H = E_{mn} / E_{mo}$ is the overall reduction of the polymer modulus under saturation conditions, E_{mn} being the maximum degradation of polymer modulus due to moisture absorption. For the Epikote 828 LV polymer, this material property was found to be $H = 0.56$, approximately. Parameter γ appearing in equation (2), equals to unity approximately, and may be determined numerically by interpolating experimental data. For the polymer under consideration, it was found $\gamma = 1.18$. Solid line in Figure 1 represents the best exponential fitting of experimental data, as expressed by equation (2) and is determined by assigning the above values for the parameters H and γ.

Assuming particles of spherical shape, a representative volume element (RVE) may be considered. This RVE consists of three separate subregions properly bonded to construct the composite, i.e., the filler, the interphase and the matrix. They have spherical shape and they are bounded by the radii r_f, r_i and r_m respectively, as shown in Figure 2. The volume fraction for each one of the phases in the composite may then be written as

$$V_f = r_f^3 / r_m^3, \qquad V_i = (r_i^3 - r_f^3) / r_m^3,$$

$$V_m = (r_m^3 - r_i^3) / r_m^3 \tag{3}$$

where V denotes volume fraction and subscripts f, i, m refer to filler, interphase and matrix respectively.

Filler and matrix are considered as homogeneous material, while the interphase material is inhomogeneous in the radial only direction. Its properties are also affected on the moisture absorption process, as described above. Next by taking into account that the moisture absorbed by the fillers is negligible, the fact that the interphase is mainly a part of the matrix phase, and the high moisture concentration in the interphase region, the variation of the interphase modulus was assumed to vary according to the relation (Anifnatis et al. 1997, Kakavas et al. 1998)

$$E_i(\rho, \lambda) = \left[E_{mo} + (\alpha E_f - E_{mo}) R(\rho) \right] F(\lambda) \tag{4}$$

where α represents the adhesion efficiency. Function $R(\rho)$ represents the radial variation of the modulus and has the form

$$R(\rho) = \frac{1 - \rho e^{1-\rho}}{1 - \rho_f e^{1-\rho_f}} \tag{5}$$

where $\rho = r / r_i$ and $\rho_f = r_f / r_i$.

Equation (4) includes moisture degradation effects, based on experimental evidence. In the same equation further, parameter α is a measure of the matrix - filler adhesion. As it is proposed in reference (Sideridis et al. 1986), parameter α may be defined as the magnitude of the discontinuity of the material moduli at the filler-interphase boundary, i.e., at $r = r_f$. According to this definition,

$$\alpha = \frac{E_{io}}{E_f} \tag{6}$$

where $E_{io} = E_i(r_f, \lambda)$. The value $\alpha = 1$ corresponds to perfect adhesion conditions, while values of α lower than unity assign for imperfect adhesion conditions. Furthermore, when $\alpha = 0$, a two phase model is obtained in which no interphase is present.

The interphase thickness $\Delta r = r_i - r_f$ is also a function of absorbed moisture as shown in Figure 3 for the same epoxy resin as it was considered above. The dependence of the interphase zone on the moisture absorption was discussed in reference (Sideridis et al. 1986), and the experimental results have shown that the interphasial region is a function of the filler volume fraction too. As the filler volume fraction increases, the thickness of the interphase increases too. For values of the moisture content close to 0.55% a minimum value of the thickness is observed. In the case of moisture absorption, the action of moisture results in partial disruption of the bonds between filler and matrix by the formation of additional cavities which could be filled with water. It has been established that the water enters at the interphase at a rate of approximately four hundred and fifty times more rapidly than in the bulk matrix itself (Sideridis et al. 1986), and consequently any absorbed moisture is concentrated preferentially into the interphase area.

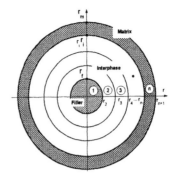

Figure 3. Description of problem by homogeneous subregions.

4 HOMOGENEOUS SUBDOMAIN APPROACH

Let us consider that the representative volume element (RVE) of the problem is discretized into n subdomains all having the shape of concentric spherical shells, as shown in Figure 2. Then, a particular subdomain, e, for e = 1, 2, ..., n, is defined between the radii r_e and r_{e+1}. Material properties within each subdomain are constant in the average sense, but different from adjacent ones, depending on inhomogenuity presented by the corresponding phase. The inner (the central) and the outer (the peripheral) suddomains represent the filler and the matrix, respectively.

In order to study the stress state inside the representative volume element, an externally applied hydrostatic pressure, p, to the RVE, is considered. Due to spherical symmetry, stresses and strains depend only on the radial abscissas. Then, the displacement field $u_e = u_e(r)$, in each subdomain e, is given by (Timoshenko & Goodier, 1951)

$$u_e = \frac{A_e}{r^2} + B_e r, \quad e = 1, 2, ..., n \quad (7)$$

where A_e, B_e are the unknown constants to be determined. The components of strain and stress within each subdomain e are given by

$$\varepsilon_{re} = \frac{d u_e}{d r}, \quad \varepsilon_{te} = \frac{u_e}{r} \quad (8)$$

$$\sigma_{re} = -\frac{2 E_e}{1 + v_e} \frac{A_e}{r^3} + \frac{E_e}{1 - 2v_e} B_e \quad (9)$$

$$\sigma_{te} = \frac{E_e}{1 + v_e} \frac{A_e}{r^3} + \frac{E_e}{1 - 2v_e} B_e \quad (10)$$

where E_e denotes averaged values of elastic modulus over subdomain, as defined by equation (4) for

the interphase or is the exact E-value for the matrix and filler subdomains; v_e denotes the Poisson ratio. Across the interfaces, between adjacent subdomains, the compatibility conditions are

$$u_{e-1}(r_{e-2}) = u_e(r_{e-1}), \quad e = 2, 3, ..., n \quad (11)$$

and the continuity conditions are simplified to the form

$$\sigma_{r(e-1)}(r_{e-2}) = \sigma_{re}(r_{e-1}), e = 2, 3, ..., n \quad (12)$$

At the outer surface of the representative volume element the equilibrium condition requires that

$$\sigma_m(r_{n+1}) = p \quad (13)$$

and at the center, the solution must be bounded, hence equation (7) yields

$$A_1 = 0 \quad (14)$$

Equations (11) to (14) provide enough conditions to solve for the unknown coefficients A_e and B_e. To this end, these equations are properly assembled to a system of linear algebraic equations of the form

$$[K]\{x\} = \{q\} \quad (15)$$

where the components of the matrix [K] contain terms of displacement and stress, as they result from equations (8) to (10); vectors appearing in equation (15) contain the unknown coefficients and the loading terms, i.e.

$$\{x\}^T = \lfloor A_1 \ A_2 \ ... \ A_n \ B_1 \ B_n \ ... \ B_n \rfloor \quad (16)$$

206

$$\{q\}^T = \lfloor 0\ 0 \ldots p \rfloor \qquad (17)$$

where superscript T means matrix transpose.

The system of equations (15) contains all the necessary information for the proposed model. In these equations the proper values of the global design variables, are also included, thus permitting analysis for any particular case. Due to the large number of equations, only numerical treatment is permitted and a proper algorithm was constructed for the determination of unknown coefficients, displacements, strain and stress fields.

5 RESULTS AND DISCUSSION

In the present study the particulate under consideration constitutes from iron spherical particles of average diameter 150 μm, modulus of elasticity 210 GPa and Poisson ratio 0.3, free of moisture absorption, embedded in Epicote 828 LV polymeric matrix. The later possess modulus of elasticity 3.4 GPa, Poisson's ratio 0.35, and is sensitive to moisture initiated degradation. The interphase is considered as inhomogeneous phase with variable thicknes, as discussed above. Each subdomain lying within the interphase has averaged values of elastic modulus, according to equation (4). The solution of the previously formulated problem yields the desired stress and strain fields for various values of the design parameters, i.e. filler volume fraction, moisture content and efficiency of adhesion. Numerical results are listed in this section in order to study the effect of these parameters on the mechanical response of the system. Results are illustrated only in the region of the interphase as in this region concentration of degradation phenomena occurs. For reasons of simplicity, all subsequent results refer to a typical filler volume fraction of $V_f = 5\%$.

Figure 4 illustrates the influence of moisture absorption and adhesion effectiveness on the strain distribution components along the interphase thickness. Radial strain component presents a strong ascending variation with radial abscissas, although a low sensitivity is observed to moisture absorption rates. Improper adhesion conditions are shown to increase radial strain and further intensify moisture affected degradation (Figure 4.a). Tangential strain component is too small as compared to radial one, and is rather insensitive to moisture changes (Figure 4.b). This strain component seems to change significantly with the degree of adhesion. The overall effect of moisture content on peak strains observed at $r = r_i$, is illustrated in Figure 5. Along with as-

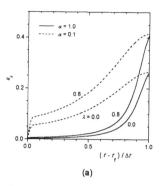

Figure 4.a. Distribution of radial strain within the interphase.

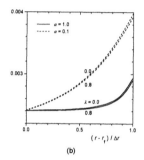

Figure 4.b. Distribution of tangential strain within the interphase.

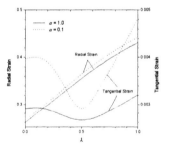

Figure 5. Peak strains with moisture absorption.

cending behavior of radial component of strain, a local minimum is observed in tangential component, as moisture increases. The later behavior probably can be attributed to the minimum interphase thickness occurring at $\rho = 0.5$. Considering both strain components, most sensitive to adhesion effectiveness is shown to be the tangential component. Imperfect adhesion further seems to soften the material structure thus producing higher strains.

Figure 6 illustrates the stress degradation process within the interphase and the influence of adhesion effectiveness. In the same Figure the normalized

207

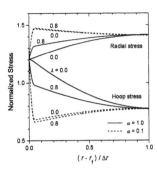

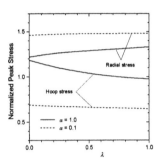

Figure 6. Distribution of stress components within the interphase.

Figure 7. Peak stresses with moisture absorption.

stress over the applied pressure magnitude p components are also shown. Both stress components present upper and lower bounds at the interphase boundaries in a reverse manner, although the radial stress component is slightly higher. The moisture content seems to enforce step stress gradients in the region very close to the filler which are further intensified when improper adhesion conditions occur. Present results show that most important moisture initiated stress degradation phenomena take place within the interphase close to the filler surface. This fact explains how relevant failures start from the interphasial region. The overall effect of moisture content on peak stresses observed very close to the radius $r = r_i$, is illustrated in Figure 7. Although these stresses are not affected by moisture content, they vary significantly when improper adhesion conditions exist.

Present results show that the moisture absorption from particulate composites plays a significant effect on material degradation process as produces high strain and stress fields. Moisture absorption produces reduction in matrix stiffness and as a result, significantly reduces interphase stiffness. Numerical results show that imperfect adhesion softens the material structure and produces step gradients in stresses. In

other words, moisture absorption is accompanied by an increase of stress gradient and increase of stresses inside the interphase. This interaction is shown to be more sensitive to adhesion efficiency between matrix and filler. Most important for the elastic behavior of the model is the presence of interphasial imperfections and the degree of filler-matrix adhesion. As a consequence, localized strain and stress concentrations take place within the interphase, initializing the degradation process. Presence of such conditions may cause severe failures in this region. Relevant failures expected to start very close to filler surface and may significantly intensified when improper adhesion conditions exist.

6 CONCLUSION

The degradation process due to moisture absorption and especially the interphasial stress and stain fields developed in particulate composites is investigated. Young modulus of epoxy matrix was determined from experimental data as a function of the moisture content. The effect of the existence of an inhomogeneous interphase and the imperfect adhesion between matrix and filler are also investigated by formulating a proper semi-analytical model. In this approach, material inhomogeneities are taken into account by considering a discretization of the problem domain into a set of homogeneous subdomains. Resulting boundary value problems were solved numerically. In this model the filler volume fraction, the value of imperfect adhesion and the moisture content are considered as design variables. Numerical results are presented for a typical particulate composite for which experimental data are available. The effect of particular values of the design variables is further illustrated and discussed. For the material examined, numerical results show a strong degradation effect on its elastic behavior that significantly increases when imperfect adhesion exists.

REFERENCES

Soutis, C. & Turkmen, D. 1997. Moisture and temperature effects of the compressive failure of CFRP unidirectional laminates, *Journal of Composite Materials*, 31: 832-849.
Shen, C. & Springer, G.S. 1976. Moisture absorption and desorption of composite materials, *Journal of Composite Materials*, 10: 2-20.
Surrel, Y. & Vautrin, A. 1989. *Journal of Composite Materials*, 23: 232-250.

Fahmy, A.A. & Hurt, J.C. 1980. Stress dependence of water diffusion in epoxy resins, *Polymer Composites*, 1: 77-80.

Marom, G. & Broutman, L.J. 1981. Moisture penetration into composites under external stress, *Polymer Composites*, 2: 132-136.

Kakavas, P.A., Anifantis, N.K., Bexevanakis, K., Katsareas, D.E. & Papanicolaou, G.C. 1995. The effect of interphasial imperfections on the micromechanical stress and strain distribution in fiber reinforced composites, *Journal of Materials Science*, 30: 4541-4548.

Anifantis, N.K., Kakavas, P.A. & Papanicolaou, G.C. 1997. Thermal stress concentration due to imperfect adhesion in fiber-reinforced composites, *Composites Science and Technology*, 57: 687-696.

Kakavas, P.A., Anifantis, N.K. & Papanicolaou, G.C. 1988. The role of imperfect adhesion on thermal expansivities of transversely isotropic composites with an inhomogeneous interphase, *Composites*, Part A, 29: 1021-1026.

Theocharis, P.S, Papanicolaou, G.C. & Contou, E.A. 1983. Interrelation between moisture absorption, mechanical behavior and extent of the boundary interphase in particulate composites, *Journal of Applied Polymer Science*, 28: 345-3153.

Sideridis, E, Theocharis, P.S. & Papanicolaou, G.C. 1986. The elastic modulus of particulate composites using the concept of a mesophase, *Rheological Acta*, 25: 350-358.

Chamis, C.C. 1987. Simplified composite micromechanics equations for mechanical, thermal and moisture-related properties, in J.W. Weeton et al. (eds), *Engineers' Guide to Composite Materials*, 3-8-3-24, ASM International, Materials Park, OH.

Timoshenko, S.P. & Goodier, J.N. 1951. *Theory of elasticity*, 2nd edition, McGraw-Hill Book Company, New York.

Recent Developments in Durability Analysis of Composite Systems, Cardon, Fukuda, Reifsnider & Verchery (eds)
© 2000 Balkema, Rotterdam, ISBN 90 5809 103 1

Modeling of hygrothermal effects in composites and polymer adhesive systems

Samit Roy & Weiqun Xu
Department of Mechanical Engineering, University of Missouri-Rolla, Mo., USA

ABSTRACT: It is now well known that Fick's Law is frequently inadequate for describing moisture diffusion in polymers and polymer composites. Non-Fickian or anomalous diffusion is likely to occur when a polymer composite laminate is subjected to external stresses that could give rise to internal damage in the form of matrix cracks. In this paper, a modeling methodology based on irreversible thermodynamics applied within the framework of composite macro-mechanics is presented. The final form for effective diffusivity obtained from this derivation indicates that effective diffusivity for this case is a quadratic function of crack density. A finite element procedure that extends this methodology to more complex shapes and boundary conditions is also presented. Comparisons with test data for a 5-harness satin textile composite are provided for model verifications.

1 INTRODUCTION

The benefits of lightweight polymer matrix composite (PMC) components to aircraft engines are now well known. Although thousands of PMC components are currently in service, barriers still exist to further implementation in more structurally critical and higher temperature applications. Most of these barriers are associated with the inability to accurately predict component lives, and therefore, component life-cycle costs. A fiber reinforced composite material with a polymer matrix will typically absorb moisture in a humid environment and at elevated temperatures. Combined exposure to heat and moisture affects a PMC in a variety of ways. First, the hygrothermal swelling causes a change in the residual stresses within the composite that could lead to micro-crack formation. These micro-cracks in turn provide fast diffusion paths and thus alter the moisture absorption characteristics of the laminate. Secondly, heat and humidity may cause the matrix to become plasticized thus causing an increase in the elongation to failure of the matrix.

Thirdly, the chemical bond at the interface between fiber and matrix may be affected which in turn would influence strength and toughness. Finally, in the event of cyclic heating and cooling with a sustained use-temperature above the boiling point of water, vaporization and out-gassing of absorbed moisture may take place leading to physical damage and chemical changes within the PMC, especially at temperatures greater than the T_g of the matrix. Continuous exposure to high moisture concentrations at the exposed surfaces of a PMC component could also lead to damage in the form of matrix cracking, dissolution, and peeling.

It is now widely recognized that cyclic moisture absorption and desorption plays a significant role in influencing the mechanical behavior, and therefore, long-term durability of polymers and PMC. Numerous diffusion models have been proposed over the years for modeling hygrothermal effects in PMC. The one most frequently used by researchers is the one-dimensional Fickian model. Unfortunately, this model tends to overestimate the moisture absorption in panels for short diffusion time (Shen 1981). Some researchers have suggested that the deviation can be explained by a two-stage Fickian process (Gurtin 1979,Carter 1978). Others claim that the diffusion process in a PMC is really non-Fickian (Shirrell 1979,Weitsman 1991). In reality, the nature of the diffusion process depends on the material and on the environmental conditions that the material is exposed to. For example, if the rate of viscoelastic relaxation in a polymer is comparable to the rate of moisture diffusion, then the diffusion is likely to be non-Fickian. In addition, the presence of strong temperature and stress gradients has been known to engender non-Fickian driving forces. The presence of damage in the form of matrix cracks could also lead to anomalous diffusion. Employing a rigorous thermodynamic approach, Weitsman (Weitsman 1987) developed a model for coupled damage and moisture transport in a transversely isotropic, fiber reinforced polymer composite. The damage entity was represented as a skew-symmetric tensor and was included in the model as an internal state variable. However, the be expressed in a polynomial form that remains invariant to coordinate transformations. Such a polynomial can be mathematically expressed as a combination of invariant terms obtained from the so-model was mathematically complex and was not amenable to a simple closed-form solution.

In this paper, theory of irreversible thermodynamics is applied within the framework of continuum mechanics to derive governing equations for diffusion in a PMC from first principles. A special form for Gibbs potential is formulated for an orthotropic material using stress, temperature, damage and moisture concentration as independent state variables. The resulting governing equations are capable of modeling the effect of interactions between complex stress, temperature, damage and moisture concentration on the diffusion process within an orthotropic material. The primary focus of this work is to model diffusion in the presence of a pre-existing state of damage; consequently damage evolution is not included in the present analysis for tractability. Because the mathematically complex nature of the governing equations precludes a closed-form solution,

a variational formulation is used to derive the weak form of the nonlinear governing equations that are then solved using the finite element method. This approach provides a significant improvement over solution methods reported in the literature for this type of problems. For model validation, the model predictions are compared with experimental data for the special case of isothermal diffusion in an unstressed 5-harness satin weave graphite/epoxy $[0/90/0/90]_s$ laminate with distributed matrix micro-cracks.

2 MODEL DEVELOPMENT

The Gibbs potential for an orthotropic material subjected to applied stress and internal damage must called irreducible integrity bases (Adkins 1959,Talreja 1994). The irreducible integrity bases for an orthotropic material are,

$$\bar{\sigma}_{11}, \bar{\sigma}_{22}, \bar{\sigma}_{33}, d_{11}, d_{22}, d_{33}, \bar{\sigma}_{23}^2, \bar{\sigma}_{13}^2, \bar{\sigma}_{12}^2, \bar{\sigma}_{23}\bar{\sigma}_{13}\bar{\sigma}_{12}, d_{23}^2, d_{13}^2, d_{12}^2$$
$$, d_{23}d_{13}d_{12}, \bar{\sigma}_{23}d_{23}, \bar{\sigma}_{13}d_{13}, \bar{\sigma}_{12}d_{12}, d_{23}\bar{\sigma}_{13}\bar{\sigma}_{12}, d_{13}\bar{\sigma}_{12}\bar{\sigma}_{23}, d_{12}\bar{\sigma}_{23}\bar{\sigma}_{13},$$
$$\bar{\sigma}_{23}d_{13}d_{12}, \bar{\sigma}_{13}d_{12}d_{23}, \bar{\sigma}_{12}d_{23}d_{13} \qquad (1)$$

where, the normalized stress

$$\bar{\sigma}_{ij} = \frac{\sigma_{ij}}{\sigma_t},$$

where σ_t is the ultimate stress in a material principal direction, the damage tensor d_{ij} is a symmetric tensor of the 2nd rank.

The chemical potential of moisture in the polymer is given by,

$$\mu = \rho_s \frac{\partial \phi}{\partial m} \qquad (2)$$

where, ϕ is Gibbs potential, ρ_s is the mass density of the polymeric solid, and "m" is the moisture concentration. It should be noted that in this treatment the moisture concentration "m" is assumed to be a scalar valued variable with the same value in all symmetry directions. Conservation of diffusing mass within a unit volume of the polymer requires,

$$\frac{\partial m}{\partial t} = -\frac{\partial f_i}{\partial X_i} \quad , \qquad i = 1,3 \tag{3}$$

where, in the absence of temperature, stress and damage gradients, moisture flux f_i for orthotropic symmetry is assumed to be of the form,

$$f_i = -D_i \frac{\partial \mu}{\partial X_i} \tag{4}$$

where D_i are the diffusion coefficients in the lamina material principal directions. It should be noted that repeated indices do not imply summation in equation (4). Combining equations (2), (3), and (4), gives the governing equation for diffusion in an anisotropic medium,

$$\frac{\partial m}{\partial t} = \frac{\partial}{\partial X_i}\left(D_i \frac{\partial \mu}{\partial X_i} \right) \qquad i = 1,3 \tag{5}$$

Please note that repeated indices imply summation in equation (5).

Special Case

Consider a laminate with intralaminar cracks oriented perpendicular to the X_1 axis, subjected to inplane uniaxial loading in the X_1 direction under isothermal conditions as schematically shown in Figure 1. For the special case of uniaxial loading, the states of stress and damage reduce to,

$$\overline{\sigma}_{22} = \overline{\sigma}_{33} = \overline{\sigma}_{13} = \overline{\sigma}_{23} = \overline{\sigma}_{12} = 0$$

and,

$$d_{22} = d_{33} = d_{13} = d_{23} = d_{12} = 0$$

Based on a definition of damage originally proposed by Talreja (Talreja 1994), it can be shown that the non-zero damage component d_{11} is given by,

$$d_{11} = \frac{\kappa(m,T) t_c^2 \delta_1}{t} \tag{6}$$

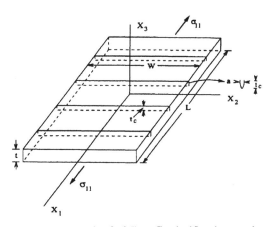

Figure 1. Schematic of a Micro-Cracked Laminate under Uniaxial Stress State

In equation (6) $\kappa(m,T)$ is an experimentally determined influence parameter that incorporates the constraining influence of moisture concentration (m), temperature (T), ply-orientation, and fiber architecture on crack opening displacement; t_c is the crack size, t is the total thickness of the laminate, and δ_1 is the crack density in the X_1-direction as depicted in Figure 1.

Using the irreducible integrity bases from equation (1) and assuming that terms beyond second order in uniaxial stress and damage can be neglected, the Gibbs potential becomes,

$$\begin{aligned} \rho_s \phi = \hat{C}_0 + \hat{C}_1 \overline{\sigma}_{11} + \hat{C}_2 d_{11} + \hat{C}_3 \overline{\sigma}_{11}^2 + \hat{C}_4 \overline{\sigma}_{11} d_{11} \\ + \hat{C}_5 d_{11}^2 + \hat{C}_6 \overline{\sigma}_{11}^2 d_{11} + \hat{C}_7 \overline{\sigma}_{11} d_{11}^2 \end{aligned} \tag{7}$$

In equation (7), the $\hat{C}_i$ are thermodynamic coefficients that could be expressed, in general, as functions of temperature (T) and concentration (m). In this manner, concentration and temperature are implicitly included in the polynomial expansion of Gibbs potential as state variables.

The chemical potential of moisture in the polymer for this special case is,

213

$$\mu = \rho_s \frac{\partial \phi}{\partial m}$$

$$= \rho_s \Big[\ \frac{\partial \hat{C}_0}{\partial m} + \frac{\partial \hat{C}_1}{\partial m}\overline{\sigma}_{11} + \hat{C}_1 \frac{\partial \overline{\sigma}_{11}}{\partial m} + \frac{\partial \hat{C}_2}{\partial m} d_{11}$$

$$+ \hat{C}_2 \frac{\partial d_{11}}{\partial m} + \frac{\partial \hat{C}_3}{\partial m}\overline{\sigma}_{11}^{\ 2} + 2\hat{C}_3 \overline{\sigma}_{11}\frac{\partial \overline{\sigma}_{11}}{\partial m} + \frac{\partial \hat{C}_4}{\partial m}\overline{\sigma}_{11}d_{11}$$

$$+ \hat{C}_4 d_{11}\frac{\partial \overline{\sigma}_{11}}{\partial m} + \hat{C}_4\overline{\sigma}_{11}\frac{\partial d_{11}}{\partial m} + \frac{\partial \hat{C}_5}{\partial m}d_{11}^{\ 2} + 2\hat{C}_5 d_{11}\frac{\partial d_{11}}{\partial m}$$

$$+ \frac{\partial \hat{C}_6}{\partial m}\overline{\sigma}_{11}^{\ 2}d_{11} + 2\hat{C}_6\overline{\sigma}_{11}d_{11}\frac{\partial \overline{\sigma}_{11}}{\partial m} + \hat{C}_6\overline{\sigma}_{11}^{\ 2}\frac{\partial d_{11}}{\partial m}$$

$$+ \frac{\partial \hat{C}_7}{\partial m}\overline{\sigma}_{11}d_{11}^{\ 2} + \hat{C}_7 d_{11}^{\ 2}\frac{\partial \overline{\sigma}_{11}}{\partial m} + 2\hat{C}_7\overline{\sigma}_{11}d_{11}\frac{\partial d_{11}}{\partial m} \ \Big]$$

(8)

Because chemical potential is in general a function of σ_{11}, d_{11}, T, and m, therefore,

$$\frac{\partial \mu}{\partial X_i} = \frac{\partial \mu}{\partial m}\cdot\frac{\partial m}{\partial X_i} + \frac{\partial \mu}{\partial T}\cdot\frac{\partial T}{\partial X_i} + \frac{\partial \mu}{\partial \overline{\sigma}_{11}}\cdot\frac{\partial \overline{\sigma}_{11}}{\partial X_i} + \frac{\partial \mu}{\partial d_{11}}\cdot\frac{\partial d_{11}}{\partial X_i}$$

For the special case of uniform stress, damage, and temperature distributions only the moisture gradient term is dominant and equation (4) can be written as,

$$D_i \frac{\partial \mu}{\partial X_i} = D_i\left(\frac{\partial \mu}{\partial m}\right)\left(\frac{\partial m}{\partial X_i}\right)$$

$$= \rho_s D_i \Big[\ \frac{\partial^2 \hat{C}_0}{\partial m^2} + \frac{\partial^2 \hat{C}_1}{\partial m^2}\overline{\sigma}_{11} + 2\frac{\partial \hat{C}_1}{\partial m}\frac{\partial \overline{\sigma}_{11}}{\partial m}$$

$$+ \hat{C}_1 \frac{\partial^2 \overline{\sigma}_{11}}{\partial m^2} + \frac{\partial^2 \hat{C}_2}{\partial m^2}d_{11} + 2\frac{\partial \hat{C}_2}{\partial m}\frac{\partial d_{11}}{\partial m} + \hat{C}_2 \frac{\partial^2 d_{11}}{\partial m^2}$$

$$+ \frac{\partial^2 \hat{C}_3}{\partial m^2}\overline{\sigma}_{11}^{\ 2} + 4\overline{\sigma}_{11}\frac{\partial \hat{C}_3}{\partial m}\frac{\partial \overline{\sigma}_{11}}{\partial m} + 2\hat{C}_3\left(\frac{\partial \overline{\sigma}_{11}}{\partial m}\right)^2$$

$$+ 2\hat{C}_3\overline{\sigma}_{11}\frac{\partial^2 \overline{\sigma}_{11}}{\partial m^2} + \frac{\partial^2 \hat{C}_4}{\partial m^2}\overline{\sigma}_{11}d_{11} + \frac{\partial \hat{C}_4}{\partial m}\frac{\partial \overline{\sigma}_{11}}{\partial m}d_{11}$$

$$+ \frac{\partial \hat{C}_4}{\partial m}\overline{\sigma}_{11}\frac{\partial d_{11}}{\partial m} + \frac{\partial \hat{C}_4}{\partial m}d_{11}\frac{\partial \overline{\sigma}_{11}}{\partial m} + \hat{C}_4\frac{\partial d_{11}}{\partial m}\frac{\partial \overline{\sigma}_{11}}{\partial m}$$

$$+ \hat{C}_4 d_{11}\frac{\partial^2 \overline{\sigma}_{11}}{\partial m^2} + \frac{\partial \hat{C}_4}{\partial m}\overline{\sigma}_{11}\frac{\partial d_{11}}{\partial m} + \hat{C}_4\frac{\partial \overline{\sigma}_{11}}{\partial m}\frac{\partial d_{11}}{\partial m}$$

$$+ \hat{C}_4\overline{\sigma}_{11}\frac{\partial^2 d_{11}}{\partial m^2} + \frac{\partial^2 \hat{C}_5}{\partial m^2}d_{11}^{\ 2} + 4d_{11}\frac{\partial \hat{C}_5}{\partial m}\frac{\partial d_{11}}{\partial m}$$

$$+ 2\hat{C}_5\left(\frac{\partial d_{11}}{\partial m}\right)^2 + 2\hat{C}_5 d_{11}\frac{\partial^2 d_{11}}{\partial m^2} + \frac{\partial^2 \hat{C}_6}{\partial m^2}\overline{\sigma}_{11}^{\ 2}d_{11}$$

$$+ 2\frac{\partial \hat{C}_6}{\partial m}\overline{\sigma}_{11}d_{11}\frac{\partial \overline{\sigma}_{11}}{\partial m} + \frac{\partial \hat{C}_6}{\partial m}\overline{\sigma}_{11}^{\ 2}\frac{\partial d_{11}}{\partial m} + 2\frac{\partial \hat{C}_6}{\partial m}\overline{\sigma}_{11}d_{11}\frac{\partial \overline{\sigma}_{11}}{\partial m}$$

$$+ 2\hat{C}_6 d_{11}\left(\frac{\partial \overline{\sigma}_{11}}{\partial m}\right)^2 + 2\hat{C}_6\overline{\sigma}_{11}\frac{\partial \overline{\sigma}_{11}}{\partial m}\frac{\partial d_{11}}{\partial m} + 2\hat{C}_6\overline{\sigma}_{11}d_{11}\frac{\partial^2 \overline{\sigma}_{11}}{\partial m^2}$$

$$+ \frac{\partial \hat{C}_6}{\partial m}\overline{\sigma}_{11}^{\ 2}\frac{\partial d_{11}}{\partial m} + 2\hat{C}_6\overline{\sigma}_{11}\frac{\partial \overline{\sigma}_{11}}{\partial m}\frac{\partial d_{11}}{\partial m} + \hat{C}_6\overline{\sigma}_{11}^{\ 2}\frac{\partial^2 d_{11}}{\partial m^2}$$

$$+ \frac{\partial^2 \hat{C}_7}{\partial m^2}\overline{\sigma}_{11}d_{11}^{\ 2} + \frac{\partial \hat{C}_7}{\partial m}\frac{\partial \overline{\sigma}_{11}}{\partial m}d_{11}^{\ 2} + 2\frac{\partial \hat{C}_7}{\partial m}\overline{\sigma}_{11}d_{11}\frac{\partial d_{11}}{\partial m}$$

$$+ \frac{\partial \hat{C}_7}{\partial m}d_{11}^{\ 2}\frac{\partial \overline{\sigma}_{11}}{\partial m} + 2\hat{C}_7 d_{11}\frac{\partial \overline{\sigma}_{11}}{\partial m}\frac{\partial d_{11}}{\partial m} + \hat{C}_7 d_{11}^{\ 2}\frac{\partial^2 \overline{\sigma}_{11}}{\partial m^2}$$

$$+ 2\frac{\partial \hat{C}_7}{\partial m}\overline{\sigma}_{11}d_{11}\frac{\partial d_{11}}{\partial m} + 2\hat{C}_7\frac{\partial \overline{\sigma}_{11}}{\partial m}d_{11}\frac{\partial d_{11}}{\partial m} + 2\hat{C}_7\overline{\sigma}_{11}\left(\frac{\partial d_{11}}{\partial m}\right)^2$$

$$+ 2\hat{C}_7\overline{\sigma}_{11}d_{11}\frac{\partial^2 d_{11}}{\partial m^2} \ \Big]\left(\frac{\partial m}{\partial X_i}\right)$$

(9)

For the special stress-free case where $\overline{\sigma}_{11} = 0$, equation (9) becomes,

$$D_i \frac{\partial \mu}{\partial X_i} = \rho_s D_i \Big[\ \frac{\partial^2 \hat{C}_0}{\partial m^2} + \frac{\partial^2 \hat{C}_2}{\partial m^2}d_{11} + 2\frac{\partial \hat{C}_2}{\partial m}\frac{\partial d_{11}}{\partial m}$$

$$+ \hat{C}_2 \frac{\partial^2 d_{11}}{\partial m^2} + \frac{\partial^2 \hat{C}_5}{\partial m^2}d_{11}^{\ 2} + 4d_{11}\frac{\partial \hat{C}_5}{\partial m}\frac{\partial d_{11}}{\partial m}$$

$$+ 2\hat{C}_5\left(\frac{\partial d_{11}}{\partial m}\right)^2 + 2\hat{C}_5 d_{11}\frac{\partial^2 d_{11}}{\partial m^2} \ \Big]\left(\frac{\partial m}{\partial X_i}\right)$$

$$= \rho_s D_i \Big[\left\{\begin{matrix} \dfrac{\partial^2 \hat{C}_0}{\partial m^2} + 2\dfrac{\partial \hat{C}_2}{\partial m}\dfrac{\partial d_{11}}{\partial m} + \hat{C}_2 \dfrac{\partial^2 d_{11}}{\partial m^2} \\ + 2\hat{C}_5\left(\dfrac{\partial d_{11}}{\partial m}\right)^2 \end{matrix}\right\}$$

$$+ \left\{\frac{\partial^2 \hat{C}_2}{\partial m^2} + 4\frac{\partial \hat{C}_5}{\partial m}\frac{\partial d_{11}}{\partial m} + 2\hat{C}_5\frac{\partial^2 d_{11}}{\partial m^2}\right\}d_{11}$$

$$+ \left\{\frac{\partial^2 \hat{C}_5}{\partial m^2}\right\}d_{11}^{\ 2} \ \Big]\left(\frac{\partial m}{\partial X_i}\right)$$

(10)

If, for conceptual simplification, it is assumed that swelling due to moisture absorption influences only the opening displacement without significantly affecting either the crack size, t_c, or the crack density, δ_1, then,

$$\frac{\partial}{\partial m}\left(\frac{\kappa t_c \delta_1}{t}\right) = \frac{t_c^2 \delta_1}{t}\frac{\partial \kappa}{\partial m}$$

(11)

Taking the definitions of damage and its derivative given respectively by equations (6) and (11) and substituting in equation (10) results in the final expression for moisture flux,

$$D_i \frac{\partial \mu}{\partial X_i} = \overline{D}_i \frac{\partial m}{\partial X_i}$$

where, each orthotropic component of effective diffusivity ($\overline{D}_i$) can be obtained as a quadratic function of the crack density given by,

$$\overline{D}_i = \rho_s D_i \left[\begin{Bmatrix} \dfrac{\partial^2 \hat{C}_0}{\partial m^2} + 2 \dfrac{t_c^2}{t} \dfrac{\partial \hat{C}_2}{\partial m} \dfrac{\partial \kappa}{\partial m} \delta_1 + \hat{C}_2 \dfrac{t_c^2}{t} \dfrac{\partial^2 \kappa}{\partial m^2} \delta_1 + \\ 2\hat{C}_5 \left(\dfrac{t_c^2}{t} \right)^2 \left(\dfrac{\partial \kappa}{\partial m} \right)^2 \delta_1^2 \end{Bmatrix} \right.$$

$$+ \left\{ \dfrac{\partial^2 \hat{C}_2}{\partial m^2} + 4 \dfrac{t_c^2}{t} \dfrac{\partial \hat{C}_5}{\partial m} \dfrac{\partial \kappa}{\partial m} \delta_1 + 2\hat{C}_5 \dfrac{t_c^2}{t} \dfrac{\partial^2 \kappa}{\partial m^2} \delta_1 \right\} \kappa \dfrac{t_c^2}{t} \delta_1$$

$$+ \left\{ \dfrac{\partial^2 \hat{C}_5}{\partial m^2} \right\} \left(\kappa^2 \dfrac{t_c^4}{t^2} \right) \delta_1^2 \left. \right]$$

$$= C_0 + C_1 \delta_1 + C_2 \delta_1^2$$

(12)

where,

$$C_0(m,T) = \rho_s D_i \frac{\partial^2 \hat{C}_0}{\partial m^2}$$

(13)

$$C_1(m,T) = \rho_s D_i \left(\frac{t_c^2}{t} \right) \left[\kappa \frac{\partial^2 \hat{C}_2}{\partial m^2} + 2 \frac{\partial \hat{C}_2}{\partial m} \frac{\partial \kappa}{\partial m} + \hat{C}_2 \frac{\partial^2 \kappa}{\partial m^2} \right]$$

(14)

$$C_2(m,T) = \rho_s D_i \left(\frac{t_c^2}{t} \right)^2 [2\hat{C}_5 \left(\frac{\partial \kappa}{\partial m} \right)^2 + \kappa \begin{Bmatrix} 4 \dfrac{\partial \hat{C}_5}{\partial m} \dfrac{\partial \kappa}{\partial m} \\ + 2\hat{C}_5 \dfrac{\partial^2 \kappa}{\partial m^2} \end{Bmatrix}$$

$$+ \kappa^2 \frac{\partial^2 \hat{C}_5}{\partial m^2}]$$

(15)

Substituting equation (12) in (5) yields the governing equation for moisture diffusion in an orthotropic

laminate subjected to zero stress and uniform intralaminar damage,

$$\frac{\partial m}{\partial t} = \frac{\partial}{\partial X_i} \left[\left(C_0 + C_1 \delta_1 + C_2 \delta_1^2 \right) \frac{\partial m}{\partial X_i} \right]$$

(16)

As defined in equations (13)-(15), the damage coefficients C_0, C_1, C_2, depend on polymer density, polymer diffusivity, moisture concentration, temperature, and the ratio t_c^2/t. These coefficients can be characterized from absorption experiments on pre-cracked laminate specimens as discussed in the "Model Verification" section.

For modeling absorption and/or desorption in a laminate in the through-thickness , i.e., in the X_3 direction, equation (16) reduces to,

$$\frac{\partial m}{\partial t} = \frac{\partial}{\partial X_3} \left[\left(C_0 + C_1 \delta_1 + C_2 \delta_1^2 \right) \frac{\partial m}{\partial X_3} \right]$$

(17)

3 FINITE ELEMENT FORMULATION

In order to extend the simple one-dimensional analytical model in equation (17) to more complex shapes and boundary conditions, a three-dimensional finite element code (NOVA-3D) was developed. The variational (weak) form of Equation (16) in three-dimensions is given by,

$$\int_{V^{(e)}} \left[u \frac{\partial m'}{\partial t} + D' \frac{\partial u}{\partial X_i} \frac{\partial m'}{\partial X_i} \right] dV - \int_{A^{(e)}} \left[u \left(D' \frac{\partial m'}{\partial X_i} \right) n_i \right] dA = 0$$

(18)

where, u is an admissible variational test function. Based on the variational statement, the diffusion boundary conditions can now be identified as,

$$\left(D' \frac{\partial m'}{\partial X_i} \right) n_i + \hat{q} = 0 \quad \text{on } A_1^{(e)} \quad \text{(specified solvent flux)}$$

215

$$m = \hat{m} \quad \text{on } A_2^{(e)} \qquad \text{(specified concentration)}$$

where, $\qquad A_1^{(e)} + A_2^{(e)} = A^{(e)}$

and n_i are the components of the unit outward normal at the boundary. Thus,

$$\int_{V^{(e)}} \left[u \frac{\partial m^t}{\partial t} + D^t \frac{\partial u}{\partial X_i} \frac{\partial m^t}{\partial X_i} \right] dV = - \int_{A_1^{(e)}} u\hat{q}\, dA \tag{19}$$

A standard finite element interpolation of the concentration field over each element is given by,

$$m(X_i, t) = \sum_{j=1}^{N} N_j(X_i) m_j(t) \tag{20}$$

where, m_j are the nodal concentrations, N_j are the interpolation functions and N is the number of nodes per element. Substituting equation (20) in (19) and employing matrix notation, equation (19) becomes,

$$\left[T^{(e)}\right]\{\dot{m}\} + \left[K^{(e)}\right]\{m\} = \{F^{(e)}\} \tag{21}$$

where, the superscript $^{(e)}$ is used to denote that the equations are satisfied over each element and

$$T_{jk}^e = \int_{V^{(e)}} (N_j N_k)\, dV$$

$$K_{jk}^e = \int_{V^{(e)}} \left\{ D^t \frac{\partial N_j}{\partial X_i} \frac{\partial N_k}{\partial X_i} \right\} dV$$

$$F_j^e = - \int_{A_1^{(e)}} N_j \hat{q}\, dA \quad , \quad i = 1,3 \text{ and } j, k = 1, N$$

The time derivative $\{\dot{m}\}$ is approximated using a standard theta-family of approximations, yielding at time t_n and t_{n+1},

$$\left[A^{(e)}\right]\{m\}_{n+1} + \left[B^{(e)}\right]\{m\}_n = \{P^{(e)}\}_n \tag{22}$$

where,

$$\left[A^{(e)}\right] = \left[T^{(e)}\right] + \theta\Delta t_{n+1}\left[K^{(e)}\right]$$

$$\left[B^{(e)}\right] = \left[T^{(e)}\right] + (1-\theta)\Delta t_{n+1}\left[K^{(e)}\right]$$

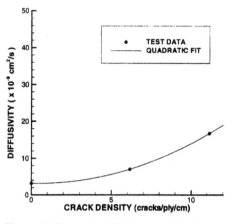

Figure 2. Change in Diffusivity with Micro-crack Density

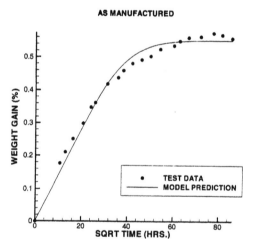

Figure 3. Predicted vs. Measured Weight Gain for $\delta_1 = 0$ cracks/ply/cm

$$\{P^{(e)}\} = \Delta t_{n+1}\left[\theta\{F^{(e)}\}_{n+1} + (1-\theta)\{F^{(e)}\}_n\right]$$

Equation (22) is solved using a value of $\theta = 0.5$, which corresponds to the Crank Nicholson scheme and is unconditionally stable. Note that for n=1, the value of the starting concentration in equation (22) is known from initial conditions.

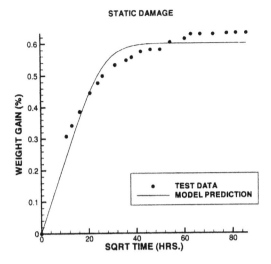

Figure 4. Predicted vs. Measured Weight Gain for δ_1 = 6.142 cracks/ply/cm

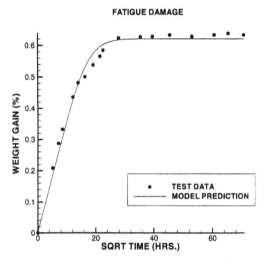

Figure 5. Predicted vs. Measured Weight Gain for δ_1 = 11.142 cracks/ply/cm

4 MODEL VERIFICATIONS

In order to characterize the model coefficients and perform preliminary model verifications, moisture weight gain data for graphite/epoxy 5-harness satin [0/90/0/90]$_s$ laminate with different micro-crack densities were obtained from hygrothermal tests performed elsewhere (Pratt & Whitney Aircraft). Specifically, test specimens were mechanically pre-cracked by uniaxial fatigue as well as static loading and then exposed to 75% relative humidity at 40°C. The crack densities reported for the specimens were 6.142 cracks/ply/cm for the static load case and 11.142 cracks/ply/cm for the fatigue load case respectively. An un-cracked specimen was also included in the test matrix as the control specimen. The specimens were not subjected to any applied mechanical stress during absorption. The effective diffusivity for each test specimen was extracted from weight gain data using standard analytical procedure (Shen 1981).

A quadratic least-squares curve-fit to the diffusivity data is shown by the solid line in Figure 2. The values of the damage coefficients defined in equation (12) were obtained using this procedure for this case and are given as C_0=3.184x10^{-09} cm^2/sec, C_1= -1.036 x10^{-10} cm^3/sec, and C_2= 1.184 x10^{-10} cm^4/sec. The corresponding equation for through-thickness diffusivity as a function of crack density is given by,

$$D_3 = 3.184 \times 10^{-09} - 1.036 \times 10^{-10}\delta_1 + 1.184 \times 10^{-10}\ \delta_1{}^2 \ \text{cm}^2/\text{sec}$$

$$(23)$$

It should be noted that although the linear damage coefficient C_1 has a negative value, the over-all value of the diffusivity is, for all practical purposes, a monotonically increasing function of crack-density as evidenced in Figure 2.

Moisture weight gain curves predicted by the model using the quadratic diffusivity-damage relation given by equation (23) combined with the diffusion governing equation in (17), and its comparison with test data are shown in Figures 3, 4, and 5. The results shown in Figure 3 correspond to an undamaged specimen. The moisture data shown in Figure 4 correspond to a micro-crack density of 6.142 cracks/ply/cm, and those in Figure 5 correspond to a micro-crack density of 11.142 cracks/ply/cm. Reasonable agreement between model predictions and test data is observed for all three cases for the duration of the tests. While these results do not conclusively prove that laminate diffusivity is a quadratic function of crack density, it does confirm that retaining only up to the quadratic terms in damage in the expansion of the Gibbs potential is an acceptable modeling approximation for the material system under consideration. As anticipated, the moisture uptake curves predicted by the finite element model for the simple one-dimensional case agree closely with the analytical model predictions and are not separately presented in Figures 3 through 5.

5 DISCUSSION AND CONCLUSIONS

A modeling methodology based on irreversible thermodynamics developed within the framework of composite macro-mechanics was presented to allow characterization of non-Fickian diffusion coefficients from moisture weight gain data for laminated composites with damage. A symmetric damage tensor based on continuum damage mechanics was incorporated in this model by invoking the principle of invariance with respect to coordinate transformations. To maintain tractability, the diffusion governing equations were simplified for the special case of a laminate with uniform matrix cracks that is subjected to a uniaxial tensile stress state. A finite element procedure that extends this methodology to more complex shapes and boundary conditions was also presented. Because of the macro-mechanics formulation employed in developing this model, the model is currently restricted to the study of overall laminate absorption / desorption characteristics in the presence of non-evolving damage states. An alternative approach that considers the more detailed physical aspects of moisture ingress into a micro-cracked laminate can be found in (Roy 1999).

The material coefficients needed to model the effect of matrix micro-cracking on laminate diffusivity were evaluated by using hygrothermal test data for a [0/90/0/90]$_s$ graphite/epoxy 5-harness satin textile composite. The moisture weight gain curves predicted by using a quadratic diffusivity–damage relation yielded good correlation with test data. While these observations do not conclusively prove that the laminate diffusivity is indeed a quadratic function of crack density, it does indicate that retaining only up to the quadratic terms in damage in the expansion of the Gibbs potential is an acceptable modeling approximation for the material system under consideration. The primary purpose of this paper is to establish a theoretical framework for using the proposed modeling approach to characterize and eventually, to predict, absorption and desorption in micro-cracked laminates. The formulation presented in this paper is deemed to provide a small yet positive step towards that goal. Additional hygrothermal tests on specimens subjected to combined stress and damage states at different temperature and humidity levels are currently underway to comprehensively evaluate the accuracy of the proposed model when these data become available.

ACKNOWLEDGEMENT

The authors are grateful to Dr. Rajiv Naik and Mr. Ron Cairo of Pratt & Whitney, and to Mr. Mike Meador of NASA Lewis Research Center for supporting this research and for supplying the test data. The authors would like to thank Mr. Uday Idnani for his assistance with the manuscript.

REFERENCES

Shen, C. H. and G.S. Springer, 1981, "Effects of Moisture and Temperature on the Tensile Strength of Composite Materials," in *Environmental Effects on Composite Materials*, G. S. Springer, ed., Lancaster, PA: Technomic Publishing Co., Inc., pp. 79-93.

Gurtin, M.E., and Yatomi, C. "On a model for Two Phase Diffusion in Composite Materials," *Journal of Composite Materials*, Vol. 13, (April 1979), pp. 126-130.

Carter, H.G., and Kibler, K.G. "Langmuir-Type Model for Anomalous Diffusion in Composite Resins," *Journal of Composite Materials*, Vol. 12, (April 1978) , pp. 118-130.

Shirrell, C. D., Leisler, W. H. and F. A. Sandow. 1979. "Moisture-Induced Surface Damage in T300/5208 Graphite/Epoxy Laminates," *in Nondestructive Evaluation and Flaw Criticality for Composite Materials*, ASTM STP 696, R. B. Pipes, ed., American Society for Testing and Materials, pp. 209-222.

Weitsman, Y. 1991. "Moisture in Composites: Sorption and Damage," in *Fatigue of Composite Materials*, K. L. Reifsnider, ed., Elsevier Science Publishers B.V., pp. 385-429.

Weitsman, Y. 1987. "Coupled Damage and Moisture Transport in Fiber- Reinforced, Polymeric Composites," *International Journal of Solids and Structures*, 23(7): 1003-1025.

Adkins, J.E., 1959. "Symmetry Relations for Orthotropic and Transversely Isotropic Materials," Arch. Rational Mech. Anal., Vol 4, pp. 193-213.

Talreja, R., 1994. "Damage Characterization by Internal Variables," Damage Mechanics of Composite Materials, Edited by R. Talreja, Elsevier Science, pp. 53-78.

Private Communications, Pratt & Whitney Aircraft, West Palm Beach, Florida.

S. Roy and T. Bandorawalla, " Modeling of Diffusion in a Micro-cracked Composite Laminate using Approximate Solutions," *Journal of Composite Materials*, Vol. 33, No. 10/1999, pp. 872-905.

Recent Developments in Durability Analysis of Composite Systems, Cardon, Fukuda, Reifsnider & Verchery (eds)
© 2000 Balkema, Rotterdam, ISBN 90 5809 103 1

Durability of adhesive joints subjected to elevated temperature aging

D. R. Veazie
Department of Engineering, Clark Atlanta University, Ga., USA

J. Qu & J. S. Lindsay
School of Mechanical Engineering, Georgia Institute of Technology, Atlanta, Ga., USA

ABSTRACT: Fracture toughness parameters for a predictive scheme to estimate the remaining life of a bonded joint are experimentally determined in this paper. To this end, double cantilever beam, end notch flexure and crack lap shear tests were performed to obtain the Mode I, Mode II, and mixed mode strain energy release rates, respectively. Testing was performed at room and elevated temperatures (177°C and -54°C) on Ti-6Al-4V titanium joints bonded with an adhesive (FM®x5) based on a polyimide developed at the NASA-Langley Research Center (LaRC™-PETI-5). To enhance the durability of the joints, a Sol Gel chemical pretreatment process was performed on the surface of the titanium adherends. Specimens cut from a bonded sheet were tested in the as-received state, as well as isothermally exposed for 5,000 hours at 177°C, and isothermally exposed to a hot/wet environment (80°C, 90%+ relative humidity). In all loading cases except Mode I, the threshold strain energy release rates were reduced as a result of 5,000 hours of aging and elevated temperature testing. Specimens tested in Mode II or Mixed Mode I & II appear to be more susceptible to degradation by environmental exposures and elevated temperature testing than by Mode I testing alone.

1 INTRODUCTION

Adhesively bonded joints posses many advantages over traditional fastening techniques such as rivets and bolts. Adhesive joints are usually lower in weight, allow the bonding of dissimilar materials, and have lower fabrication costs. Several structural adhesive joints are in service on military and commercial aircraft (Baker et al. 1984, Dunkerton & Vlattas 1998), however, these bonded joints have been limited to secondary structures partly because the environmental durability of these joints have not been extensively characterized. Since structural adhesive joint systems are composed of adherends, adhesives, and interphase regions, detailed analyses of these joints must be performed not only in their fabricated condition but also after they have been exposed to elevated temperatures and other environmental conditions.

Commercial supersonic aircraft capable of traveling at speeds in excess of Mach 2 could be subjected to temperatures between -54°C to 177°C in the wing and fuselage structures. It is believed that the key to develop advanced adhesive joints to meet future challenges of elevated temperature, environmental exposure, and durability is to understand the mechanics of the adhesive joints based on the adhesive's microstructure and the physics of interfacial adhesion. To obtain this understanding is the rational for a comprehensive study with two aims:

1. Establishing a correlation between the microstructural changes and the long term bond strength in adhesive joints by the use of interfacial fracture mechanics to characterize interfacial toughness as a function of the mode mixity at the debonding crack-tip;

2. Developing finite element computations to obtain a calibration between the applied load and the energy release rate based on the load and specimen geometry. ·

There are usually two methodologies used when analyzing a structural adhesive joint. The first methodology is based on stress analysis and it focuses on the determination of the distribution of shear and peel (or normal) stresses in the adhesive bond line under static testing conditions (Goland & Reissner 1944). This method is excellent for enabling the mechanical behavior of adhesive joint specimens to be easily understood and predicted. However, limits to this method include the fact that the stress states calculated are unique for each specimen, and the method does not take into account flaws which are present in the bond line. The second methodology relies on the concept of fracture mechanics, which is based on the theories of Griffith (1920) and Irwin (1948). This method focuses on the knowledge that flaws are present in every material and mathematically analyzes the loads at which these flaws propagate. Ripling et al. (1964a) was the first to use fracture mechanics theory for bonded joint analyses. The major advantage of fracture mechanics in analyzing bonded joints is in part due to the observation that adhesive joints generally fail by progressive crack growth.

The strain energy release rate, G, is the fracture mechanics parameter which measures the amount of energy required to extend a crack over a unit surface area of an adhesive structural joint. The critical strain energy release rate, G_c, can be composed of three major components which correspond to the modes by which fracture can occur. These are opening (Mode I), shearing (Mode II), and tearing (Mode III). For most aerospace applications, failure usually occurs by Mode I, Mode II, or a combination of both Mode I and Mode II. The most commonly used test methods to obtain G_c for these modes include 1) the Double Cantilever Beam (DCB), in which the test specimen fails due to Mode I loading, 2) the End Notched Flexure (ENF), which fail specimens due to Mode II loading, and 3) the Cracked Lap Shear (CLS), in which the test specimen fails by a combination of Mode I and Mode II. The CLS specimen most accurately resembles the loading of adhesive structural joints in aerospace service.

This research investigates the degradation mechanisms and the long term effects of time and temperature on polymer bonded joints through an experimental study. Fracture toughness parameters (G_c) are obtained for possible use in a predictive scheme to estimate the remaining life of a bonded joint. To this end, room and elevated temperature ($177^{\circ}C$ and $-54^{\circ}C$) tests are conducted. The adhesive system used for the study is comprised of Ti-6Al-4V titanium adherends bonded with an adhesive (FM®x5) based on a polyimide developed at the NASA-Langley Research Center (LaRC™-PETI-5). Room and elevated temperature Mode I, Mode II and Mixed-Mode I and II fracture toughness results are obtained from DCB, ENF, and CLS tests, respectively. Testing was performed on specimens with no environmental exposure (as-received), as well as isothermally exposed for 5,000 hours at $177^{\circ}C$, and isothermally exposed to a hot/wet environment ($80^{\circ}C$, 90%+ relative humidity). The fracture surfaces of the adhesive joints were also analyzed to determine the method by which the crack propagated (i.e. cohesive, interfacial, or both).

2 MATERIALS AND TESTING PROCEDURES

Specimens cut from a 95.3 cm. by 69.9 cm. bonded panel were provided by The Boeing Company. The unaged glass transition temperature, T_g, of the adhesive as measured by Dynamic Mechanical Analyzer (DMA) G'' peak was $250^{\circ}C$. The titanium adherend thickness was 0.25 cm., whereas the adhesive measured 0.04 cm. thick. To enhance the durability of the joints, a Sol Gel chemical pretreatment process was performed on the surface of the titanium adherends. The inclusion of non-adhesive Kapton™ film provided the crack initiation sites, and slots were incorporated in the panel for venting of volatiles during cure.

DCB and ENF tests were performed to obtain the Mode I and Mode II fracture toughness parameters, respectively. The DCB and ENF specimen configurations are identical, and consist of two adherends joined by the adhesive with a mid-plane embedded insert (Kapton™) at one end. The insert prevented bonding of a nominal 5.08 cm. region. DCB and ENF specimens used for this research were nominally 1.90 cm. wide and 26.67 cm. long, as shown in Figures 1 and 2. The CLS specimen consists of a 30.48 cm. long beam (lap) bonded to a 26.67 cm. shorter beam (strap), as shown in Figure 3.

2.1 Environmental conditioning

Specimens were subjected to one of two forms of environmental exposure prior to testing, according to temperature conditions typically experienced in commercial supersonic flight. The 'hot/dry' environment consisted of 5,000 hours of isothermal exposure at a temperature of $177^{\circ}C$. Isothermal exposure was conducted in a convection oven equipped with a digital controller. Specimens were stored inside a desiccator until the start of each test. No load was applied to the specimens during environmental exposure.

The 'hot/wet' environment consisted of 5,000 hours of isothermal exposure at a temperature of $80^{\circ}C$, and a relative humidity of greater than 90%. This was accomplished by suspending the specimens above a pool of distilled water inside a sealed glass container which was, in turn, placed inside an oven maintained at the desired temperature. Following exposure and prior to testing, specimens exposed to the 'hot/wet' environment were stored over distilled water at room temperature in sealed containers (at 90%+ relative humidity) in an attempt to maintain the moisture level (Johnson & Butkus 1998).

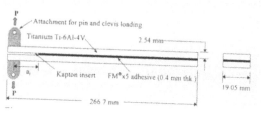

Figure 1. Schematic for the Ti-6Al-4V/FM®x5 double cantilever beam specimen.

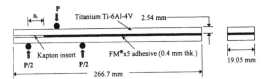

Figure 2. Schematic for the Ti-6Al-4V/FM®x5 end notch flexure specimen.

Moisture absorption by the adhesive resulted in slight weight gains by the specimens exposed to the 'hot/wet' environment. The weight gain by the specimens following 5,000 hours exposure to the 'hot/wet' environment was approximately 0.3%.

2.2 *Experimental procedures and equipment*

All of the fracture tests were performed at atmospheric pressure in laboratory air (23°C, 50% RH) on a 100 kN servo-hydraulic test frame equipped with digital controller and computer data acquisition. To reduce experimental error, several replicate tests were performed, permitting the calculation of multiple G_c values for each specimen.

Load transfer to the DCB specimens was accomplished by means of a pin-and-clevis attachment bolted to the adherends (Fig. 1). Monotonic testing, using ASTM Specifications D3433 and D5228 as guidelines, was conducted to obtain the Mode I strain energy release rate (G_{Ic}) for the DCB specimen. Crack growth was measured using a traveling long focal length microscope. A crosshead displacement rate, equal to a crack mouth opening of 1.0 mm/min was used. Deviation from linearity of the load verses displacement plot indicated the onset of crack growth. This was also confirmed by optical observations.

ENF tests were performed to obtain the Mode II strain energy release rate (G_{IIc}). This was accomplished by using a three-point bend fixture with cylindrical supports and applying load to the ENF specimen in the middle of the supports with a cylindrical surface (Fig. 2). The load was applied to the ENF specimen with a crosshead rate of 0.5 mm/min and simultaneous load-deflection data was recorded. The deflection, observed as the motion of the loading nose relative to the supports, was measured by LVDT. In this research, the span length was 10.16 cm. and the crack tip was placed halfway between an outer loading point and the middle loading point, resulting in a 2.54 cm. initial crack. The specimen was then monotonically loaded until the unstable crack propagated, since the crack-length-to-specimen-length ratio was 0.5. A deviation from linearity of the load versus displacement plot indicated the onset of crack growth in the bond line region or possible yield of adherends. This was also confirmed by visual observation through a traveling long focal length microscope. If multiple tests were possible, the length of the crack was measured and the specimen was unloaded. The test was performed again by positioning the new crack at a point midway between an outer support and the loading pin, resulting in another initial crack length of 2.54 cm.

The CLS tests closely resemble the type of Mixed-Mode (I and II) loading found in aerospace structures. The CLS specimen was loaded in tension by applying load (100 N/sec.) to both the lap, and to the strap and lap part of the specimen (Fig. 3). Axial loading of the CLS specimen causes mostly Mode II stresses at the crack tip, but Mode I stress is also caused due to the eccentricity of the load path at the crack tip. Hydraulic grips incorporating flat-faced wedges with a non-aggressive surface finish were used to allow for firm gripping of the adherends without grip-induced failures. An extensometer was mounted on the specimen during testing and placed in a position so that the gage length encompassed the point of the initial crack tip as well as a region in which the strap and lap were bonded. The extensometer served as an indicator for when the crack propagated and not as a means to measure strain in the CLS specimen. Deviation from linearity in the load versus strain plot indicated that the crack was probably growing, but the value for critical load was taken when the crack tip was visually seen growing by a traveling long focal length microscope.

3 ANALYSIS PROCEDURE

3.1 *Mode I (G_{Ic}) closed form solution*

The DCB specimen was first introduced for fracture toughness testing by Ripling et al. (1964b). For DCB specimens with two adherends of the same material and thickness, the strain energy release rate, G_{Ic}, may be calculated by using Equation 1

$$G_I = \frac{P^2}{2b} \frac{dC}{da} \qquad (1)$$

where P = load; a = crack length; b = specimen width; δ = crosshead opening displacement; and C = specimen compliance (δ/P).

By using beam theory and assuming that the DCB specimen consists of two cantilever beams with a built-in-support on the end opposite the load application point, Equation 1 reduces to

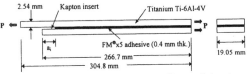

Figure 3. Schematic for the Ti-6Al-4V/FM®x5 crack lap shear specimen.

$$G_I = \frac{3P\delta}{2ba}.$$ (2)

Equation 2 can be modified to account for the relationship between specimen compliance and observed crack length using (O'Brien & Martin 1993)

$$G_I = \frac{3P\delta}{2b(a+|\Delta|)}.$$ (3)

The value Δ is the intercept of the a- axis obtained from a linear relationship between $C^{1/3}$ and a. The Δ term serves as a correction term to account for the false assumption that the cantilever beams were built end since the uncracked end of the DCB specimen is free. For the monotonic test, the critical load P corresponded to the load at which the load versus displacement data deviated from linearity.

3.2 *Mode II (G_{IIc}) closed form solution*

Determination of the applied Mode II strain energy release rate, G_{IIc}, from ENF specimens was carried out using the beam theory method. The method is discussed by Carlsson et al. (1986), and the formula for the beam theory is given as

$$G_{IIc} = \frac{9P^2a^2}{16Eb^2t^3}$$ (4)

where P = load; b = specimen width; a = crack length (2.54 cm); t = thickness of one adherend; and E = adherend elastic modulus.

Similar to the DCB tests, crack growth was observed to begin at or near the onset of nonlinearity in the load versus displacement curve. A note should be made that this analysis does not take into account friction between the adherends. In previous analyses, friction was neglected due to the smooth surface of interlaminar cracks in composites. However, this assumption may not be valid here due to the increase in G_{IIc} values as multiple tests were taken on the same specimen.

3.3 *Mixed Mode I & II (G_{Tc}) closed form solution*

The CLS specimen geometry was devised by Brussat et al. (1977), and its closed form solution is based upon beam theory, given as

$$G_{Tc} = \frac{P^2}{2b^2}\left[\frac{1}{E_s t_s} - \frac{1}{E_s t_s + E_l t_l}\right]$$ (5)

where P = load; E_l = lap elastic modulus; b = specimen width; t_s = strap adherend thickness; E_s = strap elastic modulus; and t_l = lap adherend thickness.
Note that this Equation 5 does not provide for the determination of the individual Mode I and Mode II components present in the cracked lap shear specimen. Currently, research still continues on ways to separate these two distinct strain energy release rate modes using a closed form solution.

4 RESULTS AND DISCUSSION

The effects of environmental exposure on the Mode I critical strain energy release rate (G_{Ic}) or fracture toughness from DCB tests at room and elevated temperatures (177°C and -54°C) are compared in Figure 4. The error bars represent the ±95% confidence interval. In Mode I cases, crack growth in the Ti-6Al-4V/FM®x5 specimens was cohesive (i.e. contained within the adhesive bondline), as shown in Figure 5. DCB tests showed a moderate reduction in the Mode I fracture toughness due to 5,000 hours of aging and elevated temperature testing. The most reduction was observed following 5,000 hours of the 'hot/wet' exposure.

Mode II critical strain energy release rates (G_{IIc}) are shown in Figure 6 as a function of environmental exposure and elevated temperature testing. In the Mode II cases, an interfacial failure (i.e. adhesive layer debonds from the adherend) was observed as shown in Figure 7. ENF results show a reduction of approximately 25% in the Mode II fracture toughness due to 5,000 hours of 'hot/dry' exposure, whereas a reduction of approximately 50% is observed due to 5,000 hours of 'hot/wet' exposure. Elevated temperature testing further reduced the fracture toughness of the joints by more than 50%. Prior to aging, however, no reduction was observed due to -54°C testing.

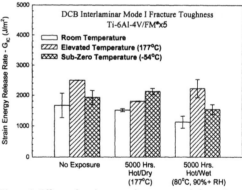

Figure 4. Effects of environmental exposure and elevated temperature testing on Mode I fracture toughness.

Figure 5. Fracture surface of DCB specimen showing cohesive failure.

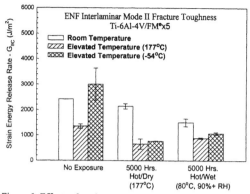

Figure 6. Effects of environmental exposure and elevated temperature testing on Mode II fracture toughness.

Figure 7. Fracture surface of ENF specimen showing interfacial failure.

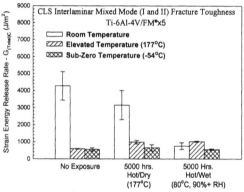

Figure 8. Effects of environmental exposure and elevated temperature testing on Mixed Mode I & II fracture toughness.

Figure 9. Fracture surface of CLS specimen showing alternating failure.

Figure 8 shows a comparison of the effects of environmental exposure and elevated temperature testing on the Mixed Mode I & II fracture toughness from CLS tests. In Mixed Mode I & II cases, an alternating failure (i.e. both cohesive and interfacial failure) was observed (Fig. 9). Critical strain energy release rate (G_{Tc}) results from CLS tests show a reduction of approximately 25% due to 'hot/dry' exposure, whereas a significant reduction of approxi-

mately 75% is observed due to the 'hot/wet' exposure. For the Mixed Mode I & II tests, elevated temperature testing significantly reduced the fracture toughness by approximately 75%.

Fractographic analyses of the failed specimens showed that pure Mode I loading of the Ti-6Al-4V/FM®x5 adhesive system yielded a mostly cohesive type of crack, whereas pure Mode II loading produced mostly an interfacial crack (Figs 5, 7). As shown in Figure 9, Mixed Mode I & II loadings produced cracks that alternated from cohesive to interfacial. In all loading cases, there was no evidence that elevated temperature testing or exposure to either 'hot/dry' or 'hot/wet' environments caused a change in the crack growth path.

The effects of Mode I versus Mode II critical strain energy release rates (G_{Ic} vs G_{IIc}) is shown in Figures 10-12 as a function of environmental exposure to 5,000 hours of the 'hot/dry' and 'hot/wet' conditions for room temperature, 177°C testing, and -54°C testing, respectively. For room temperature tests, exposure to 5,000 hours of environmental conditioning had a greater effect on specimens loaded in pure Mode II than in pure Mode I, and exposure to 5,000 hours of the 'hot/wet' environment was most detrimental (Fig. 10). Environmental conditioning had a similar effect on the Mode I and Mode II fracture toughness results at the 177°C test temperature, however exposure to 5,000 hours of the 'hot/dry' environment was shown to be most detrimental (Fig. 11). Results from -54°C testing showed that the effects of environmental conditioning on the Mode I fracture toughness were minor, however significant effects were observed for the Mode II fracture toughness. For -54°C testing, exposure to 5,000 hours of the 'hot/dry' environment was also shown to be the most detrimental (Fig. 12).

5 CONCLUSIONS

Experiments were performed to determine the effects of elevated temperature testing and environmental exposure on the fracture toughness of the Ti-6Al-4V/FM®x5 adhesively bonded system, pretreated with a Sol Gel chemical process on the titanium adherends to enhance durability. The Ti-6Al-4V/FM®x5 bonded joints were loaded in Mode I, Mode II, and Mixed Mode I & II in the as-received state, isothermally exposed for 5,000 hours at 177°C, and isothermally exposed to a hot/wet environment (80°C, 90%+ relative humidity). Tests were performed at room temperature, 177°C, and -54°C. Experimental results and established analytical methods were used to investigate the critical strain energy release rates, G_{Ic}, G_{IIc}, and G_{Tc} from double cantilever beam, end notch flexure and crack lap shear tests, respectively.

In all loading cases except Mode I, the threshold

strain energy release rates were reduced as a result of 5,000 hours of aging and elevated temperature testing. In some of the Mode II and Mixed Mode I & II cases, the reduction was as much as 75% due to elevated temperature testing or environmental exposure. In these experiments, the type of failure depended on the particular loading mode and was virtually unaffected by testing at the elevated temperature or the environmental exposures. Specimens loaded in pure Mode I experienced cohesive types of failures, whereas an interfacial failure was observed in specimens loaded in pure Mode II. For the Mixed Mode I & II loading cases, an alternating failure was usually observed, consisting of both cohesive and interfacial cracks. The test data was consistent and repeatable over the range of testing temperatures and environmental exposures.

From the critical strain energy release rate and fractographic results presented here, crack propagation due to pure Mode I loading appears to be governed primarily by the toughness property of the LaRC™-PETI-5 based adhesive FM®x5 (cohesive failures). The data for this loading mode indicates that the Mode I interfacial fracture toughness is higher than the Mode I fracture toughness of FM®x5, and that FM®x5 is somewhat resistant to the elevated temperatures and environmental conditions experienced in this study.

For pure Mode II and Mixed Mode I & II loadings however, crack propagation appeared to be governed primarily by the interfacial fracture toughness property (FM®x5 adhered to the Sol Gel chemical pretreated titanium surfaces). Because of the interfacial failures observed for this loading mode, the Mode II interfacial properties appear to be much lower than the shear properties of FM®x5, and the interface seem very susceptible to elevated temperature testing and degradation due to environmental exposure.

This investigation of a candidate system for supersonic commercial aircraft consisting of titanium

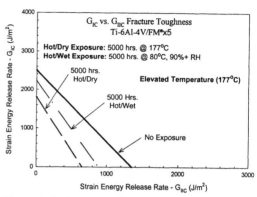

Figure 11. Environmental effects on Mode I versus Mode II fracture toughness at 177°C.

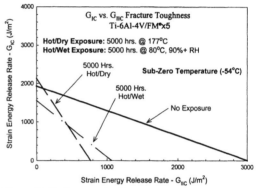

Figure 12. Environmental effects on Mode I versus Mode II fracture toughness at -54°C.

adherends bonded with a polyimide adhesive has identified the need to address the effects of elevated temperatures and environmental exposure on fracture toughness. Although the results presented here are specific to the materials, loading modes, testing temperatures, and the environments examined, the sensitivity of the fracture toughness to environmental exposure and elevated temperatures implies that a predictive scheme to estimate the remaining life of a bonded joint must account for elevated temperatures and moisture. This implication would also hold true when considering the development of accelerated test methods based upon the changes in the adhesive's microstructure and the physics of interfacial adhesion due to temperature and moisture.

6 ACKNOWLEDGMENTS

The authors gratefully acknowledge Ron Zabora of the Boeing Commercial Airplane Group for providing the materials used in this study. Funding for this

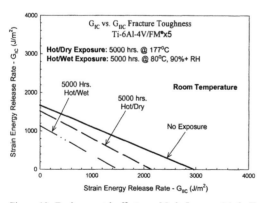

Figure 10. Environmental effects on Mode I versus Mode II fracture toughness at room temperature.

work was provided by the NASA Faculty Awards for Research (Grant NAG1-1727 monitored by Tom Gates), the NASA Graduate Student Researchers Program, and the NASA Center for High Performance Polymers and Composites (HiPPAC).

7 REFERENCES

Baker, A.A., Callinan, R.J., Davis, M.J., Jones, R. & Williams, J.G. 1984. Repair of Mirage III aircraft using the BFRP crack-patching technique. *Theoretical and Applied Fracture Mechanics* 2: 1-15.

Brussat, T.R., Chiu, S.T., & Mostovoy, S. 1977. Fracture mechanics for structural adhesive bonds - final report. *AFML-TR-77-163*. Burbank: USAF Materials Laboratory.

Carlsson, L.A., Gillespie, J.W., & Pipes, R.B. 1986. On the analysis and design of the end notched flexure (ENF) specimen for Mode II testing. *Journal of Composite Materials* 20: 594-605.

Dunkerton, S.B. & Vlattas, C. 1998. Joining of aerospace materials - an overview. *International Journal of Materials & Product Technology* 13: 105-121.

Goland, M. & Reissner, E. 1944. The stresses in cemented joints. *Journal of Applied Mechanics* 11(1): 65-77.

Griffith, A.A. 1920. Phenomena of rupture and flow in solids. *Philosophical Transactions* 221: 163-198.

Irwin, G.R. 1948. Fracture dynamics. *Fracturing of Metals*: 147-166. Cleveland: American Society for Metals.

Johnson, W.S. & Butkus, L.M. 1998. Considering environmental conditions in the design of bonded structures: a fracture mechanics approach. *The International Journal of Fatigue and Fracture of Engineering Materials & Structures* 21(4): 465-478.

O'Brien, T.K. & Martin, R.H. 1993. Round robin testing for Mode I interlaminar fracture toughness of composite materials. *Journal of Composites Technology and Research* 15(4): 269-281.

Ripling, E.J., Mostovoy, S., & Patrick, R.L. 1964a. Application of fracture mechanics to adhesive joints. *ASTM STP 360*: 5-19. Philadelphia: American Society for Testing and Materials.

Ripling, E.J., Mostovoy, S., & Patrick, R.L. 1964b. Measuring fracture toughness of adhesive joints. *Materials Research and Standards* 6: 129-134.

Recent Developments in Durability Analysis of Composite Systems, Cardon, Fukuda, Reifsnider & Verchery (eds)
© 2000 Balkema, Rotterdam, ISBN 90 5809 103 1

Strain rate and temperature effects in polymeric matrices for composite materials

K. Reifsnider, C. Mahieux, B. Walther & F. Sun
Department of Engineering Science and Mechanics, Materials Response Group, Virginia Polytechnic Institute and State University, Blacksburg, Va., USA

ABSTRACT

The effect of temperature and strain rate during deformation on the properties and performance of materials has been a long-standing subject of engineering interest, driven by the observation that significant differences in strength, stiffness and life can be induced by these effects. In polymer-based materials, the discussion of such effects usually centers on viscoelastic response and changes in stiffness. The present paper examines subjects that are related to strength, life, and to accelerated testing. Micromechanical effects are identified and discussed, and new concepts are introduced to assist in the interpretation and representation of physical behavior, over ranges of strain rate from impact to quasi-static testing, and temperature from glassy to flow behavior. In this paper, for polymeric matrices over wide ranges of strain range and temperature, theoretical models of strain rate dependent properties and temperature effects are discussed. A quantitative equivalence equation between strain rate and temperature effects is proposed. Finally, comparisons between numerical results and experimental data are shown and the applications / validations of the proposed models are discussed.

1 STRAIN RATE DEPENDENCE

The behavior of materials, especially polymeric matrices in composites under dynamic loads, is a very important issue in practical engineering, essential for any work involving polymeric composite materials and structures in aerospace and ocean applications (Sierakowski and Chaturvedi, 1997). Most of the prior research dealing with the dynamic properties of composites has concentrated on the strain rate dependent behavior of composite laminates and structures, and on macro-mechanics analysis (Ross and Sierakowski, 1973; Finn and Springer, 1991; Hull, 1991).

In order to fully understand the strain rate dependent mechanisms and to develop better impact-tolerant composites through a micro-mechanics approach, it is necessary to know the strain rate dependent properties of the constituents, i.e., the fibers and matrices. It is only in fairly recent years that tensile impact techniques for fiber bundles and matrix have been established (Harding and Welsh, 1982; Kawata et al, 1982; Xia et al, 1986.). There are some tensile impact data showing

that the strength of epoxy, widely used as a matrix in polymeric composites, increases with strain rate (Harding and Welsh, 1983; Dong et al, 1990). However, few quantitative analyses for the strength variations and rate dependent mechanisms for this material have been found so far, and few explanations of strain rate effects on polymeric matrices have been offered.

On the other hand, it is common knowledge that the mechanical properties of polymeric matrices change with strain rate as well as with ambient temperature. For very long-loading-time deformations, such as creep and relaxation, there is a time-temperature equivalence, e.g., WLF superposition principles may apply (Williams, et al, 1955). Mechanical properties of polymers at times over several decades of significance can be predicted by the data at higher temperatures and over short times. But this equivalence is not generally valid at temperatures below Tg, and few data support the application of WLF principles to the relationship between temperature and strain rate for polymeric matrices. It is only reported qualitatively that there exists an "equivalence" between strain rate in impact process and temperature, for polymeric resins (Nielsen and Landel, 1994; Hall, 1981), and for elastomers within the range of rubbery behavior (Smith and Sterdy, 1960); but, there are no quantitative descriptions regarding this equivalence.

Various investigators have discussed material property-strain rate relations and several models have been proposed (Gates, 1993; Ashby and Jones, 1986; Xia, et al, 1994). For the present case, we propose to use a relationship developed in another (related) context.

For the relationship between time to failure and strain rate, there is a simple and widely used equation for creep of metals, polymers and other materials. It is usually referred to as the Monkman-Grant equation (Rosen, 1982; Evans, 1984), as expressed by:

$$t_b \cdot \left(\dot{\varepsilon} \right)^m = C \qquad (1)$$

Here, for creep, t_b is time to failure, $\dot{\varepsilon}$ is the steady-state strain rate and C is a material constant. We have applied the Monkman-Grant equation to the impact process and fit the experimental data in work by Gates, 1993; Ashby and Jones, 1986; Xia, et al, 1994 and Xia, et al, 1993, by equation 1, treating t_b as the time to failure and $\dot{\varepsilon}$ as the constant strain rate in the impact process for impact processes under constant strain rate conditions.

Using the data sets from Gates, 1993; Xia, et al, 1987; Ashby and Jones, 1986; Xia, et al, 1994; Xia and Xing, 1996; Carrington, 1994; Rosen, 1982; Evans, 1984; Bartenev and Zuyev, 1968; Xia, et al, 1993; Hartwig, 1978; Dutta, 1993, the validity of equation 1 for fiber, matrix, and composite over the entire range of strain rates is evaluated, in Figs. 1-3.

It can be seen that the equation fits the data for epoxy at strain rates from the quasi-static to very high strain rate ranges (up to 10^3 1/sec) almost perfectly. The constants for the three conditions are:

Fiber: $C_1 = 0.0273$ $m = 0.935$
correlation = 0.999
Matrix: $C_1 = 0.0377$ $m = 0.908$
correlation = 0.999
Composite: $C_1 = 0.0419$ $m = 0.908$
correlation = 0.999

The applicability of this relationship to such a wide range of conditions is quite surprising. It raises several possibilities for predictive models when strength, strain to failure, and time to failure

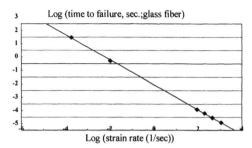

Figure 1 Variation of time to failure with strain rate for glass fibers.

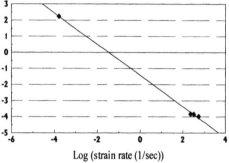

Figure 2 Variation of time to failure with strain rate for epoxy matrix.

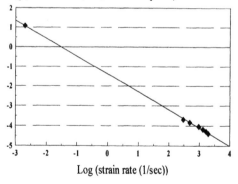

Figure 3 Variation of time to failure with strain rate for Glass-reinforced epoxy composite.

must be estimated in composites, and when testing or characterizations are available for one strain rate but applications require different strain rate conditions.

An equivalence between strain rate effects (or time to a certain state of material at various strain rates) in impact processes and the temperature effect on the same state of materials can be proposed for such matrices, similar to that for creep and relaxation. In the present paper, an empirical superposition principle was employed to calculate equivalent temperatures for different strain rates, especially for impact processes. Here, the time to break, t_b, which is a function of strain rate is used in the calculation. The time corresponding to low temperatures can be arbitrarily chosen at a convenient strain rate for experiment. For the present example, it is set to t_S, the time to break in a quasi-static test.

Then, the room temperature data for very short-time fracture (high strain rate) is converted to the data at long fracture times (quasi-static test) and low temperature, by a shift factor a_T, along the time scale.

Here,

$$a_T = \frac{t_S}{t_b} \tag{2}$$

In order to calculate Log (a_T), the following Arrhenius relationship is employed (Miyano and Kanemitsu, 1983):

$$\log(a_T) = \frac{\Delta H}{2.303G}\left[\frac{1}{T} - \frac{1}{T_0}\right] \tag{3}$$

Here, the activation energy is ΔH =219 for epoxy.

Therefore, the temperature for the quasi-static test, which is equivalent to high strain rate, can be calculated by combining equations (1) and (3) as follows:

$$T = \left[\frac{2.303G\log\left(\dfrac{t_S}{C_1(\dot{\varepsilon})^{-m}}\right)}{\Delta H} + \frac{1}{T_0}\right]^{-1} \tag{4}$$

Comparisons between the experimental data of longitudinal and shear moduli at various strain rates and the calculated values at equivalent temperatures were made for epoxies in work by Gates, 1993; Xia, et al, 1993; and Dutta, 1993. Some examples of the results are shown in Figure 4. It can be seen that the calculated results corresponding to the equivalent temperature (lower temperature) at a reference strain rate agree, quite well, with the experimental data at higher strain rates and temperatures over seven orders of strain rate. (Calculations for two reference values of the activation energy are shown.)

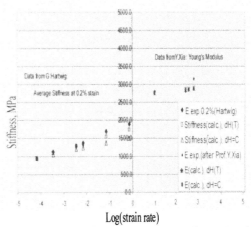

Figure 4 Predicted (from quasi-static tests) and observed stiffness values as a function of strain rate for epoxy, using two different activation energy values.

It can be seen that the calculated stiffness values match the measured values very closely over a wide range of strain rates.

2 STIFFNESS MODELING ACROSS TRANSITION TEMPERATURES

As a result of the typical use of spring and dashpot models, modulus-time relationships for polymers are often written in the form

$$E(t) = \sum_{i=1}^{N} E_i \exp\left(\frac{-t}{\tau_i}\right) \longrightarrow$$

$$E(t) = \int_{-\infty}^{+\infty} H(\tau)\exp\left(\frac{t}{\tau}\right)d(\ln\tau) \tag{5}$$

where τ_i are relaxation times and the usual distribution function, $H(\tau) = \tau E(\tau)$ has been introduced. For our situation, we use the distribution function to represent the action of secondary bonds as temperature changes, to capture the physics of instantaneous stiffness changes caused by changes in mobility at the macromolecular level. We make the argument that the distribution function should correspond to the behavior of secondary bonds under rapid loading (since we are ultimately interested in micro-fracture processes), and that we want to capture the interactions between bond failure events, especially in the region of Tg. We also invoke the exchangeability of time and temperature variables

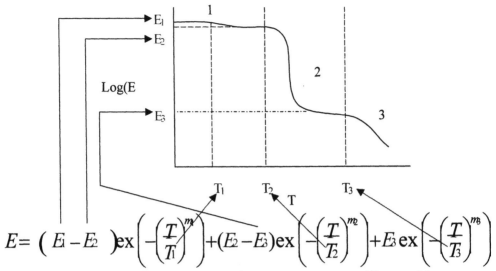

$$E= \left(E_1 - E_2 \right) \text{ex} \left(-\left(\frac{T}{T_1}\right)^{m_1} \right) + (E_2 - E_3)\text{ex}\left(-\left(\frac{T}{T_2}\right)^{m_2} \right) + E_3 \text{ex}\left(-\left(\frac{T}{T_3}\right)^{m_3} \right)$$

Figure 5 Schematic diagram of the transition temperature stiffness model.

for such instantaneous response (Aklonis and MacKnight, 1983) Finally, we require that the model of stiffness be valid over essentially all of the transitions induced by temperature, so that the representation can be used in engineering models across operating ranges that include glassy, rubbery, and even flow behavior. For certain applications, such wide operating ranges are being used. The resulting representation is shown in Fig. 5. All of the constants in the equation can be measured independently. The transition temperatures, T_1 can be measured by dynamic methods such as DMA testing. The modulus step sizes (e.g. (E_1-E_2)) can be measured by various dynamic methods as well; typically ultrasound is used (VanKrevelen, 1990). The Weibull coefficients are obtained by fitting pilot data, at this point. However, methods for obtaining them from first-principals and independent data are being developed.

We have conducted an extensive study to validate the form of the model shown in Fig. 5 (Mahieux, 1999) An example of the results of that study is shown in Fig. 6. The material in that example is PMMA. The experimental data are from (Ashby, and Jones, 1998). It can be seen that the model follows the form and function of the stiffness variations well for this amorphous linear polymer. The various stiffness values from the literature were typically measured at 1-3 Hz. Numerous examples were also compared, for cross-linked thermosets, and crystalline and amorphous thermoplastics. In all cases, the model represented the data well.

The Weibull modulus for the primary transition

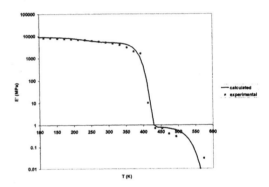

Figure 6 Comparison of model to experimental data from (Ashby and Jones, 1998).

around Tg is large, about 20 for the example in Fig. 6. High coefficients characterize a very deterministic and simultaneous process. Values of m_2 as high as 40 were found in some cases. As one would expect, the values of the Weibull coefficients associated with the glass transition region, m_2, and with the flow region, m_3, are dependent on physical variables like molecular weight, crystallinity, crosslinking, and the presence of fillers. We have studied some of those dependencies, but much more data is needed (Mahieux, 1999).

We have also studied how these transitions affect the mechanical properties and performance of polymer matrix composites (Mahieux, 1999). Only a few examples of those results will be summarized here. One surprising result has to do with the tensile strength of fiber-dominated unidirectional composites, in the fiber direction. Walther reports the surprising result that the quasi-static tensile strength of such materials can vary by 15-30 percent across the Tg of the polymer matrix (Walther, 1998). Even though the fibers are unaffected by such temperature changes, the failure process, which depends greatly on the connectivity between the fibers and the matrix, is substantially altered in some cases. Subramanian and Reifsnider developed a micromechanical model that accounts for such an "efficiency factor" η by modifying the rule of mixtures expression for the longitudinal Young's modulus to read

$$E_{11} = \frac{E_m V_m + \eta E_f V_f}{V_m + \eta V_f} \qquad (6)$$

where V_1 are the constituent volume fractions. Mahieux introduced the model described into that relationship to obtain

$$E_{11}(T) = \frac{E_m(T) V_m + \sum_{i=1}^{N} \lambda_i \exp\left(-\left(\frac{T}{T_i}\right)^{m_i}\right) E_f V_f}{V_m + \sum_{i=1}^{N} \lambda_i \exp\left(-\left(\frac{T}{T_i}\right)^{m_i}\right) V_f} \qquad (7)$$

where the coefficients λ_i are normalized material constants, generally less than unit value. The utility of this approach is shown in Fig. 7. The material is AS4/PPS with a fiber volume content of 69 percent.

The modulus of the fibers was 2.26 MPa and the values of the λ_i coefficients were 0.03, 0.05 and 0.92. The efficiency factor variation was fixed by the independently measured data for PPS, so that there were no undetermined constants in the model. The agreement with experimental data is quite good; the amorphous material and crystalline material models seem to bracket the average observed behavior, as expected.

Figure 8 shows the loading arrangement used for the second example. Unidirectional specimens with dimensions of 152.4X12.5X10.16 mm were end-loaded to produce out-of-plane bending. The material was the same AS4/PPS composite used for the study in Fig. 7. The stress-rupture of the material at fixed bend radius and temperature was successfully modeled using micromechanics representations of the local micro-buckling that initiated the specimen failure (Mahieux, 1999). Then the micromechanical model of uniaxial stiffness mentioned above was used in the micromechanical buckling relationship to estimate the stress rupture behavior as a function of temperature. Room temperature fatigue data were also collected with the same loading method to obtain a model of the fatigue response of the specimens. Finally, the time dependent (stress rupture) and cycle dependent (fatigue) models were combined using the critical element method and MRLife code developed by our research group over a number of years (Reifsnider, 1991; Reifsnider and Stinchcomb, 1986; Reifsnider, 1992; Reifsnider, 1995; Reifsnider, 1996a; Reifsnider, 1996b). An example of the type of results obtained is shown in Fig. 9

For this case, the fatigue degradation at elevated temperature is less than the stress rupture at that temperature. The models seem to represent that feature well. The most striking part of the result, however, is that the sharp change in the behavior above the Tg of the material is followed well by the model, and also seen in the experimental data.

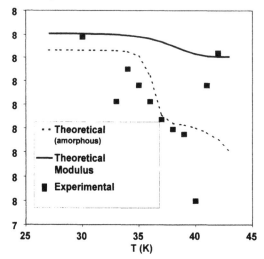

Composite tensile modulus

Figure7 Comparison of model predictions and data for composite tensile strength.

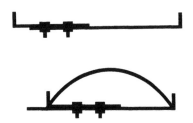

Figure 8 Loading arrangement for end-loaded bending.

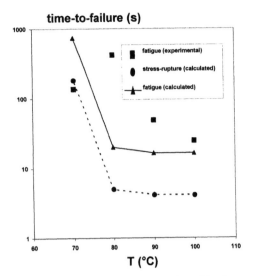

time-to-failure (s)

Legend:
- fatigue (experimental)
- stress-rupture (calculated)
- fatigue (calculated)

T (°C)

Figure 9 Predicted and observed combined stress rupture and fatigue behavior of AS4/PPS in end-loaded bending at 75 percent strain and several temperatures, across Tg.

Hence, we have predicted combined fatigue-stress rupture behavior across Tg of the composite for fiber dominated behavior, in this case.

3 CONCLUSIONS

Based upon the above discussion and data, it can be concluded that:

(1) The Monkman – Grant relationship in creep may be applicable to the description of the relationship between time and strain rate processes, not only for the time to failure and strain to failure, but also for the time to a certain material state as a function of strain rate for epoxy. This suggests that there may exist a common mechanism between creep and rate processes such as impact in this material, and also suggests that relationships of this type may be developed for similar materials.

(2) An equivalence between strain rate effects and temperature effects on material states was proposed. A quantitative equivalence equation was obtained for the longitudinal and shear moduli of epoxy, as an example. It was found that an empirical superposition approach based upon an Arrhenius equation

is an effective method for predicting the equivalent temperatures for various strain rates. Predicted values of stiffness for different strain rates (corresponding to impact conditions) based on values obtained from quasi-static testing at room temperature (obtained by predicting an equivalent lower temperature for which the stiffness should match the experimental value obtained at a rapid strain rate at room temperature) matched observed values well for epoxy over seven orders of magnitude of strain rates.

(3) A robust model of polymer stiffness as a function of temperature across all transitions was presented. The model is based on the statistical behavior of secondary bonding.

(4) The stiffness-temperature model was successfully applied to the prediction of unidirectional tensile strength, and combined fatigue-stress rupture for fiber dominated AS4/PPS composites.

Several distinctive features of this work should be emphasized. First, we have addressed the effect of temperature and strain rate on failure, specifically the strength, time to failure, and strain to failure. And second, we have developed simple concepts and equations that make it possible to calculate values of stiffness, strain to failure, and time to failure as a function of temperature and strain rate, based on independent reference data

Expected applications of these concepts are numerous. An especially important one, at the global level, is the construction of algorithms that can be used in predictive analysis codes to estimate stiffness (and, thereby, stresses and strains, strains to failure, and time to failure) as a function of strain rates and temperature for materials used in dynamic loading situations, especially impact conditions. Another global use is the "acceleration" of testing, in the sense that test results under very difficult (and expensive) test conditions (such as very high strain rates or very high or low temperatures) can be estimated from more easily obtained room temperature quasi-static test data. And at the local level, micromechanical analysis of fiber / matrix / interface behavior depends on a firm knowledge of stiffness as a function of strain rate and temperature, since events like fiber fracture are inherently high-strain-rate processes. If we have any hope of getting the local stress state correct for such events (and estimating composite strength using micromechanics), we must get the local stiffness correct under those conditions. The present approach offers an approach to that objective.

4 ACKNOWLEDGEMENTS

The authors gratefully acknowledge the support of the Air Force Office of Scientific Research (grant no. F49620-95-1-0217) for the research on high temperature polymer composites, and the National Science Foundation under grant no. DMR9120004 for support of the micromechanical modeling.

5 REFERENCES

Aklonis, J.J. and W.J. MacKnight, 1983. *Introduction to polymer viscoelasticity*. Second Edition. John Wiley & Sons, Inc.

Ashby, M.F. and D.R.H. Jones, 1986. *Engineering Materials 2*. Pergamon Press. 226-231.

Ashby, M.F. and D.R.H. Jones, 1998. *Engineering Materials: An Introduction to Microstructures, Processing and Design*.

Bartenev, G.M. and Y.S. Zuyev, 1968. *Strength and failure of visco-elastic materials*. Translated by F.F. and P. Jaray. Pergamon Press. 102.

Carrington, G., 1994. *Basic thermodynamics*. Second Edition. Oxford Science Publications. 275-280.

Dong, L., Y. Xia and B.Yang, 1990. Tensile impact testing of fiber bundles. *ICSTAD Proceedings*. India, 184-189.

Dutta, P.K., K.L. Faran and D. Hui, 1993. Influence of low temperature on energy absorption in laminated composites. *Proc. ICCM 9*. 311-320.

Evans, H.E., 1984. *Mechanisms of creep fracture*. Elsevier Applied Publishers. 20.

Finn, S.R. and G.S.Springer, 1991. Composite plates impact damage. Technomic Publishing Co., Inc.

Gates, T.S., 1993. Matrix-dominated stress/strain behavior in polymeric composites: effects of hold time, nonlinearity, and rate dependency. *ASTM STP 1206*. E.T. Camponeschi, Jr., Ed. Philadelphia. 177-189.

Hall, C., 1981. *Polymer materials*. John Wiley & Sons. 73.

Harding, J. and L.M. Welsh, 1982. Impact testing of fiber-reinforced composite materials. *Proc. ICCM 4*. Tokyo, Japan. 845-852.

Harding, J. and L.M. Welsh, 1983. A testing technique for fiber-reinforced composites at impact rates of strain. *J. of Materials Science*. 18:1810-1826.

Hartwig, G., 1978. Mechanical and electrical low temperature properties of high polymers. *Nonmetallic materials and composites at low temperatures*. Edited by A.F. Clark, R.P. Reed and G. Hartwig. Plenum Press. 33-50.

Hull, D., 1991. A unified approach to progressive crushing of fiber-reinforced composite tubes. *Composites science and technology*. 40:377-421.

Kawata, K., S. Hashimoto and N.Takeda, 1982. Mechanical behaviors in high velocity tension of composites. *Proc. ICCM 4*. Tokyo, Japan. 829-836.

Mahieux, C.A., 1999. *A systematic stiffness-temperature model for polymers and applications to the prediction of composite behavior*. Dissertation submitted to the Department of Engineering Science and Mechanics, Virginia Polytechnic Institute and State University, Blacksburg, VA.

Miyano, Y. and M. Kanemitsu, 1983. Time and temperature dependence of flexural strength in transversal direction of fibers in CFRP. *Fiber science and technology*.18:65-79.

Nielsen, L.E. and R.F.Landel, 1994. *Mechanical properties of polymers and composites*. Second Edition. Marcel Dekker, Inc. 256.

Reifsnider, K. L. Editor, 1991. *Fatigue of Composite Materials*. London: Elsevier Science Publishers.

Reifsnider, K. L., 1992. Use of mechanistic life prediction methods for the design of damage tolerant composite material systems. *ASTM STP 1157*. M.R. Mitchell & O. Buck Editors. Philadelphia, PA: American Society for Testing and Materials. 205-223.

Reifsnider, K. L., 1995. Evolution concepts for microstructure-property interactions in composite systems. *Proc. IUTAM Symp. On Microstructure-Property Interactions in Composite Materials*. Aalborg, Denmark. R. Pyrz, Editor. New York, NY: Kluwer Publishers. 327-348.

Reifsnider, K. L., 1996a. Recent advances in composite damage mechanics. *Proc. Conf. On Materials and Mechanical Testing*. European Space Agency. Noordwijk, Netherlands. SP-386. 483-490.

Reifsnider, K. L., 1996b. A micro-kinetic approach to durability analysis: the critical element method. *Progress in Durability of Composite Systems*. A.H. Cardon, K.L. Reifsnider, & H. Fukuda, Editors. Balkema, Rotterdam. 3-11.

Reifsnider, K. L. and W. W. Stinchcomb, 1986. A critical element model of the residual strength and life of fatigue-loaded composite coupons. *Composite Materials: Fatigue and Fracture*,

ASTM STP 907. H.T. Hahn, Editor. Philadelphia, PA: American Soc. for Testing and Materials. 298-313.

Rosen, S.L., 1982. *Fundamental principles of polymeric materials*. John Wiley and Sons, Inc. 242.

Ross, C.A. and R.L.Sierakowski, July 1973. Studies on the Impact resistance of composites. *Composites*. 157-161.

Sierakowski, R.L. and S.K. Chaturvedi, 1997. *Dynamic loading and characterization of fiber-reinforced composites*. John Wiley & Sons, Inc.

Smith, T.I. and P.J. Sterdy, 1960. Time and temperature dependence of the ultimate properties of an SBR rubber at constant elongations. *J. of Applied Physics*. 31, 11:1892-1898.

Van Krevelen, D.W., 1990. *Properties of polymers*. Third Edition. Elsevier. New York.

Walther, B.M., 1998. *An investigation of the tensile strength and stiffness of unidirectional polymer-matrix carbon-fiber composites under the influence of elevated temperatures*. Thesis submitted to the College of Engineering, Virginia Polytechnic Institute and State University, Blacksburg, VA.

Williams, M.L., R.F.Landel and J.D.Ferry, 1955. The temperature dependence of relaxation mechanisms in amorphous polymers and other glass-forming liquids. *J. American Chemical Society*. 77:3701-3707.

Xia, Y., B.Yang L.Dong and D.Jia, 1986. Experimental study of constitutive relation for unidirectional glass fiber reinforced resin under tensile impact. *Proc. of ISCMS*. Beijing, China. 997-1002.

Xia, Y., B. Yang and M. Li, 1987. Tensile impact response of resin matrices and their composites. *ACTA MATERIAE COMPOSITAE SINICA*. 4, 2:59-65, (In Chinese).

Xia, Y., X. Wang and B. Yang, 1993. Brittle-ductile-brittle transition of glass fiber-reinforced epoxy under tensile impact. *J. Materials Sci. Lett*. 12:1481-1484.

Xia, Y., J. Yuan and B. Yang, 1994. A Statistical model and experimental study of the strain–rate dependence of the strength of fibers. *Composites Science and Technology*. 82:499-504.

Xia, Y. and W. Xing, 1996. Constitutive equation for unidirectional composites under tensile impact. *Composites Science and Technology*. 56:155-160.

Recent Developments in Durability Analysis of Composite Systems, Cardon, Fukuda, Reifsnider & Verchery (eds)
© 2000 Balkema, Rotterdam, ISBN 90 5809 103 1

Fatigue life prediction of polymer composites with hysteretic heating

X. R. Xiao & I. Al-Hmouz
Concordia Center for Composites, Concordia University, Montreal, Que., Canada

ABSTRACT: This paper presented a simple prediction scheme which correlates the fatigue life to the thermal degradation of fatigue strength. Shifting factors similar to the time-temperature shifting in viscoelastic media were employed to account for the temperature effect on the fatigue strength and an iso-strength plot was introduced for fatigue life prediction under non-isothermal conditions. The scheme presented in this paper can predict the load frequency effect associated with hysteretic heating from limited basic material information. The load frequency effect on the fatigue life of thermoplastic composite AS4/PEEK ±45 laminate was investigated at 1 Hz, 5 Hz and 10 Hz. The fatigue life prediction for 5 Hz and 10 Hz based on SN data at 1 Hz was demonstrated. The predictions were in a reasonably good agreement with the experimental data.

1. INTRODUCTION

The load frequency is known to have a substantial effect on fatigue performance of polymers and polymer matrix composites. Literature data indicate that the load frequency has a two-sided effect on fatigue life of polymer composites, depending on the extent of temperature rise associated with hysteretic heating. When the temperature rise is not significant, fatigue life appears to increase with increasing load frequency. This has been observed in graphite/epoxy (Sun & Chan, 1979) and boron/epoxy (Reifsnider et al, 1977), where lower frequencies were found to cause greater damage as compared to that at the same number of cycles at higher frequencies. On the other hand, in materials associated with large-scale hysteretic heating, such as glass fiber reinforced epoxy (Dally & Broutman, 1967) and thermo-plastic composite AS4/PEEK Dan-Jumbo et al, 1989, Curtis et al, 1988), fatigue life decreases remarkably with increasing load frequency.

Sun and Chan (1979) has developed a model to account for the load frequency effect on fatigue life of notched composite specimens. The model proposed in ref.3 was similar to a model by Schapery (1975) for creep crack propagation in viscoelastic media and also took into the consideration of the effect of temperature rise during fatigue. It has a relatively simple form with two experimentally determined parameters. The model was in a good agreement with experimental results on graphite/epoxy which exhibited moderate temperature rises. For materials with more significant hysteretic heating, a modified model was developed (Dan-Jumbo, 1989), which relates

the stress and load frequency directly to the fatigue life. Nevertheless, more parameters are involved and have to be determined by curve fitting of experimental data. In an earlier attempt, we have proposed a model (Xiao & Al-Hmouz, 1997) based on the two models mentioned above. The model contains fewer parameters but its predictive capability is still at the same level as the other models because of the curve fitting procedure involved.

The present work is aimed at developing predictive models for the load frequency effect on the fatigue life of polymer composites associated with large hysteretic heating using the approaches based on viscoelastic theory. It is well known that the load frequency dependence behavior of polymer composites in fatigue is, to a great extent, due to the viscoelastic nature of the polymer matrices. In viscoelastic media, the time-temperature, time-stress superposition approaches are common technique in modeling the time-dependent mechanical responses such as creep and relaxation and have been proven to be effective in polymer composites. This approach has also been extended to model the fatigue behavior of polymer composites. Rotem and Nelson (1981) introduced temperature shifting factors to predict long-term fatigue strength from short-term testing at elevated temperatures. Miyano et al (1994) constructed master strength-time curves for a woven carbon/epoxy composite from the flexural fatigue data obtained at two frequencies over a wide range of temperatures. These works, however, are for the isothermal conditions, whereas the temperature effect associated with hysteretic heating is a non-isothermal one.

In this paper, a simple scheme correlating temperature rise caused by hysteretic heating with the thermal degradation of the strength and the fatigue life of the material is presented. The non-isothermal problem is addressed through an iso-strength approach. The procedure for fatigue life prediction is demonstrated and the results are compared with the data obtained on AS4/PEEK [±45]4s laminates.

2. FATIGUE STRENGTH UNDER NON-ISOTHERMAL CONDITIONS

In the analysis proposed by Rotem and Nelson (1981), the fatigue strength at a given temperature is determined by considering the cyclic stress field at each lamina, allowing progressive failure of individual plies until the final fracture of the laminate. One needs to know not only the stiffness and strength components of the unidirectional composite, the progressive damage law of the laminate, but also the nonlinearity in these properties at a given temperature. The calculation involved in non-isothermal cases will be even more tedious. In this paper, we follow the concept of temperature shifting factor at the laminate level rather than from the lamina level. The scheme presented in this paper is equally applicable to the lamina level.

We start from the isothermal conditions. Following Rotem and Nelson (1981), if the S-N curve of a composite laminate at a reference temperature T_0 is described as

$$\sigma(T_0) = F(N) \qquad (1)$$

then the S-N curve at a arbitrary temperature T will be

$$\sigma(T) = F[N, a_T(T), b_T(T)] \qquad (2)$$

where a_T and b_T are the shifting factors. b_T corresponds to a vertical shift due to the changes in the static strength with temperature; a_T corresponds to the rotation of S-N curve with temperature. The exact form of F function and the values of a_T and b_T are to be determined from experimental results. $b_T(T)$ is equal to the ratio of static strength at temperature T to that at the reference temperature T_0. $a_T(T)$ is determined by the relative change in slope of S-N curves at the two temperatures. If the slope of S-N curves over a temperature range does not change, a_T will be a unity. A series of isothermal S-N curves described by Eq.2 are shown schematically in Fig.1.

The fatigue life in a non-isothermal condition can be presented by plotting iso-strength curves in a temperature-number of cycles plot, as illustrated in Fig.2. To obtain an iso-strength curve at σ_i, a horizontal line corresponding to σ_i is drawn which

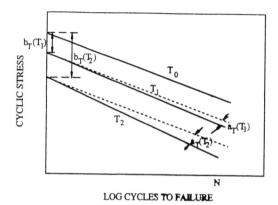

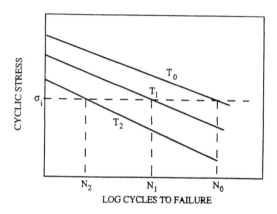

Figure 1. Schematic isothermal SN curves and shifting factors.

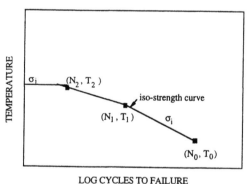

Figure 2. Constructing an iso-strength curve.

intercepts isothermal S-N curves of T_0, T_1 and T_2 at fatigue lives N_0, N_1 and N_2, respectively. These three sets of data provide three points for the iso-strength curve at σ_i. Since the isothermal S-N curves are available only at some specific temperatures, the iso-strength curve is constructed by piecewise linear segments linking these points.

Another segment of the iso-strength curve is the horizontal line, which corresponds to the temperature when the static strength decreases to the iso-strength value. The envelope formed by iso-strength curve approximates the fatigue life of a sample subjected to a cyclic stress at the iso-strength value in a non-isothermal condition.

As to be discussed in the next section, the temperature rise in fatigue caused by hysteretic heating can be predicted for a given load level and frequency. By overlaying the predicted transient temperature curve of a sample cycled at a stress level equal to the iso-strength curve value, one can predict the fatigue life under non-isothermal condition, as illustrated in Fig.3. Fatigue failure occurs when the transient temperature curve intercepts the iso-strength curve. If the temperature rises rapidly, the transient temperature curve may intercept the horizontal line, i.e., the static strength of the material thermally degrades to the applied cyclic stress level and hence the material fails as in a static test before fatigue failure. For both cases, the number of cycles corresponding to the crossing point gives the fatigue life prediction.

3. TEMPERATURE RISE DUE TO HEATING

The analysis for specimen temperature rise under cyclic loading has been discussed by a number of researchers. The present work follows the analysis presented by Hanh and Kim (1976). Considering the heat loss from the specimen to the environment, the temperature rise associated with internal heating for a smooth specimen is given by

$$\frac{dT}{dt} = \frac{q}{\rho c_p} - \frac{HA}{\rho c_p V} (T-T_0) \qquad (3)$$

where q is the heat generation rate, ρ is the density, c_p is the specific heat, H is a general heat transfer coefficient accounting for the heat loss from the specimen to environment due to all types of heat exchanges, T_0 is the environment temperature, and A and V are the surface area and volume of the fatigue test specimen, respectively.

The heating rate q may be estimated from the viscoelastic properties of the material or from the area of the hysteresis loop w by assuming that all the loss energy turns into heat

$$q=wf \qquad (4)$$

The area of the hysteresis loop may be approximated by the area of two triangles. Hanh and Kim (1976) found that the width of the triangle base is proportional to the stress amplitude $C(\sigma-\sigma_t)$ and hence

$$w = C(1-R)^2\sigma(\sigma-\sigma_t) \qquad (5)$$

where C is a parameter, R is the stress ratio of the

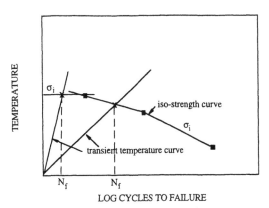

Figure 3. Fatigue life prediction for non-isothermal conditions.

fatigue test, and σ_t is the threshold stress below which no hysteretic heating occurs.

The heat transfer coefficient H can be estimated from the equilibrium condition

$$\frac{q}{\rho c_p} - \frac{HA}{\rho c_p V} (T_e-T_0) = 0 \qquad (6)$$

where T_e is the equilibrium temperature.

Having determined w and H values, the transient temperature T can be calculated by

$$\frac{T-T_0}{T_e-T_0}=1-\exp(-\frac{HA}{\rho c_p V} t) \qquad (7)$$

with

$$T_e-T_0=\frac{q}{\frac{HA}{V}} \qquad (8)$$

Eq. 7 is valid when q is constant, i.e. the area of hysteresis loop remains constant during fatigue test. To account for variations in hysteresis, $q=w(t,T)f$, Eq.3 can be solved by incremental method:

$$\Delta T_i = \frac{1}{\rho c_p} [(w(t_i, T_i)f - \frac{HA}{V} (T_i-T_0)]\Delta t \qquad (9)$$

$$T_{i+1} = T_i + \Delta T_i$$

$$t_{i+1} = t_i + \Delta t$$

4. EXPERIMENTAL

Specimens
AS4/PEEK [±45]4s laminates of 8x8 inches were molded using a WABASH hot press following ICI's procedure. The straight sided [±45]4s specimens were then cut from the laminates. The nominal dimensions of the specimens were

200x18x2 mm with 38 mm long aluminum end tabs, which follows the specification of ASTM D3479.

Fatigue tests
The fatigue tests were carried out using a MTS machine controlled by MTS TestStar II control system. Tensile-tensile fatigue tests were performed under load controlled mode with a sinusoidal load wave form. The stress ratio in fatigue was R=0.13. Fatigue tests were carried out at three load frequencies: 1 Hz, 5 Hz and 10 Hz. At each frequency, fatigue tests were conducted at three stress levels: 60%, 70% and 80% of the ultimate tensile strength. To complete the S-N curve at 1 Hz, few more fatigue tests were conducted at other stress levels. All fatigue tests were conducted without forced cooling.

During the tests, the load level was monitored from the signal of the load cell of the MTS machine while the axial strain was measured using an extensometer. The displacement of cross-head was also recorded to provide complementary strain reading. The readings from two strain measurements were in good agreement. During fatigue tests, cyclic stress-strain data from the load cell and extensometer were collected at selected cycles by the TestStar software and the hysteresis loops were displayed on a monitor and later plotted using graphic package Excel. The possible hysteresis of testing and measuring systems was examed by testing a steel specimen of similar dimensions at the highest load level at the three load frequencies. Since linear responses were recorded the hysteresis due to these systems was considered negligible. The temperatures were measured during the test using thermocouples attached to the center region of the specimen surface and recorded by a data acquisition system.

5. RESULTS AND DISCUSSION

S-N Data
The data on fatigue strength - number of cycle to failure (S-N) obtained are presented in Table 1. The values of the equilibrium or the maximum temperature recorded during fatigue tests are also listed. As seen, increasing frequency reduces the fatigue life and this life reduction effect appears to be larger at lower stress levels.

Temperature Rise
The typical temperature variations during fatigue tests are shown in Fig.4 by solid lines for the three load frequencies at three stress levels. At 1 Hz, temperature rise reached the equilibrium for fatigue tests conducted at all load levels. At 5 Hz, a peak on temperature rise curve, which is similar to that reported by Curtis et al (1988), was observed at tests conducted at 60% level. At higher stress levels, monotonic temperature rises were recorded. At 10 Hz, continuously rising temperatures were measured at all stress levels.

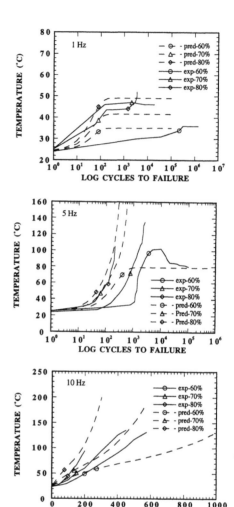

Figure 4. Temperature rise during fatigue. Solid lines are measured values, dashed lines are the predictions. (a) 1 Hz; (b) 5 Hz; (c) 10 Hz.

The temperature rises during fatigue were predicted following the procedure presented in Eq.3 to Eq.9. The parameters in Eqs.4-8 were obtained as the following. In Eq.5, σ_t was estimated to be 120 MPa, which is equal to the elastic limit measured by static tensile test. The dimensionless parameter C was determined by curve fitting. It turned out to be 2.4 x10^{-5}. Hence, the w values in Eq.4 for 60%, 70% and 80% stress levels were 0.29, 0.46 and 0.77 MN/m^2, respectively. Since temperature equilibrium was reached only in fatigue tests conducted at 1 Hz, H value was estimated from 1 Hz data from the slop of q versus T_e-T_0 plot, similar as that described by Hahn & Kim (1976). HA/V was found to be about 40,000N/m^{2}s°C and in turn H was about 80J/m^{2}s°C. For APC-2, ρ is known to be 1.6g/cm^3 and c_p is 1.1J/g°C.

Table 1.   Fatigue test results for AS4/PEEK ±45 laminate

Load Level (% of σ_u)	Frequency (Hz)	Number of Cycles To Failure	Maximum Temperature (°C)
60%	1	1,400,000*	34
		1,200,000*	32
65%	1	50,7000	-
70%	1	20,650	43
		16,000	40
		21,000	45
		18,800	46
80%	1	2,900; 3,000	45; 45
		3,850; 3,570	- ; 52
93%	1	1,500	-
60%	5	100,000	84
	186,000	84	
70%	5	2,200 1,500	127; -
		1,100; 2,600	135; 152
		2,620	127
75%	5	350	100
80%	5	200; 220	110; 97
	200	108	
60%	10	550; 575	152; 140
	1,040; 580	-	
70%	10	450; 260	136; 95;
	280; 470	-; 119	
80%	10	200; 180; 315	110; 102; 135
		120; 100	-; 90

* sample did not fail

Our experimental results shown that the hysteresis loop may either increase or decrease in area, depending upon the testing condition. At 1 Hz, the loop area appeared to decrease or first decrease and then increase with number of cycles. At 5 Hz, 60% level, the loop area is almost constant. At high stress levels and frequencies, the loop area increased with number of cycles. Eq.7, therefore, was employed for 1 Hz and 5 Hz, 60% level whereas incremental method Eq.9 was used for other conditions.

The increase in hysteresis loop area, i.e. cyclic softening may be contributed by mechanical damage and by thermal softening. Since it is rather difficult to separate the two contributions, the change in loop area was expressed as a function of temperature, i.e. w=w(T). Fig. 5 plots the ratio of loop area to initial loop area versus recorded temperature rise. The data in Fig.5 show that the tendency of increasing loop area appears to be relatively moderate at below 80°C and becomes significant at higher temperatures. The w(T) function is therefore approximated by two piecewise linear segments:

$$w = w_0 \qquad\qquad T < 80°C$$
$$w = w_0[1+\beta(T-80)] \qquad T > 80°C \qquad (10)$$

where w_0 is the initial loop area and β is a parameter to be determined by curve fitting. β is 0.025 for the data in Fig.5.

The predicted temperature rises are presented in Fig.4 by dashed lines. As seen, an equilibrium behavior is predicted for tests at 1 Hz at all stress level and at 5 Hz at 60% stress level. The predicted equlibrium value is in a good agreement with the measured equilibrium value but the predicted temperature rise occurs much faster than measured curves. Considering loop area increase at above

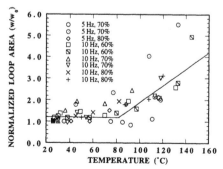

Figure 5. Variation of loop area as a function of temperature.

80°C results in an ever rising temperature curve which is in good agreement with measured temperature rise curves at these testing conditions.

Fatigue Life Prediction

As presented above, our fatigue life prediction for non-isothermal cases is based on the iso-strength plot constructed from isothermal SN curves. Ideally isothermal S-N curves should be generated experimentally. At present time, these curves are generated based on the following assumptions.

The SN curve at 1 Hz is used as the reference curve. Since the maximum temperature rise during fatigue tests at 1 Hz varied between 32°C to 46°C with an average value of about 40°C, 40°C is chosen to be the reference temperature.

The SN curve is usually modeled by a linear relation

$$\sigma = a + b \log N \qquad (11)$$

This relation fits most of the experimental SN data from intermediate to low stress region but fails when the cyclic stress is close to the static strength or the fatigue limit.

In this work, the SN curve is modeled by a four-parameter power law relation:

$$p = p_0 + \frac{1-p_0}{(1+\tau N)^n} \qquad (12)$$

In Eq.12, the cyclic stress appears in a normalized form, $p = \sigma/\sigma_u$ and $p_0 = \sigma_0/\sigma_u$, in which σ_u and σ_0 are the static strength and the fatigue limit, respectively; τ is a dimensionless parameter related to a characteristic time or number of cycle; and n is an exponent. τ and n are to be determined by curve fitting. The parameters in Eq.10 for the SN curve at 1 Hz were determined by curve fitting. It turned out to be $\tau = 0.0004$ and $n = 0.5$. The p_0 was estimated to be 0.55 based on the SN data at 1 Hz. In Fig.6, the SN curve at 1 Hz described by Eq.12 (the solid line) is compared with experimental data (symbols). As seen Eq.12 fits the data reasonably well over an entire stress range from the static strength to the fatigue limit.

The SN curves at elevated temperatures are estimated by shifting the reference SN curve,

$$p = b_T \left(p_0 + \frac{1-p_0}{(1+a_T \tau N)^n} \right) \qquad (13)$$

where a_T and b_T are the shifting factors defined earlier. In this work, the rotation shifting factor a_T was assumed to be 1. This assumption was based on literature data (Rotem & Nelson, 1981, Miyano & McMurray, 1994) which have shown that the slope of SN curves of angle-ply laminates remains almost the same over a wide temperature range.

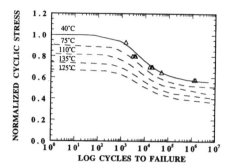

Figure 6. SN curve fitting using Eq.12 (solid line) for 1 Hz data (symbols) and the isothermal SN curves obtained by shifting the reference SN curve using Eq.13 (dashed line).

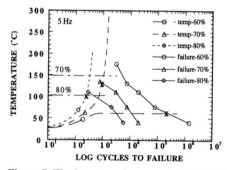

Figure 7. The iso-strength curves for 60%, 70% and 80% load levels and the fatigue life prediction for 5 Hz.

The vertical shift factor b_T was equal to the ratio of the static strength at temperature T to that at the reference temperature. For AS4/PEEK ±45 laminate, the following data are available in literature (Cogswell, 1992):

Relative Tensile Strength of ±45 AS4/PEEK Laminate

Temperature (°C)	23	120	180
Relative Strength	1	0.75	0.65

The b_T values were determined by interpolation of the above data:

T (°C)	40	75	110	135	175
b_T	1	0.9	0.82	0.75	0.67

A series isothermal SN curves generated using Eq.13 are plotted in Fig. 6, together with the reference SN curve. Following the procedure presented in Figs.1 and 2, iso-strength curves corresponding to 60%, 70% and 80% load levels are constructed in Fig.7. Fig.7 also illustrates the prediction for fatigue life at 5 Hz. The predicted transient temperature curves for the three load levels are overlaid on the iso-strength curves. The crossing points of these two groups of curves at the same load level give the fatigue life at that testing condition, as indicated by the crossing marks.

Figs.8 and 9 present the fatigue life prediction for 5 Hz and 10 Hz, respectively. The solid symbols in these two figures are the temperatures measured at the moment of fatigue failure. As seen, most of the data fall close to the predicted failure points. This supports a basic assumption in our prediction that the thermal effect is the dominant factor for the frequency dependence of the fatigue life for materials with large hysteretic heating.

As noted, the maximum temperature values in Table 1 appear to be controversy. The maximum temperatures measured on samples tested at 10 Hz somehow were lower than those measured at 5 Hz at the same load level. Also, the measured maximum temperatures were lower on samples tested at higher levels as compared to those at lower load level at the same frequency for 5 Hz and 10 Hz cases. Figs. 8 and 9 clearly indicate that the maximum temperature that might be reached in a fatigue test depends upon the number of cycles a sample can endure before failure and its value is predictable using the present scheme. The predicted maximum temperature rises are compared with the experimental average values in Table 2. As seen, the predictions are in a good agreement with the experiment. For 5 Hz cases, the predictions are within the scatter in experimental data. Lower value measured in 10 Hz tests might be due to the delay in heat transfer. Continue temperature rise after the load had been removed was noted in 10 Hz tests.

The temperature prediction capability in the present scheme has solved a key problem for models relying on curve fitting. To account for the hysteretic heating effect, the temperature rise has to be included as a variable in this type of models, either directly or indirectly. On one hand, using experimentally measured maximum temperature value will limit the predictive capability of a model. On the other hand, from analysis one can only predict the maximum equilibrium temperature but not the temperature at failure, i.e. the predicted

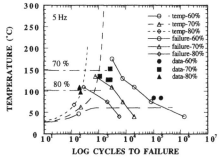

Figure 8. Fatigue life prediction for 5 Hz. The symbols are experimental data.

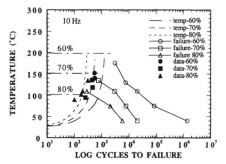

Figure 9. Fatigue life prediction for 10 Hz. The symbols are experimental data.

maximum temperature rise is always higher for a sample tested at a higher frequency and load level. Actually, this problem cannot be solved by using a single equation.

The predicted fatigue lives are compared with the test results in Table 2. As seen, the predictions are in a good agreement with the experimental results for 10 Hz at 70% and 80% levels, and for 5 Hz at at 80% load levels, i.e. within the scatter of experimental data. The prediction overstimates the fatigue life at 60% load level for about 40% for 5 Hz and 80% for 10 Hz cases while underestimate

Table 2. Comparison of predicted fatigue lives and temperature at failure with experimental results

Load Level (% of σ_u)	Frequency (Hz)	Number Of Cycles To Failure		Temperature At Failure (°C)	
		Experiment (Mean±S.D.*)	Prediction	Experiment (Mean±S.D.*)	Prediction
60%	5	143,000±4300	200,000	84	79
70%	5	2,004±679	950	135±12	130
80%	5	207±11	235	105±7	101
60%	10	686±236	1230	146±8	200
70%	10	365±110	470	117±21	140
80%	10	183±85	190	109±19	101

*S.D.=standard deviation

the fatigue life at 5 Hz, 70% level for about 50%. Nevertheless, these predictions are still in a reasonably good agreement with experiment data considering the standard deviations of the data were about ±30% in our results.

The errors in prediction may come from many sources. Firstly, it could be due to the assumptions in the isothermal SN curves, such as the linear relation for b_T value assumed in interpolation and the unit value for a_T. Secondly, the iso-strength plot is based on isothermal SN curves which does not consider the difference in damage accumulation rate at different temperatures. Finally, as mentioned earlier, the predicted temperature rise occurs faster than recorded curves which tends to lead to an underestimated fatigue life.

6. CONCLUSIONS

A simple scheme correlating the temperature rise caused by hysteretic heating with the thermal degradation of the strength and the fatigue life of the material is developed. By using shifting factors to account for the temperature effect on SN curve and by introducing iso-strength plot, the present prediction scheme requires few material information. As demonstrated, fatigue life predictions can be made with the knowledge of the static strength-temperature relation, SN curve at a reference temperature and the predicted temperature rise due to hysteretic heating. This gives the present approach a much stronger predictive edge as compared to the models relying on curve fitting. Fatigue life was predicted for AS4/PEEK [±45]4s laminates at 5 Hz and 10 Hz based on SN curve generated at 1 Hz. The predictions appear in a reasonably good agreement with the experimental results.

7. REFERENCES

Cogswell, F.N., 1992. Thermoplastic Aromatic Polymer Composites, Butterworth-Heinemann Ltd, 1992.

Curtis, D.C., D.R.Moore, B.Slater, N.Zahlan, 1988. "Fatigue testing of multi-angle laminates of CF/PEEK", Composites, 19: 446-452.

Dally J.W., L.J.Broutman, 1967. "Frequency effects on the fatigue of glass reinforced plastics", J.Composite Materials, 1:424-442.

Dan-jumbo, E., S.G.Zhou, C.T.Sun, 1989. "Load-frequency effect on fatigue life of IMP6/APC-2 thermoplastic composite laminates", ASTM STP 1044, G.M.Newaz, Ed., 113-132.

Hahn, H.T., R.Y.Kim, 1976. "Fatigue behaviour of composite laminate", J.Composite Materials, 10: 156-180.

Miyano, Y., M.K.McMurray, 1994. "Loading rate and temperature dependence on flexural fatigue behaviour of a satin woven CFRP laminate", J. Composite Materials, 28: 1250-1260.

Reifsnider, K.L., W.W.Stinchcomb, T.K.O'Brien, 1977. "Frequency effects on a stiffness based fatigue criterion in flawed composite specimens", Fatigue of Filamentary Composite Materials. ASTM STP 636, K.L.Reifsnider and K.N.Lauraitis, Eds., 171-184.

Rotem, A., H.G.Nelson, 1981. "Fatigue behaviour of graphite-epoxy laminates at elevated temperatures", Fatigue of Fibrous Composite Materials, ASTM STP 723, 152-173.

Schapery, R.A., 1975. "Deformation and failure of viscoelastic composite materials", Inelestic Behaviour of Composite Materials, C.T. Herakovich Ed., ASTM, 1975, 127-156.

Sun, C.T., W.S.Chan, 1979. "Frequency effect on the fatigue life of a laminated composite", Composite Materials: Testing and Design, ASTP STP 674, Ed. S.W.Tsai, 418-430.

Xiao, X.R., I.Al-Hmouz, 1997. "Effect of load frequency on the tensile fatigue behaviour of angle-ply AS4/PEEK", ICCM-11, Gold Coast, Australia, 1997.

Recent Developments in Durability Analysis of Composite Systems, Cardon, Fukuda, Reifsnider & Verchery (eds)
© 2000 Balkema, Rotterdam, ISBN 90 5809 103 1

Effect of marine exposure on weatherability of FRP laminates

H. Takayanagi & K. Kemmochi
National Institute of Materials and Chemical Research, Tsukuba, Japan

I. Kimpara
University of Tokyo, Japan

ABSTRACT: We fabricated eight kinds of fiber-reinforced plastic from the matrices of unsaturated polyester resin and epoxy resin with glass fiber and carbon cloth. The FRP laminates were attached to the side of a large structure floating in Tokyo Bay. The FRP laminates were removed after one and two years, and mechanical property tests were conducted to assess the degradation. The results clearly indicated that Young's modulus and the bending modulus were the smallest for the specimens exposed at the waterline. The tensile and bending strengths of the specimens exposed for one and two years exceeded the initial values, and were reduced compared to the specimens exposed for one year, respectively. As the compatibility of reinforcing fiber and matrix, the carbon fiber and glass fiber were in conformity with epoxy resin and unsaturated polyester resin on the bending strength, respectively.

1 INTRODUCTION

In the future, electrical power will be produced extensively using the force of the sea as a practical natural energy resource. We have proposed the Mighty Whale Project and have conducted sea experiments to demonstrate the multipurpose production of electrical power. New high-polymer materials will be applied to marine structures to utilize this energy resource. Durability is an important feature of these new materials (Takayanagi 1990, Kemmochi 1995). There are relatively few existing studies into the degradation caused by marine exposure.

We fabricated eight kinds of fiber-reinforced plastic (FRP) from the matrices of unsaturated polyester resin, epoxy resin with glass fiber, and carbon fiber reinforcing materials. The resultant FRP laminates were exposed on the side of a large structure floating in Tokyo Bay. The FRP laminates were removed after one and two years, and mechanical property tests were conducted to assess the degradation.

2 EXPERIMENTAL PROCEDURES

2.1 *Specimen*

Two lamination systems, MR lamination (a combination of glass roving cloth and glass chopped-strand mat) and plain woven cloth lamination, were used on the large marine structure as shown in Table 1.

Glass fiber, shown in Table 2 above the cost performance, and carbon cloth, shown in Table 3 above the specific strength and rigidity, were used for reinforcement. A resin-rich layer consisting of a surface mat was applied to both surfaces to increase the corrosion resistance. In addition, an ivory-colored, gel-coated specimen was included to increase the weather resistance. The following resins were used for matrices:

· Unsaturated polyester resin (made of Dainippon Ink & Chemicals, Co.)

    Article   :Polylite FG-283 (Neopentyliso)
    Class   :UP-HE (JIS K 6919)
    Viscosity :4.2~5.2 P (25℃)

· Epoxy resin (made of Dainippon Ink & Chemicals, Co.)

    Article   :Epoxed resin, Epiclon 857
              Curing agent, Luckamide WH-042
    Quality   :EPICLON 857

        Epoxy equivalent · · · ·200 g/eq
        Viscosity (25℃) · · · · 9.9 St
    Luckamide WH-042
        Activated hydrogen equivalent

                  · · · · 66 g/ eq
        Viscosity (25℃) · · · · 1.3 P
    Compounding ratio (Weight)
        Epoxed resin : Curing agent
        = 100 : 35

· Gel-coat resin (made of Dainippon Ink & Chemicals, Co.)

    Article   :Polyton Ivory G-19206 (Neo-pentyliso)
    Viscosity : 25.0 P (25℃)
    Thixotropy : 5.6

We assumed that the structural elements of the large marine structure, which were directly affected by the climate of the exposure site, would need to be thicker and larger for increased strength, which would restrict the molding method. We selected the non-heating cure type and hand lay-up molding methods. Spacers were placed at both edges of the specimen to achieve the design thickness. A dead weight was kept on the pressure plate until the curing was finished. A post-cure was conducted at 50℃ for eight hours.

Table 1. Lamination of FRP.

Type	MRS-UP	MRS-EP	MRG-UP	MRG-EP	C_1S-UP	C_1S-EP	C_2S-UP	C_2S-EP
Size(mm)	1000×500		1000×500		1000×500		1000×500	
Thickness(mm)	5.2		5.2		3.6		3.6	
Reinforcing fiber [*1]	S_1 ×1 ply R ×1 M ×4 R ×1 S_1 ×1		G S_2 ×1 ply R ×1 M ×4 R ×1 S_2 ×1 G		S_1 ×1 ply C_1 ×13 S_1 ×1		S_1 ×1 ply C_2 ×9 S_1 ×1	
Matrix[*2]	UP	EP	UP	EP	UP	EP	UP	EP

[*1] S_1, S_2 : Surface mat, R: Glass roving cloth, M: Chopped-strand mat, G: Gel-coat resin, C_1: Glass cloth, C_2: Carbon cloth
[*2] UP: Unsaturated polyester resin, EP: Epoxy resin

Table 2. Reinforcing glass fiber.

(1) Surface mat

Symbol	Article index	Mass per unit area (g/m^2)	Kind of binder
S_1	MF60P 104	60	Polyester
S_2	MF30P 104	30	Polyester

(2) Roving cloth

Symbol	Article index	Mass per unit area (g/m^2)	Structure	Number of ends and picks per 25 mm	
				End	Pick
R	WR570C-100	570	Plain-weave	6.5	6.0

(3) Chopped-strand mat

Symbol	Article index	Mass per unit area(g/m^2)	Kind of binder
M	MC450A-104	450	Polyester

(4) Glass cloth

Symbol	Article index	Mass per unit area(g/m^2)	Treatment	Structure	Number of ends and picks per 25 mm	
					End	Pick
C_1	WF230 N 100	203	Silane	Plain-weave	19	18

Table 3. Reinforcing carbon cloth.

Symbol	Article index	Mass per unit area(g/m^2)	Carbon fiber	Number of ends and picks per 25 mm	
				End	Pick
C_2	W-3101	197	Besfight HTA-3K	12.3	12.3

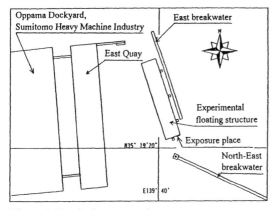

Figure 1. Detailed exposure site.

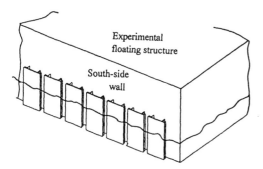

Figure 2. Schematic view for fitting up specimens.

Figure 3. Adhering condition of marine living things.

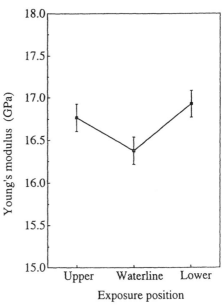

Figure 4. Relation between Young's modulus and position.

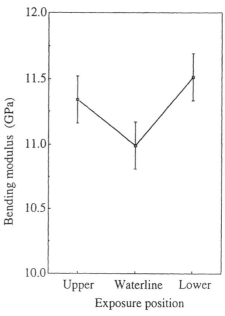

Figure 5. Relation between bending modulus and position.

2.2 Experimental method

The experimental floating structure has been 200 m off the shore of the East Quay of the Oppama Dockyard of Sumitomo Heavy Machine Industry Co. which is located in Yokosuka City, Kanagawa Prefecture. Its geographical location is situated at lat. 35 ° 19 ´ N and long. 139 ° 01 ´ E as shown in

Figure 1. The floating structure has been 300 m long, 60 m wide, 2 m deep and 0.5 m at the waterline. The mean temperature over the past 12 years was 15.4 ℃. The waves in the experimental area were low in

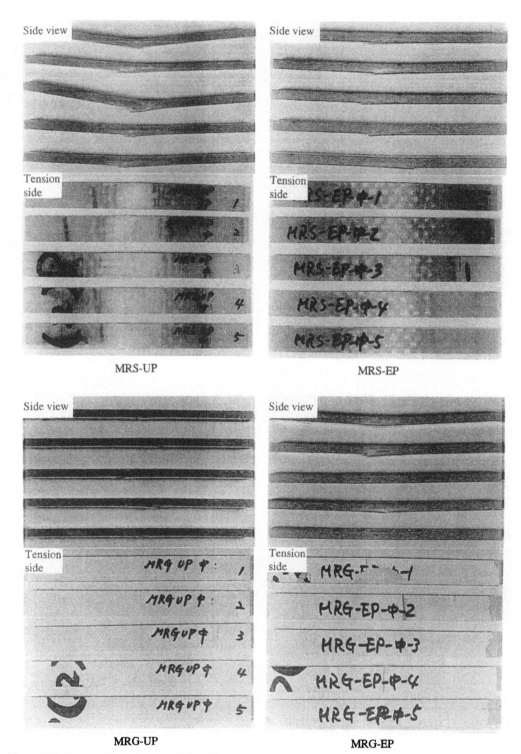

Figure 6(1). Results of bending test of MRS-UP, MRS-EP, MRG-UP and MRG-EP exposed for two years at the waterline.

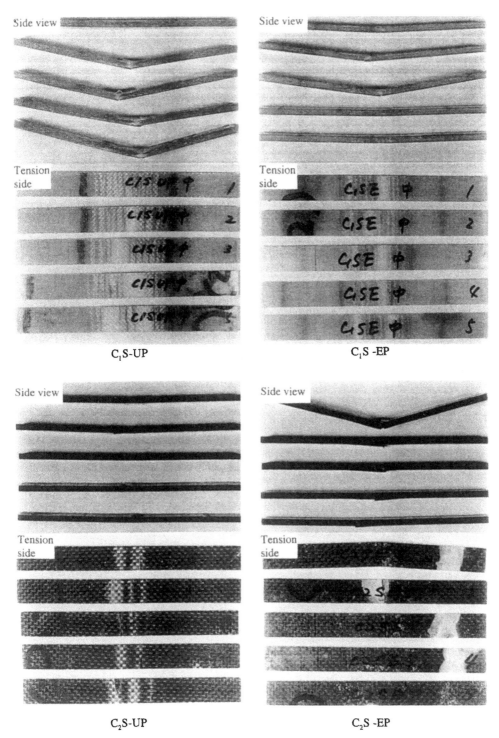

Figure 6(2). Results of bending test of C_1S-UP, C_1S-EP, C_2S-UP and C_2S-EP exposed for two years at the waterline.

height and of short period, $H_{1/3}=0.25\sim0.5$ m and $T_{1/3}=3\sim4$ sec, for 65% of the experiment.

Twenty-four specimens, 1000 mm high and 500 mm wide, were attached to the south wall of the floating structure using 8 holes 13 mm in diameter to ensure that the position (40 cm from the lower edge) agreed with the waterline. The exposure direction was inclined 18° from south to east as shown in Figure 2.

After exposure, we removed the attached seaweed and shellfish shown in Figure 3 from specimens with plastic and stainless steel spatulas. Since the gel-coat on the inside of Specimen MRG-EP was delaminated within one year, both sides of gel-coat on the narrow section of the tensile specimen were removed, and strain gages were mounted on both sides. The tensile strength, tensile modulus, and Poisson's ratio were measured using the A type specimen, based on JIS K 7054. The gel-coat on the compression side of bending test Specimen MRG-EP was removed completely. A bending test was performed under the condition that the exposure side agreed with the tension side in bending. The bending strength, bending modulus, and maximum deflection were measured, based on the four-point bending test of JIS K 7055. The support spans were 120 mm (span depth ratio, $L/t=23$) for Specimens MRS and MRG, and 96 mm ($L/t=27$) for Specimens C_1S and C_2S. After exposure, the support spans were 96 mm ($L/t=18$) for Specimens MRS and MRG, and 75 mm ($L/t=21$) for Specimens C_1S and C_2S.

3 RESULTS AND DISCUSSION

3.1 *Effect of exposure position on weather resistance*

In order to clarify the effects of exposure position, that is, the upper side of waterline, the waterline and the lower part of waterline, on the Young's modulus after two years of exposure, we used a three-way classification. Three factors were the reinforcing fiber, matrix and exposure position. There were three measurements per cell. The values exposed for two years were used as the data hereafter. Each factor was significant at the one percent level. The confidence limits at 95% confidence coefficient were shown in Figure 4. The same confidence coefficient was applied hereinafter. The order of the reinforcing fiber on the Young's modulus was C_2S, C_1S, MRS, and MRG. The order of the matrices was UP and EP. We observed that the Young's modulus of the specimens exposed at the waterline was the smallest as shown in Figure 4. We assume that the repeated force applied by waves, the weather, and their interactions caused the severe degradation at the waterline. The exposure position did not significantly affect the tensile strength.

The effect of the exposure position on the bending strength and bending modulus was significant at the one percent level. The bending modulus of the specimens exposed at the waterline was the smallest as shown in Figure 5.

The exposure position did not significantly affect the Sharpy impact value after two years of exposure.

3.2 *Effect of matrices on weather resistance of specimens exposed at different exposure positions*

The results of bending test of specimens exposed for two years at the waterline were shown in Figure 6. The failure mode of C_2S-UP was buckling of surface mat on compression side at low load-level, the failure mode of the other specimens , however, were fiber breakage on tension side.

The relation between bending strength and matrix was shown in Figure 7. Concerning to bending strength, there was a large interaction between the reinforcing fiber and matrix. The bending strengths of MRS-EP, MRG-EP and C_1S-EP were smaller than MRS-UP, MRG-UP and C_1S-UP, respectively. It was considered as the above reason that the epoxy resin was non-heating cure type and elastic.

It was seen based on the comparison among MRS, MRG and C_1S that the effect of the difference of fiber contents between the surface-mat layer and the reinforcing glass-fiber layer on the shear force of interface between them was small regardless of the kind of matrix.

The bending strength of C_2S-EP was about twice the other specimens because of the high strength and modulus of carbon fiber. The bending strength of C_2S-UP, meanwhile, was about half the C_2S-EP. It was recognized that the effect of the difference of fiber contents between the surface-mat layer and the carbon-cloth layer on the shear force of interface between them increased more extensively by using unsaturated polyester resin than epoxy resin. It was also assumed from the failure mode.

The relation between maximum deflection and matrix was shown in Figure 8. It was seen that the maximum deflection of C_2S-UP was much smaller than MRS-UP, MRG-UP and C_1S-UP.

3.3 *Effect of exposure time on weather resistance*

In order to clarify the effects of exposure time, that is, before exposure, after one year and after two years, on the Young's modulus and tensile strength, we used a three-way classification. Three factors were the reinforcing fiber, matrix and exposure time. There were three measurements per cell. The values at the waterline were used as the data for exposure hereinafter.

Each factor was significant at the one percent level except that the factor, matrix, concerning to tensile

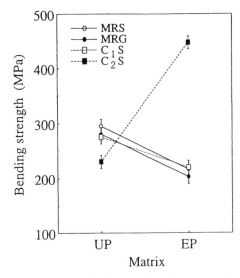

Figure 7. Relation between bending strength and matrix.

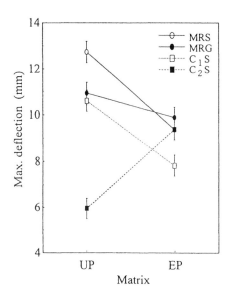

Figure 8. Relation between maximum deflection and matrix.

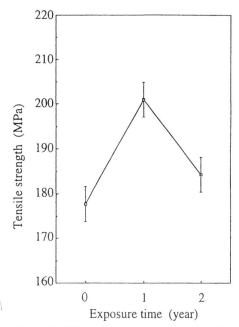

Figure 9. Effect of exposure time on tensile strength.

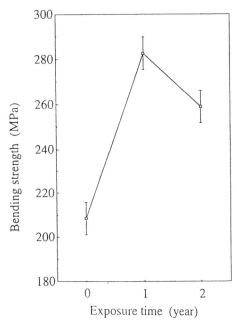

Figure 10. Effect of exposure time on bending strength.

strength was significant at the five percent level. We observed that the tensile strength of the specimens exposed for one and two years increased by 13% and 4% the initial value, respectively as shown in Figure 9. It was considered that the progress of curing during one-year exposure exceeded the degradation, and the deterioration proceeded after two years.

We observed that the bending strength of the specimens exposed for one and two years increased by 36% and 24% the initial value, respectively as shown in Figure 10.

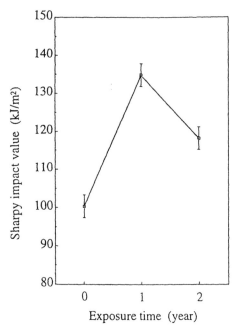

Figure 11. Effect of exposure time on Sharpy impact value.

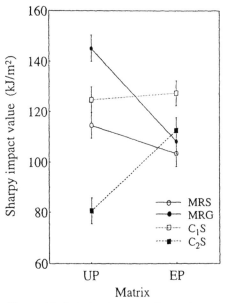

Figure 12. Relation between Sharpy impact value and matrix.

The Sharpy impact value of the specimens exposed for one and two years increased by 34% and 18% the initial value, respectively as shown in Figure 11. This may also correlate with the bending strength.

3.4 *Effect of matrices on weather resistance of specimens exposed for two years.*

The relation between Sharpy impact value and matrix was shown in Figure 12. With respect to the Sharpy impact value, there was a large interaction between the reinforcing fiber and matrix. The Sharpy impact value of C_2S-UP was less than C_2S-EP. This may also correlate with the maximum deflection.

4 CONCLUSIONS

Results were summarized as follows:
The Young's modulus and the bending modulus of the specimens exposed at the waterline were the smallest. It was assumed that the repeated force produced by waves, weather, and their interactions resulted in the degradation.

The tensile strength, the bending strength, and the Sharpy impact values of the specimens exposed for two years exceeded the initial value and decreased from those exposed for one year. It was considered that the progress of curing during one-year exposure exceeded the degradation, and the deterioration proceeded after two years.

As the compatibility of reinforcing fiber and matrix, the carbon fiber and glass fiber were fitted for epoxy resin and unsaturated polyester resin were superior on the bending strength, respectively.

5 ACKNOWLEDGMENTS

The authors would like to thank T.Takano, High Polymer Test & Evaluation Center, JCII, the Standardization Committee of Evaluation Method for New, High Polymer Material to Generate Electric Power by Surge (High Polymer Test & Evaluation Center), H.Kittaka, Nittobo FRP Research Center Co., Ltd. for the supply of specimen.

REFERENCES

Kemmochi, K., H.Takayanagi, C.Nagasawa, J.Takahashi & R.Hayashi, Advanced Performance Materials, 2, 385-394(1995).
Takayanagi, H., N.Koshizaki, K.Kemmochi & R.Hayashi, Bulletin of Industrial Products Research Institute, No.117 (1990) July.

Recent Developments in Durability Analysis of Composite Systems, Cardon, Fukuda, Reifsnider & Verchery (eds)
© 2000 Balkema, Rotterdam, ISBN 90 5809 103 1

Accelerated marine aging of composites and composite/metal joints

P. Davies
Marine Materials Laboratory, IFREMER Centre de Brest, Plouzané, France

A. Roy
CRITT Matériaux Poitou-Charente, Rochefort, France

E. Gontcharova & J.-L. Gacougnolle
Laboratoire de Mécanique et de Physique des Matériaux, ENSMA Poitiers, France

ABSTRACT: Long term predictions of the integrity of marine structures require quantitative assessment of the evolution of material properties. In order to obtain such data in a reasonable time it is necessary to accelerate tests, usually by raising temperature. This paper will present results from a study in which the validity of such tests is examined. Three composite systems (polyester and epoxy reinforced with woven and stitched glass fabrics) and a steel/composite bonded joint have been aged at sea in the Brest estuary for up to 2 years. Specimens were removed regularly for mechanical testing in tension and flexure. At the same time specimens of identical geometry were aged in the laboratory using temperatures up to 50°C to accelerate aging. There are analogies between the aging of joints in the tidal zone and in demineralized water, for which the aging is dominated by diffusion, whereas continuous immersion in sea water results in oxidation of the steel at the steel/adhesive interface The comparison of residual mechanical and physico-chemical properties has enabled the correlation between the natural and accelerated aging procedures to be established.

1 INTRODUCTION

The use of composites in the marine industry is widespread [1-3]. The materials used are mostly polyester thermosets, with some vinyl esters and epoxies are also employed. The fibre reinforcement is mostly glass, principally woven fabric or random mat, although more stitched fabrics are also being introduced. While some applications involve complete composite structures, generally produced by hand lay-up, in many cases composites are joined to other materials, mostly metals and particularly steel. These marine structures are usually designed for long service lives, tens of years, and the good reputation for corrosion resistance of composites in sea water is frequently a major selling point. However, the designer is required to guarantee that service life and the usual procedure is to apply a significant 'safety factor' for long term performance. This generally results in satisfactory behaviour but may significantly over-dimension the structure. When weight is critical, as in high speed vessels, this may not be acceptable.

Given the long experience with composites in this industry it might be thought that all necessary design data would be available from previous studies. Weitsman has recently summarised available data [4] and indeed, some long immersions have been noted, particularly associated with mine-sweeper pro-

grammes, but few results have appeared in the published literature. A brief summary of French naval tests was published [5], based on 21 years of immersion. Although few experimental details were presented this is a valuable contribution, showing a correlation between a change in mechanical behaviour observed after 15 years of natural aging at sea and 1000 hours accelerated ageing at 70°C.

The main value of long term immersion data is to validate the short accelerated tests which many laboratories are obliged to run to find answers to questions from designers both qualitatively such as 'does this new resin we want to use have as good long term aging resistance as the one we currently use?', or quantitatively 'what safety factor should we use to take account of mechanical property degradation over a 20 year lifetime ?'. This study was performed in order to address the second type of question. The aim was to take samples with the same geometry from the same batches of fibre and resin materials, produced at the same time, and characterise their mechanical behaviour using the same test procedures after different natural and accelerated aging periods and conditions.

Three composite systems (polyester and epoxy reinforced with glass fabrics) and a steel/composite bonded joint have been studied. Table 1 shows the resin and fibre combinations. The joints were pre-

Table 1. Materials tested			
Specimen	Fibre	Resin	Form
Composite 1 RM/ISO	Rovimat (Chomarat 500/300 g/m²) 5 layers	Isophthalic polyester Scott Bader 491PA	Panels 100 x 250 mm Thickness 6 mm
Composite 2 QX/EP	Quadriaxial stitched (Cotech 1034 g/m²)3 layers	Epoxy (Sicomin 1500/2505 amine)	Panels 100 x 250 mm Thickness 4 mm
Composite 3 QX/ISO	Quadriaxial stitched (Cotech 1034 g/m²) 6 layers	Isophthalic polyester Scott Bader 491 PA	Panels 100 x 250mm Thickness 6mm
Mild steel (2 mm) bonded to Composite (6 mm), (Redux 420)	Rovimat (Chomarat 500/300 g/m²) 5 layers	Isophthalic polyester Cray T7039	Lap shear specimens 25 mm wide, 25 mm overlap

Photo 1. Composite panels and bonded joints before immersion at sea.

Table 2. Aging conditions		
Specimen	Natural aging (months)	Laboratory aging (months)
Composite 1 RM/ISO	SW 3, 6, 9, 12, 18, 24	16°C DW, 9 50°C DW 1, 2, 3, 9
Composite 2 QX/EP	SW 3, 6, 9, 12, 18, 24	16°C DW, 9 50°C DW 1, 2, 3
Composite 3 QX/ISO	currently underway	
Mild steel bonded to Composite (Redux 420)	Tidal zone 1, 3, 6, 9, 12 Immersion SW 3, 12	25.C DW, 1, 3, 6, 9, 12 40.C DW, 1, 3, 6, 9, 12 50°C SW 1, 2, 3

DW: distilled water, SW: Sea water

pared by abrading and alcohol cleaning the surfaces and applying a Ciba Geigy Redux 420 two part adhesive (amine hardener) and curing for 2 hours at 70°C.

The composites were immersed at sea at a test site in the Brest estuary for up to 2 years. Panels were fixed to immersion baskets, placed at 5 metres depth, and periodically removed by divers, Photo 1.

Some joint specimens were also immersed, others were placed in the tidal zone. Table 2 shows the natural and accelerated test conditions. These tests are continuing and the QX/polyester series started after the other two materials so results will be presented later.

After aging specimens were placed in a temperature (20°C) and relative humidity (50%) controlled laboratory for 2 weeks to stabilise them, before testing them in the same laboratory. The sea aged specimens were covered with marine growth, and this was carefully removed before stabilisation. Weight gains of all specimens were recorded.

2 TESTS PERFORMED.

All the composite panels were cut in the same way, as shown in Figure 1, using a diamond-tipped cutting wheel.

Each panel provided 4 tensile specimens, 4 flexure specimens and at least 6 interlaminar shear specimens. The test procedures were as follows:

2.1 Tensile tests

Loaded up to 1 ton force, with a clip-on extensometer, unloaded, then tested to failure. Loading rate 2 mm/minute. This provides:
- tensile modulus E_t, strain at knee on stress-strain plot (indicative of damage in 90° reinforcement), ε_d, failure stress (maximum stress), σ_t

2.2 Flexure tests

Loaded to failure at 2 mm/min, three point flexure fixture, span of 16 times thickness,. providing:
- flexure modulus E_f, flexural strength σ_f

2.3 Interlaminar shear tests

Loaded to failure at 5 mm/min, three point flexure fixture, span of 5 times thickness,. providing:
- apparent interlaminar shear strength τ.

2.4 Tests on joints

The joint specimens were loaded to failure at 1 mm/minute. Tabs were bonded to the ends to allow symmetrical loading. An extensometer was placed on

the specimen, fixed above and below the bonded overlap, to measure apparent strain (including tensile, shear and rotation contributions). Some laboratory-aged joint specimens were dried before testing. These tests provide two values:
- force at damage initiation F_d , force at failure, F_r

3 RESULTS

3.1 *Composite panels*

Examples of the influence of aging on composite behaviour are shown below. Figure 2 shows tensile moduli, as presented at the previous DURACOSYS conference [6]. Here it appeared that the laboratory tests at 50°C were useful in accelerating the modulus degradation, as the forms of the degradation plots are similar in both cases.

Strength values will now be considered. The analysis of failure is much more complex than stiffness. For example, the tensile stress-strain plots for rovimat and quadriaxial reinforcement show a clear knee as damage occurs in the regions containing off-axis reinforcement. These are in the 90° weave and mat layers for the rovimat, and in the 90° and 45° layers in the quadriaxially reinforced composite. This is one of the parameters which can be correlated, but in the limited space available here only the final failure stresses will be considered. Figures 3-5 show tensile, flexure and interlaminar shear strength results, for RM/polyester and QX/Epoxy composites aged naturally and in the laboratory. Tensile strength is governed by the whole specimen state, flexural strength is dominated by the outer layers (which age most quickly), while interlaminar shear is probably most dependent on resin and interface degradation in the centre of the specimen.

3.1.1 *Correlation between accelerated and natural aging of composites*

These strength results suggest that the 50°C distilled water test is very severe in accelerating tensile and flexure strengths for both materials. Over the first year there seems to be a reasonable correlation between the forms of the changes in strength properties, (as there was for modulus), an accelerating factor in the laboratory of around four to five times (3 months in the laboratory corresponds to around a year at sea). However the polyester composite tensile and flexural strengths after 18 and 24 months at sea increase again. The interlaminar shear results for polyester give a better quantitative correlation with respect to natural aging.

There is virtually no reduction in the tensile or flexural strength of the epoxy composite while ILSS drops about 15% after 2 years. The laboratory epoxy specimens aged for 9 months lose over 30% of their ILSS and again appear very severe.

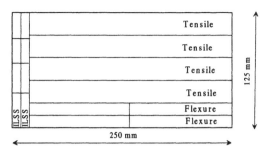
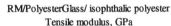

Figure 1. Specimens cut from each panel

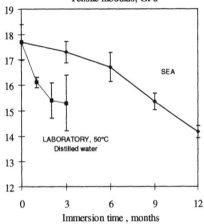

Figure 2. Tensile moduli, RM/Polyester up to 12 months at sea

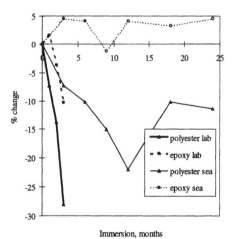

Figure 3. Tensile strength

255

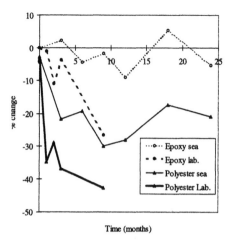

Figure 4. Flexural strength

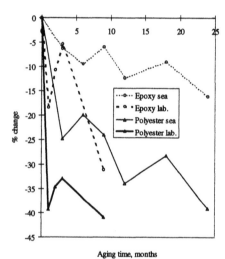

Figure 5. Interlaminar shear strength

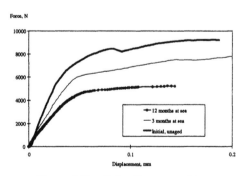

Figure 6. Tensile loading curves for
steel/composite joint

3.2 Composite/Metal bonded Joints

Examples of tensile loading curves of lap shear joints are shown in Figure 6 below.

There is again a clear knee on the curve and this is caused by the appearance of an interlaminar crack at the joint extremity in the composite. The force at which this appears, F_d, and the force at failure, F_r, are the two parameters which were used here to examine the correlation between laboratory aging and natural aging. A large number of laboratory tests were performed in order to understand the mechanisms of degradation of these joints, and more data are available elsewhere [7].

The curves in Figure 6 clearly show the change in behaviour after aging. In order to simulate this in the laboratory several series of specimens were aged in natural sea water at 50°C, and in distilled water at 25°C and 40°C. Some results for F_r are shown in Figure 7 below.

There is a significant drop in failure load after 3 months aging and the tidal zone exposure is initially more severe than full immersion. There is roughly a factor of three acceleration between tidal exposure and laboratory 50°C sea water immersion.

The comparison between different aging conditions is shown more clearly in Figures 8 and 9, which present the drops in load at the knee (Fd) and at failure (Fr) for the five conditions studied.

These figures show that after three months the influence of aging is more significant for the damage threshold value than the final failure load. All the laboratory tests result in more severe drops in failure load than those observed after sea immersion but less severe than tidal zone exposure.

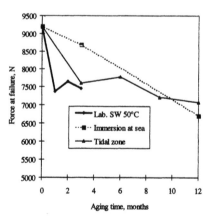

Figure 7. Tensile behaviour of aged lap shear
metal/composite

Force at damage 'knee', Fd, after 3 months aging

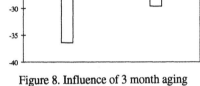

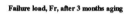

Figure 8. Influence of 3 month aging under different conditions on Fd.

Failure load, Fr, after 3 months aging

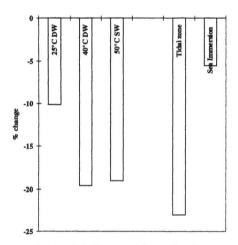

Figure 9. Influence of 3 month aging under different conditions on Fr.

4 CONCLUSION

This paper shows some of the results available after accelerated laboratory aging and tests at sea on composite and metal/composite specimens. While there is a reasonable correlation for the joints under some conditions, provided sea water is used as the accelerated aging medium, there are significant differences for the composite strengths. The detailed analysis of these results is now underway. It involves study of the failure mechanisms involved, but also a

Photo 2. Three metal/composite joints fixed to 100 x 250mm composite panel, after 12 months' immersion at sea.

Photo 3. Three joint specimens after 12 months in tidal zone.

quantitative appreciation of the factors which can cause scatter in experimental results. These include fabrication variations (assessed by physicochemical and baseline mechanical property variations), testing parameters and variability in the aging conditions. The latter are particularly complex when aging at sea is considered and to avoid seasonal effects a full 12 month cycle is a minimal requirement. The influence of biological factors cannot be neglected and Photos 2 and 3 show the appearance of joint specimens after 12 months immersion and tidal zone aging respectively.

REFERENCES

[1] Smith CS, 'Design of Marine Structures in Composite Materials', Elsevier, 1990.

[2] Davies P, in 'Durability analysis of structural composite systems', ed Cardon, 1996, p33.Balkema Publishers

[3] Shenoi RA, Wellicome JF, 'Composite Materials in Maritime Structures', Cambridge Ocean Technology series, 1993.

[4] Weitsman YJ, 'Effects of fluids on polymeric composites', Report MAES98-5.0-CM, University of Tennessee August 1998.

[5] Gutierrez J, LeLay F, Hoarau P, 'A study of aging of glass fibre-resin composites in a marine environment', Proc. 3rd IFREMER conference 'Nautical construction with composite materials', Paris 1992, IFREMER publication p338.

[6] Davies P, Choqueuse D, Mazéas F, 'Composites Underwater', Proc.DURACOSYS 97, p19-24.

[7] Roy A, Gontcharova E, Davies P, Gacougnolle J-L, 'Hygrothermal effects on failure mechanisms of composite/steel bonded joints', Proc ASTM Symp. on Time dependent and Non-linear effects in polymers & Composites', May 1998

Recent Developments in Durability Analysis of Composite Systems, Cardon, Fukuda, Reifsnider & Verchery (eds)
© 2000 Balkema, Rotterdam, ISBN 90 5809 103 1

Thermal ageing effects on G_{IIC} interlaminar fracture mode response of 2-D composites

F. Segovia, V. Amigó & Mª. D. Salvador
Departamento de Ingeniería Mecánica y Materiales, Escuela Técnica Superior de Ingenieros Industriales, Universidad Politécnica de Valencia, Spain

C. Bloem
Escuela de Ingeniería Mecánica, Facultad de Ingeniería, Universidad de Los Andes, Mérida, Venezuela

ABSTRACT: The present work is focused in the study of toughness loss on fibreglass vinylester laminate plates, which were exposed to high temperature environment. A damped exponential mathematical model on the strain energy release rate (G_{IIc}) against exposure time fits to the data collected. Laminates were done by the hand contact method. The reinforcement configuration of an E-glass-fibre plain bidirectional multiaxial 2D cloth was composed with 8 plies (0-90°/±45°), four 0-90° and four ±45° alternatively ordered. Two cured temperatures, 20 °C and 50 °C, were selected to study the effect of cure treatment on mode II interlaminar fracture toughness. The results shows a strongly dependence between G_{IIC} values and cure conditions.

1 INTRODUCTION

The polyester glass fibre reinforced composites GFRP are getting more importance as a basic material in the elaboration of rust free equipment and facilities due to the advantages over metals. These equipments some times are exposed to environments which the temperatures are higher than atmospheric one's.

The increased use of GFRP and the newer technologies has been developed a special interest on the study of the environmental effect over the static mechanical properties (Karama et al. 1993; Manrique et al. 1996; Evans & Crook 1997) or impact (Boukhili et al. 1996; Kasamori et al. 1996), or fatigue behave (Chateauminois & Vauthier 1996; Smith & Weitsman 1996; Lesko et al. 1997). Some works determine the effect on the fracture toughness due to a different kind of liquids exposition (Selzer & Friedrich 1993; Springer et al 1980; Russell & Street 1985; O'Brien et al 1986; Hooper & Subramanian 1991).

The Vinylester-Bisphenol is one of the most remarkable resins on the unsaturated polyester group. The outstanding of this resin are its high chemical stability, high strength and elevated toughness. The knowledge of the thermal effect over the statical mechanical characteristics of this resin is really old and well-known (Reinhart 1988). No data is available on fracture toughness loss of Vinylester-Bisphenol composites due to the thermal ageing, as well the thermal ageing mathematical correlation of fibreglass vinylester composites is not found.

The aim of this work is to evaluate the interlaminar fracture toughness degradation due to a elevated temperature exposition, to perform a mathematical model of it, and to correlate the relationship of fracture toughness to the thermal cured treatment of fibreglass vinylester laminates.

2 MATERIALS.

Glass fibre reinforced laminates were done employing a vinylester-bisphenol A resin (Basf A430 VE/BA). The resin's mechanical and thermal characteristics such as strain (A%) and tensile strength (TS), flexural modulus and strength (E_F, R_F), impact toughness (ρ), and glass transition temperature (T_G) are summarised on table 1. The reinforcement were a multilayered equilibrated glass fibre E multiaxial 2D cloth of 440 g/m^2.

Table 1. Vinylester-Bisphenol characteristics.

A (%)	TS (MPa)	T_G (°C)	ρ (J/cm^2)	E_F (MPa)	R_F (MPa)
6	83	130	4	4000	150

The reinforcement configuration (0°-90°/±45°) were done with 8 alternately disposed sheets, where a ±45° sheet is placed over a 0°-90° sheet.

Laminates were cured at environment temperature (20°C) and controlled temperature (50°C), the cured time in both cases was the same, 24 hours.

The end notched fracture samples ENF were done inserting between the 4th and 5th reinforcement ply a Teflon® sheet with a thickness of 50 µm.

3 EXPERIMENTAL PROCEDURE

3.1 *Polymer ageing.*

The sample ageing process was developed employing a controlled temperature chamber. The chamber's temperature was of 100°C close to the heat distortion temperature, but lower to the glass transition temperature (130°C). A thermocouple electronic device had controlled the camber's temperature.

3.2 *Interlaminar fracture mode II essay.*

The essay was performed accordingly to Davis (1993) procedure. The samples measures were of 25 mm in width, 150 mm in length. Teflon® inserts of 50 mm in length were employed to produce the pre-crack effect, on which 25 mm were the precrack length (**a**). The supports were positioned at the distance of (**2L**) 100 mm, so the relation between **a** and **L** (**a/L**) was 0.5. The fracture essay was developed at 2 mm/sec on an Instron 4202 universal testing machine.

The prior determination of the compliance (**C**) employing the equation (1) to different values of (**a**) was necessary to determinate the interlaminar fracture toughness $\mathbf{G_{IIc}}$.

$$C = C_o + m\, a^3 \qquad (1)$$

Where:
C is the compliance: The relation between δ and the load **P**.
a is the crack length.
m is the adjust constant.

The deformation energy release rate $\mathbf{G_{IIC}}$ were calculated applying the equation 2

$$G_{IIC} = \frac{3m\,P^2\,a^2}{2w} \qquad (2)$$

Where:
W is the sample's width, and
P is the load on the following criteria:
- The point where the **P-δ** relationship loss its linearity (**NL**).
- A 5% lower slope than initial **P-δ** (**C5%**)

4 RESULTS AND DISCUSSION.

The flexural characteristics show a strong loss on strength **S$_f$** and Modulus **E$_f$** due to time see table 2. If the cured temperature was of 20° C the losses reach up to 25% on modulus (E$_F$) and 15% on strength (S$_F$), but if the cured temperature was of 50° C the losses on both characteristics are 22%.

The cured temperature plays an important role because it increases those mechanical characteristics. Samples cured at 50° C show better modulus up to 7% and strength of 3% compared with 20°C samples. After 5260 hours this difference becomes higher, up to 11% on modulus and 7% on strength.

Table 2. Flexural strength characteristics.

T (°C)	t (h)	E$_F$ (GPa)	S$_F$ (MPa)
20	0	13.0	357
	5260	9.7	304
50	0	13.9	368
	5260	10.8	325

This behaviour promotes the study on the interlaminar fracture toughness evolution due to long-term heat exposed samples. Figure 1 shows the non-linear (NL) criteria of the energy release rate G_{IIC} against time. Figure 2 represents the same parameters but under the 5%higher compliance criteria (**C5%**).

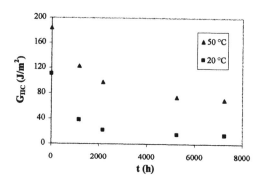

Figure 1. G_{IIC} against time. NL criteria.

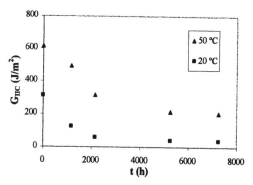

Figure 2. G_{IIC} against time. C5% criteria.

The strain energy release rate loss follows a damped exponential model against ageing time. The model could be mathematically expressed as equation (3)

$$G_{IIC} = A + B\,e^{-kt} \qquad (3)$$

Where:
A represents the asymptotic value.
B represents the difference between the initial value and the asymptotic one.
k represents the extinction constant in hours^{-1} and
t is the exposition time.

The adjusting parameters of the thermal ageing model are summarised on table 3

Table 3. Thermal ageing model's adjusting parameters.

Criteria	NL		C5%	
Cured (°C) Temperature	20	50	20	50
A (J/m^2)	14	70	39	210
B (J/m^2)	77	132	241	425
A+B(J/m^2)	91	202	280	635
k(10^{-4}h^{-1})	9	8	10	6
R^2	0.98	0.98	0.99	0.95

Whatever criteria are followed, a strong decay on the fracture toughness due to thermal effect could be observed.

Nowadays the second criterion (C5%) is widely accepted by the majority of the international researchers. There are many reasons to employ this criterion. To obtain the point where the linearity disappears is really hard to find and the difficult to visually distinguish the delamination's beginning. As many authors express the **C5%** criterion is more accurate.

Table 4. G$_{IIC}$ percent retention (%).

NL	A/(A+B)	C5%	A/(A+B)
20 °C	15	20 °C	14
50 °C	35	50 °C	33

Table 4 shows the material toughness degradability due to ageing time as a function of cured temperature.

The relation **A/(A+B)** exposes de accumulated loss beyond the asymptotic value **A**. The **A** corresponds the behaviour's value of a 7000 hours ageing. The **A+B** value predicts the model's value at 0 hours of ageing. The residual toughness on 20°C cured samples are nearly 14~15% of the initial value. While on those samples cured at 50° C the residual toughness reach values up to 33%~35% approximately. The increase of cured temperature from 20° to 50°C decreases up to twice the material's degradability enhancing its durability.

The cured temperature plays an important role on these resin's behaviour. Figure 3 represents the comparison factor between the toughness at those temperatures of cured.

Figure 3 shows that initially (0 hours (A+B)), the samples cured at 50°C behave twice better than 20°C cured samples. Taking in account the asymptotic value **A** for long-term exposition the behaviour improves up to 5 or 7 times depending on the criteria employed. This phenomena corroborates that the higher temperature of cured ameliorate the mechanical characteristics through the ageing. Figures 1 and 2 show that the samples cured at 20°C has the highest rate of ageing during the first 2160 hours, while the samples cured at 50°C nearly complete its ageing at 5260 hours.

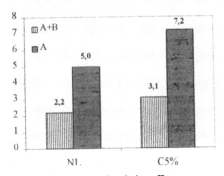

Figure 3. Cured temperature's relative effect

Some aspects of micrographed fractures were analysed and discussed to invigorate this work. So in this way would be capable to conceive some aspects or characteristics in the samples due to the thermal ageing. Figures 4 and 5 display some morphologic aspects on the fracture plane. These shapes are not the same as function of temperature.

Figure 4. Non aged sample cured at a 20 °C. x30.

The fracture of the samples cured at 20°C occurs in the 0-90° reinforcement ply. The relative weakness of 90° layer promotes the crack growing between this one and the 0° layer.

In the case of the 50°C cured samples the fracture occurs through 0-90° and ±45° layers, this phenomena clearly explains the increase on fracture toughness on those samples cured at 50°C.

It is important to indicate that the fracture occurs on mixed mode as could be seen on figure 5, the typical shear fracture are clearly defined by the apparition of hackle formations (M II), while the classical mode I fracture shows the typical cleavage morphology as river marks (M I).

The resin's relative high toughness and the equilibrated bidirectional reinforcement promote the mixed mode fracture. Observing figures 6 and 7 where the thermal ageing effect over the fracture morphology is distinguishable.

Figure 6 shows a closed packed and deep curved morphology on its fracture. While the aged sample (figure 7) shows a hackle packet reduction, an undeformed hackle morphology like facets and a lack of relief.

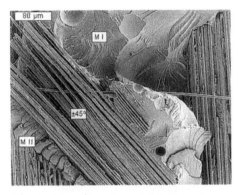

Figure 5. Non aged sample cured at 50 °C. X100.

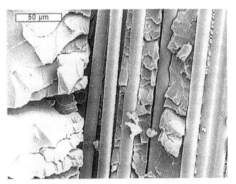

Figure 6. Non aged sample cured at 20 °C. X500.

Figure 7. Sample cured at 20°C, 2160 h aged. X500

5 CONCLUSIONS

1. The long-term exposition at high temperature (100°C) causes a high fracture toughness decrease on Vinylester glass fibre reinforced laminates.

2. The strain energy release rate G_{IIC} evolution due to ageing follows a damped decreasing exponential model.

3. The temperature of cured plays an important role over composite's toughness characteristics, a higher temperature of cured, higher initial toughness.

4. The degradability is cured temperature reverse dependent.

5. The ageing effect is higher whiles lower is the cured temperature and vice versa. The adjusting parameters of the proposed model are temperature dependent too.

6. The interlaminar fracture implies a 0-90° and ±45° plies on 50° cured samples, while on 20°C cured samples, the interlaminar fracture occurs exclusively in the 0-90° ply.

7. As a function of cured temperature the fracture interface shows dissimilar and particular features.

8. The fracture pattern is mainly mode II (shear), but some type of mode I fracture could be distinguished too.

6 BIBLIOGRAPHY

Karama M., Touratier M. y Pegorarol M. 1993. Test of acelerated aging composites material in shipbuilding. 9th ICCM proceedings vol 5 pp 585-592, Madrid.

Manrique F., Bonhome J. y Belzunce F.J. 1996. Influence of enviromental aging on mechanical properties of single lap joints. Progress in Durability Analysis of Composite Systems, pp 311-316 Balkema Editors, Rotterdam.

Evans D., y Crook M.A. 1997 Irradiation of plastics: damage and gas evolution. MRS Bulletin, 22 (4), pp 36-40.

Boukhili R., Champoux L. Y Martin S. 1996 Effect of water absortion on the low energy repeated impact of carbon/epoxy laminates. Durability Analysis of Composite Systems, pp 259-264, Balkema Editors, Rotterdam.

Kasamori M., Funada Y., Awazu K., Watanabe Y., Nakada M. y Miyano Y. 1996 Accelerated evaluation of mechanical degradation behavior of GFRP in hot water. Durability

Analysis of Composite Systems, pp 273-277, Balkema Editors, Rotterdam.

Chateauminois A. y Vauthier E. 1996 Fatigue behaviour of aged glass-epoxy composites. Progress in Durability Analysis of Composite Systems, pp 159-165, Balkema Editors, Rotterdam.

Smith L.V. y Weitsman Y.J. 1996 Sea water effects on the fatigue response of polimeric composite. Durability Analysis of Composite Systems, pp 217-223, Balkema Editors, Rotterdam.

Lesko J.J., Hayes M.D., García K., McBagonluri D. y Verghese 1997 N. Environmental-mechanical durability of glass/vinyl ester composites. DURACOSYS 97 proceedings, pp 4.10-4.13, Blacksburg, Virginia.

R. Selzer, K. Friedrich. 1993 Effects of water up-take on interlaminar fracture properties of various carbon fiber/epoxy composites. 9th ICCM proceedings vol 5 pp 875-881, Madrid.

G.S. Springer, B.A. Sanders, R.W. Tung. 1980 Environmental effects on glass fiber reinforced polyester and vinylester composites. Journal of Composite Materials, vol. 14, pp 213-233.

A.J. Russell, K.N. Street. 1985 Moisture and temperature effects on the mixed-mode delamination fracture of unidirectional graphite/epoxy. Delamination and Debonding of Materials. ASTM STP 876, pp 349-370.

T.K. O'Brien, I.S. Raju, D.P. Garber. 1986 Residual thermal and moisture influences on the strain energy release rate analysis of edge delamination. Journal of Composites Thecnology and Research, vol. 8, (2) pp 37-47.

S.J. Hooper, R. Subramanian. 1991 Effects of water and Jet fuel absortion on mode I and mode II delamination of graphite/epoxy. Composite Materials: Fatigue and Fracture, vol. IV. ASTM STP 1156, pp 318-340.

T.J. Reinhart. 1988 Composites. Engineered materials Handbook, vol. 1. ASM International, Metals Park, Ohio.

Davis P. 1993 Protocols for interlaminar fracture testing of composites, European Structural Integrity Society, Polymers & Composites Task Group.

Infrastructure applications

Recent Developments in Durability Analysis of Composite Systems, Cardon, Fukuda, Reifsnider & Verchery (eds)
© 2000 Balkema, Rotterdam, ISBN 90 5809 103 1

Durability of fiber reinforced composites in civil infrastructure : Issues, results and implications

V. M. Karbhari & S. Zhang
Department of Structural Engineering, University of California, San Diego, Calif., USA

ABSTRACT: E-glass and carbon fiber reinforced composites have been shown, through laboratory and field demonstration projects, to have significant potential for use in civil infrastructure applications. These materials provide immense advantages over conventional materials due to aspects such as lightweight, high performance-to-weight ratios, and potentially high durability. The determination of the durability of these materials for periods of use as long as 75-150 years, in uncertain environmental conditions, and for the most case under load, is a critical aspect to the further acceptance of composites by civil engineers and designers. This paper addresses issues related to durability of composites in civil infrastructure with specific emphasis on effects of alkalinity, freeze-thaw, and moisture. Implications of results on design using conventional civil design methodologies is also discussed.

1 INTRODUCTION

1.1 *Application Area Issues*

Fiber reinforced polymer (FRP) matrix composites are increasingly being considered for use in civil infrastructure in applications ranging from the seismic retrofit of columns and strengthening of slabs and beams, to use in all-composite replacement bridge decks and in new structural systems. Although the performance advantages of FRP composites make them very attractive for use in these applications, the eventual acceptance of these materials is predicated on the resolution of a few critical aspects, foremost among which are the aspects of durability and cost. It should be noted at the outset that cost is paramount in the area of civil infrastructure, especially as new materials and technologies are compared with conventional materials on an acquisition cost rather than a life cycle cost basis. Further, civil infrastructure components such as bridges are designed for lifetimes ranging from 75-150 years. The use of composites in aerospace applications has been predicated on extensive materials testing for the purposes of qualification, followed by strict adherence to prescribed specifications for autoclave based fabrication in highly controlled factory environments. These materials and processes are unlikely to find significant application in civil infrastructure due to cost and processing specific aspects. Civil applications, at present, are more likely to (a) use processes such as wet lay-up, pultrusion and resin infusion than autoclave molding, (b) fiber and resin as separate constituents rather than in the form of preimpregnated material, and (c) resin systems such as polyesters, vinylesters, phenolics and lower temperature cure epoxies rather than the higher temperature curable epoxies and thermoplastics. Further, there is likely to be extensive use of processes under ambient conditions in the field, rather than fabrication in factory controlled environments. Thus, the civil engineering environment not only brings with it new challenges for the control of quality and uniformity of composites, but also makes it difficult (if not impossible) to use the well established databases generated by DoD sponsored research (such as those for AS4/3501-6 or T300/5208 based systems). Environmental conditions likely to be faced by these composites range from extremes in temperature (heat as in Saudi Arabia, cold as in Alaska), significant variation in humidity levels, potential for routine submersion or immersion (such as for structures in flood plains), exposure to chemicals, including road salt, and widely varying load conditions. It is important that these issues be kept in mind while selecting materials and processes for the fabrication of FRP composites for civil infrastructure. This paper attempts to raise the awareness of the effect of a number of these materials-process-environment systems on the durability of the structure, while emphasizing the implication of such factors on overall design and use.

Both from the standpoint of design, and from considerations of life-cycle durability, it is important to note that the use of composites in infrastructure renewal can be classified into the different classes of rehabilitation and new structural systems. Within the scope of rehabilitation of concrete structures, it is essential that we differentiate between repair, strengthening and retrofit, terms, which are often erroneously used interchangeably, but in fact, refer to three different structural conditions (Figure 1).

In "repairing" a structure, the composite material is used to fix a structural or functional deficiency such as a crack or a severely degraded structural component. In contrast, the strengthening of structures is specific to those cases wherein the addition or application of the composite would enhance the existing designed performance level, as would be the case in attempting to increase the load rating (or capacity) of a bridge deck through the application of composites to the deck soffit. The term retrofit is specifically used as related to the seismic upgrade of facilities, such as in the case of the use of composite jackets for the confinement of columns. The differentiation is important not just on the basis of structural functionality, but also because the specifics related to the use of the material in conjunction with existing conventional materials, and its expected life, have a significant effect on the selection (or rejection) of fiber-resin combinations from a variety of alternatives. Also, this emphasizes the differentiation between configurations that are expected to be load bearing over the entire life of the structure from those in which loads may be transient and of short duration separated by long periods of no load (such as in the case of seismic retrofit).

1.2 Materials and Process Implications

At this juncture it may be worthwhile reviewing the specific characteristics of the materials systems likely to be most commonly used in civil infrastructure applications, especially vis-à-vis the generic resin systems and their characteristics. The use of processes such as wet lay-up, pultrusion, wet winding, RTM, and resin infusion require that the resins have a sufficiently low viscosity so as to be able to infuse the reinforcing structure of fibers cost-effectively. Both polyesters and vinylesters are likely to be widely used with preference being given to vinylesters because of their greater durability. The common vinylesters used to date in applications such as bridge decks, rebar, prefabricated plating, and profiles belong to the generic classification of thermosetting resins and are comprised of low molecular weight polyhydroxyether chains with reactive groups at chain ends. Styrene in monomeric form is used as a diluent in the resin in quantities between 20-60%. It is important to note however that the increase in styrene content results in (a) an increase in hydrophobicity, thereby effectively decreasing the level of moisture absorption, and (b) an increase in shrinkage to levels of 5-10% by volume which can result in significant microcracking in resin rich areas and high residual stresses in composites having high volume fractions. In comparison to polyesters, which have double bonds at about 250 g/mol level, vinylesters have reactive double bonds at about every 500-1000 g/mol. The increased distance between cross-linkages in the polymer results in a network that has greater fracture toughness. It is important to note that irrespective of the cure mechanism used, vinylesters do not completely polymerize, generally reaching a level of cure higher than 95%, with the last part of cure continuing very slowly. Incomplete cure can result due to environmental conditions, incorrect stoichiometry of resin system components, or the failure to reach a sufficient temperature of cure. This state can affect mechanical properties, moisture absorption, and susceptibility to moisture induced degradation of the resin, and the fiber-matrix interphase. Figure 2 schematically depicts the relation between degree of cure (or cross-link density) and specific characteristics in these materials.

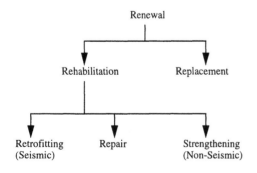

Figure 1. Types of Renewal Strategies.

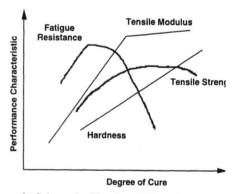

Figure 2. Schematic Showing Effect of Degree of Cure on Material Characteristics

The aspect related to change in properties with degree of time is important not just from a performance point of view for thermosetting resin based composites, but especially for composites used in civil infrastructure which in a large number of cases will be cured under ambient conditions and then either post cured over a prolonged exposure to sun light and ambient conditions, or through a short period of exposure to elevated temperatures. For components fabricated with vinylesters it is likely that the performance characteristics would change with time, as shown schematically in Figure 1, and hence it is important that the designer use values characteristic of the material in use (after a reasonable period of time), rather than values based on testing conducted immediately after fabrication. This aspect has specific relevance to the use of "knock down" factors or safety margins.

Although carbon fibers would generically be preferred on the basis of their inertness to most environmental conditions, cost and performance considerations (such as the need for higher levels of strain to failure, coefficients of thermal expansion in composite form to match that of concrete etc.) entail that E-glass fibers will be used in a majority of current applications. It should be noted further that current sizings/finishes on carbon fibers are not compatible with vinylesters and polyesters, especially when loads are to be carried in compression or shear. Although fiber-resin compatability is not as great an issue with E-glass fibers, the susceptibility of E-glass to moisture and alkali induced damage creates the necessity for the appropriate selection of the resin system to serve not just as a binder but also as a protective layer that would reduce diffusion and UV induced degradation. Due to the high potential for use in the field of processes such as wet lay-up and resin infusion in the presence of moisture, the use of anhydride cured epoxies would not be recommended.

Although changes in glass transition temperature due to moisture absorption are considered in other application areas, the effect is perhaps more striking in the civil infrastructure area due to the low initial T_g itself (due to the predominance of ambient cure, especially in applications involving structural rehabilitation). It should be noted that although the T_g has been shown to recover after drying, the actual situation in a large number of geographical areas is that a high level of humidity and/or moisture will exist continuously with very little chance of the composite decreasing in moisture content. Further, effects of continuous load on structural elements are likely to result in exacerbating the overall detrimental effect.

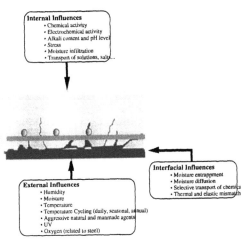

Figure 3. Systems Level Interactions

1.3 Interface Aspects

In a large number of cases, the fiber reinforced composite component will be used in contact with or adjacent to a concrete substrate. In such cases one must consider that concrete is a porous and chemically active material with pH of pore water being as high as 13.5. Existing concrete can also contain high levels of chlorides, carbonates, and sulfates, all of which can be brought to the concrete-composite interface through moisture that diffuses through the concrete. Figure 3 emphasizes the various levels of interaction that need to be considered in using composites in applications such as external strengthening, wherein a relatively thin composite layer is adhesively bonded to the surface of "active" concrete.

It is important to remember that the strength and effectiveness of assemblies depends largely on the cohesive strength of the adhesive and the degree of adhesion between the two adherends. Of equal importance to the efficiency of the adhesive, is the method of surface pretreatment of the adherends prior to the application of the adhesive. Although most engineers tend to focus on the achievement of a prescribed initial bond strength, it must be noted that although important, this metric is not as important as that of bond durability as dictated by the environmental stability of the adherend-adhesive interface. This is especially true of assemblies that are subjected to the vagaries of nature, including extremes of temperatures (often within a 24 hour cycle), moisture, sustained and cyclic loading. In addition, it should be remembered that in comparison to bonding in the aerospace environment, the degree of control possible in a civil infrastructure environment (due to site related factors) is far less. Depending on the type of retrofit, access may be difficult, substrates may not be "clean", and the adhesion may have to be done on a vertical plane or on the bottom face of a horizontal plane against the forces of gravity. While considering adhesion, four generic mechanisms of adhesion need to be considered in the investigation of bond, i.e., mechanical interlocking, diffusion theory, electronic theory and absorption theory. In the case of bonding of composites to concrete, the first is by far the dominant mechanism. Mechanical interlocking assumes that the major source of intrinsic adhesion is the interlocking of the resin into the irregularities of the concrete surface. The rough and deep surface topography created through surface abrasion and glass beading creates a structural morphology that allows the resin to penetrate into the irregularities forming a strong interfacial layer. Surface pre-treatment plays an important role in all bonding operations. However, in a civil infrastructure environment, this can often be restricted to the removal of loose and unsound material and drying of the concrete surface to the best degree possible, keeping in mind that the internal structure of concrete may be moist and will remain that way depending on geographical location, and the sealing of microporosities and large gaps on the surface may be the only practical path forward in some cases.

Although a good definition of the bond line is desirable, this may not always be possible and thick bondlines have to be designed for with lower levels of dimensional precision from those seen in the aerospace environment. Also the surface to be repaired will consist of a variety of morphologies similar to those shown schematically in Figure 4. Some of these will be surface phenomena only and hence may be removed through abrasion, whereas others may involve cracking that is internal or extends into the structure, thereby providing a path for internal flow of the resin and/or primer or providing an area which could at a later date enable transport of moisture or interfacial crack initiation. The efficiency of an adhesive bond, even if predicated by mechanical interlock at the surface morphology level, is dependent

on the condition of the surfaces to be joined. Whereas some degree of control may be possible on the composite end, very little control is possible as related to the chemical and electrochemical characteristics of the concrete. Based on the condition of the structure, concrete may be chloride or carbonate contaminated, or may contain a high percentage of alkali rich pore water. It should be noted that concrete throughout its life would have a high pH level, which necessitates the selection of adhesives and resin that are not affected by the alkali. Although the reinforcing fibers in the composite will be embedded in the resins, due care must be taken in their selection to offset long term durability concerns related to moisture diffusion, and alkali, solvent and salt penetration. In a number of cases the use of protective coatings and resin rich surface layers can slow down the diffusion of moisture containing alkalis through the resin to the interface and fiber surfaces to the point where the degradation may not be a short term concern. The choice of fiber system should in all cases be predicated on the specifics of the application, structural risks involved, stress levels likely to be faced by the retrofit system and the duration that the repair/retrofit is being designed for. Notwithstanding the degradability of glass fibers, design of retrofits using glass are distinct possibilities if the appropriate precautions in terms of materials selection and design stress levels are adhered to. Furthermore the choice of the retrofit system could also affect the state of the deteriorated area (consisting of concrete) since a change in resistivity in the system due to the addition of the repair material could isolate the area electrochemically from the surrounding region causing acceleration in corrosion of the pre-existing reinforcing steel.

1.4 Scope

Within the scope of the current paper (a) effects of alkalinity on E-glass fiber vinylester composites, (b) effects of freeze-thaw on both E-glass and carbon epoxy composites used as jackets on concrete, and (c) field level ongoing investigations, will be discussed. In each case the choice of composite reflects its actual use in the field, thereby providing a basis for actual field-level comparison rather than comparisons on the basis of model systems.

2 ALKALI EFFECTS

2.1 Origin and Rationale

The concern related to alkaline environments is due to the use of concrete in most civil structures, which means that if glass fiber reinforced composites are used in conjunction with concrete, it will be either embedded in concrete (as in the case of rebar or tendons) or will be placed around concrete (as in the

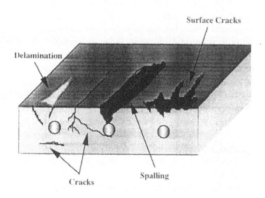

Figure 4. Common Surface Morphologies

case of column retrofit or beam strengthening). From previous research it is well known that bare glass fibers placed in concrete are degraded through loss in toughness and strength and through embrittlement. The fibers are damaged due to a combination of two processes, (a) chemical attack on the glass fibers by the alkaline cement environment, and (b) concentration and growth of hydration products between individual filaments. The embrittlement of the fiber is due to the nucleation of calcium hydroxide on the fiber surface. Cleavage cracks in the crystal can propagate in a direction parallel to the (0001) pseudohexagonal cleavage of the crystal, but rather than stopping at the fiber-crystal interface, extend into the fiber. The hydroxylation itself can cause fiber surface pitting and roughness, which act as flaws severely reducing fiber properties in the presence of moisture. This behavior is seen not just when the concrete is moist, but also after it has "set." Although concrete pore water has a high pH level, 13.5, it has been clearly shown that degradation of glass fibers is not merely due to the high pH levels, but rather to the combination of hydroxylation products, presence of moisture, and the pH, which acts as a catalyst. Although the use of a polymer matrix as a binder around glass bundles and individual filaments does provide a level of protection to the fiber from the above mentioned degradation, there is still concern related to the migration of high pH solutions and alkali salts through the polymer to the fiber surface.

2.2 Test Setup

Despite the significant amount of research that has been conducted in this area, serious questions related to procedures and effects still remain. Two specific aspects are (a) comparison of effects from fresh uncured concrete with those related to cured concrete over time, and (b) effects of different simulated alkali solutions and the validity of their equivalence to actual conditions in concrete. In this paper, discussion is restricted to the latter only. Since it is easier to simulate field conditions in a laboratory for purposes of aging and durability evaluation through the use of solutions in baths, various researchers have used different combinations and concentrations of chemicals to simulate alkaline pore water that would exist in concrete. Table 1 provides, as an example, a listing of some of these solutions with the names of the primary researchers involved. It should be emphasized at this point that the effect that is being attempted to be simulated is that of alkaline pore water in or from concrete, rather than just a solution with high pH. Unfortunately, a number of investigators have neglected this aspect and have conducted comprehensive experiments that either just simulated pH levels (with no concern to the presence of alkali salts) or simulated alkali salts without serious

thought as to which salts are truly representative of the pore water solution (and which are likely to diffuse through the polymer into the composite. In addition most experiments have been conducted using solutions that have to be renewed in terms of salt concentration over the test time period, thereby potentially using greater concentrations of salts than would be achievable through contact with concrete pore water. It should also be noted that although saturated calcium carbonate solutions are often used as an equivalent laboratory substitute for concrete pore water, pH levels of this solution vary dramatically over time, beginning at about pH 12.5 and rapidly dropping to level of 8.5-9.5 due to the interaction of the salt in solution with carbon dioxide in the air resulting in the formation of calcium bicarbonate, which has the lower pH. The continuous use of Calcium Hydroxide as a rechargent to bring up the level of pH is a classic example of adding significantly greater percentages of a salt (calcium hydroxide in this case) just to maintain pH levels. It is possible that these changes, or the presence of greater quantities of some salts, and the absence of others, would have a significant effect on the degradation and rate of chemical attack on glass fiber reinforced composites.

Table 1: Example of Alkaline Solutions used In Composite Investigations

Researchers	Details of Solution
Uomoto and Nishimura	NaOH solution at 1 mol/l at 40 C
Rostasy	Saturated solution of $Ca(OH)_2$ + 0.4N KOH for a pH13 level
Hawkins, Steckel, Bauer and Sultan	$CaCO_3$ solution at 23 C to get pH 9.5
Christensen	0.32 mol/L KOH, 0.17 mol/L NaOH and 0.07 mol/L of $Ca(OH)_2$ in distilled water
Vijay, GangaRao and Kalluri	0.2% $Ca(OH)_2$, 1% NaOH, 1.4% KOH by weight (pH 13)
Conrad, Bakis, Boothby and Nanni	Saturated solution of $Ca(OH)_2$

It should be obvious that the solutions in Table 1 (which represent an incomplete list) do not provide the same level of attack on the glass fiber, and further do not even provide the same concentration profile during diffusion. Also, it is debatable whether the effect (chemical and otherwise) of the above solutions come close to replicating, over the short- and long-term, the effects of pore water from concrete, or those associated with migration of water through concrete into adjacent glass fiber reinforced composites.

In order to simulate effects due to migration of water through concrete to the composite, alkaline solutions were prepared by placing disks of concrete of 150 mm diameter and 25 mm height in water and allowing alkaline salts to collect in the water directly from the concrete itself. For purposes of comparison concrete disks from two different sets of ordinary Portland Cement Concrete were used, the first resulting from cylinders allowed to cure for 28 days (New Concrete) and the second resulting from cylinders that were allowed to age 10 years before use (Old Concrete). To further compare effects a cementitious extract was prepared at the same time as pouring of the "New" Concrete by collecting the solution that formed after settling of the aggregate and diluting the solution by the same level of water added to the concrete disks used previously. Standard potassium buffer solution providing a pH of 10 was also used. Further deionized water by itself was also used as a comparative test solution. pH levels of these solutions (with the exception of the buffer) were measured over an extended period of time and results are shown in Figure 5.

As can be seen the pH levels of the "New" Concrete solution, "Old" Concrete solution, and Cementitious extract are different at the beginning with the extract solution having the highest pH and the "Old" concrete solution having the lowest. The two solutions that derive alkali salts and pH from actual concrete both drop to a pH level of 8.5 fairly rapidly (60-80 days) whereas the cementitious extract solution falls to a level of about 8.5 in about 250 days. It is interesting to note that if the concrete containing water solution was placed in isolation (in this case in a sealed dessicator) the pH level does not fall but stays relatively constant at a level of 12-12.5. This is in line with the previously mentioned effect of carbon dioxide on carbonates. The effect, however does raise questions related to effects arising from situations wherein the concrete in question is isolated from free atmospheric interaction, which would seemingly result in maintenance of higher pH levels as well as of salt concentrations that conceivably would be more degrading to glass fiber reinforced concretes, than others.

Glass fiber reinforced composite panels were fabricated using layers of UM2403 unidirectional fabric with Interplastics Corezyn CORVE8121 using a wet lay-up process followed by vacuum bagging at 2-3 in of Hg pressure. Panels were fabricated in configurations of 2, 4 and 8 layers of fabric with the mat side facing the same direction (towards the mold surface) in all cases. Fiber volume fraction ranged from 41.7% to 51.1% with less than 2% voids. Once cured at ambient conditions (1 month) panels were cut into tensile coupons of size 254 mm x 25.4 mm and all cut edges were sealed with the same resin system through light brushing. In all cases MEKP

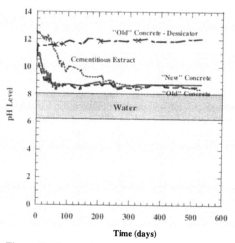

Figure 5: Change in pH Levels With Time

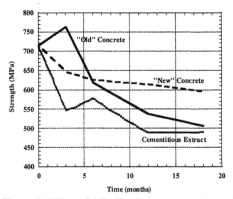

Figure 6: Effect of Alkaline Exposure on Strength of 2 Layer Samples

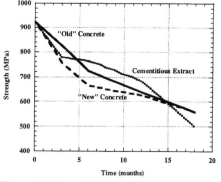

Figure 7: Effect of Alkaline Exposure on Strength of 4 Layer Samples

was used as the catalyst/initiator. After careful inspection and weighing, each coupon was placed in

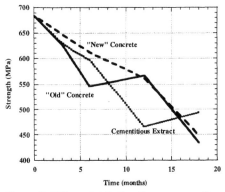

Figure 8: Effect of Alkaline Exposure on Strength of 8 Layer Samples

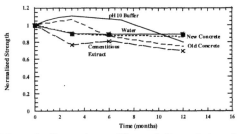

Figure 9: Comparison of Strength Degradation after Immersion in pH10 Buffer with Degradation Due to Cementitious Solutions

the appropriate solution ("New" concrete, "Old" concrete, cementitious extract, pH 10 buffer, and water and tested after periods of 3, 6, 12 and 18 months. Tests were also conducted using DMTA techniques to assess changes in glass transition temperature and storage and loss moduli.

2.3 Results

Summary effects of exposure to the various solutions on tensile strength are shown in Figures 6-8 for the 2 layer, 4 layer and 8 layer composites respectively.

In all cases the drop in strength is seen to be significant, with results at the end of 18 months being roughly comparable in all cases. Micrographic analysis shows significant roughening of the fiber surface with some cracking and pitting as well. In addition there is noticeable loss of bond in local areas between the fiber and the matrix. It is instructive to compare effects resulting from these solutions with those from a potassium based pH 10 standardization buffer, as in Figure 9. This shows that the effect due to the higher overall pH of the buffer (since the pH in the three solutions falls below 9 in 200 days) over a short time interval does not provide the same degree of degradation, emphasizing the importance of the presence of the alkali salts. However, after a length of about 180 days, degradation of the samples exposed to the buffer is rapid, approaching levels seen in the other solutions. It is significant however, that this appears to be concentrated at the interfacial level rather than that of the fiber indicating that mechanisms of degradation in the two cases, pH buffer and actual concrete solutions, is different.

Figure 10: Interface Cracking and Fiber Pitting in Samples Exposed to Cementitious Extract Solution for 18 Months

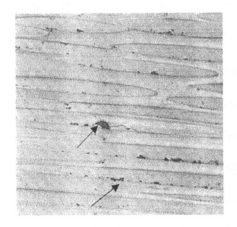

Figure 11: Fiber Level Damage and Pitting Seen in Sample Exposed to "Old" Concrete Solution for 18 Months

Figures 10 and 11 clearly show the effects of exposure in terms of fiber pitting, degradation through cracking, and interfacial deterioration. Micrographic and EDAX analysis shows that the damage at the interphase level is initiated by moisture in a highly alkaline medium, whereas that at the fiber level is due to initial formation of a calcium hydroxide crystal with notches and abrades the fiber itself. Both these effects could conceivably be aggravated further through the introduction of stress, and could be potentially negated through the use of appropriate fiber sizings/finishes and use of resin systems such as polyurethane modified vinylesters which have been shown to be more resistant to moisture and alkali related degradation.

3 FREEZE-THAW EFFECTS

Another environmental factor that is of significant concern to the use of composites in civil infrastructure, especially as related to their use in seismic retrofitting and strengthening of columns, is the effect of sub-zero environments and freeze-thaw exposure, both in terms of temperature variation and as related to synergistic effects of road salt that could diffuse into the composites during that time period.

3.1 *Test Setup*

Concrete cylinders were wrapped with carbon and glass fiber unidirectional fabrics which were impregnated using the wet lay-up process using a commercially available epoxy system following procedures used in the field for this system for column strengthening. Three distinct wrap architectures were used, (a) 3 layers of the carbon fabric with the fibers in the hoop direction, (b) 2 layers of carbon fabric, one each with fibers in the hoop direction and in the height direction with the hoop layer being the outermost (i.e., a 90/0 lay-up), and (c) 7 layers of the glass fabric with fibers in the hoop direction. The carbon fiber had a nominal modulus of $E_f = 372$ GPa and a strength of $\sigma_f = 1.52$ GPa, whereas the glass fiber had a nominal modulus and strength of $E_f = 68.95$ GPa and $\sigma_f = 1.52$ GPa respectively. The resulting fiber weight fraction in the wraps was about 50-55%. The number of glass and carbon layers in the hoop direction were chosen to provide roughly the same lateral confining pressure as determined by

$$f_l^{'} = \frac{2 f_{com} \cdot t}{(d_c - 2t)}$$ wherein f_{com} is the ultimate tensile strength of the FRP composite of thickness, t, and d_c is the specimen diameter. It should be noted that lateral confining pressure has an important effect on the level of strengthening achieved by confinement.

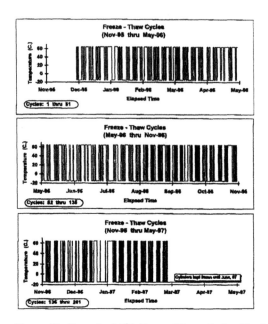

Figure 12: Record of 201 Freeze-Thaw Cycles Used For Exposure of FRP Composite Wrapped Concrete Specimens

Specimens were then divided into two groups, one set being kept under ambient conditions, and the other being exposed to 201 cycles of freeze-thaw as shown in Figure 12.

After exposure, wrapped and unwrapped cylinders were tested in direct compression to assess effects of exposure on strengthening efficiency.

3.2 *Results and Discussion*

It is important to note as we review results that there are actually three overlapping effects that must be considered (a) effect of exposure on the concrete itself, (b) effect of exposure on the FRP composite, and (c) effect of exposure on the FRP composite confined concrete acting as an integral system. It is thus expected that some important effects would only be apparent on review of damage and failure mechanisms, and overall response, rather than merely on the basis of changes in ultimate strength levels. The concrete (unwrapped) control cylinders tested after the total period gave a strength and stiffness of 60.47 MPa (8.77 ksi) and 32.73 GPa (4.75 Msi) respectively. In comparison the unwrapped cylinders exposed to the freeze-thaw cycles gave a strength and stiffness of 63.92 MPa (9.27 ksi) and 26.04 GPa (3.78 Msi) respectively. It was noted that in general there was only a slight, 5.7%, change in the average ultimate compressive strength which falls within overall scatter bounds, but the freeze-thaw cycling caused a 20.4% decrease in initial stiffness.

Response (stress-axial strain) curves for the specimens subjected to the 22.5°C exposure and those subjected to the freeze-thaw cycles shown a bilinear response characteristic of FRP composite confined concrete. In all confined specimens a definite kink point is seen which has been attributed to the initiation of full confining action with microcracking in the concrete and along the FRP composite-concrete bond interface. The specimens wrapped with the hoop based layers show greater ductility than the 90/0 configuration, which as is expected due to the effect of axial component of the wrap itself. As expected, the wrapping of concrete with FRP composite jackets results in an increase in both ultimate strength and ductility, as measured by the increase in level of axial failure strain. The effect of the confining action provided by the FRP composite jackets over the unwrapped concrete after exposure as 22.5°C and after freeze-thaw exposure in terms of changes in levels of strength, stiffness and strains, are detailed in Tables 2 and 3 respectively. In each case the results are normalized by the values for the corresponding set of unwrapped concrete cylinders.

Table 2: Effect of Confinement - 22.5°C conditions (values are normalized by those of unwrapped concrete cylinders subjected to the same exposure)

System	Strength	Stiffness	Axial Strain	Hoop Strain
Glass (0°)	1.54	1.06	2.59	5.49
Carbon (0°)	1.96	1.00	3.30	3.81
Carbon (90/0)	1.32	1.06	1.75	2.92

Table 3: Effect of Confinement – After Freeze-Thaw Exposures (values are normalized by those of unwrapped concrete cylinders subjected to the same exposure)

System	Strength	Stiffness	Axial Strain	Hoop Strain
Glass (0°)	1.42	1.22	2.21	11.79
Carbon (0°)	1.85	1.34	2.30	6.94
Carbon (90/0)	1.29	1.08	1.35	4.78

As can be seen from Table 2, the use of three layers of unidirectional carbon increased the strength by 96% over the unwrapped concrete, whereas the use of a 90/0 (axial and hoop layers) configuration resulted in an increase of 32%, which correspond closely to earlier results. It should be noted that although the glass fiber reinforced FRP composite jacket resulted in a lower level of strength increase, 54% than the all hoop carbon fiber reinforced FRP composite jacket, the higher strain capacity of the glass fibers (2.1% versus 1.5% for the carbon fibers) results in a significantly larger increase in ultimate hoop strain. Under action of axial compressive loading the confined concrete responds through increases in ultimate values of both radial expansion and axial compression strain. The radial, or hoop, strain is dominated by the response of the FRP composite jacket in being able to confine the concrete core and the confinement enabled by the FRP composite jacket is dependent on the stress level achieved in the FRP composite itself.

From Table 3 it is clear that even after freeze-thaw cycling all FRP composite wrapped systems show an increase in ultimate strength and strain over the unwrapped, but similarly exposed, concrete specimens. It is noted, however, that the strength increases are slightly lower than those noted for the set that was kept as 22.5°C, whereas there is a definite increase in system stiffness after freeze-thaw cycling in all cases. This effect can be traced to the increase in matrix stiffness and strength, which causes an increase in the axial stiffness of the overall system. This effect should be expected to be greater in systems having fibers perpendicular to the loading direction and this is born out by the very small increase, 8%, in stiffness of the 90/0 wrap system which has the layer of fibers in the axial direction in contact with the concrete. The increased matrix hardening, and resulting increase in FRP composite stiffness also leads to changes in failure modes of the confined concrete samples as is described later.

A direct comparison of the effect of freeze-thaw exposure on confining action of the FRP composite alone can be provided through the normalization of values resulting from wrapped samples after exposure to freeze-thaw conditions by those resulting from wrapped samples after exposure to 22.5°C. (This can be shown as resulting from the ratio $P_{freeze-thaw}/P_{22.5°C}$, wherein P represents the property of the system under consideration). The inclusion of effects directly attributable to the concrete core and the bond between the FRP composite jacket and the

concrete core are, however, given by normalizing the results given in Table 2 by the corresponding ones in Table 1. (This provides a simple ratio of
$$\left[\frac{P_{freeze-thaw}}{P_{22.5}}\right] x \left[\frac{Q_{22.5}}{Q_{freeze-thaw}}\right]$$ where Q represents the
corresponding property of the unwrapped concrete samples). Results on the basis of these two ratios are given in Figures 13 and 14 for strength, stiffness and strains, as determined by the two above described normalizations respectively.

Stiffness levels also are seen to change very slightly for the jackets using hoop reinforcement, whereas a

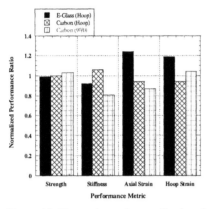

Figure 13: Comparison of normalized performance metrics on the basis of wrapped cylinders only ($P_{freeze-thaw}/P_{22.5°C}$)

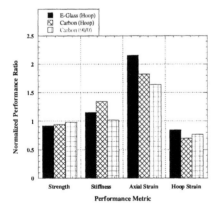

Figure 14: Comparison of normalized performance metrics on the basis of exposure effects on wrapped and unwrapped cylinders $\left[\frac{P_{freeze-thaw}}{P_{22.5}}\right] x \left[\frac{Q_{22.5}}{Q_{freeze-thaw}}\right]$

definite reduction in stiffness due to freeze-thaw exposure (11%) is noticed in the 90/0 carbon fiber reinforced FRP composite jacketed system. This again is not surprising since the effects of change in properties of the FRP composite are likely to be affected to a greater extent in the off-axis direction. As the specimens are loaded in compression the hoop elements of the FRP composite jacket are actually loaded in tension due to the dilation of concrete causing hoop directed layers to be loaded in the direction of their reinforcement whereas layers in the 90° direction (axial) are actually loaded transverse to the fiber direction, i.e. in the matrix and interphase dominated direction. Thermal cycling at negative temperatures results in smoother fracture surfaces with increased fiber-matrix debonding and decrease in tensile properties when tested transverse to the fiber direction. This weakness results in earlier failure in the axial layers and the change in damage mechanism is clearly seen on comparison of fracture surfaces of the FRP composite jacket with 90/0 architecture subjected to freeze-thaw cycling with those exposed to just 22.5°C temperature. In the case of specimens exposed to 22.5°C the failure mode is dominated by two, almost diametrically opposite, vertical splits through the jacket, with fairly "clean" fracture surfaces. In comparison, specimens subjected to the freeze-thaw exposure show a very smooth fracture surface, but a jagged surface for the outer, hoop, layer with significant splitting between hoop directed fibers, and pullout across the fiber surface.

In contrast, the overall strengths do show a slight decrease, with the reduction being greater for the 0°, or hoop based, systems, when reviewed on the basis of inclusion of the unwrapped concrete specimens as norms. It is noteworthy that when compared on this basis the two hoop wrapped systems both show an increase in stiffness, 15% and 34% for the glass and carbon systems respectively. It should be remembered at this point that the unwrapped concrete specimens subjected to freeze-thaw exposure showed a 20.4% decrease in stiffness in comparison to the specimens exposed to 22.5°C, emphasizing the effect of exposure on the concrete itself.

The effect of freeze-thaw exposure on the failure mode of the FRP composite wrap in the case of the 90/0 carbon architecture was explained earlier. There are also effects due to this exposure on the two hoop oriented systems. In the case of the carbon fiber reinforced, hoop oriented, composite systems subjected to 22.5°C exposure failure was through the formation of two primary cracks, roughly diametrically opposite each other. Fracture surfaces are jagged. Damage in the form of hoop matrix splitting, fiber fracture and interlayer delamination or separation was restricted to a local region on either side of the line of fracture. In comparison, in the case of specimens subjected to freeze-thaw exposure, the longitudinal splits were accompanied by splits in the hoop direction which often resulted in the line of failure being stepped rather than roughly vertical. This is attributed to effects of freeze-thaw exposure on the matrix in similar fashion as that elucidated in the earlier explanation for the 90/0 lay-up. The hoop splits were seen to be fairly widespread extending 60-100 mm on average on either side of the vertical failure line. Further, failure in the vertical direction was accompanied by very clean local FRP composite surfaces on the inner side with almost no concrete adhesion close to the crack face. This is due to the transfer of high levels of fracture energy from fiber fracture to the locally adjacent interface areas causing local failure of the matrix-concrete bond. It should be noted, however, that failure away from the vertical surfaces was within the concrete with good bond integrity remaining between the inner resin-rich layer and the concrete itself.

In the case of the jackets using glass fibers in the hoop direction specimens exposed to 22.5°C failed through the formation of bands spread wide with widths of 76 - 100 mm (3-4 inches) with fracture taking place individually across the height of these bands. Smaller length cracks were formed in the hoop direction as a result of the initial fracture but did not traverse along the entire circumference. Axial cracking and pullout accompanied failure of the larger hoops. In contrast, specimens exposed to the freeze-thaw conditions show the formation of numerous hoop bands of 12-19 mm width (0.5-0.75 inches) with fracture across individual hoops. Fracture, subsequent to hoop splitting, was accompanied by minor delamination and tearing. The transition from the larger width bands seen in specimens exposed to 22.5°C to the smaller bands in greater number seen in specimens exposed to freeze-thaw exposure is due to the decrease in tensile strength in the off-axis direction and formation of weaker fiber-matrix interfaces resulting in formation of more bands. Also the freeze-thaw exposure causes some matrix microcracking in resin-rich regions leading to local zones of weakness.

From investigation of fracture surfaces of samples after failure, and from some samples extracted after being loaded short of failure it would appear that the differences in damage modes caused due to differences in exposure are initiated from local zones of weakness caused by changes in materials (resin, interply, fiber-resin bond, and composite) characteristics which initiate in mechanisms fairly early in the loading process. Since most fracture/failure mechanisms are energy based, it is instructive to examine differences in system-level energy due to exposure conditions at different points in the loading cycle. Based on the initiation of small microcracks, minor crazing, and some very local fiber-matrix splitting either at or just before the kink point in the bilinear curve of the confined concrete specimens, energy (defined as the area under the stress-strain response curve) levels were computed for all specimens. Table 4 shows the results at the kink point and at ultimate failure for each of the three classes of jackets after normalization of results obtained from freeze-thaw cycling by those obtained from exposure to 22.5°C conditions. It can clearly be seen that the freeze-thaw exposures result in significantly higher energy levels at the kink point, but insignificant changes at failure except for the 90/0 wrap where effects are modified due to the presence of fibers in both hoop and axial directions. This provides further verification for the significant role played by the matrix and fiber-matrix interface, as described previously, in increasing performance levels till initiation of damage, after which the overall failure energy does not change significantly on a systems level basis, although system ductility measured after the kink-point decreases due to a combination of concrete stiffness degradation, and increased matrix-hardening and brittleness in the FRP composite.

It is emphasized that although no investigations were carried out using aqueous solutions or salt, during or prior to the thaw cycling, based on diffusion of moisture in composites and the subsequent damage mechanisms, it is expected that ingress of moisture and salts in these cracks and disbonds could cause further degradation of composite characteristics

Table 4: Comparison of FRP Composite Jacketed Specimens on the Basis of Energy Response. (Values obtained as a result of freeze-thaw cycling are normalized by those obtained after 22.5•C exposure)

System	Normalized Energy Till Kink Point	Normalized Energy At Failure
Glass (0)	1.36	1.01
Carbon (0)	1.52	0.99
Carbon (90/0)	1.43	1.15

through increased microcrack formation and reduction in cohesive energy as well as further matrix embrittlement. Overall comparison of the wrapped specimen behavior shows that although there is an insignificant change in ultimate strength of the specimens after freeze-thaw exposure, the stiffness is seen to increase as compared to specimens exposed to 22.5°C. Further, the carbon fiber reinforced specimens show a decrease in hoop strain capacity after freeze-thaw exposure resulting in more catastrophic failure modes. These effects need to be considered in further testing as well as in current design practices through the selection of appropriate resin types, curing conditions, and safety factors to account for material changes and the susceptibility for a more catastrophic mode of failure with higher energy at damage initiation, which could have significant effects on overall life-cycle, especially in the presence of moisture and aqueous ingress into the FRP composites.

4 FIELD INVESTIGATIONS

Acknowledging that results from laboratory exposure can at best approximately simulate the field environment, it is also necessary to acknowledge that the results accruing from laboratory exposures are generally more severe than those from the field. It is hence critical that results be assessed in that light and as far as possible laboratory results should be compared to results from the field.

4.1 Seismic Retrofit

As part of a Civil Engineering Research Foundation Project aimed at the evaluation of composite wrapping schemes for the retrofit of columns, durability of composites is being assessed through the testing of rings of 508 mm diameter using the NOL burst procedure. The NOL ring burst test has been identified as an efficient test that combines both materials and structural aspects for the determination of design properties and durability of the FRP jacket since it simulates the pressure on the jacket during dilation of concrete. Specimens from 4 different systems are being tested after exposure to laboratory simulated conditions and after field exposure at three sites in the United States representative of some of the variations in climate. Results from burst, short-beam shear, DMTA, and micrographic analysis will be correlated to provide a baseline for use of these systems in the future by the state departments of transportation. Figures 15-17 show a range of failure modes resulting from the testing of three different systems, as examples.

Figure 15: Tensile Failure in a Well-Consolidated and Cured Specimen Resulting From the Use of a Tow-Preg Based Heat Cured System

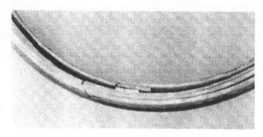

Figure 16: Failure Initiated At Adhesive Level in a System Using Prefabricated Adhesively Bonded Sections

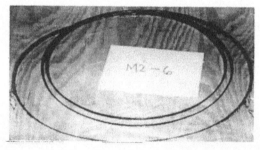

Figure 17: Failure Through Layer Separation Due to Poor Wet-Out/Interply Bonding in a Wet Lay-up Based System

5 CONCLUSIONS

Although fiber reinforced composites have significant potential for use in civil infrastructure, there are still considerable questions related to durability of materials and systems that need to be resolved in light of use of lower cost resin systems, non-autoclave based ambient or moderate temperature cure processes, and use in the field in a changing environment. Investigations are needed at the systems rather than simply at the coupon level.

Recent Developments in Durability Analysis of Composite Systems, Cardon, Fukuda, Reifsnider & Verchery (eds)
© 2000 Balkema, Rotterdam, ISBN 90 5809 103 1

Durability of reinforced concrete beams repaired by composites under creep and fatigue loading

E. Ferrier & P. Hamelin
Material Mechanical Engineering Laboratory, Lyon-1 University, France

ABSTRACT: The external bonding of fiber reinforced plastic to concrete structures is an effective method so as to increase the structural capacity of such structures. But, for the design of the reinforcement, one has to take into account the long-term behavior of each material. If the influence of natural exposure on the durability of reinforced structure begins to be well identified [1], only a few works study the long-term behavior under creep and fatigue loads [2]. The use of composite glued on the concrete support requires the identification of the adhesive layer behavior. This adhesive layer is usually an epoxy polymer. This paper deals with the fatigue and creep effect on reinforced structures. Then, the influence of the differed behavior of the polymer is taken into account for prediction of the lifetime of composite reinforced concrete beams.

1 INTRODUCTION

Because of their strength and specific stiffness, composite materials present a significant interest in the conception of bearing structures by an increase of the dead load and of the bearing capacity. They are particularly well adapted in order to strengthen shapes of complex geometry. It is necessary to point out that our repair technique aims at evaluating components (fiberglass, carbon fibers, epoxy polymers) developed in the industrial field (Fig. 1). To design composite reinforcement for a RC beam under fatigue or creep load, one has to obtain the mechanical behavior of each material and above all, of the concrete/composite adhesive layer

2 LIFETIME DESIGN METHOD

The analysis of a reinforced concrete beam strengthened by a composite material plate can be described by a classical method using the theory of steel reinforced concrete beams. Concerning the mechanical behaviors of the material, we consider that the behavior of the composite is linear up to failure, that the mechanical behavior of steel is elastic-plastic and finally, that the mechanical behavior of concrete in compression is deduced from the french design code (Fig. 2). The moment - curvature diagram up to failure can be built step by step. This can be done by initializing, first, the deformation of the concrete and, in a second time, by recalculating the position of the neutral axis taking into account the loss of effort in the composite plate after slipping. The balance of forces is so satisfied. When the internal force balance is done, the values of curvatures and moments can be calculated. The maximal moment is evaluated by taking into account failure criteria, such as the yielding of steels, the failure of the composite in tension or the failure of the concrete in compression.

$\Sigma F_{ex} = 0$

$$N_B + N_A + N_T + N_P R \ 0 \qquad (2.1)$$
$$N_b = \int \sigma_b \, da_b$$

$\Sigma M = 0$

$$M = M_b + M_a + M_t + M_C A \ M_{ap} \qquad (2.2)$$
$$M_b = \int \sigma_b \, Y \, da_b$$

Experimental test results show that there are different premature failure modes of the structure in addition to the previously stated failure criteria.

These failure criteria specific to the strengthened structure allow evaluating values of maximum moments and curvatures.

This calculation method uses standard principle in predicting differed properties of steel reinforced concrete beams.

This method described by Triantafillou and Plevris [4] uses the differed properties of each material of the beam. It means that the behavior of steel, concrete, adhesive and composite has to take into account.

The differed properties of steel and concrete under creep and fatigue loading have been assessed by several authors [5], [6], [7]. But only a few researches have been done on the composite and on the adhesive layer properties.

Using a tensile shear test for the adhesive, we have done this identification. When all the mechanical properties of the adhesive layer under creep and fatigue are assessed, the evolution of beam curvature is calculated. This calculation is done by using mechanical properties for a moment or at the end of N cycles. Using the software developed by Material Mechanical engineering Laboratory, the calculation is done using two hypotheses. As a first step of calculation, the adhesion between concrete and composite is perfect, then, the sliding between the two materials is taken into account using the identification of the adhesive behavior (Fig. 3).

In the second case, the loss in strength in the composite plate is calculated. The first step of the design process is assessed with a perfect adhesion hypothesis.

Then, the "sliding effect" between the composite and the structure is estimated using the shear stress distribution in the adhesive layer and the mechanical relation between shear stress and strain obtained by tests (Fig. 5). For the second hypothesis, the shear stress in the adhesive layer has to be calculated. Several authors propose analytic solutions to calculate this shear stress in case of beam. We have compared those methods with finite element and experimental solutions [8]. All the solutions show a localization of high stress at the end of the composite plate. We have chosen the solution given by the analytical calculation method assessed by Täljsten [9].

This calculation method has been verified by static tests on beams reinforced by composite plate.

The durability of a reinforced concrete beam is influenced by the differed properties of the adhesive layer. Because of the fatigue, creep or aging of the

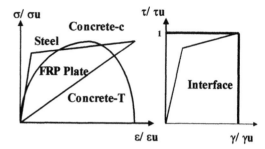

Fig. 2: Mechanical law of each material

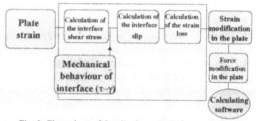

Fig. 3: Flow-chart of the slipping calculation method [8]

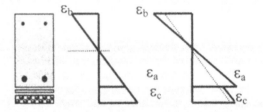

First hypothesis: perfect adhesion

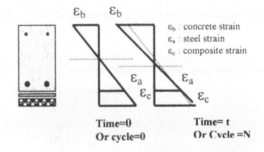

ε_b : concrete strain
ε_a : steel strain
ε_c : composite strain

Time=0   Time= t
Or cycle=0   Or Cycle =N

Second hypothesis: sliding effect in the adhesive layer

Fig. 4: Differed behavior calculation method

polymer, the loss in strength in the composite plate increases the deflection of the reinforced concrete beam.

So, it is necessary to assess the differed properties of the adhesive layer.

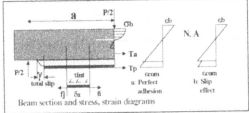

Fig. 5: Slip effect in the plate

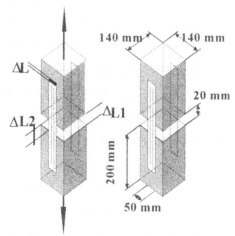

Fig. 6: Test principle

3 INDENTIFICATION OF CREEP AND FATIGUE ADHESIVE LAYER PROPERTIES

3.1 Test device

We have developed a specific tensile-shear test. Composite reinforcements are made of two carbon fabric strips of 425 mm x 60 mm. The strips assemble two concrete blocks (Fig. 6). We have carried out a set of tests with fatigue cycles on the concrete/composite interface and a set of creep tests with the same device. The first test has been conducted until failure in static (speed: 1 mm/min) with two composites (B and C; table 1).

The second is conducted on polymers (A, B, C; table 1) to estimate the influence of time-temperature-loading on the viscoelastic behavior. The third study is done under fatigue loading to assess the shear behavior of polymer B and C.

Table 1. Composite mechanical properties

	Composite A	Composite B	Composite C
E (MPa)	65000	50000	56000
R (MPa)	600	550	430
Strain to failure (%)	0.8	1.05	0.07
Tg (°C)	46.7°C	51.8°C	83.7°C

The two faces are instrumented as indicated in the figure 7. The global slipping of the four composite plates is measured.

The global shear modulus is approximately calculated with the relation 3.1:

$$G = \frac{\tau_{moy}}{\Delta l2 - \Delta l1} \cdot s \qquad (3.1)$$

with: τ_{moy}: mean shear stress (MPa),
Δl_2: average displacement in the adhesive layer (mm),
Δl_1: composite displacement (mm),
s: thickness of the adhesive layer

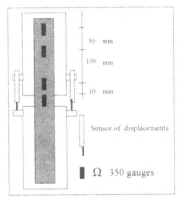

Fig. 7: Measurement device

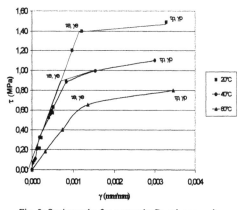

Fig. 8: Static results for composite B under several temperatures

The average shear strain measured (Fig. 8) with the device developed with this test is compared with analytic and numerical solutions.

3.2 Shear stress and strain calculation

Two methods (finite element analysis (ANSYS) and analytic solution) have been used to be compared with the experimental results. The finite element model of the test device uses volume elements (Solid 45) for concrete and adhesive layer and plane element (plane 42) for the composite. The nodal solution allows calculating the average shear stress and strain. The analytic solution is given by the equation (3.2) (Fig. 9).

$$\tau(x) = \frac{G}{s} \cdot \left[U_c(x) - U_B(x) \right]$$

$$\varepsilon_c = \frac{dU_C(x)}{dx} \; and \, \varepsilon_B = \frac{dU_B(x)}{dx}$$

$$\varepsilon_c(x) = \frac{N_c(x)}{A_c \cdot E_c}$$

(3.2)

$$\varepsilon_b(x) = -\frac{N_B(x)}{A_B \cdot E_B}$$

$$\frac{d^2\tau(x)}{dx^2} = \frac{G}{s} \cdot \left[\frac{d^2U_c(x)}{dx^2} - \frac{d^2U_B(x)}{dx^2} \right]$$

$$\frac{d^2\tau(x)}{dx^2} = \frac{G}{s} \cdot \left[\frac{dN_c(x)}{A_c \cdot E_c \cdot dx} + \frac{dN_B(x)}{E_B \cdot A_B \cdot dx} \right]$$

$$\tau(x) = A_1 \cdot sh(w \cdot x) + B_1 \cdot ch(w \cdot x)$$

With limit conditions (3.3), the constants are assessed.

$$\varepsilon_B(-\frac{L_C}{2}) = \frac{N_B}{E_B \cdot A_B}$$

(3.3)

$$\varepsilon_c(\frac{L_c}{2}) = \frac{N_B}{2 \cdot E_C \cdot A_C}$$

$$\frac{N_B \cdot G}{E_B \cdot A_B \cdot s \cdot w} = A_1 \cdot \sinh\left(w \cdot \left(-\frac{L}{2}\right)\right) + B_1 \cdot \cosh\left(w \cdot \left(-\frac{L}{2}\right)\right)$$

$$\frac{N_C \cdot G}{E_C \cdot A_C \cdot s \cdot w} = A_1 \cdot \sinh\left(w \cdot \left(\frac{L}{2}\right)\right) + B_1 \cdot \cosh\left(w \cdot \left(\frac{L}{2}\right)\right)$$

$$A_1 = \frac{N \cdot G}{2 \cdot s \cdot w \cdot \cosh\left(w \cdot \frac{L}{2}\right)} \cdot \left(\frac{1}{E_B \cdot A_B} + \frac{1}{2 \cdot E_C \cdot A_C}\right)$$

$$B_1 = \frac{N \cdot G}{2 \cdot s \cdot w \cdot \sinh\left(w \cdot \frac{L}{2}\right)} \cdot \left(-\frac{1}{E_B \cdot A_B} + \frac{1}{2 \cdot E_C \cdot A_C}\right)$$

$$\tau(x) = \frac{N \cdot G}{2 \cdot s \cdot w \cdot \cosh\left(w \cdot \frac{L}{2}\right)} \cdot \left(\frac{1}{E_B \cdot A_B} + \frac{1}{2 \cdot E_C \cdot A_C}\right) \cdot sh(w \cdot x)$$

$$+ \frac{N \cdot G}{2 \cdot s \cdot w \cdot \sinh\left(w \cdot \frac{L}{2}\right)} \cdot \left(-\frac{1}{E_B \cdot A_B} + \frac{1}{2 \cdot E_C \cdot A_C}\right) \cdot ch(w \cdot x)$$

With :

εb, εc: strain in concrete and composite
Ub, Uc: displacement in concrete and in composite
G: shear modulus (MPa), s: polymer thickness (mm)
N: load (N)
Eb: concrete modulus (MPa), Ec: composite modulus
L: anchorage length
Ac: composite area, Ab: concrete area

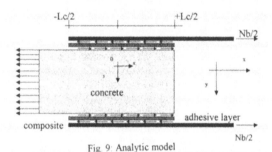

Fig. 9: Analytic model

Table 2. Comparison between experimental and theoretical solution

	Experimental	Analytical	Numerical
τ_{mean} **(MPa)**	1	0.91	0.90
γ_{mean} **(m/m)**	0.00083	0.00076	0.00075

The theoretical result verifies the static test. The measurement device chosen for the slipping evaluation allows assessing the shear strain in the adhesive layer for the elastic part with a suitable accuracy (table 2). For the following studies, this measurement device will be used to follow the shear strain evolution. With the double lap joint test, the creep, fatigue and aging of the adhesive layer are studied.

Then, using the calculation method described above, the differed properties of reinforced concrete beams are assessed.

3.3 Creep

The creep test allows setting up the creep of the composite reinforcement under combined effects of stress and environment. Three composites have been tested (A, B, C).

The main difference between the entire composite is the glass transition temperature. The parameters of the rheological model (equation 3) are identified conducting several tests with several temperatures. Those tests are described in other publications [7]. The main result of this study is the fact that the behavior under coupled effect of creep and temperature is identified and modeled using rheological model. The compliance of the rheological model is obtained using mathematical relation given by Laplace Carson (Fig. 11). The rheological parameter depends on the temperature creep. The dependence of temperature on the parameter is fit with experimental results (Fig. 10). The figure 12 assesses the influence of temperature on the rheological parameter. The experimental results allow assessing the rheological parameter given by table 3. So the strain caused by the creep is obtained for any time and temperature.

$$\gamma(t,T) = \tau_0 \cdot D(t,T)$$

$$D(t,T) = f(T) + g(T) \cdot e^{\frac{-t((a_d+b_d \cdot T)-((a_d-ai)+(b_d-bi)))}{\eta_1 \cdot (a_i+b_i \cdot T)}} \quad (3)$$
$$+ g(T) \cdot e^{\frac{-t((a_d+b_d \cdot T)-((a_d-ai)+(b_d-bi)))}{\eta_1 \cdot (a_i+b_i \cdot T)}}$$

with : f(T), g(T), ai, bi: rheological parameters.

The result of the test assessed in dry environment shows an important creep for the composite A when the temperature increases (Fig. 10.a). The main conclusion of this part is that the Tg effect should be limited to 55°C or upper to limit the creep in the adhesive layer. The strain in the adhesive layer is obtained for any time and temperature (20°C-60°C); the differed properties of the reinforced concrete beams are so studied.

3.4 Fatigue

For this part, only composites B and C have been tested. This choice has been done because of bad results obtained on the first polymer during creep tests.

In order to evaluate the performance of a composite repair, we propose to determine the behavior under fatigue loading of concrete-composite adhesive layer. The test allows determining the loading level to apply to a structure to guaranty the durability of the reinforcement. The double lap joint test sets up the fatigue life and shear modulus evolution for several levels of shear stress in the interface. The number of cycle at failure allows setting up the time life of the interface and the increase in displacement due to the adhesive layer fatigue. This measurement allows setting up the shear modulus evolution. The test measurement device is the same as the one described in the first part.

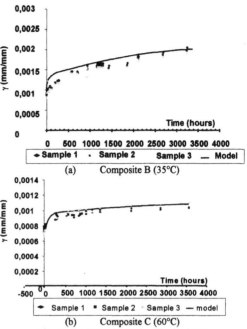

(a)   Composite B (35°C)

(b)   Composite C (60°C)

Fig. 10: Creep results on the adhesive layer

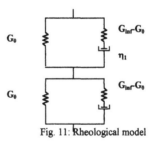

Fig. 11: Rheological model

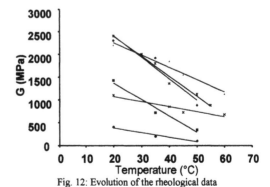

Fig. 12: Evolution of the rheological data

Table 3. Rheological parameters

	a_i	a_d	b_i	b_d	η_1	η_2
Polymer A	3280	583	-44	-10	2.10^6	2.10^6
Polymer B	3356	1290	-47.3	-10.9	$2.3.10^6$	$2.4.10^6$
Polymer C	2795	2092	-26.9	-36.1	$3.4.10^5$	$2.7.10^6$

With the static result on the double lap joint test, different levels of shear stress to be applied to the adhesive layer are assessed.

The main observations of the fatigue test on the both polymers are:

Composite B

Four levels of shear stress have been applied (100 %, 80 % 60 %, and 45 %). For each test two double lap joint samples are tested. For all the test series, the failure occurs for the same number of cycles.

For the test with 80 % (0.10-1.10 MPa) and 60 % (0.10-0.90 MPa) of the failure load, the strain in the bonding zone increases respectably for 1 000 cycles and 100 000 cycles.

The result with a load of 45 % (0.10 MPa-0.60 MPa) shows a low strain increasing on the composite bonding zone during cyclic loading, on this last sample the failure did not occur until 1 million cycles.

Composite C

Three levels of shear stress have been applied (100 %, 80 % and 60%). For each test, two double lap joint samples are tested. For the test with 80 % (0.10 MPa-1.10 MPa) and 60 % (0.10 MPa-0.90 MPa) of the failure load, the strain in the bonding zone increases respectively for 10 000 cycles and 500 000 cycles.

This low increase of strain in the bonding zone of the composite plate is explained by the fact that during the fatigue test, the debonding of the composite plate occurs suddenly just before the failure. The failure occurs with a partly cohesive delamination in the adhesive layer and with a diagonal failure in the concrete at the beginning of the adhesive joint.

The test allows to assess the shear modulus (G) of the adhesive layer in function of number of cycles (N) and the maximum fatigue stress curve ($\Delta\tau$)– number of cycles to failure (N_f).

$$\Delta\tau = m.\log(N_f)+n \qquad (2)$$
$$G = b.\log(N)+a \qquad (3)$$

The coefficients m, n, b, a, depend on mechanical composite characteristics and concrete surface treatment.

Table 4. Fatigue parameters

	m	n	B	a
Composite B	-0.07	0.98	-150	1200
Composite C	-0.04	0.98	-60	1200

A difference is noticed between the two polymers. The polymer C shows a better behavior under fatigue loading. This result confirms the creep tests assessed on the material B and C.

The main difference between the two composites is the polymer. On the composite C, the improve of the lifetime under fatigue loading shows that the choice of an appropriate polymer for long term durability of the reinforcement is necessary. Polymers with high Tg should be chosen because all of those tests have been done in 20°C air temperature. The test assessed in higher temperature should decrease the lifetime of the adhesive joint.

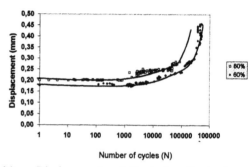

(a)   Displacement evolution in the adhesive layer

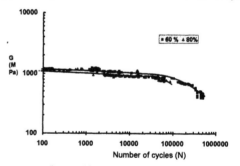

(b)   Shear modulus evolution

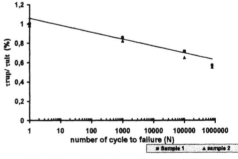

(c) Stress in function of number of cycles to failure

Fig. 13: Test fatigue results on the adhesive layer

The experimental results are used for the calculation of the slipping effect in the adhesive layer in case of reinforced concrete beams. Furthermore the shear stress in the adhesive is limited to the lifetime prediction given by the result on the double lap joint test. Thus, it is possible to design a steel reinforced concrete beam strengthened by composite with taking into account the fatigue behavior of the adhesive. The fatigue of the carbon composite is neglected because of the low stress developed in the composite and because of the good behavior of carbon fabric under cyclic loading.

4 DURABILITY OF RC BEAM REINGORCED BY COMPOSITIES

With the experimental results on creep, fatigue and aging, the differed properties of RC beams reinforced by composite are studied. This research is done with the design method described above but also with experimental tests on beams in the case of fatigue loading.

4.1 Theoretical study of creep effect on the reinforced concrete beams

The software described in the first part is used to model the behavior of a RC beam reinforced with several polymers. The creeping result on the three polymers is used to study the effect of temperature, loading on the mid span displacement. For this study, the result given by the bibliography is used for the concrete creep. This relation between time and differed strain in the concrete is given by the french research center CEB-FIP [10]. This mechanical law takes into account the instantaneous and differed plasticity of the concrete. The creeping data is assessed with a good accuracy such as:

$$\phi(t, t_0) = \beta_a(t_0) + \phi_d \cdot \beta_d(t - t_0) + \phi_f \lfloor \beta_f(t) - \beta(t_0) \rfloor$$

with
β_a : coefficient for loading time and level
β_d : function for differed elasticity
β_f : function
ϕ_f : coefficient for differed elasticity
ϕ_d : coefficient for differed plasticity
Creeping strain is then calculated by:

$$\varepsilon_{\phi(t)} = \sigma(t_0) \cdot \frac{\phi(t, t_0)}{E_{c28}}$$

$\sigma(t_0)$: stress applied at (t_0)

With all those results the long-term properties of the beam are assessed under permanent loading with a hypothesis of constant temperature.

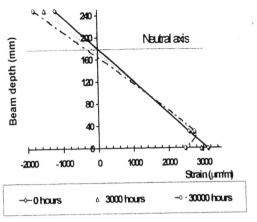

Fig. 14: Strain evolution of a reinforced concrete beam section (composite A, 20°C)

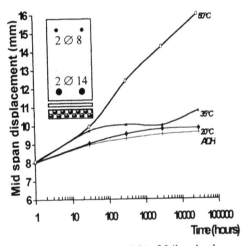

(a) Composite A 60 % of failure load

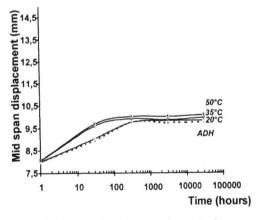

(b) Composite B 60 % of failure load

Fig. 15: Evolution of deflection under creep loading

285

The calculation is done for several temperatures and also with the hypothesis of perfect adhesion. The result shows that the creep adhesive layer behavior can not be neglected in all the cases.

The increase in strain in the polymer creates an augmentation of the curvature of the beam.

This phenomenon is caused by a loss in strength in the composite. The concrete creep also increases the curvature by strain augmentation in steel and composite. All those phenomenon are coupled and modify the differed behavior of the beam (Fig. 15.a, 15.b).

The theoretical study is assessed on a steel reinforced beam (150 mm x 250 mm x 2000 mm) reinforced with composite A, B and C.

The evolution of the deflection is calculated for a load corresponding to 60 % of the failure load. The conclusion of this calculation is that the adhesive layer does not affect the behavior of the beam when the temperature is low. In this case, the difference between the perfect adhesion hypothesis and the sliding effect can be neglected (Fig. 13.b). But for the composite A, when the temperature increases, this sliding effect in the polymer becomes very important in the global behavior of the structure. With a polymer with a low Tg (<50°C), the variations of temperature during permanent loading increases the creep in the adhesive layer and so the curvature or deflections of the beams. With this kind of polymer, the loading should be limited to avoid prematured failure.

4.2 Theoretical and experimental effects of fatigue on the reinforced concrete beams

Theoretical study

The analysis of a reinforced concrete beam strengthened by a composite material plate can be described by a classic method using mechanical behavior law of each material.

The method of non-linear calculation allows the determination of moment-curvature diagrams using the typic hypothesis of reinforced concrete theory in conjunction with the slipping phenomenon of the plate.

Using results from fatigue tests on composite and on adhesive layer, the design of a RC beam reinforced with FRP is determined.

The fatigue response and life prediction of concrete under compression loading is given by Jakobsen [3] for ultimate strength and by Ballgame and Shah [4] for Young modulus (Fig. 8). For steel, Kashani [5] has shown that a steel framework under tensile cyclical loading of 0.95 steel yielding stress has a

loss in Young modulus of 3-5 % for 20.10^3 cycles.

The main research on the fatigue of steel has shown that under the yielding stress the evolution of mechanical properties can be neglected. Taking into account the fatigue response of each material, the evolution of modulus, strength of each material is assessed for any cycle of loading. The mechanical law given by the bibliography (steel and concrete) or by the test (adhesive layer and composite) gives all the differed properties.

The table 5 gives the evolution of the mechanical characteristics during fatigue loading. This result shows that the differed behavior is estimated with a good accuracy.

Table 5. Evolution of mechanical characteristics of each material

		Steel	Concrete	Composite	Adhesive layer	
10^4 cycles	E (MPa)	199000	25000	42750	G (MPa)	1000
	R (MPa)	475	25	560	τ (MPa)	1
10^6 cycles	E (MPa)	178000	15500	40500	G (MPa)	860
	R (MPa)	425	20	550	τ (MPa)	0.80

Experimental study

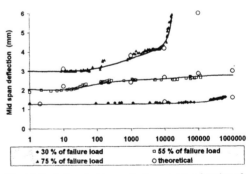

Fig. 16: Evolution of mid span displacement as a function of number of cycles

In the case of loading, the calculated result has been compared with the experimental result (Fig. 14).

The main conclusion is that, with a fatigue loading under a temperature of 20°C, the behavior of the structure is not really affected by the fatigue of the adhesive layer. In this case, the composite (carbon

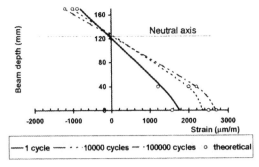

Fig. 17: Strain evolution in a reinforced concrete beam section (55 % of failure load)

Fig. 18: Experimental device of fatigue tests on beams (L2M laboratory University Lyon 1)

epoxy) and the adhesive layer show a good behavior. The main evolution of the deflection is caused by concrete cracking.

The experimental result shows that the durability of the concrete beams can be modeled with a good accuracy, although in the non-linear behavior of the beam (75 % of failure load Fig.14), the modeling of the differed beam properties is more difficult. This can be explained by the concrete cracking and steel anchorage damaging.

The result obtained shows that the yielding of steel of reinforced concrete beam should not be accepted in the design because of the fatigue failure of steel.

4.3 Theoretical effect of aging on the reinforced concrete beams

Aging of polymer

To study the effect of aging on the polymer, a test has been conducted on both polymers B and C. The aging is obtained by 40°C water immersion. The consequence for the polymer is a strength decrease of young modulus. The figure 19 shows the evolution of mechanical properties as a function of time aging.

Theoretical study of adhesive layer aging effect on the reinforced concrete beams

The software used in this study allows taking into account the result given by the aging on the polymer. The calculation is done on the same beam as the creeping study. For the stage of this aging, the decrease of the mechanical properties of the adhesive does not damage the global behavior of the reinforced structure. The increase in displacement is very low: 3 %.

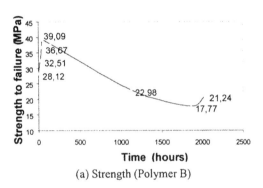

(a) Strength (Polymer B)

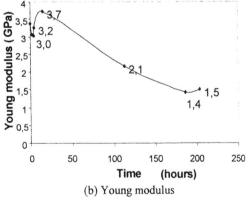

(b) Young modulus

Fig. 19: Polymer mechanical properties evolution as a function of aging (polymer B).

287

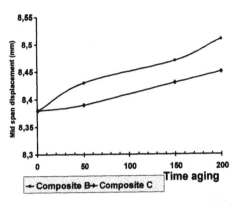

Fig. 20: Influence of the mechanical evolution on a RC beam (composite B and C 60 % of failure load).

This result has to be confirmed with higher aging level on the polymer.

Once again, the difference between the two polymers is noticed. The polymer with higher Tg has a better behavior under aging (Fig.20).

5. CONCLUSIONS

The main conclusions of this study are :

Creep :

The mechanical characteristics of the polymer at the interface between concrete and carbon fibers are evaluated by performing thermo-stimulated creep tests. Maxwell rheological model may be used to assess the creep function identification. The result shows that, beyond a limit, the creep behavior is non-linear.

The validity of the rheological model for an adhesive layer can be justified by the fact that average shear stress in a flexural behavior of the polymer joint is very low.

The determination of the long-term composite reinforcement shear behavior law allows the long-term behavior prediction of a repaired beam by composite fibers. This prediction can be done with non-linear calculation by taking into account the rheological behavior of each material (concrete, composite), but also by taking into account the shear behavior law identified in this study. The long-term slipping effect in the composite plate is carried out with the method of Hamelin and Varastehpour.

This method uses major calculation rules of the reinforced concrete, by taking into account the composite fabric slippage from its support. The design method taking into account the bending creep

in flexural behavior between concrete beams and the composite plate allows designing a safety reinforced structure with high durability.

Fatigue :

This study has shown that:

- the average shear stress for a double lap joint is 0.80 MPa for a lifetime of 10^6 cycles under a loading of 1 s^{-1},
- the average shear stress for carbon epoxy composite plate is 550 MPa for the same lifetime.

The design of a reinforced concrete beam with FRP can be done using $\Delta\tau$ –N to failure curve for double lap joint and $\Delta\sigma$ – N to failure curve for composite plate. The result of the test and the design settle the fact that the adhesive joint and the composite plate are strong enough for a fatigue loading of 10^6 cycles (1 Hz).

The material properties that limit the loading in the case of the fatigue test, is concrete and steel strength. Using the software and mechanical behavior law evolution under cyclic loading the evolution of the mid span displacement of a RC beam with FRP is calculated. Tests on several beams designed with this method verify this design.

The limit load for flexural loading is 26 KN or 60 % of the failure load for a non reinforced concrete beam. Other effects like concrete cracking, normal stress at the end of the plate, or loss in strength in the adhesive layer may modify the classic flexural behavior of a beam. The calculation of the shear stress distribution along the adhesive joint is assessed using the Täljsten analytic solution to take into account that specific behavior.

The safety factor to be applied for a long-term behavior of a reinforced concrete beam can be assessed with those laws and with the software. Suitable range of loads for concrete, steel, composite and interface are assessed to predict the life under fatigue and creep load of the reinforced concrete beams. Flexural tests on RC beams with FRP show that, with suitable length of anchorage determined with the first test and with suitable range loads, the long life of reinforced concrete structure can be predicted with reasonable accuracy.

The identification of concrete-composite shear behavior law with various environmental conditions allows taking into account the combined effect of time – temperature – loading of a reinforced concrete structure.

The same identification can be done for time temperature and moisture.

This effect can be more or less predominant according to the structure exposure (building or civil engineering structure).

The authors believe that the use of high modulus carbon fiber should improve mechanical behavior of beam, by decreasing composite strength and beam curvature evolution. The material properties that limit the loading in case of fatigue test, are concrete and steel strength. Tests on several beams designed with this method verify this design.

The research assessed in this paper shows that polymer with higher Tg than 55°C should be chosen for composite reinforcement in civil engineering structures.

BIBLIOGRAPHY

[1] V. M. Karbhari, *Effect of environmental exposure on the external strengthening of concrete with composites-short term bond durability*, Journal of reinforced plastics and composites, vol 15 – december 1996.

[2] M. Shahawy, T.E Beitelman, *Fatigue performance of RC beams strengthened with CFRP laminates*, CDCC, Sherbrooke (Québec) 1998.

[3] H. Varastehpour, P. Hamelin, *Analysis and study of failure mechanism of RC beam strengthened with FRP plates*, Second international conference on composite material for bridge and structure, Montreal (Canada), 11-15 August 1996.

[4] N. Plevris, T.C Triantafillou, *Time dependent behavior of RC members strengthened with FRP laminates*, Journal of structural engineering, vol 120, n°3, p 1016-1042, March 1994.

[5] A.K. Jakobsen, *Fatigue of concrete beams and columns*, bulletin n°70-1, NTH Institut for beton konstruksjoner, Trandheim, sept 1990.

[6] B. Balagru and S.P Shah, *A method of predicting crack width and deflections for fatigue loading*, Fatigue of concrete structures ACI publications SP-75.

[7] E. Ferrier, P. Hamelin, *Influence of time-temperature-loading on carbon epoxy reinforcement for concrete construction*, FRPRCS-4 ACI, Baltimore, Novembre 1999.

[8] I. M'Bazaa, *Renforcement en flexion de poutres en béton armé à l'aide de lamelle en matériaux composites: optimisation de la lamelles*, Mémoire ES sciences Université de Sherbrooke (Canada), 1995.

[9] B. Täljsten, *Strengthening of beams by plates bonding*, Journal of materials in civil engineering, vol 9, n°4, p 206-212, November 1997.

[10] CEB, *Comportement dans le temps du béton*, CEB-FIP 1978.

Experimental aspects

Recent Developments in Durability Analysis of Composite Systems, Cardon, Fukuda, Reifsnider & Verchery (eds)
© *2000 Balkema, Rotterdam, ISBN 90 5809 103 1*

Performance of pultruded FRP reinforcements with embedded fiber optic sensors

A. L. Kalamkarov, D. O. MacDonald, S. B. Fitzgerald & A. V. Georgiades
Department of Mechanical Engineering, Dalhousie University, Halifax, N.S., Canada

ABSTRACT: Fiber optic strain sensors are successfully embedded in glass and carbon fiber reinforced polymer (GFRP and CFRP) tendons during pultrusion. The specific application is the use of the smart composite reinforcements for strain monitoring in innovative bridges and structures. To verify the operation of the optic sensors embedded in the smart pultruded rods, mechanical tests were conducted and the output of the fiber optic sensors was compared to that of an extensometer. These mechanical tests were performed at room temperature as well as under conditions of low and high temperature extremes. The reliability assessment of the fiber optic sensors further entailed the study of their fatigue and creep behaviour as well as their performance when the rods in which they are embedded are placed in severe environments (e.g. alkaline solutions) which may simulate conditions encountered in concrete structures wherein the composite rods may be used as prestressing tendons and rebars.

1 INTRODUCTION

Fiber optic sensing techniques are currently being investigated for monitoring the strains in critical structures in civil and marine engineering, including for example bridges and hydroelectric dams. Fiber optic strain sensors have many advantages over traditional strain gages including resistance to EMI losses, inherent corrosion resistance, minimal need for cabling, a very small size which implies little or no disturbance to the substance being monitored, and the ability to make absolute strain measurements. Two popular types of fiber optic gages are currently the Fabry Perot sensor and the Bragg Grating sensor.

Fiber optic sensors have been used to monitor the state of structures made up of many different materials including steel, aluminum, concrete and composites. The fiber optic sensors can be surface mounted on a structure or test sample in much the same manner as traditional foil gages. A suitable adhesive is used to bond the sensor to the substrate. The characteristics of the adhesive are very important as it must effectively transfer strain from the substrate to the sensor.

A relatively new field of research involves the production of smart composite materials in which the fiber optic sensing element and accompanying lead are embedded inside the composite material during its fabrication. The embedded sensor is thus well protected from harsh external environments as well as from rough handling during the construction

phases of large projects. Examples of composite structures which can benefit from embedded fiber optic strain sensors include CFRP tendons on cable stayed suspension bridges, GFRP bars for concrete reinforcement in corrosive environments, and GFRP or CFRP prestressing tendons. It has been shown that very effective strain-transfer characteristics from the host composite material to the sensing element are achievable using embedded fiber optic sensors. For example, GFRP and CFRP pultruded smart tendons have shown fiber optic strain measurements that closely match corresponding readings from traditional externally mounted devices such as extensometers, under both static and dynamic loading conditions (Kalamkarov et al., 1998a).

While fiber optic sensors and smart composite materials have shown promise in replacing or strategically complimenting traditional materials and strain gages, there is not a great deal of data available which reflect upon their long term behaviour which, as is the case with all materials, is significantly affected by the surrounding environment. For civil and marine applications, the composite materials and associated fiber optic sensors will encounter, in addition to mechanical stress, external conditions of high and low temperature, humidity, and chemical ion exposure.

To fully understand environmental effects on smart composite structures, one must consider both the individual components and the overall system. The manufacturers of Fabry Perot strain sensors specify that the sensors are reliable over a temperature range of -100°C to +125°C (RocTest, 1997). It has been shown that the decay of reflectivity in an optical fiber is fairly small at temperatures up to 300°C, and that losses are negligible over a life-span of 50 years if the temperature during that period never exceeds 80°C (Erdogan et al., 1994). If the decay of reflectivity becomes substantial due to some adverse combination of moisture exposure and temperature, there may be some drift of the signal from sensitive devices such as Fabry Perot and Bragg Grating sensors. However, specialized fiber optic sensors can be manufactured to operate at elevated temperatures, although the high associated costs limit the use of such sensors to mainly aerospace applications. For example, Wang et al. (1994) demonstrated and tested a sapphire fiber-based polarimetric sensor which was operational at temperatures in excess of 1000°C.

It has also been documented (Habel et al., 1994) that optical fibers and their protective coatings are susceptible to attack by exposure to certain chemicals. The authors studied the effect of highly alkaline solutions (typical in concrete) on the integrity of several coating materials normally applied to the bare optical fiber. It was discovered that at pH values of 11-14, polyimide coatings were seriously degraded, whereas acrylate coatings were only slightly affected. Fluorine thermoplastic coatings showed no evidence of degradation when exposed to alkaline solutions. In conclusion, neither acrylate nor polyimide coated optical fibers were considered suitable for direct contact with cement and concrete mixtures. This conclusion can be extrapolated to the fiber optic sensors as well. However, if embedded in a protective composite material, these sensors could be incorporated in concrete structures and thus monitor the health of such structures.

In addition to the behaviour of the sensors themselves, the effects of temperature, humidity and other environmental factors on the commonly used fiber optic coating materials must also be assessed. It has been documented that the structural integrity of many polymeric materials used as fiber optic coatings is often jeopardized when exposed to elevated temperatures. A review by Leka and Bayo (1989), discusses the findings of several researchers. It was discovered that acrylate coated optical fibers could not sustain temperatures higher than 85°C, whereas polyimide coated optical fibers could withstand temperatures as high as 385°C. Kalamkarov et al. (1998b) reported similar findings with respect to acrylate and polyimide coated optical fibers. Gunther et al. (1994), demonstrated the ability to coat optical fibers with high-temperature nickel-based super alloys for use in temperatures approaching 1000°C which would be encountered in advanced aerospace and energy applications.

In considering the overall system, one ought to look at the behavior of the composite material over extended periods of time. Unlike the actual fiber optic sensors, the environmental performance of composite materials has been extensively characterized for many years in published journals and books. For example, it is known that moisture is absorbed into a composite through diffusion into the matrix (Mallick, 1988). Typically, the absorption rate increases initially, but reaches a saturation level after several days. The absorbed moisture can cause dimensional changes in the composite as well as a reduction in its glass transition temperature. Although moisture absorption may have a negligible effect at room temperature, it causes a drastic reduction in the properties of a composite (elastic modulus for example) at elevated temperatures. Interlaminar shear strength is also strongly dependent on moisture absorption. Contrary to composite materials, the overall effects of absorbed moisture on an embedded sensor has not been characterized to this date. Elevated temperatures in the absence of moisture can also degrade the composite properties. The matrix dominated properties such as the off axis strength in a unidirectional composite are affected as the temperature approaches the glass transition temperature for the polymer matrix. The strength and stiffness of a composite material are not adversely affected by low temperature exposure, although it may be more prone to impact damage. Again, the overall effects of temperature extremes on an embedded fiber optic sensor have not been characterized to this date.

A comprehensive reliability assessment of the embedded fiber optic sensors should envelop more than the effects of environmental factors. The fatigue behaviour of the smart tendons should also be examined if the fiber optic sensors are to successfully replace foil gages or extensometers as strain-monitoring devices.

It is known that metallic materials, which are normally ductile in nature, can fail in a brittle fashion when subjected to repeated cyclic stresses. In general, composite materials are very resistant to fatigue damage and are used as a replacement for traditional metals in highly fatigued environments. One such common application was the replacement of the steel leaf springs with graphite epoxy composite in the rear suspension of a Ford passenger van. Composites are fatigue resistant because a high fracture energy is needed to propagate a crack in a direction perpendicular to the reinforcing fibers.

For the current research involving the placement of fiber optic strain sensors within the pultruded composite tendons, it is useful to determine if the sensors themselves can withstand the effects of repeated cyclic sinusoidal loading. It will be of interest to determine if the sensors still perform adequately and provide a strain readout after enduring a large number of stress cycles, and if the sensors retain their accuracy and repeatability. It is also of interest to determine if the embedded sensors adversely affect the normal fatigue performance of the host composite material. Some related work has been previously published. Badcock and Fernando (1994) stated that an embedded optical fiber sensor did not affect the S-N curve of a graphite epoxy laminate under tension-tension fatigue with a stress ratio of 0.1. However, there was some discrepancy in the fatigue properties under tension-compression fatigue loads. These authors also reported that when the data from their embedded sensor was compared to that from an extensometer, there was good agreement up until 100,000 cycles, after which the sensor data began to develop some scatter. They attributed the scatter to the debonding of the sensor from the composite matrix. Friebele et al. (1996) published test data obtained from embedded sensors within a pultruded C-channel component. They carried out a range of fatigue test protocols including cyclic loading from 0 to 4000 Kg within a period of 6 seconds. They found that there was good agreement between the embedded optical sensors and externally bonded strain gages. However, testing was carried for a short duration of 100 cycles only. As well, Yaniv et al. (1994) also described the tension fatigue properties of a glass-epoxy composite with an embedded fiber optic sensor. Their tests were conducted at a stress ratio of 0.1 and a frequency of 0.1 Hz. Again, good correlation was reported for the data obtained from the embedded sensor and an externally affixed extensometer. These data were also reported for a low number of cycles.

Another important issue which must be examined in relation to the reliability analysis of fiber optic sensors embedded in smart composite materials is their creep behaviour. It is known that both composites and more traditional materials such as steel can exhibit creep behaviour, although it is often related to elevated service temperatures.

Sullivan (1994) offered a concise review of the measurement of composite creep properties. As composites are viscoelastic materials, primarily as a consequence of a polymeric matrix, they are particularly susceptible to undesirable deformations. The magnitude of the viscoelastic deformations under a given loading condition is dependent upon such factors as the stress level, loading frequency, temperature, and humidity. The creep properties of individual composites must be understood at the design stages of a project, with consideration to both the load and the environmental conditions.

Tuttle and Brinson (1986) had conducted work to determine the creep properties of graphite-epoxy composites. Their goal was to measure the short term viscoelastic behaviour of the material through a laboratory creep test, and compare these data with analytical results from a numerical model. While this numerical model was successful in predicting creepage within 10% of experimental data, the authors stressed that it was necessary to link the viscoelastic deformations to damage and fracture mechanisms within the composite. Gates et al. (1986) also conducted similar work in which short term creep tests were used to obtain material parameters. These parameters were then used to predict the long term creep behaviour of the composite material by means of a mathematical model. Long term creep tests were subsequently used to compare with the predicted values.

Two types of tests are usually performed to obtain data reflecting the creep properties of composite materials; creep tests and stress relaxation tests. To conduct a creep test, a sample is subjected to a constant stress while the strain is measured as function of time. Conversely, in a stress relaxation test, the sample is subjected to constant strain and the stress is measured as a function of time. Kaci (1995) studied the stress relaxation behaviour of kevlar fiber reinforced prestressing cables, and compared the results to corresponding data pertaining to standard steel prestressing cable. It was observed that after 120 hours, the stress relaxation percentage of Kevlar was about four times that of traditional steel. After 1000 hours however, this ratio is reduced from 4 to 1.5, and after an even longer period the trend was reversed with the relaxation percentage of steel now exceeding that of kevlar.

However, no references in the literature were found which describe the use of embedded fiber optic strain sensors to monitor the creep behaviour of composite materials. It would be of interest to be able to characterize the long term creep behaviour of the embedded fiber optic sensors and determine their suitability for monitoring long term service conditions.

2 MATERIALS AND EQUIPMENT

Central to the current study was the fabrication of carbon and glass FRP tendons using pultrusion. 9.5 mm-diameter rod stock was produced using a customized pultrusion machine (Kalamkarov et al., 1997a,b, 1998a,b and 1999). The reinforcing fibers (E-glass or carbon rovings) were pulled from a creel system into a shaping die. The die is equipped with three temperature zones, each with its own PID temperature controller. Prior to entering the die, the fibers are wet out in a dip-type wet bath and then distributed evenly over the rod's cross section by a series of specially machined high density polyethylene cards. One roving traveled a straight path through the cards to the center of the rod. This center roving was used to carry the optical fiber in experiments where a fiber optic sensor was required to be embedded in the composite rod. The pulling force was maintained by a set of counter rotating wheels which provided consistent pulling speeds.

The fiber glass rovings used are continuous E-glass filaments formed into a single end reinforcement, free from catenary and treated with 0.45 nominal wt.% sizing which is a silane based and compatible with most resin systems. The carbon rovings used had a standard epoxy-based sizing. The rods were produced using a urethane modified bisphenol-A based vinyl ester resin system known for its good mechanical properties and excellent processability. Two types of organic peroxide catalysts were used to cure the resin, di-peroxydicarbonate and tert-butyl peroxybenzoate. Adequate release from the die was achieved by using an internal lubricant. The 9.5 mm diameter carbon rods were pulled with 22 ends of rovings giving a volume fraction of 62.5%, while their glass counterparts were pultruded with 26 ends giving a volume fraction of 64%. Finally, die temperatures of 120°C, 150°C, and 120°C at the three zones, as well as a pulling speed of around 25 cm per minute were found to produce good quality FRP rods.

3 FIBER OPTIC SENSORS

In this study, the first fiber optic sensors that were embedded during the pultrusion of carbon fiber reinforced rods were of the Fabry Perot type. These sensors and the demodulating equipment are currently "off the shelf" items (RocTest Ltd., 1997). The Fabry Perot sensor has been developed to use a broadband light source as opposed to laser light. It is highly sensitive and can make precise, linear, and absolute measurements. Two multimode fibers are inserted and fused into a larger glass capillary tube with an overall diameter of 200-250 microns. The ends of the fibers which are inserted into the capillary are polished and contain a semi- reflective coating. The distance between the fused locations defines the gauge length of the sensor. The sensor is designed such that a predefined gap exists between the two polished optical fiber ends within the capillary tube. Hence, some of the light introduced to the sensor reflects from the end of the lead-in fiber, while some travels through the air gap and reflects from the second polished fiber end. In the reflective mode of operation, both reflections are transmitted back through the lead-in fiber to a detector. As external forces are applied to the sensor, the length of the air gap changes and hence, so does the phase difference between the two reflections. Several demodulation techniques are available to evaluate this phase difference and relate it to strain. One such device which uses a Fizeau interferometer to aid in the measurement of the Fabry Perot cavity

length is described in more detail by Belleville and Duplain (1993).

The other type of fiber optic sensors used was of the Bragg Grating type. Bragg Grating sensors are based on creating a pattern of refractive index differentials directly onto the material of the fiber core. This may be achieved by directing two laser beams operating in the ultraviolet, into the fiber from the side. An interference pattern results with alternating bright and dark fringes. At the zones of constructive interference, permanent optical damage is induced at sites occupied by germanium atoms as a result of the intensity of the ultraviolet light. This changes the refractive index of the glass material and creates a periodic pattern in the fiber which resembles a diffraction grating. Fiber gratings selectively reflect certain wavelengths and transmit others (Electrophotonics Corp., 1996). Which wavelengths are transmitted and which ones are reflected depend on both the refractive index of the core material as well as the spacing of the pattern. Changes in temperature or pressure will change the refractive index of the core material and hence cause a change in the wavelengths of peak reflection (or transmission). The presence of mechanical strain along the length of the fiber will have a similar effect since it will change the grating spacing. Measurements of these wavelength shifts provide the basis of operation of Bragg grating sensors (Electrophotonics Corp., 1996).

4 EXPERIMENTAL AND DISCUSSION

To assess and characterize the overall behavior of the embedded fiber optic sensors, mechanical testing of the pultruded tendons was carried out in ordinary laboratory conditions by applying various proof loads to the tendons while continuously monitoring strain via the embedded optical sensors and a standard extensometer clipped to the pultruded rod.
The smart FRP tendons were subjected to two basic wave forms in order to evaluate their performance and begin their reliability assessment. The first waveform was a trapezoidal waveform whereby the load was ramped from a low value (typically 100 N) to a peak value of about 3000 to 11000 N, at a slow rate of 90 N/sec. The load was held at this level for 20 seconds and then ramped back down to the initial value at the same rate. The second waveform to which the smart tendons were subjected was a sinusoidal one. The frequency was one cycle per minute (0.0167 Hertz), and a typical range through which the load was cycled was from 400 to 5000 N.

Figure 1 shows the results from a sinusoidal test performed on a GFRP tendon. The data are plotted as microstrain vs. time. The plot shows that the profile of the strain output from the extensometer follows very closely the profile of the corresponding

Microstrain

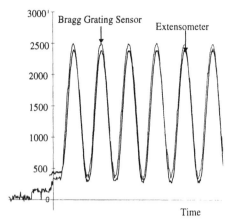

Figure 1: Strain vs. time plot from extensometer and Bragg Grating Sensor in a glass tendon subjected to a sinusoidal load (5000 N)

sensor readings. In addition there is a very high degree of conformance between the strain readings from the two devices over the entire load range. At the peak loads, the discrepancy is less than 100 microstrain (about 5%), which is quite reasonable given the resolution of the extensometer and the Bragg Grating sensor. Many more tests (trapezoidal and sinusoidal) were performed on both GFRP and CFRP tendons with embedded Bragg Grating sensors (Kalamkarov et. al., 1998a). The results of the experiments indicate that the strain output from the sensors was accurate and consistent and agreed well with that from extensometers.

Figure 2 shows the results from a trapezoidal test performed on a GFRP tendon this time with an embedded Fabry Perot sensor. This microstrain vs.

Microstrain

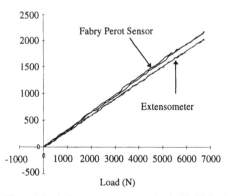

Figure 2. Strain from extensometer and embedded Fabry Perot sensor in a glass FRP tendon subjected to a trapezoidal load.

Microstrain

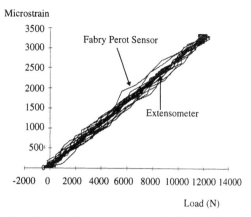

Figure 3. Strain from extensometer and embedded Fabry Perot sensor in a CFRP tendon subjected to a sinusoidal load at 60°C.

load graph illustrates that there is a good agreement between the Fabry Perot sensor and the extensometer. The same conclusion can be reached from many other experiments performed on CFRP and GFRP tendons with embedded Fabry Perot sensors (Kalamkarov et. al., 1998a).

One of the primary objectives of the present research was the study of the behaviour of the fiber optic sensors when the tendons in which they are embedded are exposed to both low and high temperature extremes. The experiments entailed subjecting the GFRP and CFRP tendons to sinusoidal and trapezoidal load waveforms of about 11 KN magnitude inside a temperature chamber. The temperature in the chamber was varied from -40°C to +60°C in increments of 20°C.

Figure 3 shows a microstrain vs. load graph for a GFRP smart tendon with an embedded Fabry Perot sensor subjected to a sinusoidal load waveform at 60°C. The sensor strain output is compared to that from an extensometer. As Figure 3 indicates, there is a remarkable agreement between the two strain monitoring devices over the entire load range.

Figure 4 shows a microstrain vs. time graph for a CFRP tendon with an embedded Fabry Perot sensor at -40°C. In this case however, the sensor data were not compared to the extensometer data but rather to strain values calculated using an experimentally determined value of 144.2 GPa for the tensile modulus of the carbon tendon (Kalamkarov et al., 1998b). It can be seen that there is a good agreement between the sensor output and the theoretical data, with a maximum discrepancy of about 9% at the peak load.

As was mentioned earlier, another major objective of the research was to study the fatigue behaviour of the fiber optic sensors, by examining the performance of the smart FRP reinforcements under conditions of cyclic loading. To this end, glass and carbon FRP

Microstrain

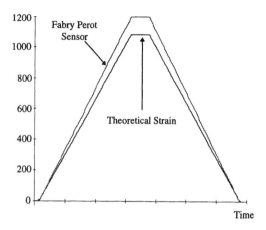

Figure 4. Strain vs. time plot from theory and Embedded Fabry Perot sensor in a CFRP tendon subjected to a trapezoidal load at -40°C.

Microstrain

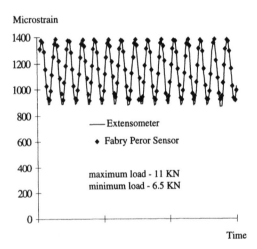

Figure 5. Strain from extensometer and embedded Fabry Perot sensor in a CFRP tendon subjected to tension-tension fatigue for 70,000 cycles.

Microstrain

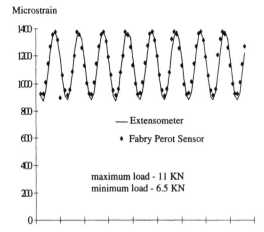

Figure 6. Strain from extensometer and embedded Fabry Perot sensor in a CFRP tendon subjected to tension-tension fatigue for 140,000 cycles.

Microstrain

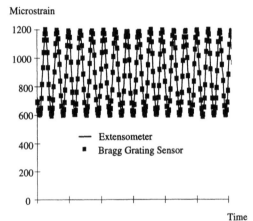

Figure 7. Strain from extensometer and embedded Bragg Grating sensor in a CFRP tendon subjected to tension-tension fatigue for 350,000 cycles.

tendons with embedded Fabry Perot and Bragg Grating sensors were subjected to a sinusoidal load waveform of 1 Hz frequency and ranging in magnitude from 6.5 KN to 11 KN (a stress ratio of 0.6). Testing was carried out for a duration of 140,000 to 350,000 cycles. The strain values from the embedded sensors were compared to those from externally mounted extensometers.

Figure 5 shows the results after 70,000 cycles of load applied to a CFRP tendon with an embedded Fabry Perot sensor. Evidently, there is an excellent degree of conformance between the extensometer data and the sensor data after 70,000 cycles.

The fatigue test to which Figure 5 pertains was continued beyond the 70,000 cycles to examine any potential changes in the performance of the Fabry Perot sensor. The results are shown in Figure 6. Even after 140,000 cycles the output of the sensor is consistent and accurate and conforms extremely well with that from the extensometer.

Similar fatigue tests were performed on CFRP and GFRP tendons with embedded Bragg Grating sensors. Figure 7 shows the results from such a test performed on a CFRP tendon. Notice that this test was carried out for 350,000 cycles. As for the Fabry Perot sensor, the strain output of the Bragg Grating sensor remains unaffected by the application of many load cycles.

Microstrain

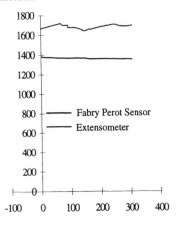

Time (hours)

Figure 8. Carbon FRP tendon with a sustained 11.5 KN load for 350 Hours.

Microstrain

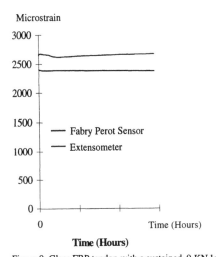

Time (Hours)

Figure 9. Glass FRP tendon with a sustained 9 KN load for 140 Hours.

Microstrain

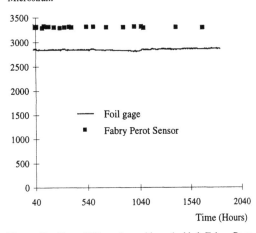

Time (Hours)

Figure 10. Glass FRP tendon with embedded Fabry Perot sensor (in an alkaline solution) with a sustained load of 11 KN.

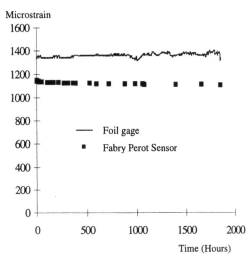

Time (Hours)

Figure 11. Carbon FRP tendon with embedded Fabry Perot sensor (in an alkaline solution) with a sustained load of 11 KN.

In addition to fatigue, another objective of the research is to characterize the long term creep behaviour of the embedded fiber optic sensors and assess their suitability for monitoring long term service conditions. The goal is not only to characterize the composite tendons themselves, but rather to characterize the interaction of the tendon and the embedded sensor under conditions of sustained load levels.

Both carbon and glass reinforced smart tendons were chosen for short term creep testing. Both tendons contained an embedded Fabry Perot sensor. The applied load was 9 KN (for a period of 140

hours) for the case of the glass tendon, and 13.5 KN (for a period of 350 hours) for the case of the carbon tendon. The results of these tests are shown in Figures 8 and 9. It can be seen that the output from the Fabry Perot sensors is practically constant over the entire duration of the tests, and thus it may be concluded that they do not exhibit short term creep behaviour at the given experimental conditions. They have the potential for significant benefit in the long term monitoring of strain levels in composite prestressing cables.

Longer term testing was subsequently performed with the object of gaining more detailed insight into the long term behaviour of the tendons and the

299

sensors. This long term testing was conducted in a caustic environment which may simulate conditions encountered in concrete structures wherein the composite rods may be used as prestressing tendons.

Both glass and carbon FRP tendons with embedded Bragg Grating and Fabry Perot sensors were chosen for this testing. The tendons were located inside an environmental chamber through which an alkaline solution was circulated by a pump. The solution (pH. of 12.8) was composed of 0.32 mol/L KOH, 0.17 mol/L NaOH and 0.07 mol/L $Ca(OH)_2$. The tendons were also subjected to a constant load of about 11 KN. Testing was carried out for a period of 2.5 months. Figures 10 and 11 show the results pertaining to the tendons with the embedded Fabry Perot sensors.

In Figures 10 and 11 the strain from the embedded Fabry Perot sensors was plotted versus that from externally bonded foil gages. It can be seen that there is a good agreement between the optical sensors and the gages. More important however is the fact that the combination of the highly alkaline solution and sustained load has no effect on the behaviour of the embedded sensors within the duration of the tests.

The experiments pertaining to the FRP tendons with the embedded Bragg Grating sensors are still on-going.

5 CONCLUSIONS

The present study is concerned with the pultrusion of smart FRP materials in which the fiber optic sensing element and accompanying optical fiber are embedded inside the composite material during its processing.

Mechanical testing was carried out in order to assess the overall behaviour of the smart FRP tendons and compare the performance of the embedded optical sensors to that of traditional strain monitoring devices such as extensometers. Glass and carbon FRP tendons with embedded Fabry Perot or Bragg Grating sensors were subjected to trapezoidal and sinusoidal load inputs at room temperature. In all cases there was a very good agreement between the sensor and the extensometer.

GFRP and CFRP tendons were subsequently subjected to sinusoidal and trapezoidal load waveforms under conditions of high (up to 60°C) and low (up to -40°C) temperature, and the strain output from the embedded Fabry Perot fiber optic sensors was compared to that from externally mounted extensometers as well as to theoretical strain values. It was determined that the strain output from the Fabry Perot sensor showed an excellent agreement with the corresponding output from the extensometer. Thus, it can be concluded that the performance of embedded Fabry Perot sensors was

not affected by ambient temperatures falling within the range of -40°c to +60°C.

Fatigue testing of the smart tendons entailed subjecting them to 140000-350000 cycles of load varying in magnitude from 6.5 KN to 11 KN. The output from the embedded Fabry Perot and Bragg Grating sensors showed an excellent conformance with that from the extensometers. Thus, many cycles of applied load had no effect on the performance of the sensors.

Other tests indicated that embedded Fabry Perot sensors do not exhibit either short-term or long-term creep behaviour even when the tendons in which they are embedded are tested in alkaline environments. Thus the sensors have the potential for significant benefit in the long term monitoring of strain levels in composite prestressing cables, tendons and rebars.

ACKNOWLEDGMENT

This work is supported by ISIS-CANADA, the Canadian Network of Centres of Excellence on the Intelligent Sensing for Innovative Structures, through the Project T3.4 on Smart Reinforcements and Connectors.

REFERENCES

Badcock, R.A. & G.F. Fernando 1995. Fatigue damage detection in carbon fiber reinforced composites using an intensity based optical fiber sensor. *SPIE* 2444: 422-431.

Belleville, C. and G. Duplain 1993. White-light interferometric multimode fiber-optic strain sensor. *Optics Letters*, 18(1): 78-80.

Electrophotonics Corporation. 1996. User Manual for FLS 3000, Concord, Ontario, Canada

Erdogan, T., Mizrahi, V., Lemaire, P.J., and D. Monroe 1994. Decay of ultraviolet-induced fiber Bragg Gratings. *Journal of Applied Physics*, 76(1): 73-80.

Friebele, E.J., Askins, C.G., Putnam, M.A., Heider, P.E., Blosser, R.G., Florio, J., Donti, R.P., and J. Garcia 1996. Demonstration of distributed strain sensing in production scale instrumented structures. *SPIE* 2721: 118-124.

Gates, T.S., Veazie, D.R. and L.C. Brinson 1997. Creep and physical aging in a polymeric composite: Comparison of tension and compression. *Journal of composite materials* 31(24): 2479- 2505.

Gunther, M.F., Zeakes, J. S., Leber, D. E., May, R. G., and R. O. Claus 1994. Sputtered metallic coatings for optical fibers used in high temperature environments. *SPIE* 2191: 2-12.

Habel, W.R., Höpcke, M., Basedau, F., and H. Polster 1994. The Influence of concrete and alkaline solutions on different surfaces of optical fibers for sensors. *SPIE* 2361: 168-179.

Kaci, S. 1995. Experimental study of mechanical behavior of composite cables for prestress. *Journal of Engineering Mechanics*, 121(6): 709-716.

Kalamkarov, A.L, Fitzgerald, S., and D.O MacDonald 1997a. On the processing and evaluation of smart composite reinforcement. *SPIE* 3241: 338-346.

Kalamkarov, A.L., MacDonald, D.O., and P. Westhaver 1997b. On pultrusion of Smart FRP composites. *SPIE* 3042: 400-409.

Kalamkarov, A.L., Fitzgerald, S.B., MacDonald, D.O., and A.V. Georgiades 1998a. Smart pultruded composite reinforcements incorporating fiber optic sensors. *SPIE* 3400: 94-105.

Kalamkarov, A.L, Liu, H.Q., and D.O. MacDonald 1998b. Experimental and analytical studies of smart composite reinforcements. *Composites Part B: Engineering* 29B(1): 21-30.

Kalamkarov, A.L., Fitzgerald, S.B., and D.O. MacDonald 1999. The use of Fabry Perot fiber optic sensors to monitor residual strains during pultrusion of FRP composites. *Composites Part B: Engineering* B30(2): 167-175.

Leka, L.G. and E. Bayo 1989. A close look at the embedment of optical fibers into composite structures. *Journal of Composites Technology & Research*, 11(3): 106-112.

Mallick, P.K. 1988. *Fiber Reinforced Composites, Materials, Manufacturing, and Design.* Marcel Dekker Inc., New York.

RocTest Ltd. 1997. *Technical Information, Equipment, and Manuals.* St-Lambert, PQ, Canada.

Sullivan, J.L. 1991. Measurement of composite creep. *Experimental Techniques*: 32-37

Tuttle, M.E. and H.F. Brinson. 1986. Prediction of the long-term creep compliance of general composite laminates. *Experimental Mechanics* 26(1): 89-102.

Wang, A., Zhang, P., May, R.G., Murphy, K.A., and R.O. Claus 1994. Sapphire fiber-based polarimetric optical sensor for high temperature applications. *SPIE* 2191: 13-22.

Yaniv, G, Zimmermann, B and K. Lou 1993. Development of an optical fiber time domain sensor for monitoring static and fatigue strains in composite laminates. SPIE 1918: 377-387.

Recent Developments in Durability Analysis of Composite Systems, Cardon, Fukuda, Reifsnider & Verchery (eds)
© 2000 Balkema, Rotterdam, ISBN 90 5809 103 1

AE measurement in acid corrosion

M. Maeda & H. Hamada
Kyoto Institute of Technology, Japan

Y. Fujii
Seikow Chemical Engineering and Machinery Limited, Amagasaki, Japan

ABSTRACT: Environmental creep test for E glass and C glass reinforced plastics were performed. Acoustic emission technique was used to detect the fracture propagation in this study. As a result, it was indicated that the durability of GFRP depended on the types of glass fiber. C glass reinforced plastic had much longer lifetime than E glass one. Furthermore, it was cleared that the fracture behaviors of both were different. Acid stress corrosion was occurred in E glass specimen, and interface was damaged in C glass specimen.

1 INTRODUCTION

Glass fiber reinforced plastics (GFRP) are widely used as an anti-corrosive material in the chemical process industry for application such as pipe work, reaction vessel, storage tanks, pumps, fans, scrubbers, etc. However, serious accident occasionally happens when it is used under loads for long period and a few cases have been reported. For example, Norwood L.S. and Hogg P.J.1984 reported a case study of GFRP tanks damaged by acid. Furthermore, Trevett, A. 1996 reported some cases caused by acid stress corrosion even in much lower level than the standard. To prevent from such severe accidents, it is determined to design tanks less than about 10% of the tensile strength. The acid stress corrosion of GFRP is related to the corrosion of the glass fiber as reinforcement. It has been considered that matrix resin works as resistance in composites and prevents from that the glass fiber is attacked by acid. However, once the acid reaches to the glass fiber, for example through cracks, it could be damaged. E glass normally used in GFRP is easily damaged by acid, and it is well known that C glass resists in acid condition. It is suggested that types of glass fiber greatly related to durability of GFRP laminates. However, no result on GFRP using C glass was reported. Moreover, durability of GFRP is normally evaluated by long term testing. It was suggested that the damage propagation in early period greatly affect the lifetime. To estimate their lifetime by short term testing should become a useful method in industry. Fujii, Y 1996 reported the technique to use acoustic emission during environmental creep test. In this study, creep test with acoustic emission (AE) monitoring was performed to determine the fracture mechanism of E and C glass reinforced plastics and relationship between fracture behavior in early period and lifetime was considered.

2 MATERIALS AND EXPERIMENTALS

Glass fiber reinforced vinyl ester laminates used in this study were fabricated by a hand lay-up method. Vinylester resin (R806; Showa high polymer Co. Ltd.) was used as matrix. The glass fiber reinforced vinylester laminates used in this study were fabricated by a hand lay-up method. Four types of GFRP were used in this study. E type woven cloth(YEM2103-N7, Mie Textile Co.)and C types woven cloth(WF230C100BS6, Nittobo Co.) reinforced composites were made of 12 plies, 3mm thick and E type chopped strand mat(ECM-450-193, Central Glass) and ECR type chopped strand mat(ECRM723, OCF) reinforced composites were made of 3 plies, 3mm thick. E type glass is used for electric product and corroded by acid. On the other hand, C type glass is used for chemical product and anti-corrosive material. Each specimen was named E mat, C mat, E cloth, and C cloth respectively. All laminates were post cured at 100℃ for 1 hour. Dumbbell shaped specimens were prepared accord-

Table 1. Mechanical property of specimens

	Elastic Modulus(GPa)	Tensile Strength(MPa)
E cloth	16.8	271
C cloth	16.2	208
E mat	7.54	97.1
C mat	8.00	103

ing to the ASTM D638. Polyvinylchloride (PVC) end tabs were glued to the specimen ends using an epoxy adhesive to reduce the noise of AE. The static mechanical property of each specimen is shown in Table 1.

The details of the creep test specimen are shown schematically in Figure. 1. The creep testing machine used in this study was a lever action type. The load on the specimen was varied by changing the dead weight. A glass tube with a rubber stopper was used to hold environmental solution. 5wt% nitric acid was used as an environmental solution. The strain gauge was mounted non-immersed position on the specimen for measuring longitudinal tensile strain. Each specimen was initially tested in air for 2hrs until the acoustic emission activity reached a steady state and then tested in environment until the specimen ruptured. Several specimens were tested at different creep stress levels. AE signals were detected using a piezoelectric transducer with a resonant frequency of 150KHz. The threshold level, that AE signals were not detected for one hour without load, was 56mv. The sensor was mounted on the specimen. Only the ring down counts were measured and recorded using an X-T recorder. All specimens were tested at room temperature. The fracture surfaces were observed by scanning electron microscopy (SEM). In this study, creep strain generally measured during the test was not monitored. Acid stress corrosion is due to the micro fracture occurred specific parts in the specimen, so that it was very difficult to consider the fracture behavior from the strain of whole specimen.

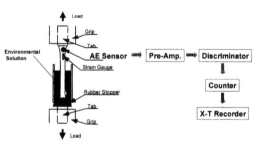

Fig. 1. Schematic illustration of environmental creep test and AE monitoring

3 RESULT AND DISCUSSION

3.1 Glass cloth reinforced plastics

Figure 2 shows the AE rate for E and C cloth specimens at 70MPa. AE rate increased drastically after acid solution added for E cloth specimen. On the other hand, C cloth was not affected by adding acid solution. E cloth was failed at 19hr but C cloth had a longer lifetime. Figure 3 shows the

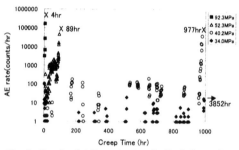

Fig. 2 AE rates for E and C cloth specimens

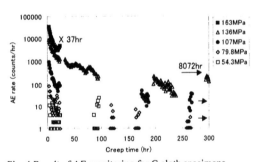

Fig. 3 Result of AE monitoring for E cloth specimens

Fig. 4 Result of AE monitoring for C cloth specimens

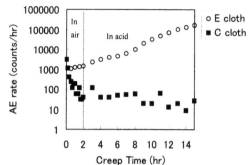

Fig. 5 E glass fiber after immersion for 8hr

result of AE monitoring for E cloth specimens. The AE rates in early period increased with increasing and the lifetimes indicated in the Figure decreased with increasing the load level. Figure 4 shows the result of AE monitoring for C cloth specimen. AE rate did not change with adding acid solution. Furthermore, the AE rate became less. Figure 5 shows the E glass fiber after immersion. Spiral cracks were found. It was cleared that the acid solution reached and attacked only E glass fiber, which is easily attacked by acid in E cloth. On the other hand, AE activity did not increase for a short period, when the reinforcement was not damaged in C cloth. The AE counts detected on C cloth specimen at low stress level would be caused by interfacial damage. Figure 6 and 7 show SEM photographs of fracture surface for E cloth and C cloth specimens respectively. From SEM observation, flat fracture surface, which was one evidence of the acid stress corrosion, was observed in E glass specimens. It was suggested that AE rate sensitively detected the propagation of the acid stress corrosion. On the other hand, the fiber-pull-out fracture was observed in C cloth specimens. From this point, it was also clear that the resources of AE counts in C cloth specimen would be damages in interface. Relationship between stress and lifetime is shown in figure 8. It was clear that the lifetime increased with decreasing the stress in both specimens and C cloth specimens had much longer lifetimes. The fracture mechanisms were different in E and C cloth specimens and it should have caused that C cloth specimens had longer lifetime compared with E cloth specimen at same stress level. Relationship between AE rate in early period and lifetime is shown in figure 9. Correlation between AE rate and lifetime was clearly found and the lifetime could be predicted from this figure. It means the possibility to predict lifetime of specimen by AE monitoring.

3.2. Glass chopped strand mat reinforced plastics

Figure 10 shows the result of AE monitoring for E mat specimens. AE rate increased drastically after acid solution added. The AE rates in early period increased with increasing and the lifetimes indicated in the figure decreased with increasing the load level. Figure 11 shows the result of AE monitoring for C mat specimen. AE rate did not change with adding acid solution. Furthermore, the AE rate became less. It was cleared that the acid solution attacked only E glass fiber, which is easily attacked by acid. On the other hand, AE activity did not increase, when the reinforcement was not damaged. The AE counts detected on C mat specimen at low stress level would be mainly caused by interfa-

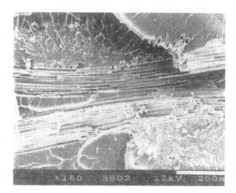

Fig. 6 SEM photograph of E cloth fracture surface
Stress; 34.0MPa Lifetime; 3852hr

Fig. 7 SEM photograph of C cloth fracture surface
Stress;137MPa Lifetime; 8072hr

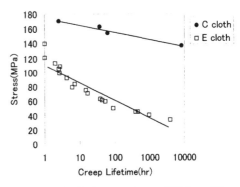

Fig. 8 Relationship between stress and lifetime of E and C cloth

cial damage. Figure 12 and 13 are SEM photographs of both specimens. The flat fracture surface, which was one evidence of the acid stress corrosion, was observed in E mat specimens. It was suggested that AE rate sensitively detected the propa-

305

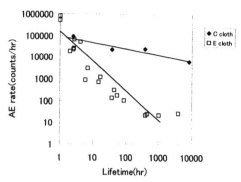

Fig. 9 Relationship between AE rate and lifetime of E and C cloth

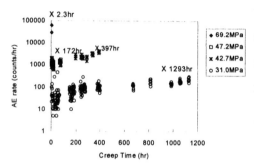

Fig. 10 Result of AE monitoring for E mat specimen

Fig.13 SEM photograph of C mat fracture surface
Stress; 52.1MPa Lifetime; 4296hr

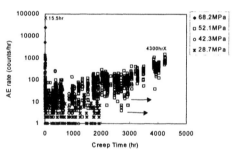

Fig.11 Result of AE monitoring for C mat specimens

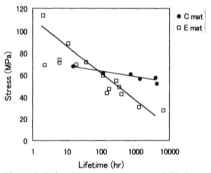

Fig.14 Relationship between stress and lifetime of E and C mat

gation of the acid stress corrosion. On the other hand, the fiber-pull-out fracture was observed in C mat specimens. From this point, it was also clear that the resources of AE counts in C mat specimen would be damages in interface. The AE rate increased and the lifetimes decreased with increasing the load. The relationship between stress and lifetime is shown in figure 14. The fracture mechanisms were different in E and C mat specimens and it should have caused that C mat specimens had longer lifetime compared with E mat specimen at same stress level. However, E mat had longer lifetime than C mat at high stress level over 60MPa. Relationship between AE rate in early period and lifetime is shown in figure 15. Correlation between AE rate and lifetime was clearly found and the lifetime could be predicted from this figure. It means the possibility to predict lifetime of specimen by AE monitoring.

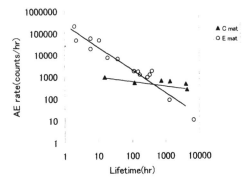

Fig. 15 Relationship between AE rate and lifetime of E and C mat

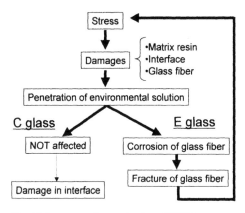

Fig. 16 Stress corrosion mechanism of GFRP

Jones, F.R.,. Rock, J.W & Wheatley, A.R., 1983, Stress Corrosion Cracking and its Implications for the Long-Term Durability of E glass Fber Composites *Composites*, **14**,3, 262-269

Norwood, L.S. & Hogg, P.J., 1984, GRP in Contact with Acidic Environments-a case study, *Composite Structures* **2**, , 1-22

Trevett A., 1996, Rescky receptacles,*Chem. Eng. (GBR)***620** 27-30

4 CONCLUSION

The difference of fracture mechanisms between E glass and C glass reinforced plastics were found and summarized in figure 16. Stress caused damages mainly in matrix resin at first and environmental solution penetrated through the damages. E glass fiber was easily attacked by acid, and it was detected sensitively by AE monitoring. On the other hand, as C glass was not corroded by acid, damage was not propagated further and had longer lifetime than E glass one. However, interface around C glass fiber was damaged. The correlation between AE rate during creep test and lifetime was found and the possibility to predict the lifetime by monitoring AE was suggested.

5 REFERENCES

Fujii, Y., Ramakrishna, S. & Hamada, H. 1996, Estimation of Durability of GFRP Laminates Under Stress-Corrosive Environments Using Acoustic Emission, ASTM STP, 1996, 190-202

Recent Developments in Durability Analysis of Composite Systems, Cardon, Fukuda, Reifsnider & Verchery (eds)
© 2000 Balkema, Rotterdam, ISBN 90 5809 103 1

Experimental characterization and analytical prediction of microscopic deformation in interlaminar-toughened laminates with transverse cracks

Nobuo Takeda
Department of Aeronautics and Astronautics, University of Tokyo, Japan

L. Neil McCartney
Center for Materials Measurement and Technology, National Physical Laboratory, Teddington, UK

Shinji Ogihara
Department of Mechanical Engineering, Science University of Tokyo, Chiba, Japan

ABSTRACT: Predictions using a stress transfer model proposed by McCartney for calculating the stress and displacement distributions in a symmetric laminate having cracks in 90° plies are compared with experimental results. A ply refinement technique is used where each ply of the laminate is subdivided into layers having the same properties in order that through-thickness variations of the stress and displacement components can be taken into account. The transverse crack opening displacement and shear deformation near ply interfaces are calculated and compared with the corresponding experimental results. Cross-ply laminates made of two material systems are used in the experiment. One is interleaved CFRP, T800H/3631-FM300, with epoxy resin (FM300) layers about 100μm thick between 0° and 90° plies. The other is toughness-improved CFRP, T800H/3900-2, with selectively toughened interlaminar layers about 30μm thick at all the ply interfaces. It is found that the new model with ply refinement technique is more accurate than a previous approximate analysis.

1 INTRODUCTION

The prediction of the initiation and multiplication of matrix cracks in composite laminates has received a great deal of attention. In particular, transverse cracking in the 90° ply of cross-ply laminates has been extensively studied and many models have been proposed to predict transverse cracking. Shear-lag analyses (Bailey et al 1979, Laws & Dvorak 1988, Han et al. 1988, Lim & Hong 1989, Lee & Daniel 1990, Takeda & Ogihara 1994a,b) and variational mechanics (Hashin 1985, Nairn 1989) were used for the prediction without checking the accuracy of stress field and the corresponding displacement field.

The shear-lag analysis usually neglects both out-of-plane stress components and the variation of in-plane normal stress in through-thickness direction. Furthermore, the shear-lag analysis uses a so-called shear-lag parameter which cannot be determined by experiments.

To overcome the disadvantage of the shear-lag analysis, Hashin (1985) developed a variational stress analysis of a cracked cross-ply laminates. The approach considers the out-of-plane stress components. However, the variation of in-plane normal stress in through-thickness direction was still neglected. In the variational mechanics, the corresponding displacement fields were not derived.

McCartney (1992) analyzed stress transfer between 0° and 90° plies in cracked cross-ply laminates. The solutions were determined to satisfy the stress-strain-temperature relations either exactly or in an average sense. The advantage of this analysis is that the corresponding displacement field can be derived, whose accuracy can be checked by some experimental techniques. In this analysis, the variation of in-plane normal stress in through-thickness direction was again neglected.

Ogihara et al. (1998) extended the McCartney's analysis to consider the effect of interlaminar resin layers between 0° and 90° plies. The displacement fields obtained by the analysis was compared with the experimental results by the micro-line/grid method. Because of the simplicity of the extension procedure of the analysis, there remained some discrepancy between the experimental data and the predictions .

Recently, McCartney (1995) developed a new stress transfer model for predicting the stress and displacement distributions in a multiple-ply cross-ply laminate having a uniform distribution of cracks in some or all the 90° plies. By making use of the ply refinement technique, where each ply of the laminate is subdivided into layers, through-thickness variations of the stress and displacement components can be predicted.

In the present study, a comparison is performed between this new model and the experimental results of the measurement of displacement fields in interlaminar-toughened CFRP cross-ply laminate in order to check the validity of the analytical method. The transverse crack opening displacement and the shear deformation along the ply interfaces near the transverse crack tips are measured using the micro-line/grid methods. Once the accuracy of the analysis is assured, it can be used with confidence to predict the evolution of transverse cracking.

2 EXPERIMENT

2.1 *Materials and laminate configurations*

Two material systems were supplied by Toray Inc. One was interleaved CFRP, T800H/3631-FM300, with epoxy resin (FM300) layers about 100μm thick between 0° and 90° plies. The other was toughness-improved CFRP, T800H/3900-2, with selectively toughened interlaminar layers about 30μm thick at all the ply interfaces. The interlaminar layers have tough and fine thermoplastic resin (polyamide) particles dispersed in the base epoxy resin.

T800H is a high strength carbon fiber. The 3631 is a modified epoxy system with improved toughness compared with a conventional TGDDM/DDS epoxy system. The fiber volume fraction was about 45% for T800H/3631-FM300 and about 55% for T800H/3900-2. The low fiber volume fraction for T800H/3631-FM300 is due to the insertion of the thick FM300 resin films. The cross-ply laminate configuration was $(0/90_4/0)$.

2.2 *Preparation of micro-lines and micro-grids*

The micro-lines or micro-grids were printed using the photo-lithography technique on specimen edge surfaces (Takeda et al. 1997, 1998a,b,c). First, the surface of the specimen was polished, then coated by the photo-resist, or photo-chemical reactive resin. The specimen was heated to cure the resist, and the surface was exposed to the light through the photo-mask, or the glass plate with micro-lines or grids. The exposed part of the resist was then removed in the developer, and vacuum-evaporated metal was deposited on the surface. Finally, the remaining resist was removed in the solvent to prepare the micro-lines or grids on the specimen surface.

2.3 *Measurement of microscopic deformation*

Tensile load is applied on a specimen in an SEM (scanning electron microscope) with a servo-hydraulic loading machine and a specimen heating unit. The specimen was loaded at 20°C until the transverse crack interval became uniform, and both the images of local area near the crack tip and a complete view of the crack were photographed. The temperature was then raised to 80°C and the crack was photographed again. This procedure was repeated at 120°C and 160°C. The load was kept constant to investigate the effect of the temperature, that is, the effect of thermal residual

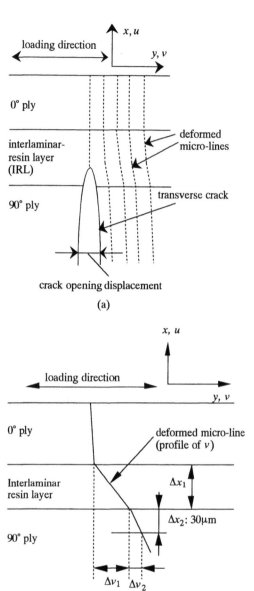

Figure 1. Schematic illustration of microscopic deformation around the transverse crack. (a) Schematic of deformed micro-line. (b) Definition of shear deformation.

stress on microscopic deformation. The crack opening displacement (COD) of the transverse cracks and the interlaminar shear deformation near the transverse crack tip were measured from these photographs (Takeda et al. 1998c).

Figure 1 (a) shows a schematic illustration of deformation around a transverse crack in the 90° ply of a cross-ply laminate. Rectangular (x and y) coordinates were selected as shown in Figure 1, and u and v denote displacements in x and y directions respectively. For example, if the micro-lines are printed parallel to the x axis, they deform, as shown in Figure 1, after transverse cracking. In this case, v can be measured as a function of x and y.

In the present study, the COD of a transverse crack was measured as a function of x. In addition, to evaluate the shear deformation, $\partial v/\partial x$ was measured as a function of y near the transverse crack tip. The quantity $\partial v/\partial x$, averaged in the interlaminar resin layer (IRL), and at the 90°/IRL interface on the 90° ply side, were measured. Detail of the definition of the shear deformation used in the present study is shown in Figure 1(b).

3 ANALYSIS

3.1 Outline of stress transfer mechanics with ply refinement

The problem under consideration concerns the in-plane deformation of a symmetric multilayered laminate constructed of 2N+2 perfectly bonded layers which can have any combination of orientations provided that laminate symmetry is preserved (McCartney 1995). As symmetry about the central plane of the laminate is assumed, it is necessary to consider only the right hand set of N+1 layers as shown in Figure 2. The y-direction defines the longitudinal or axial direction, the z-direction defines the transverse direction and the x-direction defines the through-thickness direction. The locations of the N interfaces are specified by $x=x_i$, $i=1...N$. The central plane of the laminate is specified by $x=x_0=0$ and the external surface by $x=x_{N+1}=h$ where $2h$ is the total thickness of the laminate. The thickness of i th layer is denoted by $h_i=x_i-x_{i-1}$. The orientation of the i th layer is specified by the angle ϕ_i between the y-axis and the fiber direction of this layer. The region to be considered occupies the region $x<h$, $y<L$, $z<W$ of the laminate, and is such that the faces of the laminate are stress-free. Transverse cracks in some or all the 90° plies are assumed to form only on the planes $y=\pm L$ resulting in a crack spacing $2L$, and to span entirely through the thickness and across the width of the ply, W.

Two fundamental assumptions were made in the analysis. One was that generalized plane strain conditions prevail in the laminate. The other was that the shear-stress components σ_{xy} and σ_{xz} are linear functions of x in each ply element. By making the above two fundamental assumptions, it is possible to derive an analytical representation for the stress and displacement fields, for any symmetric multiple-ply cross-ply laminate containing uniform distributions of cracks in the 90° plies, that satisfy

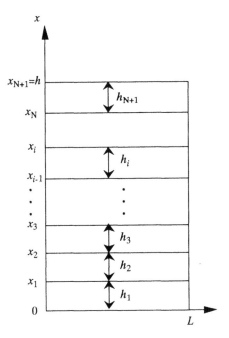

Figure 2. Schematic edge view of the right hand quadrant of a general symmetric laminate indicating coordinate axes and the geometrical interpretation of the layer parameters h_i and x_i, $i=1...N+1$.

exactly the equilibrium equations, the compatibility equations, three stress-strain relations, and the interface continuity conditions. By satisfying the remaining stress-strain relation on averaging it through thickness of each ply, the stress transfer problem is reduced to solving a system of ordinary differential equations that can be solved numerically.

One advantage of the present analysis is that the ply refinement techniques can be used where each ply of the laminate is subdivided into layers having the same properties in order that through-thickness variations of the stress and displacement components can be taken into account. Such through-thickness variations are of importance in the neighborhoods of transverse cracks where the ply refinement enables the effect of the stress singularity to be modeled accurately.

3.2 Application of the analysis

In T800H/3631-FM300 (0/90₄/0) laminate, the thickness of the T800H/3631 plies is 135μm while that of FM300 layer is 100μm. The FM300 layer and the region near a crack tip were divided into thinner elements. Three cases were calculated. In case 1, the transverse crack tip is assumed to stop at the 90°/FM300 interface. In case 2, the transverse crack penetrates into the FM300 layer at the depth of

20μm which was observed in the experiment. In case 3, the transverse crack fully penetrates into the FM300 layer and the tip stops at the FM300/0° interface. The laminate stress was set to be σ=216MPa and the crack interval $2L$=1.70mm to meet the experimental conditions.

In the T800H/3900-2 (0/90$_4$/0) laminate, the thickness of the base composite plies is 170μm and that of the polyamide particle-dispersed layers is 30μm (at 0°/90° and 90°/90° interfaces) or 15μm (at the center and the outer surface). The transverse crack was assumed to run through the thickness of the polyamide particle-dispersed layer at 0°/90° interface. The laminate stress was set to be σ=160MPa and the crack interval $2L$=2.10mm to meet the experimental conditions.

The thermomechanical properties used in the analysis are shown in Table 1. Temperature-dependent Young's moduli of T800H/3631 and T800H/3900-2 in Table 1 are taken from Ref. 17. Shear moduli, Poisson's ratios and thermal expansion coefficients were assumed to be independent of temperature, and room temperature values were used (Takeda et al. 1998c). All the properties of interlaminar-toughened layer material are assumed values. The stress-free temperature is assumed to be 180°C for both material systems.

4 RESULTS AND DISCUSSION

Figure 3 shows an example of the analytical prediction of the axial displacement in y-direction, v, as a function of the distance from the central line, x, along the transverse crack surface in T800H/3631-FM300 (0/90$_4$/0) at 20°C (Case 1). The data shown are those for $y=L$ (on the transverse crack surface). The solid line shows the average displacement in each ply element. The broken line shows the solution for the axial displacement which satisfies the continuity condition at the element interfaces. Because the transverse crack tip was assumed to stop at the 90°/FM300 interface, the average displacement in the FM300 layer and 0° ply is constant because of the symmetry. The continuous solution shows some small oscillations near the transverse crack tip because the solution cannot satisfy the symmetry condition exactly. The condition is satisfied in an average sense for each ply element.

Table 1. Material Properties Used in the Analysis.

T800H/3631-FM300

T800H/ 3631		FM300	
E_A (GPa)	169 -(0.111×T)	E_i (GPa)	2.65 -(0.010×T)
E_T (GPa)	9.62 -(0.010×T)	v_i	0.38
G_A (GPa)	4.50	α_i (/°C)	60.0×10^{-6}
v_A	0.349		
v_T	0.490		
α_A (/°C)	0.10×10^{-6}		
α_T (/°C)	35.5×10^{-6}		

T800H/3900-2

Base Composite		Polyamide Particle-Dispersed Layer	
E_A (GPa)	132 -(0.090×T)	E_i (GPa)	2.71 -(0.0105 ×T)
E_T (GPa)	8.17 -(0.007×T)	v_i	0.38
G_A (GPa)	4.50	α_i (/°C)	60.0×10^{-6}
v_A	0.349		
v_T	0.490		
α_A (/°C)	-1.73×10^{-6}		
α_T (/°C)	34.7×10^{-6}		

T: degrees centigrade (°C)
E_A : longitudinal Young's modulus, E_T : transverse Young's modulus, G_A : out-of-plane shear modulus, v_A : in-plane Poisson's ratio, v_T : out-of-plane Poisson's ratio, α_A : longitudinal thermal expansion coefficient, α_T : transverse thermal expansion coefficient, subscript i denotes interlaminar resin layer (assumed isotropic)

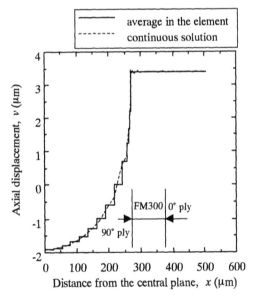

Figure 3. Axial displacement profile on the transverse crack surface in T800H/3631-FM300 (0/90$_4$/0) at 20°C (Case 1). (σ=216MPa, $2L$=1.70mm)

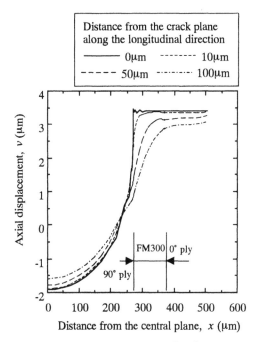

Figure 4. Displacement in y-direction, v, as a function of the distance from the central plane, x, in T800H/3631-FM300 (0/90$_4$/0) at 20°C (Case 1). (σ=216MPa, 2L=1.70mm)

Figure 4 shows the analytical results for the axial displacement distribution at different y-values for T800H/3631-FM300 (0/90$_4$/0) at 20°C (Case 1). Four examples are shown, that is, the distance from the crack plane along the longitudinal (y-) direction are 0μm (y=L), 10μm (y=L-10μm), 50μm (y=L-50μm) and 100μm (y=L-100μm). It can be seen that as the distance from the crack tip increases, the displacement at the central plane increases whereas the displacements in FM300 layer and 0° ply decrease.

Figure 5 shows the analytical results of axial displacement profile on the transverse crack surface for both the average in the ply elements and the continuous solution for T800H/3900-2 (0/90$_4$/0) at 20°C. The transverse crack is assumed to run through the thickness of polyamide particle-dispersed layer at 0°/90° interface and the average displacement in the uncracked 0° ply is constant as required by the symmetry condition. It can be seen that the derivative of the displacement is very large in the polyamide particle-dispersed layer at the 0°/90° interface.

Figure 6 shows the axial displacement distribution at different y-values for T800H/3900-2 (0/90$_4$/0) at 20°C. Similar tendency to Figure 6 can be seen. As shown in Figures 5-8, the displacement

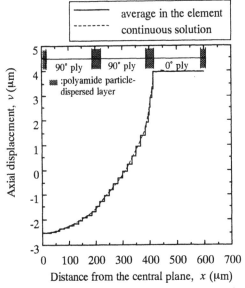

Figure 5. Axial displacement profile on the transverse crack surface in T800H/3900-2 (0/90$_4$/0) at 20°C. (σ=160MPa, 2L=2.10mm)

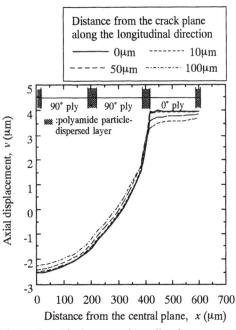

Figure 6. Displacement in y-direction, v, as a function of the distance from the central plane, x, in T800H/3900-2 (0/90$_4$/0) at 20°C. (σ=160MPa, 2L=2.10mm)

fields can be obtained anywhere in the laminate by using the analysis.

Figure 7 shows the comparison between the analytical predictions and the experimental results for the COD in T800H/3631-FM300 $(0/90_4/0)$ at various temperatures. Experimental results show that the temperature dependence of COD is very small. This can be explained qualitatively as follows. As temperature increases, the tensile thermal residual stress decreases which results in smaller COD. At higher temperature, the shear modulus of FM300 becomes lower, which may result in larger COD. The small temperature dependence of COD is expected to be a result of these two effects.

In Figure 7, analytical predictions are shown for case 1 (20°C, 80°C, 120°C and 160°C), case 2 (20°C) and case 3 (20°C). Unfortunately, the agreement is not very good. Two main reasons are considered. One is the temperature-dependent material properties are not well known especially for FM300. The other is that, in practice, the crack tip penetrates into FM300 layer at some depth. For more quantitative evaluation, these problems have to be considered.

Figure 8 shows the comparison between the same experimental results and the previous analytical prediction (Takeda et al. 1998c) without ply refinement techniques (this is called interlaminar layer

model in the present paper). The cases (a) and (b) shown in Figure 10 correspond to the assumptions that the crack tip stops at the 90°/FM300 and the FM300/0° interfaces, respectively. The agreement between the predictions and experiment is poor and the effect of the ply refinement is seen by comparing results with those in Figure 7, which implies the validity and necessity of the refinement techniques for accurate analysis.

Figure 9 shows the comparison between the analytical predictions of the new model with ply refinement techniques and the experimental results for the COD in T800H/3900-2 $(0/90_4/0)$ laminate. The agreement is rather good compared with T800H/3631-FM300 laminate (Fig.7). The analytical predictions show larger temperature dependence, that is, smaller COD at higher temperature and larger COD at lower temperature. This may be due to the inaccuracy of temperature-dependent properties of polyamide particle-dispersed layer. The properties were assumed because they cannot be measured experimentally. Another explanation for the discrepancy is plastic deformation in the polyamide particle-dispersed layer, especially at higher temperature. Because the analysis is linear elastic, the non-linear behavior of the material is not considered. However, the agreement is good at lower temperature (at 20°C and 80°C) where the

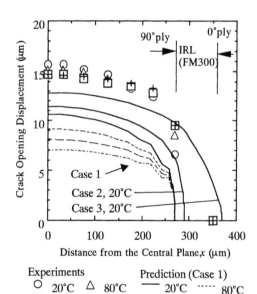

Experiments

	Prediction (Case 1)	
O 20°C △ 80°C	—— 20°C	······ 80°C
□ 120°C + 160°C	--- 120°C	···· 160°C

Figure 7. Crack opening displacement in T800H/3631-FM300 $(0/90_4/0)$. Comparison between experimental results and analysis with the ply refinement techniques. (σ=216MPa, $2L$=1.70mm)

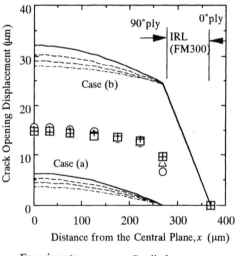

Experiments

	Prediction	
O 20°C △ 80°C	—— 20°C	······ 80°C
□ 120°C + 160°C	--- 120°C	···· 160°C

Figure 8. Crack opening displacement in T800H/3631-FM300 $(0/90_4/0)$. Comparison between experimental results and the previous analysis (interlaminar layer model (Takeda et al. 1998c)). (σ=216MPa, $2L$=1.70mm)

material can be assumed to be linear elastic, which implies the validity of the analysis.

Figure 10 shows a comparison between predictions by the new analysis and the experimental results for the shear deformation, $\partial v/\partial x$, in T800H/3631-FM300 (0/90$_4$/0). Both the average shear deformation in the FM300 layer and the shear deformation at the 90°/FM300 interface in 90° ply side are shown as functions of the distance form the transverse crack tip. The shear deformation at 90°/FM300 interface is well predicted by the analysis. The agreement of average shear deformation in the FM300 layer is not very good especially at higher temperature which may be due to the inaccuracy of the temperature-dependent properties of FM300 and the possible effect of plastic deformation. Figure 11 shows the comparison between the same experimental results and the previous analytical predictions (interlaminar layer model (Takeda et al. 1998c)). The agreement is much poorer and it can be seen that the new model can predict the effect of the singularity behavior near the transverse crack tip (Fig.11).

Figure 12 shows the comparison between the prediction of the new analysis and the experimental results for the shear deformation, $\partial v/\partial x$, in T800H/3900-2 (0/90$_4$/0). Again, both average shear deformation in the polyamide particle-dispersed layer and shear deformation at the 90°/polyamide particle-dispersed layer interface in 90° ply side are shown as functions of the distance form the transverse crack tip.

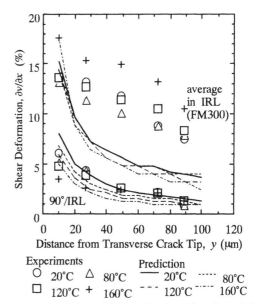

Figure 10. Shear deformation as a function of the distance from the crack tip along longitudinal direction in T800H/3631-FM300 (0/90$_4$/0). Comparison between experimental results and analysis with the ply refinement techniques. (σ=216MPa, 2L=1.70mm)

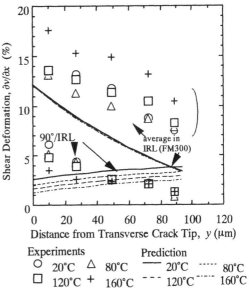

Figure 11. Shear deformation as a function of the distance from the crack tip along longitudinal direction in T800H/3631-FM300 (0/90$_4$/0). Comparison between experimental results and the previous analysis (interlaminar layer model (Takeda et al. 1998c)). (σ=216MPa, 2L=1.70mm)

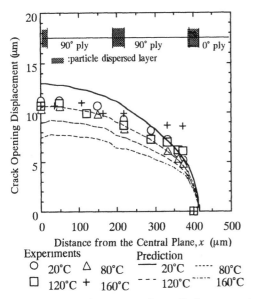

Figure 9. Crack opening displacement in T800H/3900-2 (0/90$_4$/0). Comparison between experimental results and analysis with the ply refinement techniques. (σ=160MPa, 2L=2.10mm)

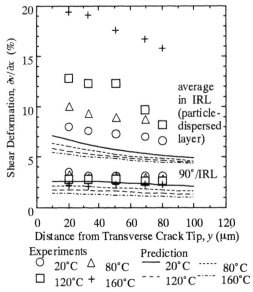

Experiments
○ 20°C	△ 80°C
□ 120°C	+ 160°C

Prediction
—— 20°C	····· 80°C
- - - 120°C	ˉˉˉ 160°C

Figure 12. Shear deformation as a function of the distance from the crack tip along longitudinal direction in T800H/3900-2 (0/90₄/0). Comparison between experimental results and analysis with the ply refinement techniques. (σ=160MPa, $2L$=2.10mm)

The shear deformation at the 90°/polyamide particle-dispersed layer interface are well predicted by the analysis. The agreement of average shear deformation in polyamide particle-dispersed layer is not very good especially at higher temperature which may be due to the inaccuracy of the temperature-dependent properties of polyamide particle-dispersed layer and the possible effect of plastic deformation. For more quantitative evaluation of the microscopic deformation, it will be necessary to establish temperature-dependent material properties of the interlaminar resin and the consideration of non-linear material behavior such as plastic deformation.

From the results obtained in the present study, it has been proved that the new analysis has sufficient accuracy within the range the material behavior that is linear elastic. The experimental techniques (the micro-line/grid methods) used in the present study are useful tools for evaluating the mechanical analysis. By combining the analytical procedure with accurate microscopic experimental verification, it will be possible to develop a validated analytical procedure to predict matrix cracking in laminated composites.

5 CONCLUSIONS

A new stress transfer model proposed by McCartney for predicting the displacement fields in a composite laminate with transverse cracks in 90° plies were compared with experimental results. The ply refinement technique was used in the analysis where each ply of the laminate was subdivided into layers having the same properties to consider through-thickness variations of the displacements. The transverse crack opening displacement and shear deformation near ply interfaces measured by using the micro-line/grid methods were used for the evaluation of the analytical predictions. Cross-ply laminates made of two material systems with interlaminar resin layers were used in the experiments. It was found that the ply refinement techniques were most effective for an accurate analysis. The new analysis was proved to be accurate enough for good correlations with experimental results in the range where the material behavior is linear elastic. Based on the present analysis, it will be possible to develop a more precise methodology to predict transverse cracking in composite laminates.

REFERENCES

Bailey, J.E., P.T. Curtis & A. Parvizi 1979. On the transverse cracking and longitudinal splitting behavior of glass and carbon fiber reinforced epoxy cross ply laminates and the effect of Poisson and thermally generated strain. *Proc. R. Soc. Lond.* A 366:599-623.

Han, Y.M., H.T. Hahn & R.B. Croman 1988. A simplified analysis of transverse ply cracking in cross-ply laminates. *Comp. Sci. Tech.* 31:165-177.

Hashin, Z. 1985. Analysis of cracked laminates: a variational approach. *Mechanics of Materials* 4:121-136.

Laws, N. & G.J. Dvorak 1988. Progressive transverse cracking in composite laminates. *J. Comp. Mat.* 22:900-916.

Lee, J.W. & I.M. Daniel 1990. Progressive transverse cracking of crossply composite laminates. J. Comp. Mat. 24:1225-1243.

Lim, S.G. & C.S. Hong 1989. Prediction of transverse cracking and stiffness reduction in cross-ply laminated composites. *J. Comp. Mat.* 23:695-713.

McCartney, L.N. 1992. Theory of stress transfer in a 0°-90°-0° cross-ply laminate containing a parallel array of transverse cracks. *J. Mech. Phys. Solids* 40:27-68.

McCartney, L.N. 1995. A recursive method of calculating stress transfer in multiple-ply cross-ply laminates subject to biaxial loading. *NPL Report DMM(A)* 150.

Nairn, J.A. 1989. The strain energy release rate of composite microcracking: a variational approach. *J. Comp. Mat.* 23:1106-1129.

Ogihara, S., N. Takeda & A. Kobayashi, 1998.

Analysis of stress and displacement fields in interlaminar-toughened composite laminates with transverse cracks. *Adv. Comp. Mat* 7:151-168.

Takeda, N. & S. Ogihara, S. 1994a. *In-situ* observation and probabilistic prediction of microscopic failure processes in CFRP cross-ply laminates. *Comp. Sci. Tech.* 52:183-195.

Takeda, N. & S. Ogihara, S. 1994b. Initiation and growth of delamination from the tips of transverse cracks in CFRP cross-ply laminates. *Comp. Sci. Tech.* 52:309-318.

Takeda, N., H. Niizuma, S. Ogihara & A. Kobayashi 1995. Experimental evaluation of thermal residual stress in CFRP cross-ply laminates. *Materials System* 16:73-78 (in Japanese).

Takeda, N., H. Niizuma, S. Ogihara & A. Kobayashi 1997. Application of micro-line/grid methods to temperature-dependent microscopic deformation and damage in CFRP laminates. *Exp. Mech.* 37:182-187.

Takeda, N., S. Ogihara, S. Suzuki & A. Kobayashi 1998a. Evaluation of microscopic deformation in CFRP laminates with delamination by micro-grid methods. *J. Comp. Mat.* 32:83-100.

Takeda, N., S. Ogihara, K. Nakata & A. Kobayashi 1998b. Characterization of microscopic failure process and deformation in glass/nylon6 composites by micro-grid methods. *Composite Interfaces* 5:305-321

Takeda, N. & S. Ogihara 1998c. Micromechanical characterization of local deformation in interlaminar-toughened CFRP laminates. *Composites Part A* 29:1545-1552.

Recent Developments in Durability Analysis of Composite Systems, Cardon, Fukuda, Reifsnider & Verchery (eds)
© 2000 Balkema, Rotterdam, ISBN 90 5809 103 1

On the use of acoustic emission and scanning electron microscopy to investigate fatigue damage in plain-woven fabric composites

S. D. Pandita, G. Huysmans, M. Wevers & I. Verpoest
Department of Metallurgy and Materials Engineering, Katholieke Universiteit Leuven, Belgium

ABSTRACT: Acoustic emission as a non-destructive inspection technique was used to evaluate the damage development of woven fabric composites subjected to tensile-tensile fatigue loads. This technique offers some advantages like continuously damaged monitoring and is not corrosive to the fatigued specimen. In order to gain good interpretation, complementary data such as residual stiffness or microscopic observation using scanning electron microscopy (SEM) are required. Woven fabric composites were subjected to tensile fatigue load with fatigue load ratio R of 0.1 and a frequency of 3 Hz. Damage development at fatigue load S_{max} of 0.5 and 0.7 was discussed.

1 INTRODUCTION

Woven fabric composites are a class of textile composites that are widely used for many structural applications. The use of textile fabrics has generally offered a lower cost to composite manufacturing and higher damage tolerance. Woven fabrics are produced by interlacing two or more yarns on a loom. The mechanical properties of woven fabric composites, such as strength and stiffness, are strongly determined by the weave parameters (weave geometry, yarn size, yarn spacing, and yarn crimp, fiber type), and laminate parameters (fiber orientations and overall fiber volume fraction), and the inherent material properties of fiber and matrix (Naik, 1995, and Vandeurzen, 1998).

During service, composite structures can be subjected to fatigue loads. The fatigue load creates fatigue damage. The major damage types in fiber reinforced composite materials are fiber fractures, matrix cracks, matrix-fiber debonding, and delamination. Those damage types were also detected in woven fabric composites that were subjected to fatigue loads. In general, the accumulation of fatigue damage is accompanied by a reduction in stiffness and strength, and an increase in the size of the hysteresis loop. Residual stiffness and hysteresis change are not easily monitored during the fatigue test, unless the test is stopped or the fatigue tests are run on a sophisticated fatigue machine where data processing is possible during the measurement. Acoustic emission (AE) is the only non-destructive test technique capable of detecting all of the above mentioned damage types (Wevers, 1997) and of providing information of damage accumulation during fatigue loading.

However, the major disadvantage of AE is that great skill is needed in order to correlate AE information with damage mechanisms in a material. The interpretation of AE signals coming from different damage types is not straightforward. In order to facilitate AE interpretation, researchers often compare AE data with other damage characterization techniques (ultrasonic C-scan, radiography and scanning electron microscopy). Microscopic techniques such as scanning electron microscopy (SEM) have often been used to complement acoustic emission. The main disadvantage of SEM is that it is a destructive technique. It means that an investigated specimen can only be subjected to a certain damage level or a specific number of fatigue cycles. The specimen then needs to be cut and analyzed either at the edge or on a cross-section of the specimen.

This paper focuses on using the acoustic emission technique and scanning electron microscopy to investigate tensile-tensile fatigue damage in plain-woven fabric composites. Woven fabric composites were subjected to tensile-tensile fatigue loads along one of its on-axis directions.

2 EXPERIMENTAL

A balanced plain woven fabric, having similar yarn properties in both warp and weft direction, was used. Because of this symmetry, fatigue tests were performed in one of these two directions. Fatigue tests on woven fabric composites were performed in the weft direction. E-glass woven fabrics (R/420) were supplied by the company Syncoglas in Belgium. The woven fabrics were impregnated by epoxy resin film F533 from Hexcel and cured in an autoclave at a temperature of 125 degrees Celsius and a pressure of 3 bar. The fiber volume fraction of the woven fabric composite is 50 %.

The woven fabric composites were subjected to tensile fatigue loading at a stress ratio R ($\sigma_{min.}/\sigma_{max.}$) of 0.1 and a frequency of 3 Hz. The fatigue specimens had a width of 25 mm, a thickness of 2 mm, and a length of 230 mm. Aluminum end tabs with length of 40 mm, were attached at both ends of the specimen to avoid failure around the gripping device during the tests. The gauge length of the specimen was 150 mm. Two AE sensors were placed at the middle of the specimen. The area of the AE investigation was 30 % of the total area of specimens. Figure 1 shows the fatigue set-up.

Fatigue damage development in the fatigued specimens was continuously evaluated by acoustic emission. Damage that occurs in the specimen creates an ultrasonic waveform. This ultrasonic waveform was then amplified and counted by an AE device. The waveform parameters (amplitude and events) were explored in order to learn the fatigue damage. A research strategy was made to optimize the fatigue damage investigation, as shown on figure 2. The change in the AE amplitude or events spectrum determines the time points to interrupt the fatigue test and then analyze the fatigued specimen microscopically. Location of a fatigue damage can be estimated by analyzing the difference of the arrival time of an ultrasonic signal reaching both AE sensors.

3 RESULTS AND DISCUSSION

3.1 *Static Tensile results.*

Before investigating the fatigue behavior of a particular material, its static behavior must be discussed first. Figure 3 shows the static tensile test result of the investigated woven fabric composites. The tensile test was performed at a strain rate of 2 mm/s, which is much slower compared to the cyclic displacement rate in a fatigue test whose frequency is 3 Hz. At 50 % of the ultimate tensile stress, damage has been created in the specimen because the stiffness of the specimen has already decreased to 75 % of the initial stiffness, as shown on figure 4. Therefore, a fatigue load S_{max} of 0.5 has been selected for the fatigue tests.

Figure 5 shows the S-N curve of woven fabric composites subjected to tensile-tensile fatigue

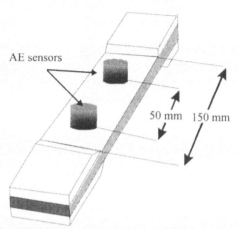

AE sensors

50 mm   150 mm

Figure 1. Fatigue Specimen with AE sensors mounting.

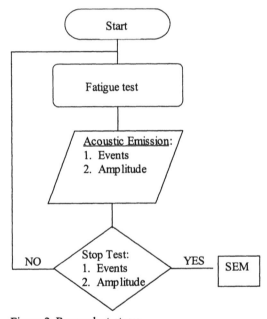

Figure 2. Research strategy.

loading. For the investigation of the damage mechanisms of woven fabric composites, two fatigue loads were selected with maximum fatigue loads S_{max} of 0.5 and 0.7. These fatigue loads were selected based on the related fatigue life. Fatigue life at fatigue load S_{max} of 0.5 σ_{ult} is 65.000 cycles and at S_{max} of 0.7 σ_{ult} is 2.700 cycles. On the other hand, at S_{max} less than 0.4 σ_{ult}, fatigue life is higher than 10^6 cycles.

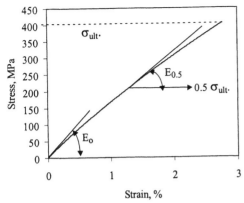

Figure 3. Static tensile result

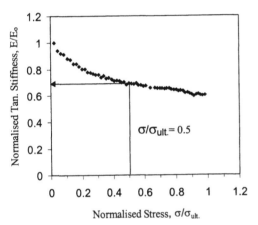

Figure 4. Normalized tangential stiffness plotted against normalised stress.

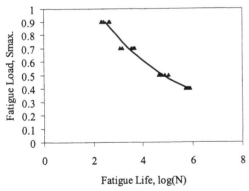

Figure 5. S-N curve

3.2 *Fatigue Loads S_{max} of 0.5.*

Based on the acoustic emission evaluation (figures 6 and 7), the fatigue damage development at the fa-

tigue load S_{max} of 0.5 consists of 4 stages. The first stage is characterized by the absence of AE activities. From the static test, stiffness at the tensile stress of 50 % of the ultimate stress decreased about 25 % from the initial stiffness. Therefore, at the fatigue load S_{max} of 0.5, the stiffness in the fatigue test was expected having a similar decrease. However, the stiffness in the fatigue test was only dropped to 95 % of the initial stiffness when the first AE event was detected. Compared to the static results, the stiffness degradation during fatigue tests is less pronounced. It indicates that the mechanical response and damage in woven fabric composites in a static test is different than in the dynamic test and hence the damage development is strain rate dependent. The different response might also be contributed to the rheologycal effects of the viscous-elastic matrix. At higher strain rate, the viscous-elastic matrix behaves more elastically. Figures 8 and 9 show the hysteresis loop and the residual stiffness at a fatigue load S_{max} of 0.5 σ_{uts} respectively. Stiffness is still around 90 % of the initial stiffness after 50.000 cycles.

Although the stiffness in the fatigue test decreased up to 95 % of the initial value, no AE events were detected. There are two justifications for this observation. First, the AE events of fatigue generated introduced outside the investigated area, will be filtered out. The second reason is that the AE intensity of the initial fatigue damage is too weak. Small fatigue damage phenomena such as fiber-matrix debonding near the edge or inset of the specimen may result in low AE intensity.

In the second stage, AE events related to fatigue damage were detected increasingly when the number of fatigue cycles increased. A complementary fatigue test on another sample was performed and stopped at 15.000 cycles, where from the first test the second

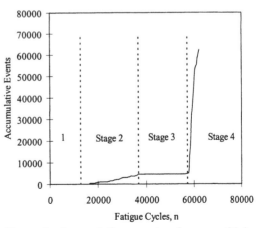

Figure 6. Accumulative events of woven fabric composites subjected to fatigue load S_{max} of 0.5

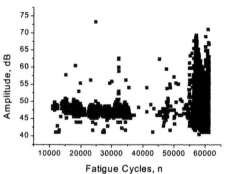

Figure 7. AE Amplitude of woven fabric composites subjected to fatigue load S_{max} of 0.5

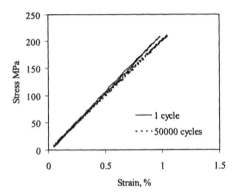

Figure 8. Fatigue profile at S_{max} =0.5

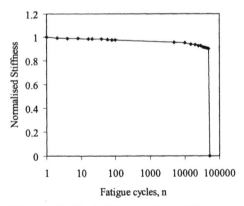

Figure 9. Residual stiffness at S_{max} =0.5

Figure 10. Transverse bundle failure in a transverse yarn (15.000 cycles)

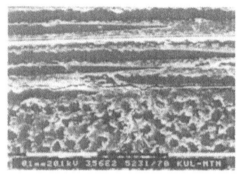

Figure 11. A transverse crack deflected by a longitudinal yarn leading to a meta-delamination. (45.000 cycles)

stage was expected to happen. The specimen was then cut and analyzed microscopically at its cross-section. From the SEM inspection, the fatigue damage at 15.000 cycles consists of transverse bundle cracks in the fatigued specimen. The AE amplitude of the AE events at the second stage is mostly between 45 and 50 dB. Another test was stopped and analyzed by SEM at 45.000 cycles, which is close to the end of the second stage. The fatigue damage in the second stage did not only consist of transverse cracks, but also of transverse cracks deflected into longitudinal yarn debond ,called as meta-delaminations (Fujii , 1993). Up to 45000 cycles, the amplitude of the AE events did not show a significantly different value. Most of the amplitudes are still between 45 and 50 dB. As the meta-delamination process consists of fiber-matrix debonds and matrix cracks that are also present with a transverse crack but in different direction, the AE waveform of the meta-delamination process is similar to the AE waveform of the transverse bundle failures. Few AE events have amplitudes higher than 60 dB. These events might result from premature fiber failures. Figures 10 and 11 show a transverse bundle failure and a meta-delamination respectively.

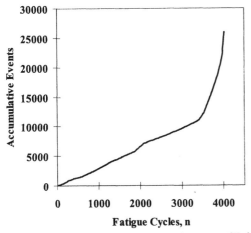

Figure 12. Accumulative events of woven fabric composites subjected to fatigue load S_{max} of 0.7

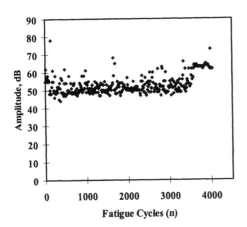

Figure 13. AE amplitude of woven fabric composites subjected to fatigue load S_{max} of 0.7

The growth of the AE event rate stabilises after all transverse bundles are damaged by transverse cracks and meta-delaminations. This condition characterized the onset of the third stage. Some events are still detected, resulting from either the propagation of transverse cracks into the matrix areas or friction phenomena occurring at the meta-delaminations sites. However, the accumulative events in this stage are much lower than the accumulative events in the second stage.

The last stage is the final failure that takes place locally and is indicated by a heavily increasing amount of AE events. The final failure is dominated by fiber fractures. The AE amplitudes associated with fiber fracture are higher than 60 dB.

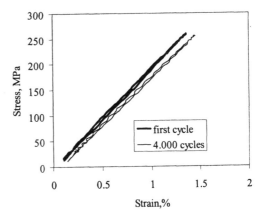

Figure 14. Fatigue profile at S_{max} =0.7

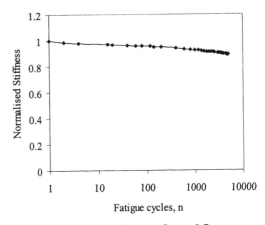

Figure 15. Residual stiffness at $S_{max} = 0.7$

3.3 Fatigue Loads S_{max} of 0.7

At the maximum fatigue load ratio S_{max} of 0.7 σ_{uts}, the fatigue damage development stages are not well separated. Figures 12 and 13 show the AE measurements. The fatigue damage only consists of the initial damage stage (transverse cracks and meta-delaminations) and fiber failure respectively. The initial failures occurred in whole material volume. AE events increased almost linearly as a function of the number of fatigue cycles. Till the point before final failure, the AE events were growing with a constant rate (ratio of events to fatigue cycles). However, at the final failure, the AE activities increased tremendously.

From the mechanical point of view, we expected that the stiffness of the fatigued specimen would have a similar decrease as during a normal static test. Figure 14 shows the fatigue profile. The stiffness of woven fabric composites subjected fatigue load S_{max} of 0.7 σ_{uts} is still around 90 percent before final failure, depicted in figure 15. This phenomenon might be due the different damage development and/or the rheologycal properties of the constituent materials as a function of the loading rate. The ultimate stress should be measured at a similar test rate as in the fatigue test. In other words, the fatigue behavior of woven fabric composites is frequency dependent. The frequency of the fatigue test obviously affects the loading rate.

4. CONCLUSION

The combination of acoustic emission and SEM was used to analyze tensile fatigue damage development in woven fabric composites. The geometry of transverse cracks is not linear in the fatigued woven fabric composites and they are difficult to be filled by liquid penetrants. Interpretation of AE results is however not straightforward, which can be a drawback of this technique.

At the fatigue load S_{max} of 0.5, the fatigue damage developed in 4 clearly distinct stages. In the first stage, AE events were not detected. The onset of the second stage starts is characterized by the first detection of AE events. Transverse cracks and meta-delaminations were observed in the second stage. The third stage is a transition stage in which the AE event rate stabilizes. All transverse bundles have been damaged. Some AE events resulting from the propagation of the transverse cracks into matrix areas and the friction in meta-delamination sites were detected. The final failure was terminated by fiber fractures. However, at a fatigue load S_{max} of 0.7, the fatigue damage development is more continuous and can therefore not be subdivided into separated stages.

ACKNOWLEDGEMENTS

This text presents research results of the Belgian programme on Inter-university Poles of Attraction, funded by the Belgian state, Prime Minister's Office, Science Policy Programming. The scientific responsibility is assumed by its authors. S.D. Pandita is financed through grants of the Governmental Agency for Cooperation with Developing Countries (ABOS-VLIR), Concerted Research Action (GOA) and Fonds voor Wetenschapelijk Onderzoek (FWO).

REFERENCES

Fujii, T., Amijima, S., Okubo K. 1993. Microscopic Fatigue Processes in Plain Weave Glass Fiber Composites. Composites Science and Technology. 49: 327-333.

Naik, R.A. 1995. Failure Analysis of Woven and Braided Fabric Reinforced Composites. Journal of Composite Materials. 29: 2334-2363.

Vandeurzen, P. 1998. Structrure-Performance Modelling of Two Dimensional woven Fabric composites. Doctoral Thesis, Katholieke Universiteit Leuven, Belgium.

Wevers, M. 1997. Listening to the sound of materials: acoustic emission for the analysis of material behavior. NDT&E International. 30: 99-106.

Recent Developments in Durability Analysis of Composite Systems, Cardon, Fukuda, Reifsnider & Verchery (eds)
© *2000 Balkema, Rotterdam, ISBN 90 5809 103 1*

Damage identification of composite material during stress rupture using acoustic emission technique

A. Rotem
Faculty of Mechanical Engineering, Technion-IIT, Haifa, Israel

ABSTRACT: T300/5208 carbon epoxy laminated composite specimens, which were manufactured 19 years ago, and kept in a conventional normal environment, were tested for creep durability and Acoustic Emission behavior. It was found that the mechanical behavior as a whole, was retained, and degradation is minimal if any. The creep history under the load and the Acoustic Emission while under loading, was monitored continuously. Effects of different deformation modes, pure shear and in plane shear combined with interlaminar shear, were separated by choosing special kinds of specimens. It was found that the AE signature is different for the different modes of creep, but remains the same as 19 years ago. The strength and creep behavior are also about the same as 19 years ago. AE activity accelerates as creep rate increased and unconnected, AE accelerate towards the end of the creep life. The Pulse Height Analysis of the two shear stresses, are different in their slopes. Thus these characteristics can serve as sensors for creep rupture.

1 INTRODUCTION

Composite materials are used in a variety of structures that are designed to serve for many years. However, there is not enough knowledge on their behavior after a long period of time. This research has two goals: the first one is to determine whether the material retain its mechanical properties many years after manufacturing and the second is, what is the mechanical durability of such a material under constant load. The material under consideration is graphite fibers in an epoxy matrix. Polymeric materials are more vulnerable for aging conditions than graphite fibers and therefore the experimental program is aimed to examine the matrix behavior. As aging is primarily crystallization of the polymeric matrix which cause degradation of strength and deformation, especially by shear, the experimental program was chosen to reflect this behavior.

2 EXPERIMENTAL PROGRAM

Two kinds of specimens were used, unidirectional and ±45° angle ply. Two types of specimens were produced from the unidirectional laminate and one from the angle ply. A 10° unidirectional specimen was used to impose primarily shear stress when loaded in the axial direction (shear to transverse stress ratio is 6.54/1). An axial unidirectional specimen with 2 opposite slits was used for imposing

Mode II stress condition along the fibers. The specimens are shown in figure 1. Also shown is the ±45° angle ply specimen which has the combination of in plane and interlaminar shear.

Specimens were mounted in a special apparatus, shown in figure 2, where they were loaded by weights through a lever to a stress level close to their ultimate strength. Creep under the load was measured by LVDT and Acoustic Emission sensor captured the released stress waves, as deformation and cracking occurred. The details of the specimen mounting is shown in figure 3. The AE signals were fed into a dedicated computer as well as the LVDT output. Creep and AE activity versus time were logged. The AE activity was further processed to give the pulse height distribution (occurrence).

3 EXPERIMENTAL RESULTS

Specimens were loaded close to their ultimate strength in order to finish the experiment in a reasonable time. The ultimate strength was measured by an Instron testing machine, where the specimens were loaded by tension. The results fall within the statistical scatter of the strength measured 19 years ago (Rotem 1981). Creep tests were done with loading of 85% to 95% of the ultimate. The creep measured in microstrain and the summation of all AE events occurring during the creep up to fracture are plotted in a single graph. AE sources in compos-

ite materials may be cracking or plastic deformations, where elastic energy is released from the
stressed medium and propagate as elastic wave. This
wave is captured by the AE sensor mounted on the
specimen. However, different stress release mechanisms delivers different strain waves with different
amplitudes content (rotem 1979). Therefore each
degradation mechanism emit unique burst of strain
amplitudes that might be called "AE signature".
Figure 4 shows a graph of pulse height analysis
taken while loading in tension a specimen of $\pm 45^{\circ}$
angle ply, some 18 years ago (Rotem 1984). This
specimen was made from the same laminate which
was used now in this research. The graph describes
the AE counts above a certain amplitude level, in a
normalized way. Certainly, at the extreme left all the
events are counted (N=TN) and none at the right side
(N=0). The amplitude is normalized with respect to
the threshold measuring level Vt. As seen in the
figure, the graph decline almost linearly. Each mode
of energy release mechanism has its unique amplitude content and therefore a characteristic slope.
Therefore these slopes may serves as identifiers to
separate between the various failure modes, fracture
and deformation. The steeper the slope is, there are
less high level amplitude in the AE. Comparing the
AE slopes may gives information on the failure
mechanism.

In this research the same material was used, for
testing three types of creep in shear. Specimens of
unidirectional fibers, cut 10° off axis were loaded in
tension. A typical test result, shown the creep and
total AE, is shown in figure 5. The ratio of shear to
normal transverse stress is 6.5 and therefore the
mechanism of creep is shear dominated. AE is
pacing the creep and indicate on the plastic deformation occurring in the material while creeping.
Figure 6 is the pulse height analysis of this specimen. Unlike in the previous case (19 years ago), the
normalization of the amplitude is done by the maximum possible value. Therefore the normalized value
of the amplitude here is smaller than one. The slope
of the PHA is B=-1.412, which indicate on moderate
combination of high and low amplitude. The second
kind of test was of the Mode II fracture mechanics
test. The specimen is unidirectional with slits perpendicular to the fibers direction. The stress raiser is
very high, given by the fracture mechanics formula
(Carlsson 1989),

$$\tau_{xy} = \frac{K_{II}}{\sqrt{2\pi x}} + \tau_{xy}^{\infty} \qquad (1)$$

$$K_{II} = \tau_{xy}^{\infty} \sqrt{\pi a} H \qquad (2)$$

where τ_{xy} is the shear stress along the fibers, x is
the distance from the crack tip (starts at the slit), ∞
mark the undisturbed stress, a is the crack length and

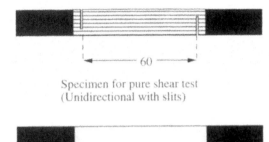

Specimen for pure shear test
(Unidirectional with slits)

Specimen for inplane and interlaminar shear
(Angle ply and off axis laminate)

Figure 1. View of specimens used in the test

Figure 2. Creep apparatus (general view)

Figure 3. Specimen mounting on the apparatus

H is a geometry factor. In our case, the stress raiser
is about 5.2. There is almost no deformation in this
test and thus no creep was recorded. Figure 7 shows
the AE accumulation while load was applied until
failure. Here the activity is mainly from crack formation and propagation and therefore the high am
plitude content in the AE is greater. This is seen in

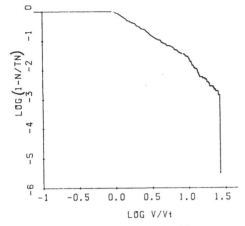

Figure 4 PHA of ±45o specimen taken 18 years ago

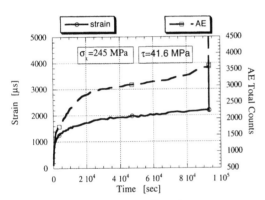

Figure 5 Creep and TAE from 10° specimen

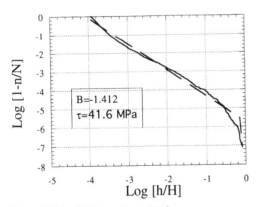

Figure 6 Pulse height analysis of 10° specimen

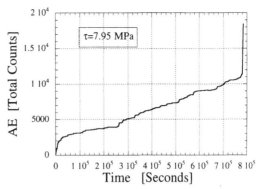

Figure 7 Accumulated AE from slotted unidirectional specimen.

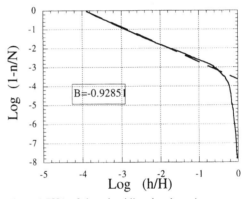

Figure 8 PHA of slotted unidirectional specimen

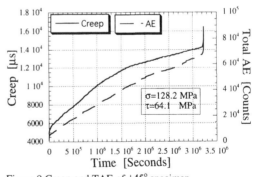

Figure 9 Creep and TAE of ±45° specimen

figure 8, where the slope of the PHA is smaller than in figure 6, which indicate on higher content of high level amplitudes.

The third shear stress test was done on ±45° specimen. Here the shear stress is exactly half the axial stress and the normal stress, transverse to the fibers, is 0.15 of the shear stress and therefore it is also shear dominated situation. However, because of the lamination, there is also interlaminar shear stress

327

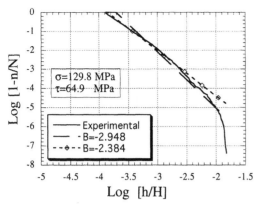

Figure 10 PHA of ±45° specimen

present near the free edges of the specimen. The pulse height analysis reflect this characteristic. Figure 9 shows the creep and the accumulated AE while testing the ±45° angle ply laminate. The total creep up to fracture comes to 1.4%, which is much higher than in the other to cases. This indicate on a large deformation activity in two regions, the inplane and interlaminar shear, that produces AE signals of relatively low level amplitudes. It has two outputs: first, the intense AE activity cause the accumulated signals to pace the strain and second, the pulse height distribution graph has a steeper slope than in the previous cases.

This is seen in figure 10, where two slopes are fitted to the experimental results, because the line is not straight enough. However, both lines has a much steeper slope than in the previous cases, that indicate on low content of high amplitude signals.

4 DISCUSSION

The material used in this research, graphite epoxy composite T300/5208, was manufactured and tested for mechanical strength 19 years ago. The strength testing which was done recently, gave the same results within a statistical error, $\sigma_x = 148 \pm 10$ Mpa for the ±45° specimens. These results indicate that the material, especially the polymeric matrix, remained stable along all these years. The matrix did not undergo any crystallization process during this period, while the material stayed in a room conditions, without any loading. This excellent characteristic does not imply that the same thing would happen if the material will be under load or stored in rough environment like direct exposure to the sun or chemicals. However, high load certainly causes creep and eventually failure.

The durability under load that causes dominated shear stress found to be quite good. On loads of approximately 90% of the ultimate, the creep time to failure was in the order of 24 days. Strain to failure of about 1.4% for the ±45° specimen also indicate on the conservation of the ductility of the matrix material. This behavior of the shear deformation was found in all the types of experiments used in this research. Detection of the AE when the specimen was under load, enable to monitor the stress waves outputted from the deformation process or crack formation and propagation. Because these processes differ in the amount of energy released, the amplitude levels of the emitted stress waves also differ. This difference enable one to identify the creep process by the content of the amplitudes in the AE signals. More high level amplitudes points out on crack formation process and vice versa, more low level amplitudes points out on plastic deformation process. The amplitude level content may be given by pulse height analysis, where the slope of the presentation indicate on the amplitude level content. Indeed, the unidirectional specimens with the slits, that serves to impose Mode II crack opening, gave the most shallow slope, which indicate on high content of high level amplitudes. On the other hand, the ±45° specimen, where creep is mainly by shear plastic deformation, which produces low level amplitudes, the slope is very steep.

5 CONCLUSSION

The durability, in the sense of mechanical behavior, of the composite material T300/5208 graphite epoxy was retain for over 19 years. Strength and deformation tests, then and now, gave statistically the same results. These "old" laminates behaved as "young" when tested under dead load for creep resistance. Therefore, 19 years in a casual conditions does not degrade the mechanical behavior. Casual conditions means, normal fluctuations of temperature and humidity in a conventional room without air condition, but without exposure to the sun or rain.

Acoustic emission technique can serve very effectively to monitor shear strain creep and crack formation. Pulse height analysis of the amplitudes enable the identification of the creep process, whether it is plastic deformation or crack formation. The identification depends on the slope of the PHA.

6 ACKNOLEDGEMENT

This research was supported in part by the fund for the promotion of research at the Technion.

REFERENCES

Carlsson, L.A. & J.W. Gillespie Jr. 1989. Mode II interlaminar fracture of composites. In K. Frie drich (ed.), *Application of fracture mechanics to composite materials*: 113-158. Amsterdam: Elsevier science publishers B.V.

Rotem, A. & E. Altus 1979. Fracture modes and acoustic emission of composite materials. *J. Testing and Evaluation*. 7(1):33-40.

Rotem, A. & H.G. Nelson 1981. Fatigue behavior of graphite/epoxy laminates at elevated temperatures. ASTM, STP 723: 152-173.

Rotem, A. 1984. Fracture modes identification of composite materials by A.E. analysis. *Composites Technology Review*. 6(4): 145-158.

Recent Developments in Durability Analysis of Composite Systems, Cardon, Fukuda, Reifsnider & Verchery (eds)
© *2000 Balkema, Rotterdam, ISBN 90 5809 103 1*

High-frequency dielectric spectroscopy as NDE technique for adhesively bonded composites structures

P. Boinard & R. A. Pethrick
Department of Pure and Applied Chemistry, University of Strathclyde, Glasgow, UK

W. M. Banks
Department of Mechanical Engineering, University of Strathclyde, Glasgow, UK

ABSTRACT: This paper discusses the application of high-frequency non-destructive techniques to the ageing study of carbon fibre reinforced plastics (CFRP) adhesively bonded composite structures. The effect of changes in the surface alignment and subsequent bulk orientation of carbon fibres on the dielectric propagation has been investigated. The ingress of moisture in the raw materials and in the joint structure is presented. The high-frequency time domain response (TDR) analysis allows the integrity of the structure to be explored and a good correlation is shown between TDR analysis and gravimetric results. This study indicates that the application of high frequency dielectric measurements is applicable to CFRP bonded structures. The dielectric study not only indicates a new way to assess the state of such a structure but is also producing new insights into the application of TDR measurement to non-isotropic materials.

1 INTRODUCTION

The integrity of power and signal cables and related structures is typically assessed by the use of time domain electrical measurement techniques. In the past fifteen years development of vector network analysers operating in the frequency range 300 kHz to 3 GHz has allowed accurate measurements of the electrical permittivity and loss in wave-guide structures.

During the past decade, the application of high frequency dielectric spectroscopy to the assessment of aluminium adhesively bonded structures has been investigated. The structure resembles a wave-guide in which the adhesive layer is the dielectric. It has been established that information on the quality and integrity of bonded structures can be obtained (Pethrick et al. 1995,1996 & 1997a to c). Its potential as a non-destructive technique to monitor ageing and degradation of adhesive joint structures during exposure to a harsh hot and humid atmosphere has been demonstrated (Pethrick et al. 1997a). Time domain analysis allowed identification of defects present in the bond line (Pethrick et al. 1996-1997a, b), whereas frequency domain analysis allows the ingress of water in the bond line (Pethrick et al. 1997c).

Degradation of bonded structures occurs when significant amounts of moisture are absorbed by the adhesive. Once the water penetrates the bond line, the adhesive properties can be altered in a reversible manner by plasticisation or in an irreversible manner by hydrolysis or crack and craze formation (Comyn 1991, Shaw 1993). Generally, degradation at the adhesive-adherent interface occurs by displacing the adhesive or by changing the adherent surface chemistry. The diffusion of water in the adhesive tends also to modify the mechanical failure mechanism of the joint from a cohesive failure in the adhesive layer to an adhesive failure at the adhesive-adherent interface (Matthews et al. 1991, Kinloch 1987).

High frequency dielectric spectroscopy techniques proved successful in the study of ageing of aluminium-epoxy resin-aluminium bonded structures since the conducting aluminium adherent generates a wave-guide structure for the propagation of electromagnetic waves. However, the aerospace industry increasingly uses adhesively bonded composite materials in aircraft primary and secondary structures. There is, therefore, a requirement for the development of non-destructive techniques to assess the integrity of these bonded composite structures. This paper explores the potential of high frequency dielectric spectroscopy as a non-destructive evaluation (NDE) method for adhesively bonded composite structures. This paper also addresses how electromagnetic wave propagation may occur in composite structures.

2 EXPERIMENTAL METHODS

2.1 Materials

The carbon fibre reinforced plastic (CFRP) adherents were manufactured from Hexcel Composites Ltd unidirectional carbon fibre pre-impregnated film, trade named 914C-TS(6K)-5-34%. The pre-impregnated film contained 66% by weight of high-tensile surface treated carbon fibres. The adhesive system used for bonding the CFRP plates was a 3M structural epoxy system trade named Scotch-Weld Brand AF-163-2U, consisting of a nylon woven fibre supported epoxy resin.

2.2 Manufacturing process

Carbon fibre pre-preg layers were used to produce a series of different lay-up designs using a vacuum bag procedure in an autoclave. Curing was achieved at a temperature of 170°C under a pressure of 7 bar. A post-cure at 190°C for 4 hours was performed. Joints were manufactured by joining two carbon fibre reinforced plastic (CFRP) plates with a lay-up of adhesive films. The assemblage, placed in an autoclave chamber, was cured using a vacuum bag system at a temperature of 125°C and a pressure of 2 bar for 1 hour. The final system was a single lap joint made of 150 mm length, 50mm width and 1.8 mm thick CFRP plates bonded over a 10 mm wide overlap by a 1 mm thick adhesive.

2.3 Gravimetric measurements

As soon as manufactured, all the samples were placed prior to their ageing in a dessicator at room temperature in order to avoid absorption of moisture. Gravimetric measurements during ageing were performed by removing the samples from a water bath held at a constant temperature of 60°C and rapidly blotted and weighed using an electronic balance (Mettler AJ100) with an accuracy of ±0.1 mg. The times for the weighing experiments were assumed sufficiently short not to influence the values of the mass measured.

2.4 Dielectric measurements

Dielectric measurements were carried out in reflection mode over the frequency range 300kHz to 3GHz using a Hewlett Packard 8753A Network analyser. The system was calibrated using three independent standards whose reflection coefficients are known over the frequency range of interest: short, open circuit and a matched load at 50 Ω. Details of the measurement technique and theory have been presented elsewhere (Pethrick et al., 1995 & 1997b). The ratio of reflected to incident waves ρ for such wave-guide lines is given by

$$\rho(t) = \rho_i(t)\delta(t) - \left[(1 - \rho_i)^2 \rho_T \sum_0^N (\rho_i \rho_T)^{N-1} \delta(t - NT) \right]$$

(1)

where ρ_i and ρ_T are the reflection coefficients at the input and end of the joint respectively, t is the time, T is the total transit time, N is the number of transits in a time t and δ is the Kronecker delta. The reflection coefficients ρ_i and ρ_T are defined by

$$\rho_i = \frac{Z_i - Z_0}{Z_i + Z_0}$$

(2)

$$\rho_T = \frac{Z_T - Z_i}{Z_T + Z_i}$$

(3)

where Z_0, Z_i and Z_T are the characteristic impedance of the system and the joint and the impedance of the line termination respectively. The latter is equal to infinity for a free end, therefore ρ_T equals one. Time domain data were obtained by using the network analyser's inverse Fourier transformation over the frequency range.

3 RESULTS AND DISCUSSION

3.1 Effect of orientation of the carbon fibre layers on the TDR

When good contacts exist between the fibres of CFRP laminates, the composite adherent structure is able to transmit electric current polarising the adherent/adhesive interfaces. In the first attempt, joints were designed so that the unidirectional fibres were in the direction of the wave propagation (Figure 1, top). The limitation of this design was that the water diffused perpendicularly to the mechanical testing direction, resulting in a limited understanding of the effects of water on the degradation of the mechanical properties of the bonded structure.

A design, which eliminates this restriction, was required. In the new design investigated, the unidirectional fibres were orientated in the same direction as the water diffusion and mechanical testing directions, therefore perpendicular to the direction of the electromagnetic wave transmission (Figure 1, bottom).

Results for unidirectional carbon fibre plate joints are presented in Figure 2 and several reflection peaks were observed (solid line). This is a typical TDR spectrum. When all the fibres are perpendicular to the dielectric wave propagation direction, the dissipation of the impulse energy in the first millimetres of the joint prevents the development of secondary reflection peaks. There is enough energy to percolate through the thickness of the plate, as shown by the presence of a first upward peak, but

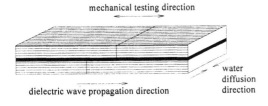

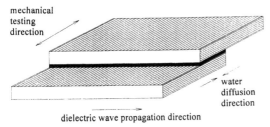

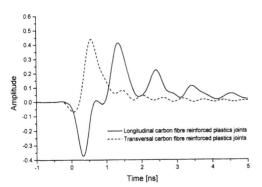

Figure 1. Previous joint design (top) and new joint design concept (bottom).

Figure 2. TDR of longitudinal and transversal CFRP joints.

the high resistivity of the laminate matrix prevents propagation of the pulse along the joint.

The next design investigated consisted of improving the conductivity of the adherent in order to propagate the signal along the joint without significantly affecting the mechanical testing and involved orientation of the fibres in the propagation direction.

During high frequency dielectric spectroscopy, an electric field appears between the CFRP plates producing a magnetic field, which propagates a transverse electro-magnetic (TEM) wave at the adherent/adhesive interface. This part of the theory has not been previously explored since only isotropic materials such as aluminium have been studied. By using composite materials, such as carbon fibre prepregs, there is the possibility to modify the adherent/adhesive interface by changing the orientation of the fibre of the laminae at the interface.

For this purpose, five different lay-ups following the sequences $[0_{14}]$, $[0_2/90/0_2/90/0]_S$, $[0/90_3/0/90_2]_S$, $[90_2/0/90_2/0/90]_S$ and $[90/0_3/90/0_2]_S$ were manufactured. The differences between these lay-ups, which modify the overall dielectric property of the adherent, are designated in the direction of the last laminae at the interface and the percentage of fibres in each direction. The percentage of fibres in the 0°direction in each of the lay-ups is 100%, 74%, 26%, 26% and 74%, respectively. It should be noted that the directions of the carbon fibres are given relative to the dielectric wave propagation direction. A 90° fibre direction means that the fibres are perpendicular to the dielectric wave propagation (Figure 1, bottom). Figure 3a presents the TDR spectrum of the $[0_{14}]$, $[0_2/90/0_2/90/0]_S$ and $[0/90_3/0/90_2]_S$ lay-up carbon fibre plate-adhesive joint (respectively 100%, 74% and 26% of the carbon fibre at 0°) with the direction of the fibre of the laminae at the interface being at 0°. Figure 3b presents the TDR spectrum of the $[90_2/0/90_2/0/90]_S$ and $[90/0_3/90/0_2]_S$ lay-up carbon fibre plate-adhesive joint (respectively 74% and 26% of the carbon fibre at 0°) with the direction of the fibre of the laminae at the interface being 90°.

The sequence of the layers does not have an influ-

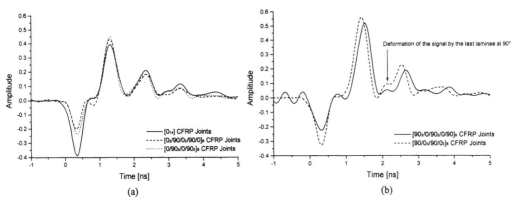

Figure 3. TDR of different carbon fibre lay-up: (a) laminae at the interface at 0° and (b) laminae at the interface at 90°.

ence on the response and, as seen in Figure 3b, the presence of some layers at 90° does not significantly disturb the response when comparing the $[0_{14}]$ response with the other laminate responses.

From these results, the $[90_2/0/90_2/0/90]_S$ configuration was selected for the adherent plates of the new joint design. This implies 14 layers of carbon fibres, 10 at 90° and 4 at 0°, representing 74% of fibres at 90° and 26% at 0°. Figure 4 presents a comparison between the TDR of the initial design and the new design.

The lower amplitude in the impulse (downward) peak indicates a better balance between the impedance of the measuring system and the joint structure. The higher amplitude of the first response (upward) peak corresponds to an improved polarisation of the CFRP-adhesive interfaces. It is believed that the percolation between the fibres at 90° minimises the energy required for the electrical wave to reach the adherent-adhesive interfaces, consequently optimising the polarisation of the adhesive. The displacement of the peaks towards longer response time is consistent with the high resistivity of the CFRP matrix, which decreases the velocity of the TEM wave.

3.2 Moisture ingress study

3.2.1 Carbon fibre layer orientation effect on water absorption

Figure 5 presents the results for different bulk orientation of carbon fibre layers in the composite adherent. The two designs presented are following the sequence $[0_2/90/0_2/90/0]_S$ and $[90_2/0/90_2/0/90]_S$. A significant difference between the plates is the amount of fibre ends exposed on each side of the plates. CFRP plates from the second design have more fibre ends in the length of the plates and therefore present a greater proportion of fibre ends to the ageing environment than plates from the first design. The $[90_2/0/90_2/0/90]_S$ plates would consequently be more sensitive to water absorption by capillary mechanism.

As shown in Figure 5, the initial rate of water uptake is higher for the second design. This confirms the concern raised previously in section 3.1 regarding the effect of the amount of fibre ends exposed to the ageing medium. However, after long time exposure, the two designs present the same rate of absorption and the same amount of water absorbed. This indicates that the matrix is at an equilibrium value.

Figure 5 also shows that all the plates are following a non-Fickian behaviour. The weight of the plates does not reach any equilibrium but steadily increases. This behaviour has been previously observed in carbon fibre–epoxy matrix systems (Cai and Weitsman, 1994). The increase in weight may be generated by the formation of clusters inside the

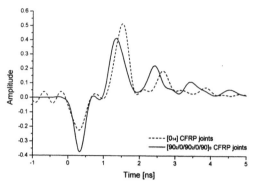

Figure 4. TDR of $[0_{14}]$ and $[90_2/0/90_2/0/90]_S$ CFRP joints.

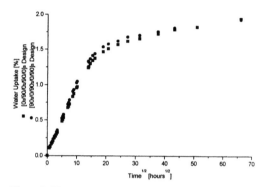

Figure 5. Water uptake in different CFRP plate designs, (■) $[0_2/90/0_2/90/0]_S$ and (●) $[90_2/0/90_2/0/90]_S$.

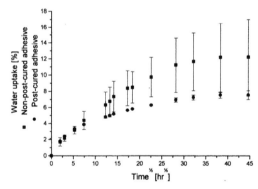

Figure 6. Water uptake in (■) non-post-cured and (●) post-cured adhesive.

material, due to disbonding of the fibre/matrix interface.

3.2.2 Effect of adhesive post-cure on water absorption

Figure 6 presents the water uptake profile for non-post-cured and post-cured adhesive. The large deviation around the value shown by the non-post-cured adhesive confirms the poor quality of the three-dimensional network of the epoxy resin, which results in a significant variation of the water absorption value. After a post-cure at 145°C for 8 hours, the water absorption of the adhesive was significantly decreased and more consistent results were obtained from sample to sample, as shown by the smaller deviation around the mean.

3.2.3 Moisture absorption in joint structure

Figure 7 presents the results for the joint structure, the CFRP adherents alone and the adhesive. The amount of water present in the adhesive has been estimated by subtracting the amount of water present in the joint to the amount of water present in the CFRP plates. The latter has been calculated from gravimetric results of CFRP plates exposed to identical ageing conditions. From these results, it is possible to dissociate the amount of water in the adherent from the amount of water absorbed by the adhesive.

The adhesive in the joint absorbed less than the raw material. This is due to the difference in the absorption configuration: in the previous experiment all the surfaces of the adhesive sheet were in contact with the water whereas in the joint structure, only the edges are exposed to water.

3.3 Effect of water on the TDR of adhesively bonded composite structures

TDR measurements were performed during ageing and are presented in Figure 8. The time difference between two neighbouring peaks is related to the effective permittivity $\bar{\varepsilon}$ of the material between the wave-guides (Pethrick et al., 1997a) by

$$\bar{\varepsilon} = \left(\frac{c}{2 \cdot l / \Delta t} \right)^2 \qquad (4)$$

where c is the velocity of light in vacuum, l is the physical length of the joint, Δt is the time difference between the impulse peak and the first response peak and $\bar{\varepsilon}$ is the average dielectric permittivity. Figure 8a shows the displacement of the response peaks towards longer response time values. This is caused by the evolution of the dielectric permittivity of the adhesive due to the absorption of water which decreases the electrical wave velocity, consequently increasing the response time. Figure 8b presents the evolution of the average dielectric permittivity as a function of exposure time.

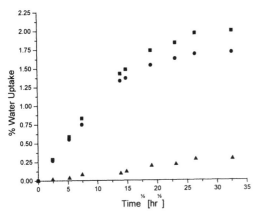

Figure 7. Water uptake of (■) the joint structure, (●) the CFRP adherents and (▲) the adhesive.

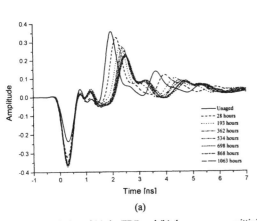

(a)

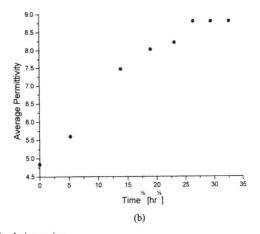

(b)

Figure 8. Evolution of (a) the TDR and (b) the average permittivity during ageing.

335

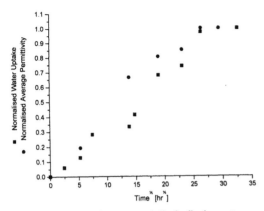

Figure 9. Correlation between normalised adhesive water uptake and average dielectric permittivity.

3.4 Correlation between water uptake and TDR of adhesively bonded composite structures

In order to compare the time domain results with the gravimetric results, normalisation of the data was performed using Equation 5

$$Data_{Norm} = \frac{Data_t - Data_0}{Data_\infty - Data_0} \qquad (5)$$

where $Data_t$ is the value at time t, $Data_0$ is the initial value, $Data_\infty$ is the final value at equilibrium and $Data_{Norm}$ is the normalised value. A correlation coefficient of 0.96 between the TDR results with the gravimetric results shown in Figure 9 confirms that water penetration in adhesively bonded structure could be assessed by high frequency TDR measurement.

4 CONCLUSION

This study has shown the effect of the carbon fibre orientation on the high-frequency time domain analysis. It has been established that non-isotropic adherents, when well designed, provide good waveguide for the TEM wave propagation. Correlation between the TDR and gravimetric data validated the use of TDR to identify the change in the dielectric permittivity of the bond line and to relate it to the water uptake in the structural adhesive.

This study on adhesively bonded composite structure has shown the potential of high frequency dielectric TDR as a non-destructive evaluation method to assess the integrity of the bond line present in aircraft primary and secondary structures.

5 ACKNOWLEDGEMENT

One of the authors (PB) wishes to thank the Non Destructive Evaluation Branch, Materials Directorate of the US Air Force for the provision of a maintenance grant in support of this study (grant No. F49620/97/1/0530), British Aerospace for the provision of materials and Dr D. Hayward at the Department of Pure and Applied Chemistry, University of Strathclyde, for his high frequency dielectric spectroscopy knowledge and advice.

REFERENCES

Comyn J., 1981, 'The relationship Between Joint Durability and Water Diffusion', Developments in Adhesives – Part 2, Applied Science Publishers, London, pp. 279-314.

Kinloch A.J., 1987, Adhesion and adhesives: Science and Technology, Chapman and Hall, London.

Matthews F.L., Kinloch A.J. and John S.J., 1991, 'Measuring and Predicting the Durability of Carbon Fibre/Epoxy Composite Joints', Composites, Vol. 22, No. 2, pp. 121-126.

Pethrick R.A., Hayward D., Joshi S.B., Li Z-C., Jeffrey K. and Banks W.M., 1995, 'High-Frequency Dielectric Investigations of adhesive bonded structures', Insight, Vol. 37, No. 12, pp. 964-968.

Pethrick R.A, Dumoulin F., Hayward D., Li Z-C. and Banks W.M., 1996, 'Non-Destructive Examination of Composite Joint Structures', Journal of Physics Part D, Applied Physics, Vol. 29, pp. 233-239.

Pethrick R.A., Hayward D., Gilmore R. and Li Z-C., 1997a, 'Investigation of Moisture Ingress into Adhesive Bonded Structures Using High-Frequency Dielectric Analysis', Journal of Materials Science, Vol. 32, pp. 879-886.

Pethrick R.A., Joshi S., Hayward D., Gilmore R. and Li Z-C., 1997b, 'High-Frequency electrical Measurements of Adhesive Bonded Structures – An Investigation of Model Parallel Plate Waveguide Structures', Non-Destructive Testing & Evaluation International, Vol. 30, No. 3, pp. 151-161.

Pethrick R.A., Hayward D., Gilmore R., Yates L.W. and Joshi S., 1997c, 'Environmental Ageing of Adhesively Bonded Joints: I. Dielectric Studies', Journal of Adhesion, Vol. 62, pp. 281-315.

Shaw S.J., 1993, 'Epoxy Resin Adhesive', Chemistry and Technology of Epoxy Resins, Blackie Academic & Professional, Glasgow, pp. 206-255.

Weitsman Y. and Cai L-W., 1994,' Non-Fickian Diffusion in Polymeric Composites', Journal of Composite Materials, Vol. 28, No. 2, pp. 130-154.

Recent Developments in Durability Analysis of Composite Systems, Cardon, Fukuda, Reifsnider & Verchery (eds)
© *2000 Balkema, Rotterdam, ISBN 90 5809 103 1*

Influence of the shape of a striker on the mechanical responses of composite beams

L.Guillaumat
LAMEFIP-ENSAM Esplanade des Arts et Métiers, Talence, France

ABSTRACT: The aim of this study is to analyze the mechanical responses of laminated composite beams under low velocity impact loading using several shapes of the stricker end to simulate the accidental fall of a tool. The impact tests have been performed using a drop weight set-up. The composite is made of glass fibres reinforced isophtalic polyester resin. Whatever the striker end the mechanical results are very similar and suggest that the shape of the impactor has no influences because the area of the end is small against the area of the beam. On the other hand, the damage induced is different according to the shape of the impactor.

1 INTRODUCTION

Design of structures is often made with a deterministic approach using the mean value of each variable of the system. However, this kind of design is not always efficient for industrial structures because of the variability of both mechanical behaviour and solicitation. The three most important types of loading are impact, fatigue and vibrations. The impact solicitation applied on a structure can be due, for instance, to an accidental shock such as the fall of a tool during maintenance or manufacturing. Impact loading can be defined in terms of the geometry, mass and speed of the impacter, the size of the composite plate, and the boundary conditions (Bernard & Lagace 1989), (Chun-Gon & Eui-Jin 1992), (Davies & Choqueuse & Pichon 1994), (Nemes & Simmonds 1990), (Mines & Worrall & Gibson 1994), (Shih & Jang 1989). The influences of the induced damage on the reliability of the affected piece are not always well recognized and rarely used to design the structure.

This paper deals with some mechanical responses of laminated composite beams under low velocity impact loading using different shapes of the end of the impactor in order to simulate the fall of several types of tools such as hammer, screwdriver and so on.

2 EXPERIMENTAL METHODS

The impact tests have been performed using a drop weight set-up (Figure 1). The device is composed of a metallic tower which supports two columns. These ones guide the drop weight during the fall. A concrete framework is used to support the specific set-up which hold the specimen.

An electric motor is used with a magnet to raise the mass. The drop height is determined by the position of a first infra-red captor which stops the motor at the desired level. A second captor located on the set-up close to the specimen activates a rebound apparatus to prevent multiple impacts on the structure. The contact load between the striker and the composite specimen as a function of time is measured by means of a piezoelectric captor which is attached to the drop weight. Two laser captors provide the striker displacement and the deflection at the centre of the composite panel versus time. The tests have been filmed by a high speed camera system Camsys+ (from 2,000 up to 11,000 frames/sec). All the results were recorded using numerical oscilloscopes and after they were transferred in a computer.

The composite structure is put on two steel opposite supports in bending condition. Two flexible sheets made of metal with a rubber end maintain it (Figure 2). These boundary conditions allow rotations.

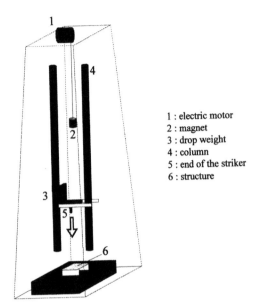

Figure 1. Drop tower.

1 : electric motor
2 : magnet
3 : drop weight
4 : column
5 : end of the striker
6 : structure

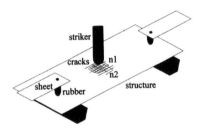

Figure 2. Boundary conditions.

Figure 3. Ends of the striker.

We have chosen five shapes for the end of the impactor (Figure 3).

The first one which is hemispherical simulates for instance the back of a handle tool. The geometry of the second one is flat. The third one simulates the end of a screwdriver. The fourth one simulates a

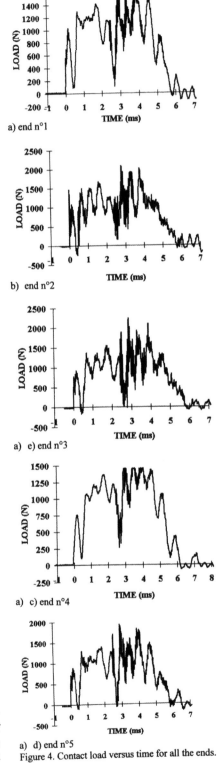

a) end n°1

b) end n°2

a) e) end n°3

a) c) end n°4

a) d) end n°5

Figure 4. Contact load versus time for all the ends.

338

pointed thing and the fifth simulates a part of a adjustable spanner for example.

Moreover, the total mass of the drop weight was about 1 kg and the velocity was 3.8 ms^{-1} corresponding to a height of 0.75m.

The striker is 20 mm in diameter.

3 MATERIAL

The composite material is made of glass fibres reinforced isophtalic polyester resin.

Several plates of about 1 m^2 were made by the hand lay-up process. All the plates are constituted of 8 layers which give a total thickness of about 3.5 mm. The stacking sequence was [0$_3$,90]s.

The specimens of this study have been cut using a band-saw from these plates and their size was 350 mm long and 66 mm wide. All the edges were polished with glass-paper.

4 RESULTS

The plot of the contact load between the striker and the structure versus time for the five ends are given (Figure 4).

For the results corresponding to the ends number 1 and 4 we have had a saturation of the signal which explains that some peaks are cut.

We can notice that the mechanical results for all the tests are very similar (Table1, Figure 4) and suggest that the shape of the impactor has no influences on the mechanical responses of the composite beams.

The maximum load for each test was measured using only the filtered signal as explain below.

Although the surface of the contact between the striker and the structure is different due to the shape of the end, it seems that this surface is too small in keeping with the surface of the beam to change the mechanical responses. Consequently, the kinetic energy of the drop weight has the same effect (for a given panel) whatever ends used in this study. The

Table1. Mechanical responses of the impact tests for different shapes of the striker.

End (Figure 3)	1	2	3	4	5
Deflection (mm)	6.2	6.5	5.9	6.2	5.8
Time of conctact (ms)	6.6	6.7	6.7	6.7	6.8
Maximum load (N)	1204	1206	1271	1255	1290

mass and the velocity of the drop weight and the dimensions of the composite structure remain the main parameters.

The films given by the high speed camera show that the undulations on the curves load versus time (Figure 4 - Figure 5, curve 1) can be attributed to the eigen modes of the panel induced by the dynamic solicitation during the shock. Effectively, firt when the striker touches the beam we can observe a local deformation which propagate up to the edges. After that, the bending of the panel begins. But, during this bending the edges which are not maintained (parallels to the span) have a sinusoïdal movement suggesting the existence of eigen modes.

To analyse only the vibratory response of the system we have first to filtered the measured signal at a given frequency to obtain only the bending response of the composite structure (Figure 5, curve 2). This frequency is linked to the time of contact between the striker and the panel because it is the inverse. In fact, the time of contact is about 6 ms and this corresponds to a frequency close to 200 Hz. After that,

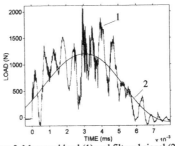

Figure 5. Measured load (1) and filtered signal (2) versus time.

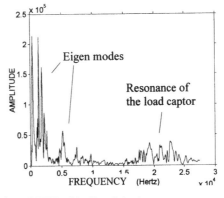

Figure 6. F.F.T. of the filtered signal.

339

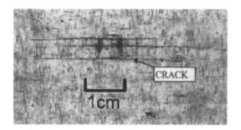

e) end n°5
Figure 7. Damage induced by each end.

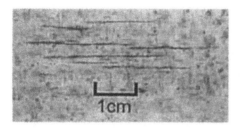

Table2. Eigen modes frequencies of the composite beams for each geometry of the striker end.

End number	1	2	3	4	5
Frequency 1 (Hz)	350	350	320	350	400
Energy 1 rate (%)	46	30	30	45	29
Frequency 2 (Hz)	1350	1350	1350	1350	1400
Energy 2 rate (%)	23	15	21	20	23
Frequency 3 (Hz)	1850	1950	1850	1870	1900
Energy 3 rate (%)	21	26	28	18	25
Frequency 4 (Hz)	2300	2320	2250	2300	2250
Energy 4 rate (%)	5	13	12	4	6
Frequency 5 (Hz)	5000	5100	5200	5200	5200
Energy 5 rate (%)	4	2	5,5	3	5

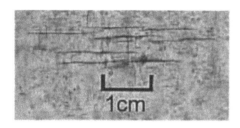

d) end n°4 : lower face

the difference between the curve 2 and 1 is calculated; It contains the frequencies of the eigen modes of the composite beam, the eigen modes of the drop weight and the resonance of the piezoelectric sensor. A F.F.T. of this difference is done to obtain all the frequencies (Figure 6).

Now, it was interesting to calculate the energy corresponding to each peak of the F.F.T. to compare the influence of all the components of the system (plate, drop weight, load sensor) on the mechanical responses.

In fact, the energy spectrum density (e.s.d.) was used because it is easier to analyse its result. The surface of each peak of the e.s.d. is calculated and it is normalised by dividing by the total surface of the e.s.d.

Consequently, we have the energy rate for each frequency. The eigen modes concerning only the composite structure are given in the Table 2.

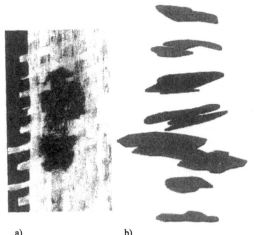

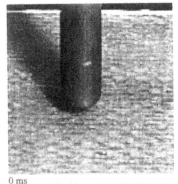

0 ms

+ 2 ms

a)   b)

Figure 8. 3D reconstruction of the delamination.
a) a photograph of a baked layer, b) draw of the 3D delamination.

First, taking into account that the boundary conditions prevent a vertical movement of the composite sample on the level of the supports and that the striker position during the contact prevent a relative movement of the panel at this location we can notice that the structure was able only to provide five main eigen modes. Moreover, the frequency 1and 2 represent from 40% up to 70% of the total energy of the system.

We can observe further frequencies in the spectrum which are attributed to the load sensor. Effectively, the resonance frequency given by the manufacturer is about 20,000 Hz and this value is close to the measured ones (Table 3).

On the other hand, the damage induced in each panel is different according to the shape of the impactor end (Figure 7).

For all the ends except for the pointed one the visible damage was only matrix cracking. We can observe long cracks parallel to the span and short cracks parallel to the width. The pointed end induces matrix cracking too, delamination and the beginning of a perforation. Obviously, this last kind of damage is more severe and could modify the behaviour of impacted panels submitted to a mechanical solicitation such as fatigue.

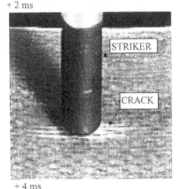

STRIKER

CRACK

+ 4 ms

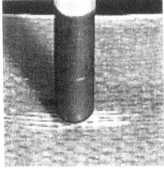

Table3. Vibratory responses of the load sensor.

Frequency (Hz)	18,000	20,000	23,000
Energy rate (%)	1.8	3.2	4.0

+ 4.5 ms

Figure 9. Evolution of the matrix cracking versus time.

341

A 3D analysis of the delamination has been realised using the de-ply technique. It was the same material but loaded with an another couple mass-velocity (close to the one used in this study) and with an hemispherical end. This technique allowed direct observation of delaminations at each interface in the composite (Freeman 1980) (Harris 1989). We used a solution of zinc iodide to penetrate the delaminations. The laminate is exposed to temperatures in the range of 150°C for a period of 1-2 h in order to vaporise the solvent. During this period the zinc iodide crystallises and is deposited on the delamination surfaces. In order to realise a partial pyrolysis of the matrix the composite is baked at about 350°C for about 1h. After that, all the laminae can be separated with a sharp blade and examined under a microscope (Figure 8a). From all the observations of each interface of the laminate we can reconstruct a 3D picture of the delamination. This picture (Figure 8b) shows that the delamination exist nearly at all the interfaces.

Moreover, the delamination area increases towards the back surface of the laminate and is the largest in the furthest interface from the impact surface. It propagated in the fibre direction of the layer below the interface. These phenomena are well known (Gao & Kim 1998) (Kim & Jun 1992). We can assume that the damage which has been observed in this previous study is close to the one induced here.

From the films of the high speed camera it is possible to reconstruct the chronological evolution of the damage. For example, the Figure 9 shows the growth of the matrix cracking. The time which is below each photograph gives the time between the current photograph and the previous.

5 CONCLUSION

First, this study has shown that there are no main influences of the geometry of the striker end on the global mechanical responses because its surface is small with regard to the surface of the impacted composite structure. We can only observe that the damage induced by the shock depends on the shape of the end. The more severe damage is due to the pointed end which creates a delamination and a beginning of perforation.

The experimental device used allows to obtain a lot of information about the mechanical responses of the impacted structure (load of contact between the panel and the impactor, deflection, velocity of the drop weight, ...). Moreover, from the films of the high speed camera we can identify a few eigen modes and the chronological evolution of the damage.

These impacted beams will be loaded under fatigue solicitation in order to quantify the influence of an accidental shock on the use of the structure.

REFERENCES

Bernard, M.L. & Lagace, P.A. 1989. Impact resistance of composite sandwich plates. Journal of Reinforced Plastics and Composites, vol 8, 432-445.

Chun-Gon, K. & Eui-Jin, J. 1992. Impact resistance of composite laminated sandwich plates. Journal of Composite Materials, Vol 26, n°15, 2247-2261.

Davies, P. & Choqueuse, D. & Pichon, A. 1994. Influence of the foam core on composite sandwich static and impact response. European Conference on Composites Testing and Standardisation, Sept 13-15, Hamburg Germany, 513-521.

Freeman, S.M. 1980. Characterization of lamina and interlaminar damage in graphite/epoxy composites by the de-ply technique. A.S.T.M. S.T.P., 787, 50-64.

Gao, S.L. & Kim, J.K. 1998. Three-Dimensional Characterization of impact damage in CFRPs. Impact response and dynamic failure of composites and laminate materials, Part 1 : impact damage and ballistic impact, Editors Kim and Yu, Trans tech publications, 35-54.

Harris, C.E. 1989. Damage evaluation by laminate de-ply. Manual on experimental methods for mechanical testing of composites, Soc. Exp. Mechanics, CN, USA, 147-149.

Kim, C.G. & Jun, E.J. 1992. Measurement of impact delamination by de-ply technique. Exp. Tech., 26-28.

Mines, R.A.W. & Worrall, C.M. & Gibson, A.G. 1994. The static and impact behaviour of polymer composite sandwich beams. Composites, Vol 25, n° 2, 95-110.

Nemes, J.A. & Simmonds, K.E. 1990. Low velocity impact response of foam-core sandwich composites. Journal of Composite Materials, Vol 26, n° 4.

Shih, W.K. & Jang, B.Z. 1989. Instrumented impact testing of composite sandwich panels. Journal of reinforced plastics and composites, vol 28, 270-298.

Experimental characterization of damage progress in high temperature CFRP laminates

S. Ogihara & A. Kobayashi
Department of Mechanical Engineering, Science University of Tokyo, Chiba, Japan

N. Takeda, S. Kobayashi & M. Oba
Department of Aeronautics and Astronautics, University of Tokyo, Japan

ABSTRACT: Transverse cracking under tensile loading in carbon/BMI (bismaleimide), G40-800/5260, and carbon/epoxy, T800H/3900-2, composite laminates with toughened-interlaminar layers is investigated experimentally. Laminate configurations are $[0/90]_s$, $[0/90_2]_s$, $[\pm45/90]_s$ and $[\pm45/90_2]_s$ for G40-800/5260, while $[0/90]_s$ and $[\pm45/90]_s$ for T800H/3900-2. Damage mechanics analysis is used to predict transverse cracking based on both the energy and stress criteria. The present analysis can be used as a characterization method of transverse cracking resistance of a material, which will be helpful in ranking materials.

1 INTRODUCTION

Carbon fiber reinforced plastics (CFRP) are remarkable for their high specific modulus and strength. CFRP is a candidate for the primary structure of supersonic transport plane (SST). Surface temperature of SST can reach -55°C at taking off and landing, while it reaches 150°C-180°C on a supersonic cruise of Mach 2.2-2.4. In this point of view, matrix resins are expected to have good heat resistance. Bismaleimide (BMI) is one of the high temperature resins. BMI has higher heat-proof quality than the conventional epoxy resins. However, damage tolerance of CFRP with BMI matrix is not well clarified. In the present study, microscopic damage progress in carbon/BMI composite laminates under static tension is investigated experimentally. Transverse matrix cracking which will be an important factor in designing laminated composites is focused.

Many investigations have been conducted on transverse cracking in laminated composites which consist of experimental observation and analytical modeling. Masters & Reifsnider (1982) observed transverse crack behavior by using the replica technique. It was shown that transverse cracks saturated before the final fracture. Takeda & Ogihara (1994) conducted in-situ observation of damage process in CFRP cross-ply laminates. The interaction between the transverse cracks and delamination was explained by using a modified shear-lag analysis which consider the delamination effect. Nairn (1989) derived variational stress analysis considering the thermal residual stresses for cross-ply laminates. McCartney (1992) analyzed stress transfer between 0° and 90° plies in cracked cross-ply laminates. The solutions are determined to satisfy the stress-strain-temperature relations either exactly or in an average sense. These analyses can be applied only to cross-ply laminates.

Gudmundson & Zang (1993) derived the thermoelastic properties of composite laminates containing matrix cracks. In the analysis, the concept of damage mechanics analysis was used. The applicability of the analysis is not limited to cross-ply laminates. Ogihara et. al. (in press) derived the energy release rate associated with cracking using this analysis and modeled the transverse crack behavior.

In the present study, microscopic damage progress in carbon/BMI laminates is investigated experimentally. Based on the experimental results and damage mechanics analysis, transverse crack behavior is modeled to evaluate the transverse cracking resistance. Furthermore, results of carbon/BMI composites are compared with the results of carbon/epoxy laminates with toughened-interlaminar layers. The methodology described in the present paper will be a procedure for the selection of SST structural materials and damage tolerance design.

2 EXPERIMENTAL PROCEDURE

Material systems used are G40-800/5260 and T800H/3900-2. G40-800/5260 is a bismaleimide-based CFRP and is a candidate for SST structural material. It is reported that G40-800/5260 material has high compressive strength and CAI strength at high temperature. T800H/3900-2 is carbon/epoxy composite with toughened-interlaminar layers. It is also reported that T800H/3900-2 material has higher hot-wet compressive strength and CAI strength (Odagiri et al. 1988, 1991) than conventional carbon/epoxy composites. Delamination suppression around an open hole under tensile loading was also shown (Takeda et al. 1998).

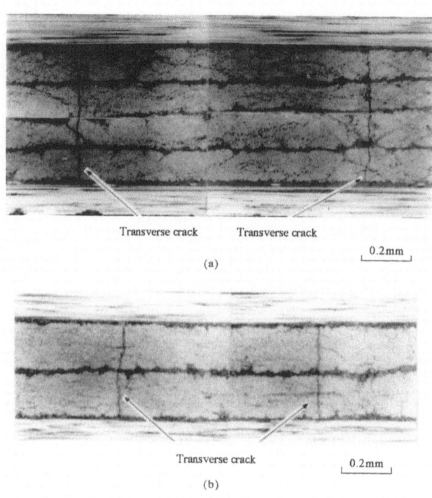

Transverse crack   Transverse crack

0.2mm

(a)

Transverse crack

0.2mm

(b)

Figure 1   Microscopic damage in (a) G40-800/5260 [0/90$_2$]$_s$ (ε=1.3%) and (b) T800H/3900-2 [0/90]$_s$ laminates (ε=1.3%).

Transverse crack

10mm

Figure 2   Microscopic damage in G40-800/5260 [0/90]$_s$ laminate (X-ray radiography, ε=1.3%).

The size of the specimen are 150 mm long, 20 mm wide for G40-800/5260, while 25 mm for T800H/3900-2. 35 mm long GFRP tabs are glued on the end of the specimens. $[0/90]_s$ and $[\pm45/90]_s$ laminates are prepared for both material systems to clarify the difference in the microscopic damage behavior. For G40-800/5260, $[0/90_2]_s$ and $[\pm45/90_2]_s$ laminates are also prepared to investigate the effect of laminate configurations on the microscopic damages. Edges of specimens are polished for microscopic observation. Laminate strain is obtained by the strain gauge glued on the center of the specimen. Tensile tests are conducted at room temperature. The cross-head speed is 0.5 mm/min. Tensile testing machine is periodically stopped and the edge of the specimens is observed by using an optical microscope. The observed area is 50mm long at the center of the specimen. The number of the transverse cracks per unit length is defined transverse crack density. Internal damage progress is observed by soft X-ray radiography (Ogihara et al. 1997). Iodozinc (ZnI), a dye penetrant opaque to X-rays, is applied along the specimen edge. Specimens are exposed to X-rays for 150 s at 20 kV and 2 mA. An X-ray point-source unit positioned 570mm away from the specimen is used.

3 ANALYSIS

In modeling transverse crack behavior in composite laminates, there are two important stages. The first is selection of stress transfer mechanics in laminated composites with transverse cracks. In previous studies, shear-lag analysis (Masters & Reifsnider 1982, Takeda & Ogihara 1994), variational analysis (Nairn 1989) and approximate elastic analysis (McCartney 1992) have been developed. But application of these analyses is limited only to cross-ply laminates containing transverse cracks in the 90° plies. In the present study, damage mechanics analysis is used which can be applied to laminates with arbitrary laminate configurations. The second stage is selection of transverse cracking criterion. In the early period, the stress criterion was often used. However, to explain the constraint effect (transverse crack onset stress (strain) becomes larger in the case of thin 90° plies), the energy criterion was developed. Recently, in many papers, experimental results are compared with analytical predictions based on only the energy criterion, because thin 90° plies are usually used. The authors have shown that both the energy and the stress criteria must be considered when predicting transverse cracking in laminates with various 90° ply thicknesses (Ogihara et al. 1998). In the present study, both the energy release rate associated with transverse cracking and the average stress of 90° ply calculated by using the damage mechanics analysis are considered. The damage mechanics analysis is outlined in the following.

Gudmundson and Zang (1993) derived in-plane compliance matrix, $\mathbf{S}_{II}(\rho^k)$, of a composite laminate containing transverse cracks as

$$\mathbf{S}_{II}(\rho^k) = \left\{ \mathbf{S}_{II}(0)^{-1} - \sum_{k=1}^{N} v^k \rho^k (\mathbf{A}^k)^T \sum_{i=1}^{N} \boldsymbol{\beta}^{ki} \mathbf{A}^i \right\}^{-1}$$

(1)

where, v^k is volume fraction of ply k (90° ply), ρ^k is normalized transverse crack density of ply k (=transverse crack density × the thickness of ply k), $\mathbf{A}^k$ is the matrix defined by compliance matrix and normal vector of crack surface of ply k, $\boldsymbol{\beta}^{ki}$ is the matrix associated with average crack opening displacement of ply k. Young's modulus of the laminate, $E(\rho^k)$, is derived from $\mathbf{S}_{II}(\rho^k)$.

Energy release rate associated with transverse cracking considering thermal residual stress can be derived using $E(\rho^k)$ (Ogihara et al. in press). Now, a new transverse crack is assumed to occur at the midway between the two existing transverse cracks. When normalized transverse crack density becomes ρ^k from $\rho^k/2$, the relation between the laminate stress, σ, and the energy release rate, $G(\rho^k)$ is derived as

$$G(\rho^k) = \frac{\sum_{i=1}^{N} a^i}{\rho^k} (\sigma - \sigma_T)^2 \left(\frac{1}{E(\rho^k)} - \frac{1}{E(\rho^k/2)} \right)$$

(2)

where a^i is the thickness of ply i, σ_T is a parameter to consider the effect of thermal residual stress.

In the energy criterion, transverse cracks are assumed to occur when the energy release rate reaches a critical value G_c. Therefore, the relation between the laminate stress, σ, and the normalized transverse crack density, ρ^k can be expressed as

$$\sigma(\rho^k) = \sqrt{\frac{G_c \rho^k}{\sum_{i=1}^{N} a^i} \left(\frac{1}{E(\rho^k)} - \frac{1}{E(\rho^k/2)} \right)^{-1}} + \sigma_T$$

(3)

Gudmundson and Zang (1993) also derived the relation between the average stress in each ply and the normalized transverse crack density. In our previous paper (Ogihara et al. in press), we derived the relation between the laminate stress, σ, and the average stress of 90° ply (ply k) in the tensile direction, $\sigma^{k(a)}$, for the case that only the ply k is the cracking ply as

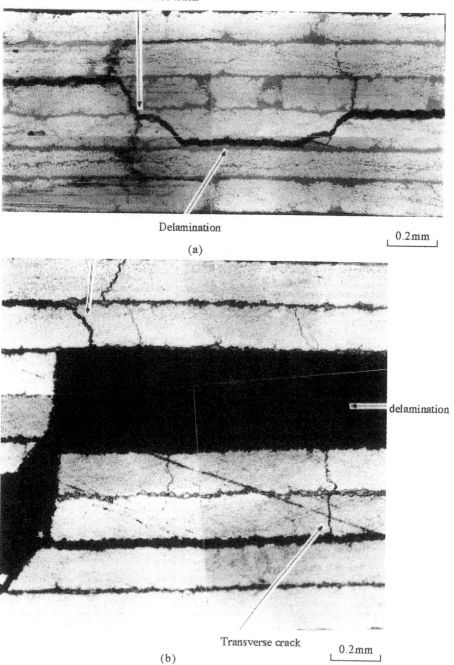

Transverse crack

Delamination

0.2mm

(a)

delamination

Transverse crack

0.2mm

(b)

Figure 3   Microscopic damage in (a) G40-800/5260 [±45/90]$_s$ (ε=1.6%) and (b) T800H/3900-2 [±45/90] laminates (ε=1.1%).

346

$$\sigma\left(\rho^k\right) = \frac{\left[\dfrac{\sigma^{k(a)}}{1-\rho^k\beta^{kk}_{(1,1)}\left(S_\Pi\right)^{-1}_{(1,1)}}+\left[\left\{\alpha_{90}-\alpha_1\left(\rho^k\right)\right\}\left(S_\Pi\right)^{-1}_{(1,1)}+\left\{\alpha_0-\alpha_2\left(\rho^k\right)\right\}\left(S_\Pi\right)^{-1}_{(1,2)}\right]\Delta T\right]}{\left(S_\Pi^k\right)^{-1}_{(1,1)}S_{\Pi(1,1)}+\left(S_\Pi^k\right)^{-1}_{(1,2)}S_{\Pi(1,2)}} \qquad (4)$$

where S_Π^k is compliance matrix of ply k, α_0 is longitudinal thermal expansion coefficient of unidirectional composite, α_{90} is transverse thermal expansion coefficient of the unidirectional composite and $\alpha_2(\rho^k)$ is transverse thermal expansion coefficient of the laminate. In the average stress criterion, transverse crack is considered to onset when average stress, $\sigma^{k(a)}$ reaches a critical average stress, σ_c. In the present analysis, both the energy and the average stress criteria are considered. This method leads the precise prediction of transverse crack behavior in various laminate configurations.

4 RESULTS AND DISCUSSION

In $[0/90_n]_s$ $(n=1, 2)$ laminates, the first microscopic damage observed is transverse cracks in 90° ply in both G40-800/5260 and T800H/3900-2. Figure 1 shows edge view of the damaged specimens ((a) G40-800/5260 $[0/90_2]_s$, $\varepsilon=1.3\%$, and (b) T800H/3900-2 $[0/90]_s$ $\varepsilon=1.3\%$, where ε is laminate strain). Transverse cracks onset and grow in both the thickness and width directions instantaneously as shown in Figure 2 (G40-800/5260 $[0/90]_s$, X-ray radiography, $\varepsilon=1.3\%$). No delamination is observed in all the cross-ply laminates.

In $[\pm45/90_n]_s$ laminates, first microscopic damage observed is transverse cracks in 90° ply in both material systems. In G40-800/5260 laminates, matrix cracks in -45° ply are observed at the tips of transverse cracks in 90° ply. On the other hand, in T800H/3900-2 laminates, matrix cracking in -45° ply at the tips of transverse cracks in 90° ply is suppressed by the toughened-interlaminar layers. Figure 3 shows microscopic damages in (a) G40-800/5260 $[\pm45/90]_s$, ($\varepsilon=1.6\%$) and (b) T800H/3900-2 $[\pm45/90]_s$ laminates ($\varepsilon=1.1\%$). Matrix cracks in -45° ply do not grow in the width direction extensively as shown in Figure 4 (T800H/3900-2 $[\pm45/90]_s$, $\varepsilon=1.4\%$). Delamination at -45/90 and 90/90 interfaces are also observed in both material systems. Delamination occurs at free edge and grows in the loading and width directions.

Figures 5 and 6 show the relation between transverse crack density and laminate strain in G40-800/5260 and T800H/3900-2 laminates, respectively. Transverse crack behavior in the same laminate configurations in the two material systems is compared. First cracking strains and transverse crack densities at final fracture are larger in G40-800/5260 laminates. In G40-800/5260 $[\pm45/90]_s$ laminates, crack density in -45° ply sometimes becomes larger than that in 90° ply. This may be due to delamination at -45/90 interface reduces stress

in 90° ply. In this case, ±45° ply carry more load and little matrix cracks in +45° ply are also observed. In T800H/3900-2 $[\pm45/90]_s$ laminates, crack density in -45° ply is smaller than that in G40-800/5260 laminates, because matrix cracks in -45° ply at the tips of transverse cracks in 90° ply are suppressed by the toughened-interlaminar layers. Delamination was also observed in T800H/3900-2 $[\pm45/90]_s$ laminates, however, matrix crack in 45° plies are not observed until the final laminate fracture.

First cracking strain becomes larger as the thickness of 90° plies becomes thinner. Transverse crack density growth rate becomes larger as the thickness of 90° plies becomes thinner. This results in larger transverse crack density just before the final fracture in the laminates with thinner 90° plies. First cracking strain becomes larger as the stiffness of adjacent plies becomes larger. Transverse crack density growth rate becomes larger as the stiffness of the adjacent plies becomes larger. That is, the larger stiffness of adjacent plies of 90° plies has a similar effect of thinner 90° plies.

Figures 5 and 6 show the predictions of transverse crack behavior in 90° ply based on the damage mechanics analysis. In the present study, a transverse crack is assumed to occur when both the energy and average stress criteria are satisfied. In other words, the prediction which gives lower crack density at the same strain is considered to be analytical prediction. The critical energy release rates assumed are 360 J/m^2 and 300 J/m^2 for G40-800/5260 and T800H/3900-2, respectively. The critical average stresses assumed are 120 MPa and 90 MPa for G40-800/5260 and T800H/3900-2, respectively. Analytical predictions based on the average stress criterion gives smaller transverse crack density for G40-800/5260 $[0/90_2]_s$, $[\pm45/90_2]_s$, and T800H/3900-2 $[0/90]_s$, $[\pm45/90]_s$ laminates. This indicates that predictions based on the average stress criterion are regarded as the analytical prediction for these laminates. Predictions are in good agreement with experimental results. But predictions of $[\pm45/90]_s$ laminates for both material systems have some mismatch in higher strain region. This higher strain region corresponds to delamination growth which is not considered in the present analysis.

5 CONCLUSION

Microscopic damage progress in bismaleimide-based CFRP, G40-800/5260 laminates and interlaminar-toughened epoxy-based CFRP, T800H/3900-2 laminates are investigated experimentally. The larger stiffness of adjacent plies of 90° plies and thinner 90° plies has a similar

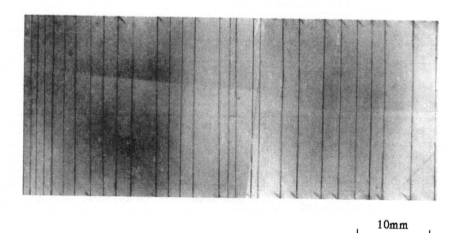

Figure 4   Microscopic damage in T800H/3900-2 [±45/90]$_s$ laminates (ε=1.4%).

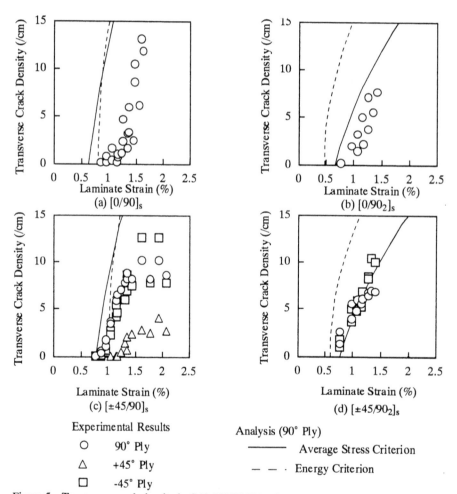

Figure 5   Transverse crack density in G40-800/5260 laminates as a function of laminate strain.   Experimental results and analytical predictions.

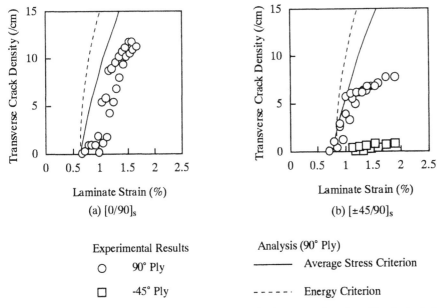

Experimental Results

○   90° Ply

□   -45° Ply

Analysis (90° Ply)

——— Average Stress Criterion

- - - - - Energy Criterion

Figure 6   Transverse crack density in T800H/3900-2 laminates as a function of laminate strain.   Experimental results and analytical predictions.

effect on the transverse crack behavior in 90° plies. Transverse crack behavior was modeled by using damage mechanics analysis. Analytical predictions considering both energy and average stress criteria are in good agreement with the experimental result.

REFERENCES

Gudmundson, P. & W. Zang 1993. An analytic model of composite laminates containing transverse matrix cracks. *Int. J. Solids Structures.* 30:3211-3231.

Masters, J.E. & K. L. Reifsnider 1982. An investigation of cumulative damage development in quasi-isotropic graphite/epoxy laminates. *ASTM STP* 775:40-61.

McCartney, L.N. 1992. Theory of stress transfer in a 0°-90°-0° cross-ply laminate containing a parallel array of transverse cracks. *J. Mech. Phys. Solids.* 40:27-68.

Nairn, J.A. 1989. The strain energy release rate of composite microcracking. *J. Comp. Mat.* 23:1106-1129.

Odagiri, N. T. Muraki & K. Tobukuro 1988. Toughness improved high performance Torayca prepreg T800H/3900 series, *Proc. 33rd Int. SAMPE Symp.* 272-283.

Odagiri, N. H. Kishi & T. Nakane 1991. T800H/3900-2 toughened epoxy prepreg system: toughening concept and mechanism. *Proc. ASC 6th Tech. Conf.* 46-52.

Ogihara, S. N. Takeda, S. Kobayashi & A. Kobayashi 1997. Effect of stacking sequence on strength and failure process in quasi-isotropic CFRP laminates with toughened interlaminar layers. *Proceedings of the 11th Int. Conf. Comp. Mater.* V:552-561.

Ogihara, S. N. Takeda & A. Kobayashi 1988. Transverse cracking in CFRP cross-ply laminates with interlaminar resin layers. *Adv. Comp. Mat.* 7:347-363.

Ogihara, S. N. Takeda, S. Kobayashi & A. Kobayashi in press. *International Journal of Damage Mechanics*

Takeda, N. & S. Ogihara 1994. In-situ observation and probabilistic prediction of microscopic failure processes in CFRP cross-ply laminates. *Comp. Sci. Tech.* 52:183-195.

Takeda, N. S. Kobayashi, S. Ogihara & A. Kobayashi 1998. Experimental characterization of microscopic damage progress in quasi-isotropic CFRP laminates. *Adv. Comp. Mat.* 7:183-199.

Recent Developments in Durability Analysis of Composite Systems, Cardon, Fukuda, Reifsnider & Verchery (eds)
© 2000 Balkema, Rotterdam, ISBN 90 5809 103 1

Initiation versus growth criteria for transverse matrix cracks

P.Gudmundson
Department of Soil Mechanics, Royal Institute of Technology (KTH), Stockholm, Sweden

ABSTRACT: Through a specially developed experimental setup, tensile test cross-ply specimens are loaded in combined tension and bending. Transverse matrix cracks are then initiated from the tensile edge of the specimens and as they grow through the width into lower strained areas, the cracks arrest before reaching the opposite edge. As the tensile and bending load are further increased already existing cracks may reinitiate their growth and new cracks are initiated from the tensile edge. The initiation and growth of transverse matrix cracks are registered by optical images at regular load intervals. The experimental data are then analyzed in terms of energy release rates and local strain states. It is found that the energy release rate controls the crack growth whereas the local strain state may be a better parameter for prediction of crack initiation. More details about the experimental setup and the theoretical analyses are presented in Gudmundson & Alpman (1999).

1 INTRODUCTION

The conditions for initiation and growth of transverse matrix cracks in glass or carbon fibre reinforced composite laminates have been investigated by several researchers over the last twenty years (Garrett & Bailey 1977a, b, Parvizi et al. 1978, Bailey et al. 1979, Flaggs & Kural 1982, Harrison & Bader 1983, Boniface & Ogin 1989). Almost exclusively, tensile test specimens have been applied. Experimental data have been presented in the form of transverse crack density versus applied loading for various laminate layups, ply thicknesses and material combinations. These investigations have provided information about conditions for initiation of cracks, but it is not obvious how to extract information about crack growth criteria, since the cracks are results of initiation and subsequent growth. For small ply thicknesses one can argue that initial flaws exist which span the ply thickness and hence the initiation is a result of crack growth from these flaws. Crack initiation data for small ply thicknesses have successfully been interpreted in this way. In order to perform direct measurements of crack growth conditions, a well defined initial crack must exist and conditions for growth of this crack should be considered. The effects of notches on the transverse layer have been investigated by Boniface et al. (1997). They found that the notches had a significant influence on the conditions for crack initiation in thick transverse plies, whereas the effect was smaller for thin plies.

These observations indicate in a nice way the differences between initiation and growth criteria.

2 RESULTS

The present investigation is aimed at well controlled crack growth measurements for determination of a relevant crack growth criterion. An experimental technique is developed that makes it possible to introduce transverse matrix cracks which start from one edge of a specimen and arrest before reaching the other edge. This is achieved through a combination of extension and bending loading of tensile specimens, see Figure 1. As the external loading increases, already existing matrix cracks continue to grow in a step-wise fashion and new cracks initiate. Matrix crack positions and lengths are registered through optical images which are captured at regular external load intervals.

The local stress and strain state as well as energy release rates of individual matrix cracks are determined from a combined global and local analysis. The average stress and stress state as a function of the loading N in the beam described in Figure 1 is modelled by beam theory including the stiffening effect of the axial force. For each load level, the average matrix crack length is determined from the optical images. The decrease in beam stiffness in the cracked part of the specimen is estimated from the model described by Adolfsson & Gudmundson

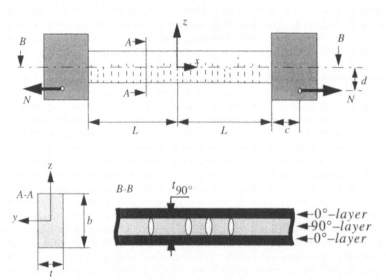

Figure 1. Experimental setup. The dashed lines in the upper figure symbolize transverse matrix cracks.

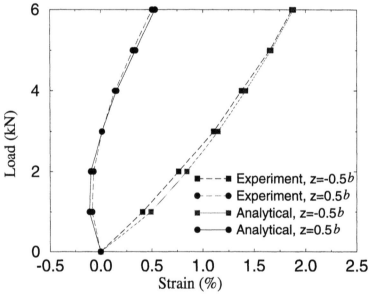

Figure 2. Comparison between measured and calculated strains for the $[0_2, 90]_s$ specimen. The position of measured and calculated strains correspond to the top ($z = 0.5b$) and the bottom edge ($z = -0.5b$) in the mid of the beam according to Figure 1.

(1997). To check the accuracy of the beam model for prediction of the global strain state, strain gages were mounted on the edges $z = 0.5b$ and $z = -0.5b$ in the middle of the beam ($x = 0$). In Figure 2 the measured strains are compared to the beam model predictions. It can be observed that a quite good agreement is obtained.

Based on extensive finite element calculations model for determination of local energy release ra for individual matrix cracks was developed (Gι mundson & Alpman 1999). It was found that 1 stress intensity factor of a matrix crack is controll by the lengths of the crack under consideration a the two neighbouring matrix cracks. If the crack

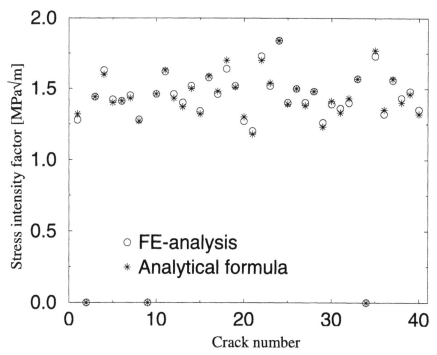

Figure 3. Comparison between approximative estimates of stress intensity factors and three-dimensional finite element calculations for the $[0_2, 90]_s$ laminate.

longer that its two neighbours, the interaction effect from the neighbours can be neglected. In the case that one or two of the neighbouring cracks are longer, correction factors depending on the distance between the cracks are derived based on extensive finite element calculations. For some load levels, the above mentioned simplified method for determination of individual stress intensity factors were compared to full three-dimensional finite element calculations of the whole specimen. In Figure 3 comparisons between the approximative analytical formula based on correction factors for determination of stress intensity factors and results from full three-dimensional finite element calculations are presented. It is observed that an acceptable accuracy is obtained by use of the approximative method. All specimens, load levels and corresponding crack states determined from the optical images are then analysed by use of the beam model for determination of the global strain state and by the derived formula for calculation of stress intensity factors (energy release rates).

An analysis of the conditions for reinitiated growth of already existent matrix cracks makes it possible to evaluate crack propagation criteria. Also conditions for initiation of new cracks can be analysed from the experiments. In this way, initiation and growth can be separated in an unambiguous way. Since several transverse matrix cracks appear in each specimen and crack growth can be observed at different load levels, a single macroscopic specimen generates lots of data. Experiments on $[0_2, 90]_s$ and $[0_2, 90_2]_s$ glass/epoxy cross-ply laminates are presented. About 40 transverse matrix cracks at 15 load levels were analysed for each specimen. This resulted in more than 400 crack growth observations and about 40 crack initiations for each specimen.

3 CONCLUSIONS

The results indicate that the energy release rate is a controlling parameter for crack growth whereas the strain state controls the crack initiation. A small transverse ply thickness dependence was however observed. The critical energy release rates for the $[0_2, 90]_s$ and $[0_2, 90_2]_s$ were found to be 151 ± 45 N/m and 124 ± 28 N/m respectively. The initiation strains were determined as $0.74\pm0.12\%$ and $0.62\pm0.05\%$. The present investigation clearly demonstrate that separate criteria for initiation and growth of transverse matrix cracks must be applied in evaluations of test data as well as in applications.

REFERENCES

Adolfsson E. & P. Gudmundson 1997. Thermoelastic properties in combined bending and extension of thin composite laminates with matrix cracks. *Int. J. Solids Structures*. 34:2035-2060.

Bailey, J.E., P.T. Curtis & A. Parvizi 1979. On the Transverse Cracking and Longitudinal Splitting Behaviour of Glass and Carbon Fibre Reinforced Epoxy Cross Ply Laminates and the Effect of Poisson and Thermally Generated Strain. *Proc. R. Soc. Lond.* A366:599-623.

Boniface, L. & S.L. Ogin 1989. Application of the Paris Equation to the Fatigue Growth of Transverse Ply Cracks. *J. Comp. Mat.* 23:735-754.

Boniface, L., P.A. Smith, M.G. Bader & A.H. Rezaifard 1997. Transverse Ply Cracking in Cross-Ply CFRP Laminates - Initiation or Propagation Controlled? *J. Comp. Mat.* 31:1080-1112.

Flaggs, D.L. & M.H. Kural 1982. Experimental Determination of the In Situ Transverse Lamina Strength in Graphite/Epoxy Laminates. *J. Comp. Mat.* 16:103-116.

Garrett, K.W. & J.E. Bailey 1977a. Multiple Transverse Fracture in 90° Cross-Ply Laminates of a Glass Fibre-Reinforced Polyester. *J. Mater. Science*. 12:157-168.

Garrett, K.W. & J.E. Bailey 1977b. The Effect of Resin Failure Strain on the Tensile Properties of Glass Fibre-Reinforced Polyester Cross-Ply Laminates. *J. Mater. Science*. 12:2189-2194.

Gudmundson, P. & J. Alpman 1999. Initiation and Growth Criteria for Transverse Matrix Cracks in Composite Laminates. *Comp. Science Techn*. In press.

Harrison, R.P. & M.G. Bader 1983. Damage Development in CFRP Laminates Under Monotonic and Cyclic Stressing. *Fibre Science and Technology*. 18:163-180.

Parvizi, A., K.W. Garrett & J.E. Bailey 1978. Constrained Cracking in Glass Fibre-Reinforced Epoxy Cross-Ply Laminates. *J. Mater. Science*. 13:195-201.

Viscoelasticity – Creep analysis

Recent Developments in Durability Analysis of Composite Systems, Cardon, Fukuda, Reifsnider & Verchery (eds)
© 2000 Balkema, Rotterdam, ISBN 90 5809 103 1

Assessment of time-temperature superposition principle as a basis for a long term behaviour prediction of CFRP under bending load

Rui Miranda Guedes

Department of Mechanical Engineering and Industrial Management (DEMEGI), Faculty of Engineering, University of Porto (FEUP), Portugal

ABSTRACT: This paper presents and discusses experimental flexural creep test results, in a controlled environment, for two different laminate composites. The purpose of the present work was to make comparisons between a tape and a woven CFRP laminate in a long-term perspective, giving guidelines to choose the appropriated CFRP laminate as a structural element for special structures for supporting delicate and precise radiation detection elements. These structures need to be highly stable under environmental conditions. The need to extrapolate experimental results induced the application of Time-Temperature-Superposition Principle (TTSP). Unfortunately there is no experimental demonstration of a real equivalence between behaviour at higher temperatures for a short period of time and long-term behaviour under lower temperatures. Therefore this work is also a contribution to assess the potential application of Time-Temperature Superposition Principle (TTSP) as a basis for a long-term behaviour prediction.

1 INTRODUCTION

Simple bending creep tests using cantilever beams are used in order to assess the potential application of Time-Temperature Superposition Principle (TTSP) as a basis for a long-term behaviour prediction. The TTSP is usually applied for acceleration of creep tests and it as been proved to be well suited to characterise the influence of temperature over the mechanical properties. Unfortunately there is no experimental demonstration of a real equivalence between behaviour at higher temperatures for a short period of time and long-term behaviour under a lower temperatures. Much effort shall be done to reduce the time and experimental work to describe creep behaviour. Some existing European standards test methods for prediction of long-term behaviour of GFRP pipes, for example, imply test data acquisition up to 10000 hours, more than one year. The future of FRP applications depend on a reliable short term creep characterisation, i.e. of a few weeks, for a confident extrapolation over a long period of time, i.e. 10 to 50 years.

This research was done at the INEGI (Institute of Engineering and Industrial Management) in the context of the characterisation of advanced composite materials to build support structures for particle detectors. These structures should present a high dimensional stability, small deviations from initial position will lead to large errors in signal detection of particles. The detectors will make part of the new particle accelerator called LHC (Large Hadron Collider). This is a big project of the European Laboratory for Particle Physics (CERN) with new challenges in many engineer fields, especially in the application of new materials.

Two types of extrapolation were tested at the linear viscoelastic domain. The first one used experimental data obtained at 50°C for a minimum period of time fitted to a Findley model (Simple Power Law). Another approach applied the TTSP to obtain the creep master curve. The curves were obtained from dynamic experimental data at frequencies from 0.01Hz up to 100Hz and at different temperatures from 23°C up to 135°C.

The creep extensions were measured, close to the section of maximum moment, using electric strain gauges, one on top surface (tension) and other on bottom surface (compression). The signal from an unloaded specimen was used to compensate temperature effects. The loads were selected in order to ensure a maximum stress level lower than 10% of the rupture stress σ_r.

The linear viscoelastic behaviour is usually observed at relatively low stress levels, hence the proposed maximum stress level lower than 10% of σ_r is expected to maintain the behaviour in the linear domain.

2 MATERIAL

A brief presentation of the experimental test specimens and test apparatus is given. A set of 50x200mm2 specimens was sectioned from composite plates fabricated at INEGI using the unidirectional UC 125 RNA and the twill woven CC 194 RNA with the same resin system (RNA) and reinforced with T300 carbon fibres. The geometry of test specimens for each stacking sequence was as follows (B= width, h= thickness):

Table 1. Geometry of test specimens from plate UC125 RNA [0°/90°]$_{4s}$.

Specimen	B2	B3	B4
h (mm)	2.30	2.28	2.26
B (mm)	49.36	49.76	49.48

Table 2. Geometry of test specimens from plate CC194 RNA [0°]$_{4s}$.

Specimen	C4	C5	C6	C7
h (mm)	2.00	2.02	2.01	2.02
B (mm)	48.91	49.22	48.28	47.26

The stacking sequence of the laminates and the geometry of test specimens were imposed by the CERN.

For carbon fibre composites the time dependency exhibited by the fibre dominated mechanical properties E_{11}, v_{12} is negligibly small, when compared to the highly time dependent matrix dominated properties. Analysing the staking sequences used in this work, no significant creep should be expected due to the existence of fibres in the loading direction. So if the deformations are constant then an internal stress transfer should occur, i.e., the non-0° plies should exhibit stress relaxation.

3 STATIC BENDING PROPERTIES OF THE TEST SPECIMENS

The layer elastic properties had been already determined at the INEGI. The table 3 summarises the results obtained for the two laminates.

The 3 point bending flexural tests were used to determine the flexural modulus (E_b) and the flexural stress strength (σ_r). The geometry of the tests

Table 3. Layer elastic properties of the laminates.

Specimen	E_1 (GPa)	E_2 (GPa)	v_{12}	G_{12} (GPa)
UC125 RNA	92.05	11.69	0.25	3.74
CC194 RNA	53.57	53.57	0.058	3.82

follows the ASTM D790M standard except for the rate definition. In these tests it was used a rate of 5mm/min instead of 12mm/min as recommended by the standard. The measured properties were as follows:

Table 4. Flexural properties of the test specimens UC125 RNA [0°/90°]$_{4s}$.

Specimen	B1	B2	B3	Average
E_b (GPa)	46.76	46.72	48.72	47.40
σ_r (MPa)	846	848	854	849

Table 5. Flexural properties of the test specimens CC194 RNA [0°]$_{4s}$.

Specimen	C1	C2	C3	Average
E_b (GPa)	41.94	42.50	42.16	42.20
σ_r (MPa)	717	727	731	725

4 DYNAMIC-MECHANICAL-THERMAL-ANALYSER (DMTA) TEST RESULTS

The characterisation of the mechanical properties of polymeric materials on a DMTA analyser gives the complex compliance (S^*) as a function of angular frequency ω. In order to characterise the viscoelastic material behaviour in a very broad range of frequencies (or times), it is necessary to combine measurements at several temperatures applying the time-temperature superposition principle (TTSP). According to this principle, a given property measured for short times must be identical with one measured for longer times at a lower temperature, except that the curves are shifted parallel to the horizontal axis, matching to form a master curve.

The dynamic test specimens were cutted out from plates with three different stacking sequences. The dynamic tests were carried out in a beam cantilever apparatus, as shown in Figure 1. The specimens had a thickness of 2.2mm, width of 10.0mm and the distance between the support and the load application of 22.0mm.

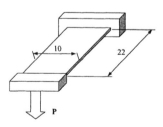

Figure 1. Geometry of DMTA test specimens.

A brief description of the methodology used to obtain the creep master curves (Guedes & Marques, 1998a) is given. For each temperature level, the frequencies varied from 0.1Hz to 100Hz. The maximum imposed deflection was 64μm and the temperature levels were within the range of 20°C to 135°C. Following the frequency-time transformation procedure, the short-term compliance curves in time domain were obtained. Then the data was shifted for each temperature to a reference temperature, T_{ref}, using the time-temperature superposition principle (TTSP) to build up the master creep compliance curve as shown in Figure 2.

Many authors already discussed the inaccuracy of the simple power law to model the whole master curve. The Simple Power Law is defined as follows:

$$S(t) = S_0 + S_1 \left(\frac{t}{\tau_0} \right)^n \tag{1}$$

where S_0 = time independent initial compliance; S_1 = coefficient of the time dependent compliance; n = material constant; and τ_0 = unit reference time.

Many polymers, such as the crosslinked polymers, deform in an asymptotic form, from glassy to rubber like behaviour. In the literature there are many expressions of the time-dependent compliance that can describe the viscoelastic phenomena over a broad range. In this study the response function (Qin, 1996) chosen to fit the master curve was given by:

$$\frac{S(t)}{S_0} = 1 + \frac{(S_\infty - S_0)}{S_0} w(t) \tag{2}$$

where $w(t)$ = the Cole-Cole function

$$w(t) = \frac{1}{1 + \left(\dfrac{\tau_0}{t} \right)^n} \tag{3}$$

Table 6. Parameters of the master compliance curve

Parameters	n	τ_0	$(S_\infty - S_0)/S_0$
UC125	0.32	5×10^5	13.468
CC194	0.32	12×10^5	15.149

where S_0 = time independent initial compliance; S_∞ = time independent compliance at infinity time; n = material constant; and τ_0 = unit reference time.

The master curves were fitted to the Cole-Cole function but it was found that the curve fitting didn't have a good agreement in the longer time range. The curve fitting and the fitting results are shown in Figure 2 and in table 6.

The discrepancy in the longer time range was related to the shape of the compliance master curves. Although the master curves shown a decrease of the creep at longer times, it was clear that they did not reach a limit value as the Cole-Cole function predicts. Nevertheless, if the working temperatures are lower than 50°C than the longer time range, 10^6 to 10^{14} hours, represents very long times, i.e., 114 and 11.4 billions of years, respectively. Hence, in this case, it is perfectly reasonable to use the Simple Power Law to predict long-term behaviour in a human time scale, i.e., 50 years.

5 EXPERIMENTAL RESULTS FOR THE CREEP TESTS

Due to limitations on temperature control device inside the laboratory, an environmental chamber was used. Inside the chamber the temperature and the relative humidity was maintained steady: 23°C and Hr. 50% for the first test and 50°C and Hr. 50% for the second test, which is still going on.

In Figure 3 the test apparatus is shown inside the environmental chamber. The maximum applied

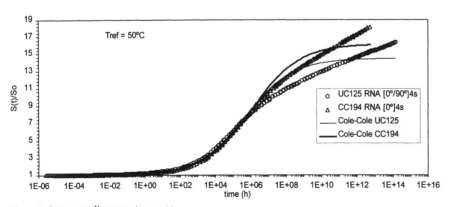

Figure 2. Creep compliance master curves.

Figure 3. Test apparatus inside the environmental chamber.

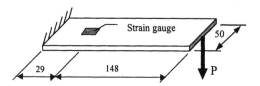

Figure 4. Geometry of DMTA test specimens.

stresses are lower than 10% of the rupture stress. At this stress level the material was expected to behave as linear viscoelastic.

A close view of test apparatus with the respective geometry is shown in the Figure 4.

The creep extensions were measured using strain gauges. For the bending tests, the tensile and compression extensions were recorded close to the section of maximum bending moment, as shown in Figure 3.

The creep compliance was defined as:

$$S(t) = \frac{\varepsilon(t)}{\sigma_0} \qquad (4)$$

where $S(t)$ = creep compliance; t = time; $\sigma_{0'}$ = applied stress; and $\varepsilon(t)$ = measured strain.

For a better comparison of all test specimens the relative creep compliance was used instead. The relative creep compliance was defined as:

$$S_R = \frac{S(t)}{S_0} \qquad (5)$$

where S_R = relative creep compliance; and S_0 = initial compliance.

After almost 1800 hours of bending creep test of specimens made of UC 125 RNA at 23°C and 50%Hr, no significant creep strains were detected within the strain gauges accuracy, i.e. ±20µε.

These results motivated the decision to increase the temperature to 50°C instead of increasing the loading stress, with two purposes: accelerate the viscoelastic behaviour and avoid the non-linear behaviour.

In Figures 5 and 6 the creep test results are plotted with the simple power law predictions (Findley) and the master creep curve (DMTA. All experimental results are in tension except the B2(C) and C5(C) that are in compression. The Findley or the simple power law was already defined by Equation (1).

These experimental results, representing more than 2500 hours, revealed an increase of the creep compliance around 6.7% in average for the CC194 and around 2.5% in average for the UC125. Surprisingly the creep tests pointed that the CC194 has a larger increase of the creep compliance than the UC125. For continuous carbon fibre composites, the time dependency exhibited by the fibre dominated mechanical properties E_{11}, v_{12} is negligibly small, when compared to the highly time dependent matrix dominated properties. Then, for these staking sequences no significant was expected creep due to the existence of fibres in the loading direction.

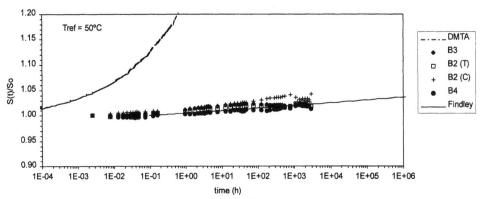

Figure 5. Relative creep compliance for the UC125 RNA [0°/90°]4s test specimens.

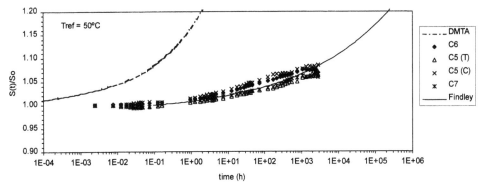

Figure 6. Relative creep compliance for the CC 194 RNA $[0°]_{4s}$ test specimens.

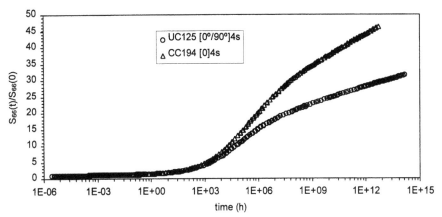

Figure 7. Relative shear creep compliance master curves.

At this stage some conclusions can be addressed. The results showed a large discrepancy between the DMTA and the bending creep, apparently having distinct behaviours. In fact each test involved specimens with different dimensions but the scale factor, alone, could not explain so large deviations. On the other hand the Findley equation prediction for the CC194 specimens, based on the first 100 hours, shown a good agreement with the experimental data until the 1000 hours were reached. After this time the creep data shown a large decrease of the creep rate, apparently diverging from the

power law. As for the UC125 specimens the predictions of the Findley are in good agreement with the experimental data, certainly due to the very low creep level presented. Having this in mind a new model was developed.

6 ANALYSIS OF EXPERIMETAL RESULTS USING A NEW MODEL

The new model approach was developed after a careful analysis of the DMTA test apparatus. It was not difficult to conclude that the beam cantilever should be considered a short beam. Therefore it become obvious that the DMTA results were related not only with the compliance but also with the shear compliance of the laminate. For that very reason the properties of the viscoelastic matrix could be determined approximately.

The Poisson's ratio v^m of the viscoelastic matrix was considered constant and equal to 0.25. The creep compliance $S^m = S_{11}^m = S_{22}^m$ and the shear creep compliance S_{66}^m of the matrix were calculated

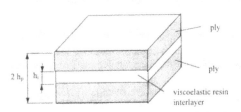

Figure 8. Laminate detail of the viscoelastic resin interlayer.

361

using some approximations. In this case it was considered that the DMTA test measured, directly, the shear creep compliance of the laminate admitting that the influence of the creep compliance of the laminate could be ignored. The Equation (6) gives the relation between the creep compliance, obtained directly from DMTA tests, and the shear creep compliance of the laminate. The Equation (6) was obtained by comparison of beam cantilever solution with and without the shear correction.

$$S_{66}(t) = \frac{1}{6}\left(\frac{L}{h}\right)^2\left(S'(t) - \frac{1}{E_b}\right) \quad (6)$$

where L = the length of the beam; h = the thickness of the laminate; $S'(t)$ = the creep compliance obtained from the DMTA and E_b = the bending modulus of the laminate considered linear elastic.

The shear creep compliance master curves obtained, using Equation (6), are plotted in Figure 7.

The micromechanics formulae based on the Halpin-Tsai equations had been adapted and used for viscoelastic analysis (Beckwith, 1975) (Horoschenkoff, 1990). Based on such analysis, assuming the fibres linear elastic and the woven composite as a multidirectional laminate consisting of the fibre angles, the relationship between shear creep compliance of the matrix, $S_{66}{}^m(t)$, and the laminate shear creep compliance, $S_{66}(t)$, was obtained as follows:

$$S_{66}{}^m(t) = \frac{1}{2(1-v_f)}\Big\{\big(S_{66}(t) - S_{66}{}^f\big)\cdot(1+v_f) +$$
$$\big[(S_{66}(t))^2\cdot(1+2v_f+v_f{}^2) +$$
$$S_{66}(t)\cdot S_{66}{}^f(2-12v_f+2v_f{}^2) +$$
$$\big(S_{66}{}^f\big)^2\cdot(1+2v_f+v_f{}^2)\big]^{1/2}\Big\} \quad (7)$$

where v_f = the volume fraction of the fibre; $S_{66}{}^f$ = fibre shear compliance considered linear elastic and equal to $22GPa$ (Cox, 1997).

Finally using a relation applied successively elsewhere (Horoschenkoff, 1990) the creep compliance of the matrix was determined by the following equation:

$$S^m(t) = \frac{S_{66}{}^m(t)}{2(1+v^m)} \quad (8)$$

A nonlinear viscoelastic analysis of the carbon fibre matrix interphase was already presented by another researcher (Sancaktar, 1990). The present approach considers a viscoelastic interlayer between each ply of the laminate, with a thickness to be determined, i.e., a resin-rich interlayer zone with the same properties of the resin, as shown in Figure 8.

Introducing the geometry and properties of the laminates into the LAMFLU program (Guedes & Marques, 1998b, 1999), it was possible to determine the resin interlayer thickness, unknown at this stage, for the UC125 and CC194 laminates using the first 100 hours of the experimental results. The results for the interlayer thickness, see Figure 8, are presented on table 7.

Table 7. Thickness of the plies and the resin interlayers.

Dimensions (mm)	h_p	h_i	h_i/h_p (%)
UC125	0.1438	0.0296	21
CC194	0.2500	0.1086	44

The values determined for the interlayer thickness reveal that the twill woven laminate (CC194) has a larger resin interlayer, i.e. 44% of the ply thickness, than the tape laminate (UC125).

The woven composites compared to tape laminates, with the same volume fractions of in-plane fibres, have usually slightly lower in-plane stiffness because of the tow waviness. Therefore, axial shear stresses are higher under nominally aligned loads (Cox & Flanagan, 1997). The tow waviness and the higher shear stress could justify why CC194 has a larger resin interlayer thickness in the present model.

The analysis of delamination in angle-ply composites lead some researchers (Wang, 1980) to model the composite laminate as an assembly of anisotropic homogeneous plies bonded by thin resin interlayers. The interlaminar resin layer was considered an isotropic material with a uniform thickness of one-tenth of the individual ply thickness as observed under microscope (Wu, 1986).

In the present case no microscopic observations were made but it was clear that the present model was a crude simplification of the real resin interlayer of the CC194 laminate. Probably the interlayer thickness that was determined for the present model reflected the influence rather the real geometry of the viscoelastic interlayer.

The predictions made by LAMFLU (Model) compared reasonably well with the experimental results plotted in Figure 5 and 6. In Figures 9 and 10 the predictions of the present model (Model) and of the simple power law (Findley) are plotted with the averaged experimental data (Exper.).

The model predicts a limit to the creep deformation and the experimental data appears to follow the same trend. The evolution of the transient creep compliance for the averaged experimental data and the model can be observed in Figures 11 and 12.

The laminate plies were considered linear elastic for the CC194 and the UC125. In fact the UC125 is

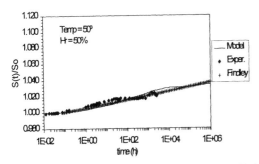

Figure 9. The averaged creep compliance of the UC125 RNA [0°/90°]₄ₛ test specimens with the model and Findley predictions.

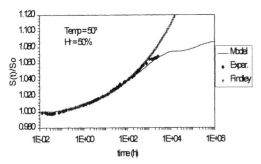

Figure 10. The averaged creep compliance of the CC194 RNA [0°]₄ₛ test specimens with the model and Findley predictions.

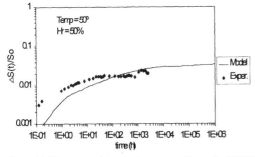

Figure 11. The averaged transient creep compliance the UC125 RNA [0°/90°]₄ₛ test specimens with the model predictions.

not really unidirectional, i.e. contains glass fibres in the weft directions, about 20% in weight. Therefore the viscoelastic interlayers suffer creep and stress relaxation simultaneously and as a consequence originate load transfer to the adjacent layers. After a certain period of time an equilibrium state is reached and the creep deformation is arrested.

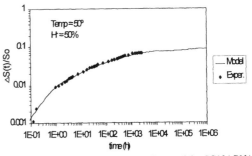

Figure 12. The averaged creep compliance of the CC194 RNA [0°]₄ₛ test specimens with the model predictions.

7 CONCLUSIONS

One simple explanation for CC 194 behaviour, which came out initially, was found on the reinforcement tissue itself. The overlap of the transverse and longitudinal fibres (twill woven) could produce plies with fibres not stretched. The tensile stresses and the viscoelastic matrix then promoted the stretching of the fibres. This phenomenon should not happen in unidirectional reinforcements. Nevertheless this paradigm did not seem satisfactory to explain the experimental results.

The creep master curves obtained from the dynamic mechanical and thermal analysis (DMTA) were very similar for both laminates UC125 RNA [0°/90°]₄ₛ and CC 194 RNA [0°]₄ₛ. The master curves shown an increase of the creep compliance (to a limit value?) between 16 and 19 times of the initial value. Further the master curves indicate, as shown in Figure 2, that it takes an absurd time to reach this limit at 50°C (1 millions of years).

One the other hand the comparison of creep master curves and creep test results have shown a large discrepancy. A new interpretation of the DMTA results was done. This allowed the determination of the matrix viscoelastic properties using the micromechanic analysis. The introduction of a viscoelastic interlayer between plies into the laminate model resulted into a new approach. The developed model explained the experimental results and permitted a long-term prediction of the creep bending response of the UC125 RNA [0°/90°]₄ₛ and CC 194 RNA [0°]₄ₛ laminates based on Time Temperature Superposition Principle master curves.

ACKNOWLEDGEMENTS

The author appreciated the Fundação Calouste Gulbenkian grant. The author would like to acknowledge the contribution for this work of his colleagues Mário Vaz (DEMEGI) and Paulo Nóvoa (INEGI).

REFERENCES

Cox, B. & Flanagan, G. 1997. Handbook of Analytical Methods for Textile Composites. NASA Contractor Report 4750.

Beckwith, S.W. 1975. Viscoelastic Characterization of a Nonlinear, Glass/Epoxy Composite Using Micomechanics Theory. JANNAF Structures and Mechanical Behaviour Working Group Meet., San Francisco, CA.

Guedes, R.M., Marques, A.T. & Cardon, A. 1998a. Creep or Relaxation Master Curves Calculated from Experimental Dynamic Viscoelastic Function. *Science and Engineering of Composite Materials* 7(3): 259-267.

Guedes, R.M., Marques, A.T. & Cardon, A. 1998b. Analytical and Experimental Evaluation of Nonlinear Viscoelastic-Viscoplastic Composite Laminates under Creep, Creep-Recovery, Relaxation and Ramp Loading. *Mechanics of Time-Dependent Materials* 2: 113-128.

Guedes, R.M., Marques, A.T. & Cardon, A. 1999. Creep/Creep-Recovery Response of Fibredux 920C-TS-5-42 Composite under Flexural Loading. *Applied Composite Materials* 6: 71-86.

Horoschenkoff, A. 1990. Characterization of the Creep Compliances J_{22} and J_{66} of the Orthotropic Composites with PEEK and Epoxy Matrices Using the Nonlinear Viscoelastic Response of the Neat Resins. *Journal of Composite Materials* 24: 879-891.

Lahiri, J., Devi, G.R., Subrahmanyam, S.V. & Balakrishna, K. 1988. Establishing the criteria for selecting suitable interlayer material for double-layered filament-wound CFRP structures. *Composites* 19(5): 376-382.

Qin, Y. 1996. Nonlinear Viscoelastic-Viscoplastic Characterization of a Polymer Matrix Composite. PhD thesis submitted to Free Univeristy of Brussels (VUB).

Sancaktar, E. & Zhang, P. 1990. Nonlinear Viscoelastic Modelling of the Fiber-Matrix Interphase in Composite Materials. *Journal of Mechanical Design* 112: 605-619.

Wang, S.S. 1980. An Analysis of Delamination in Angle-Ply Fiber-Reinforced Composites. *Transactions of the ASME* 47: 64-70.

Wu, C.M.L. 1986. Nonlinear Analysis of Edge Efferets in Angle-Ply Laminates. *Composites & Structures* 25(5): 787-798.

Recent Developments in Durability Analysis of Composite Systems, Cardon, Fukuda, Reifsnider & Verchery (eds)
© 2000 Balkema, Rotterdam, ISBN 90 5809 103 1

Viscoelastic contact problems of fiber reinforced polymers

Thomas Neff & Otto Brüller
Department of Mechanics and Material Testing, Technical University of Munich, Germany

ABSTRACT

The relative new class of materials – the fiber reinforced polymers (FRP) – exhibits a couple of outstanding properties. These materials have a low density and high strength and they can be designed to have different moduli and strength in different directions. Nevertheless, problems occur if structural parts made of metal connected by screws or pins should be substituted by parts made of FRP. Because of economical reasons, in most cases, especially in the case of series production, the original fixtures should be maintained. To fix a given component using screws or pins, one has to drill holes into the material. With fiber reinforced materials, as expected, the fibers will be cut and the strength of the component decreases considerably. The aim of the present work is to describe the behavior of the contact metal pin – fiber reinforced material under mechanical loading.

1. THEORETICAL CONSIDERATIONS

The well known correspondence principle in viscoelasticity can be applied only if there are no changes of the contact surfaces between the almost rigid pin and the viscoelastic fiber reinforced plastic. In such cases the possibility exists to reduce a viscoelastic (time-dependent) contact problem to an elastic (time-independent) one. In the present case the correspondence principle can be used if the contact area does not change with time. This restriction can only be fulfilled when the diameter of the pin and that of the hole in the FRP do not differ and remain constant. In practice the situation is different: pins may be smaller than the hole and this new situation demands a new theoretical development to be done in the next future.

2. EXPERIMENTAL

The used material were 16 layer (0°, 90°, ±45°) carbon fiber composites. The matrix was an epoxy resin (Martens Plus EP). The rectangular specimens (thickness: 2.4 mm) were loaded by a pin of ¼" (6.35 mm) which was inserted in a hole of the

specimens having the same nominal diameter. Both diameters, of the pin and of the hole, are within a tolerance of 1μm. Fig. 1 shows the schematical arrangement of the testing setup. The strain of the specimen was measured using self made strain gauge based extensometers (Fig. 2).

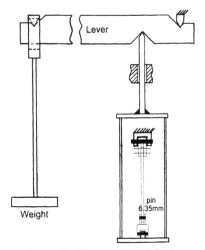

Figure 1: The experimental setup

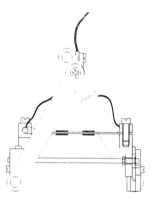

Figure 2: Extensometer

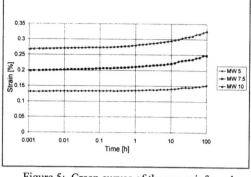

Figure 5: Creep curves of the non reinforced polymer (rectangular specimens)

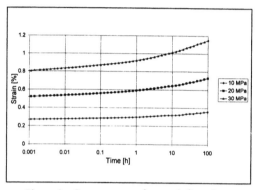

Figure 3: Creep curves of non reinforced polymer (shoulder specimens)

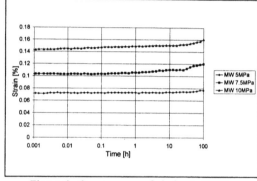

Figure 6: Creep curves of pin loaded FRP specimens at 5, 7.5, and 10 MPa

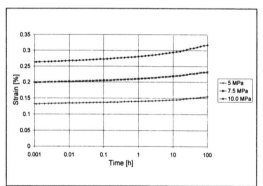

Figure 4: Creep curves of pin loaded non-reinforced polymer specimens

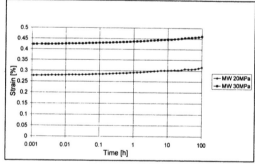

Figure 7: Creep curves of pin loaded FRP specimens at 20 and 30 MPa

3. RESULTS

As a first step, the creep behavior of the pure resin has been investigated. Figs. 3, 4, and 5 show the results. Fig. 3 presents the result of standard creep tests conducted on shoulder specimens of rectangular cross-section. In Fig. 4 the results of creep tests carried out on specimens made of the same non-reinforced polymer loaded via a pin–hole connection are shown. One can see that the creep strain of the shoulder specimens at the end of the measurement (after 100 h) is higher than that of the pin-loaded specimens. This unexpected behaviour points to a size-effect. So in a second step rectangular specimens in the same size as the holed ones were usually loaded. Fig. 5 shows the results of the rectangular specimens.

It is seen that the influence of pin loading is very small as compared with conventional loading. The deformation of the hole itself is as small as it can be neglected in relation to the creep strain of the polymer. In the case of pin loaded tests, the working stress was choosen as the load necessary to produce stresses of 20, 30, and 40 MPa on the projected area of the hole. The working stresses in the remainder of the cross-section were about 5, 7.5, and 10 MPa, respectively. The tests conducted on FRP specimens have shown that in this case the creep strain is about one decade lower. Because of this fact the loading of the reinforced specimens was increased in additional test series to 20 and 30 MPa. Although theoretically the influence of the contact area of the pin–specimen was supposed to be substantial, the results show practically no influence. These creep curves are shown in Fig. 6 and 7.

CONCLUSIONS

The behavior of a metal pin – FRP plate connection depends primarily on the matrix of the material. Using a duroplastic matrix, the loading type, i.e. with or without pin, has no influence on the behavior of the material. In a next step the behavior of thermoplastic polymers will be tested using the same procedure. In addition, different types of layer combinations have to be tested at different temperatures and loads.

LITERATURE

Johnson, K.L., (1985), *Contact Mechanics*, Cambridge University Press, Cambridge, UK

Kloiber, R., (1997), *Kleinflächige Lasteinleitung bei kohlefaserverstärktem Epoxydharz*, Student Project, Technical University of Munich

Götz, M., (1999), *Kriechverhalten von Epoxydharz*, Student Project, Technical University of Munich

Creep and stress relaxation of composite cylinders

J.T.Tzeng
US Army Research Laboratory, Weapons and Materials Research Directorate, Aberdeen Proving Ground, Md., USA

ABSTRACT: Stress relaxation and creep of composite cylinders are investigated based on anisotropic viscoelasticity. The analysis accounts for ply-by-ply variation of material properties, ply orientations, and temperature gradients through the thickness of cylinders subjected to mechanical and thermal loads. Fiber-reinforced composite materials generally illustrate extreme anisotropy in viscoelastic behavior. Accordingly, viscoelastic characteristics of composite cylinders are quite different from those of isotropic cylinders. Over a period of time, viscoelastic effects of the composite can result in a drastic change of stress and strain profiles in the cylinders, which is critical for the structural durability of composite cylinders. The developed analysis can be applied to composite pressure vessels and flywheels.

1 INTRODUCTION

Composite materials are currently used for lightweight pressure vessels and highly efficient rotors for energy storage. For both applications, prestresses are built in during fabrication of the cylinders through a "press-fit" procedure to enhance the mechanical performance. Composite overwrap pressure vessels are prestressed; therefore, the liner is in a state of hoop compression and the composite overwrap is in tension. For the other application, the rotors are subjected to a radial compression prior to operation. Accordingly, centrifugal force resulting from the rotation of the rotors generates tensile stresses in the radial and circumferential directions. Since the composite rotors are mainly reinforced in circumference (filament-wound cylinders), the radial tensile stress is critical to the ultimate performance of the rotors. Accordingly, it is essential to design and build the rotors with radial precompression. However, polymer matrix composites generally creep over a long period of time, especially at an elevated temperature (Ferry 1961). The associated stress relaxation in the composite will result in the loss of the prestresses and lead to a potential failure. The objective of this investigation is to develop an analytical method to study the viscoelastic behavior of thick-walled composite cylinders. The analysis can be applied to the design of flywheel machinery and composite pressure vessels.

To date, activities in the research of viscoelasticity have mainly concerned isotropic materials including the studies by Muki & Sternberg (1961),

Schapery (1964), Williams (1964), and Christensen (1982). These basic theories of viscoelasticity were then extended to the area of heterogeneous and anisotropic materials for a variety of applications. Hashin (1965) used the effective relaxation moduli and creep compliances to define the macroscopic viscoelastic behavior of linear viscoelastic heterogeneous media and its implementation in viscoelastic modeling. The general formulation of linear viscoelastic boundary value problems of composite materials, including the thermal viscoelastic problems for thermorheologically simple materials, and the applications of the correspondence principle were examined by Schapery (1967). Rogers & Lee (1964) investigated the viscoelastic behavior of an isotropic cylinder. In addition, finite element packages, such as ABAQUS, ANSYS, and DYNA3D, are not suitable for the viscoelastic analysis of composite cylinders because of the lack of anisotropic viscoelastic elements.

2 VISCOELASTIC FORMULATION

In the following research, the linear quasi-static viscoelastic behavior of a thick, laminated composite cylinder with an elevated temperature change is studied. The analysis accounts for ply-by-ply variation of properties, temperature changes, and fiber orientations. The thick cylinder is assumed to be in the absence of thermomechanical coupling and to be in a state of generalized plane strain such that all the stress and strain components are independent of the axial coordinate (Tzeng & Chien 1994, Chien

& Tzeng 1995). Moreover, due to the nature of axisymmetry, all the stress and strain components are also independent of the circumferential coordinate. The mechanical responses of this thick composite cylinder will, therefore, only have to satisfy the governing equation in the radial direction. Invoking the Boltzmann superposition integral for the complete spectrum of increments of anisotropic material constants with respect to time, the thermoviscoelastic constitutive relations of the anisotropic composite cylinder can be derived in integral forms. Since the thick composite cylinder is subjected to a constant elevated temperature, and boundary conditions are all independent of time, formulations of the linear thermal viscoelastic problem can have forms identical to those of the corresponding linear thermoelastic problem by taking advantage of the elastic-viscoelastic correspondence principle. In other words, all of these integral constitutive equations reduce to the algebraic relations, which are very similar to those developed for thermoelastic media when they are Laplace-transformed by means of the rule for convolution integrals. The thermoelastic analysis can thus be used to derive the transformed thermal viscoelastic solutions in the frequency domain.

The Boltzmann superposition integral of stress σ_{ij} $(i, j = 1, 2, 3)$ and strain ε_{ij} $(i, j = 1, 2, 3)$ relation for an isothermal viscoelastic problem with a constant temperature increase, ΔT, and the thermal expansion coefficient $\alpha_{kl}(T)$ is

$$\sigma_{ij}(t) = \int_0^t C_{ij}^{kl}(T, t - \tau) \frac{\partial \varepsilon_{kl}(\tau)}{\partial \tau} d\tau$$
$$- \beta_{ij}(T, t)\Delta T,\tag{1}$$

where $C_{ij}^{kl}(T, t)$ is the relaxation modulus dependent on temperature T and time t,

$$C_{ij}^{kl}(T(t), t) = C_{ij}^{kl}(T_0, \lambda(t))$$
$$\lambda(t) = \frac{t}{a_T(T(t))},$$

Here, T_0 is the base temperature, and a_T is the temperature shift factor. $\beta_{ij}(T, t)$ is given by $B_{ij}(T, t) = C_{ij}^{kl}(T, t) \cdot \alpha_{kl}(T)$. It is often desirable to use the inverse form of the constitutive relation (1),

$$\varepsilon_{ij}(t) = \int_0^t A_{ij}^{kl}(T, t - \tau) \frac{\partial \sigma_{kl}(\tau)}{\partial \tau} d\tau$$
$$+ \psi_{ij}(T, t)\Delta T,\tag{2}$$

where $R_{ij}(T, t)$ is the tensor product of the creep compliance $A_{ij}^{kl}(T, t)$ and the thermal creep coefficient ϕ_{kl}. Since the elevated temperature change, ΔT, is constant above some reference value in time,

the relaxation moduli and creep compliance are evaluated at that reference temperature, regardless of whether or not the material is thermorheologically simple, by employing the temperature shift-factor.

The Laplace transform of a function $f(t)$ is defined as

$$\bar{f} = \bar{f}(s) = \int_0^\infty e^{-st} f(t) dt,\tag{3}$$

where s is the Laplace transform variable. Applying (3) with the convolution rule to (1) and (2) reduces the integral constitutive equations to the following algebraic relations:

$$\bar{\sigma}_{ij} = \tilde{C}_{ij}^{kl} \bar{\varepsilon}_{kl} + \bar{\beta}_{ij} \frac{\Delta T}{s},\tag{4}$$

and the inverse form

$$\bar{\varepsilon}_{ij} = \tilde{A}_{ij}^{kl} \bar{\sigma}_{ij} + \bar{\psi}_{ij} \frac{\Delta T}{s},\tag{5}$$

respectively, where $\tilde{C}_{ij}^{kl} = s\bar{C}_{ij}^{kl}$, $\tilde{A}_{ij}^{kl} = s\bar{A}_{ij}^{kl}$, $\bar{\alpha}_{ij} = s\bar{\alpha}_{ij}$, $\tilde{\phi}_{ij} = s\bar{\phi}_{ij}$, $\bar{\beta}_{kl} = \tilde{C}_{ij}^{kl} \cdot \bar{\alpha}_{ij}$, and $\bar{\psi}_{kl} = \tilde{A}_{ij}^{kl} \cdot \tilde{\phi}_{ij}$. Furthermore, it can be shown that $[\tilde{A}_{ij}^{kl}] = [\tilde{C}_{ij}^{kl}]^{-1}$.

Consider a filament-wound axisymmetric thick composite cylinder consisting of N layers with the axial coordinate z, the radial coordinate r, and the circumferential coordinate θ, as shown in Figure 1. The composite cylinder has the inner radius a, the outer radius b, and the length L. Accordingly, there is a corresponding thermoelastic problem with the transformed displacement components $\bar{u}, \bar{v}$, and $\bar{w}$ in the axial direction, the circumferential direction, and the radial direction, respectively, in each layer. The axisymmetric character of the thick composite cylinder along with the assumption of the state of generalized plane strain leads to a simplified displacement field, which reflects the circumferential independence and only radial dependence of $\bar{w}$,

$$\bar{u}(r, \theta, z) = \bar{u}(r, z), \bar{v}(r, \theta, z) = \bar{v}(r, z), \text{ and}$$
$$\bar{w}(r, \theta, z) = \bar{w}(r).\tag{6}$$

Since each layer of the thick laminated cylinder is cylindrically monoclinic in reference to the global coordinates, there is no coupling between transverse shears and other deformations. Accordingly, the vanishing shear traction boundary conditions and interface continuity conditions generate zero out-of-plane shear traction and shear strains for each layer. Moreover, owing to the absence of torsional deformation, the transformed displacement components $\bar{u}$ and $\bar{v}$ become

$$\bar{u} = \bar{\varepsilon}^0 z \text{ and } \bar{v} = 0,\tag{7}$$

370

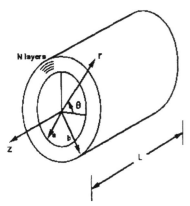

Figure 1. Cylindrical coordinate system in a laminated composite cylinder.

where the constant quantity $\bar{\varepsilon}^0$ has the physical interpretation of transformed axial strain of a layer. In fact, $\bar{\varepsilon}^0$, according to the present formulation, also represents the transformed axial strain of the entire composite cylinder. The calculation of $\bar{\varepsilon}^0$ requires the knowledge of end boundary conditions and will be given later. Likewise, solving for $\bar{w}(r)$ requires the information of transformed strain components, the constitutive equations, as well as the equilibrium equations.

The previously transformed displacement field gives the transformed strain components in cylindrical coordinates:

$$\bar{\varepsilon}_{rr} = \frac{d\bar{w}(r)}{dr}, \bar{\varepsilon}_{\theta\theta} = \frac{\bar{w}(r)}{r}, \bar{\varepsilon}_{zz} = \frac{d\bar{u}}{dz} = \bar{\varepsilon}^0, \text{and}$$

$$\bar{\varepsilon}_{\theta r} = \bar{\varepsilon}_{zr} = \bar{\varepsilon}_z = 0. \tag{8}$$

The unabridged form of the constitutive equation (4) for each layer in cylindrical coordinates with the radial coordinate r normal to the plane of symmetry is expressed as:

$$
\begin{Bmatrix} \bar{\sigma}_{zz} \\ \bar{\sigma}_{\theta\theta} \\ \bar{\sigma}_{rr} \\ \bar{\sigma}_{\theta r} \\ \bar{\sigma}_{zr} \\ \bar{\sigma}_{z\theta} \end{Bmatrix} =
\begin{bmatrix} \tilde{C}_{11} & \tilde{C}_{12} & \tilde{C}_{13} & 0 & 0 & \tilde{C}_{16} \\ \tilde{C}_{12} & \tilde{C}_{22} & \tilde{C}_{23} & 0 & 0 & \tilde{C}_{26} \\ \tilde{C}_{13} & \tilde{C}_{23} & \tilde{C}_{33} & 0 & 0 & \tilde{C}_{36} \\ 0 & 0 & 0 & \tilde{C}_{44} & \tilde{C}_{45} & 0 \\ 0 & 0 & 0 & \tilde{C}_{45} & \tilde{C}_{55} & 0 \\ \tilde{C}_{16} & \tilde{C}_{26} & \tilde{C}_{36} & 0 & 0 & \tilde{C}_{66} \end{bmatrix}
\begin{Bmatrix} \bar{\varepsilon}_{zz} \\ \bar{\varepsilon}_{\theta\theta} \\ \bar{\varepsilon}_{rr} \\ \bar{\varepsilon}_{\theta r} \\ \bar{\varepsilon}_{zr} \\ \bar{\varepsilon}_{z\theta} \end{Bmatrix} - \frac{\Delta T}{s}
\begin{Bmatrix} \tilde{B}_{zz} \\ \tilde{B}_{\theta\theta} \\ \tilde{B}_{rr} \\ 0 \\ 0 \\ \tilde{B}_{z\theta} \end{Bmatrix}. \tag{9}
$$

Furthermore, from the previous discussions, it can be shown that two of the three equilibrium equations are satisfied automatically. The only nontrivial equilibrium equation is the one in the radial direction:

$$\frac{\partial \bar{\sigma}_{rr}}{\partial_r} + \frac{\bar{\sigma}_{rr} - \bar{\sigma}_{\theta\theta}}{r} = 0. \tag{10}$$

Substituting (6), (7), and (8) into (9), the transformed stress components $\bar{\sigma}_{rr}$ and $\bar{\sigma}_{\theta\theta}$ are obtained in terms of the transformed radial displacement $\bar{w}$. Incorporating the resulting $\bar{\sigma}_{rr}$ and $\bar{\sigma}_{\theta\theta}$ functions with (10) gives a nonhomogeneous Euler differential equation of $\bar{w}$ for a layer

$$r^2 \frac{d^2\bar{w}}{dr^2} + r\frac{d\bar{w}}{dr} - \bar{\lambda}^2\bar{w} = \frac{r}{\tilde{C}_{33}}$$

$$\left[\frac{\Delta T}{s}\left(\tilde{B}_{rr} - \tilde{B}_{\theta\theta} \right) - \left(\tilde{C}_{13} - \tilde{C}_{12} \right)\bar{\varepsilon}^0 \right], \tag{11}$$

where

$$\bar{\lambda}^2 = \frac{\tilde{C}_{22}}{\tilde{C}_{33}}. \tag{12}$$

Solving (11) for $\bar{w}$ yields

$$\bar{w} = \bar{A}_1 r^{\bar{\lambda}} + \bar{A}_2 r^{-\bar{\lambda}} + \tilde{w}_p, \tag{13}$$

where

$$\tilde{w}_p = \tilde{f}_1 \bar{\varepsilon}^0 r + \tilde{f}_3 r$$

$$\tilde{f}_1 = \frac{\tilde{C}_{12} - \tilde{C}_{13}}{\tilde{C}_{33} - \tilde{C}_{22}}$$

$$\tilde{f}_3 = \frac{\tilde{S}}{\tilde{C}_{33} - \tilde{C}_{22}}$$

$$\tilde{S} = \frac{\Delta T}{s}\left[\tilde{B}_{rr} - \tilde{B}_{\theta\theta} \right] \tag{14}$$

and $\bar{A}_1$ and $\bar{A}_2$ are coefficients to be determined from boundary and continuity conditions.

Finally, it is understood that the initial condition of the original thermoviscoelastic problem is displacement-free state of rest. The boundary condition is of free traction and, hence, of free transformed traction on both inner and outer circular surfaces:

$$\bar{\sigma}_{rr} = \bar{\sigma}_{\theta r} = \bar{\sigma}_{zr} = 0 \text{ at } r = a, b. \tag{15}$$

On both end surfaces, stress resultants are zero:

$$\sum_{k=1}^{N} \int_{r_i}^{r_o} \bar{\sigma}_{zz} r dr = \bar{\sigma}_{zr} = \bar{\sigma}_{z\theta} = 0 \text{ at } z = 0, L. \tag{16}$$

The r_i and r_o are inner and outer radii, respectively, of the kth layer. The continuity conditions at each interface between two adjacent layers require continuous radial traction and continuous radial displacement at any instant as shown in Figure 2. Thus, when written in the transformed form, they become

$$\overline{\sigma}_{rr,o}^{(k)} - \overline{\sigma}_{rr,i}^{(k+1)} = 0 \qquad (17)$$

and

$$\overline{w}_{,o}^{(k)} = \overline{w}_{,i}^{(k+1)}, \qquad (18)$$

where $k = 1,...,N - 1$, and subscripts i and o denote inner and outer surfaces, respectively.

Accordingly, the formulation accounts for ply-by-ply variations of material properties and temperature change. The matrix-form numerical solution procedure with parallel computing techniques resolved the complexity and time-consuming calculation procedures in Laplace transform of a multilayered composite cylinder [10].

3 RELAXATION OF THERMAL STRESSES

The time-dependent thermal viscoelastic behavior of a 100-layer AS-4/3502 (graphite/epoxy) composite cylinder subjected to a temperature increase $\Delta T = 150°$ C is examined. Accordingly, initial residual stress built up in the cylinder due to the ΔT. The composite cylinder has an inner radius $a = 3.5$ in, an outer radius $b = 4.1$ in, and a thickness of each layer $h = 6.0 \times 10^{-3}$ in. Stacking sequence is given as $[0/30/60/90]_{25}$ from inside out, with the $0°$ direction coinciding with the axis of the cylinder. The creep properties of an AS-4/3502 graphite/epoxy composite with a fiber volume fraction of 0.67 were measured at different temperatures by Kim and Hartness (1987). The study shows that an increase of compliance with time, due to creep behaviors of material, was found at elevated temperatures. A least-squares curve fitting was used to express the transverse and shear creep from the original AS-4/3502 data in power law forms as follows:

$$S_{22}(t) = \left[1.7051 (t)^{0.1954} + 1\right] S_{22}^0, \qquad (19)$$

and

$$S_{66}(t) = \left[11.3076 (t)^{0.2771} + 1\right] S_{66}^0, \qquad (20)$$

where

$$S_{22}^0 = 7.5328 \times 10^{-7} \text{ / psi and}$$
$$S_{66}^0 = 1.3834 \times 10^{-6} \text{ / psi.}$$

The compliance in the fiber direction, $S_{11} = 5.9 \times 10^{-8}$/psi, and Poisson's ratios, $\nu_{12} = \nu_{13} = 0.3$, $\nu_{23} = 0.36$, are assumed to be time-independent. Thermal expansion coefficients of the composite in three principal directions are $\alpha_{11} = -0.5 \times 10^{-6}$/°C and α_{22}

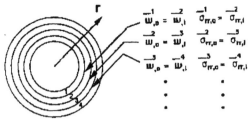

$\overline{w}$: Radial Displacement   o : outside surface

$\overline{\sigma}_{rr}$: Radial Stress   i : inside surface

Figure 2. Continuity of radial stresses and diplacments at the interfaces of layers

$= \alpha_{33} = 40.0 \times 10^{-6}$/°C, where the negative value indicates shrinkage with temperature increase.

Figures 3 and 4 show radial displacement and radial stress profiles across the thickness of the cylinder at three instants, instantaneous (initial stress), 2 years, and infinite time. The radial traction and displacement satisfied the continuity conditions at every interface of layers at all instants. The radial displacement, $w(t)$, will reach a steady state over a long period of time (infinite time) because of the creep behavior of composites. In fact, the radial displacement of most layers approaches to a constant value, except at the innermost and outermost portions of the cylinder. The free traction boundary at the surface of cylinders causes the gradients in the radial displacements. A similar phenomenon is also observed in the radial stress profile, which approaches to a constant over a long period of time. This long-term creep characteristic reflects the power law form, equations (19) and (20), of the creep compliance. The "saw" shaped radial stress distribution is the result of a variation of fiber orientations through the thickness of the cylinder. The radial stress is continuous, but the stress gradient is not. Accordingly, the stress profile illustrates the "saw" shape.

The hoop stress, $\sigma_{\theta\theta}(t)$, through the thickness of the cylinder is illustrated at three instants in Figure 5. There exist two distinct values (discontinuity) of $\sigma_{\theta\theta}(t)$ across each interface of two adjacent layers due to the various fiber orientations through the thickness. The hoop stress profile also shows a trend of relaxation over a period of time. The hoop stresses in $60°$ and $90°$ layers show a fairly steep gradient across the cylinder thickness initially. However, the gradient gradually disappears as time approaches infinity.

4 RELAXATION OF MECHANICAL STRESSES

In the following study, creep and stress relaxation of a composite cylinder subjected to internal pressure is

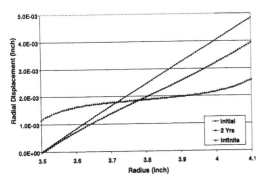

Figure 3. Radial displacement profiles in a cylinder subjected to thermal loads.

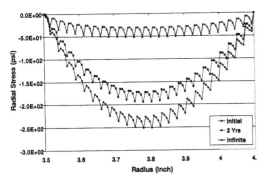

Figure 4. Radial stress profiles in a cylinder subjected to thermal loads.

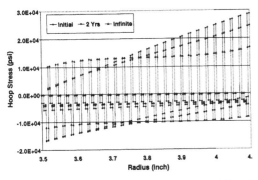

Figure 5. Hoop stress profiles in a cylinder subjected to thermal loads.

investigated. The calculation is performed using a 0.6-in-thick composite cylinder with a lay up construction of [0/30/60/90]$_{25}$. Calculations have been performed using similar basis graphite/epoxy composite materials. The material properties are the same

as described in the previous section. A pressure load of 1,000 psi was applied at the inner radius of the cylinder. Some selected displacement and stress profiles from the model prediction are illustrated and discussed in the following sections.

Figure 6 shows the radial displacement through the thickness of the cylinder at three instants, instantaneous (initial stress), 2 years, and infinite time. The radial displacement profile clearly illustrates the creep behavior of the cylinder. The inner radius of the cylinder increases over a period of time because of the application of inner pressure. The outer radius actually shrinks down because of creep characteristics. The radial strain, which is equal to the gradient of radial displacement, increases over a period of time.

Figure 7 illustrates the relaxation of radial stresses through the thickness at three instants. The radial stress at the inner radius is 1000 psi, equivalent to the pressure applied. The radial stress is zero at the outer surface of the cylinder since it is traction free. The curve is constructed by connecting the stress value at all the interfaces of the layers. The "saw shape" of the curve is due to the variation of fiber orientations through the thickness. Significant relaxation occurs over a period of time as shown in the radial stress profiles.

The hoop stress profiles at three instants are illustrated in Figure 8. The stress is not continuous from layer to layer due to the change of fiber orientation. The gradient of the stress profile is mainly due to curvature of the cylinder. The 90° layers have higher stresses because they are stiffer in the circumference direction. Accordingly, they carry more loads. The viscoelastic effect is quite interesting, as observed in the hoop stress profiles that change over a period of time. The hoop stress at the inner radius increases while it decreases at the outer radius. The integration of hoop stress through the thickness should be balanced with the inner pressure applied if a free body is taken from the cylinder. Since the stress gradient increases over a period of time, the hoop stress will also increase at the inner radius of cylinder.

5 CONCLUSIONS

An analysis has been developed for viscoelastic behavior of laminated composite cylinders with ply-by-ply variation of anisotropic viscoelastic properties, which cannot be studied using an isotropic model. Stress relaxation and creep are properly determined in a cylinder subjected to thermal and mechanical loads. The anisotropic viscoelastic behavior of the composite causes interesting characteristics in cylinders, which are critical for the durability of the structure. Creep and stress relaxation could exist in the fiber direction even though the fiber dominant properties are

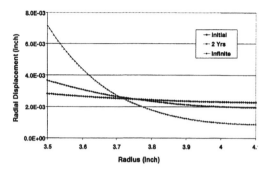

Figure 6. Radial displacement profiles in a cylinder subjected to internal pressure.

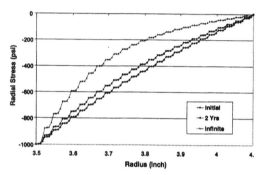

Figure 7. Radial stress profiles in a cylinder subjected to internal pressure.

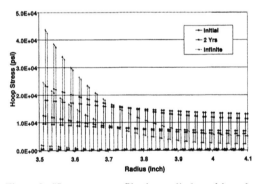

Figure 8. Hoop stress profiles in a cylinder subjected to internal pressure.

elastic. This is mainly due to the contribution of the Poisson's effects of viscoelastic transverse and shear properties. Viscoelastic characteristics are critical to the service life-cycle of applications such as pressure vessels and composite rotors designed with built-in prestress to achieve desired mechanical performance.

REFERENCES

Chien, L.S., & J.T. Tzeng. 1995. A thermal viscoelastic analysis for thick-walled composite cylinders. *J. of composite materials*. 29 (4) 525–548.

Christensen, R.M. 1982. *Theory of viscoelasticity, an introduction*, 2nd ed., New York: Academic Press, Inc.

Ferry, J.D. 1961. *Viscoelastic properties of polymers*, New York: John Wiley & Sons, Inc.

Hashin, Z. 1965. Viscoelastic behavior of heterogeneous media. *J. of applied mechanics*. 32: 630–636.

Kim, R.Y., & J.T. Hartness. Time-dependent response of AS-4/PEEK Composite. *Proc. of the 19th Inter. SAMPE Conference, October 1987*: 468–475.

Muki, R., & E. Sternberg 1961. On transient thermal stresses in viscoelastic materials with temperature dependent properties. *J. of applied mechanics*: 193–207.

Rogers, T.G., & E.H. Lee 1964. The cylinder problem in viscoelastic stress analysis. *Quarterly of applied mathematics*. 22: 117–131.

Schapery, R.A. 1964. Application of thermodynamics to thermomechanical, fracture, and birefringent phenomena in viscoelastic media. *J. of Applied Physics*. 35 (5): 1451–1465.

Schapery, R.A. 1967. Stress analysis of viscoelastic composite materials. *J. of composite materials*. 1: 228–267,

Tzeng, J.T., & L.S. Chien 1994. A thermal/ mechanical model of axially loaded thick-walled composite cylinders. *J. of composites engineering*. 4 (2) 219–232.

Williams, M.L. 1964. Structural analysis of viscoelastic materials. *AIAA Journal*. 2 (5): 785–808.

Recent Developments in Durability Analysis of Composite Systems, Cardon, Fukuda, Reifsnider & Verchery (eds)
© *2000 Balkema, Rotterdam, ISBN 90 5809 103 1*

Study of the effect of fiber orientation on the nonlinear viscoelastic behaviour of continuous fiber polymer composites

S. P. Zaoutsos & G. C. Papanicolaou
Composite Materials Group, Department of Mechanical and Aeronautical Engineering, University of Patras, Greece

ABSTRACT: The effect of fiber angle on the nonlinear viscoelastic response of a unidirectional fiber polymeric system is examined. Based on Schapery' s nonlinear viscoelastic equation and a model for the description of the nonlinear parameters, the nonlinear behavior of carbon/epoxy composite off axis specimens is studied. In order to achieve a complete characterization of an orthotropic system. The stress dependence of the parameters on the nonlinearity, introduced in the model is achieved and applied through a generic function developed and already confirmed [1,2,3] in the isotropic case. Verification of the model in the case of orthotropy is performed through a series of creep tests in composite specimens at various fiber orientations. The estimated values indicate an increase in the nonlinearity as the applied stress increases, while satisfactory predictions of the nonlinear parameters can be achieved with the use of the proposed model.

1 INTRODUCTION

Polymer matrix composites with continuous fiber reinforcement are frequently used in a variety of engineering applications. The mechanical response of these systems is time dependent due to the viscoelastic nature of the matrix, so as a consequence, time dependent properties have to be assumed when using these materials in engineering design. In addition, this behavior becomes highly nonlinear with respect to the applied loading. The nonlinearities in this behaviour have been studied the last two decades and a certain number of principles have been proposed. Much attention has been paid to the characterization of nonlinear creep behavior through multiple integral representations such as that of Green and Rivlin[4] and Pipkin and Rodgers[5] which require a rather multiple and complex experimental program for their application. Latest attempts [6,7] have been focused on Schapery's model where in comparison with previously developed single integral models seems to be more accurate and easy to handle for numerical treatment. For example in a study of Smart and Williams [8] it is evident that the model predicts well the response while the other models like MSP [9,10] or GRSN [11] fail to predict. This model is also more convenient for numerical treatment in comparison to other models [12,13]. Efforts for modeling composite systems have also been performed with satisfactory results.

A few efforts have been focused on the fiber orientation other than the transverse one, effect [14,15]. The aim of the current work is the study of the influence of fiber orientation on the nonlinear response of a composite system based on the nonlinear Schapery' s model and the nonlinear factor introduced in the above constitutive equation. The experimental results are also compared with the predictions occurring from a generic function, which is modified in order to include the fiber angle effect.

2 THEORETICAL BACKGROUND

Starting from the uniaxial isothermal case, the time dependent strain response of an isotropic material where a constant stress is applied, can be given as:

$$\varepsilon(t) = D_0\sigma_0 + \int_0^t \Delta D(t-\tau)\frac{d\sigma}{d\tau}d\tau \qquad (1)$$

where D_0 is the *initial, time-independent, component of the compliance* and $\Delta D(\psi)$ is the *transient, time-dependent, component of compliance.*

For the nonlinear behavior, according to Schapery's constitutive viscoelastic equation, the strain response of the system can be expressed as:

$$\varepsilon(t) = g_0 D_0\sigma_0 + g_1 \int_0^t \Delta D(\psi-\psi')\frac{d(g_2\sigma_0)}{d\tau}d\tau \qquad (2)$$

where ψ and ψ' are the so-called *reduced times* defined by :

$$\psi = \int_0^t \frac{dt'}{a_\sigma} \text{ and } \psi' = \psi(\tau) = \int_0^\tau \frac{dt'}{a_\sigma} \tag{3}$$

and g_0, g_1, g_2 and a_σ are *stress dependent nonlinear material parameters*.

Based on these considerations each of the factors is modeled using a law of the following form:

$$G_i = \begin{cases} 1 & \text{for } \sigma < \sigma_c \\[2mm] \dfrac{1-K_i}{1+\dfrac{\sigma-\sigma_c}{\sigma_u-\sigma}e^{\left(\frac{\sigma-\sigma_c}{\sigma_u-\sigma}\right)}} + K_i & \text{for } \sigma > \sigma_c \end{cases} \tag{4}$$

where σ_u is the ultimate tensile stress, σ_c is the critical stress from linear to nonlinear transition and K_i is the maximum value for G_i when stress tends to σ_u.

It has to be noted that the K_i parameters included in (4) have different values, depending on the parameter of interest.

In a previous analysis this model has been applied and verified in a 90° degrees carbon epoxy composite for the uniaxial response [3]. At this angle the response of the material is matrix dominated and represents the transverse properties of the composite system. The above approach was also used in the case of isotropic materials systems with satisfactory results.

The fiber orientation in each ply influences the global time dependent response of the laminate. For the case of linear viscoelastic behavior the problem can be treated by means of Boltsmann's superposition principles in combination with classical laminate theory

We now consider a unidirectional reinforced ply of a composite laminate as the basic element in a composite polymeric system. This element has to be first characterized in order to achieve the characterization of the overall time dependent mechanical response of the laminate.

In case where uniaxial loading in the x-direction of an off axis specimen is considered we may write according to Eq.2:

$$\varepsilon(t) = g_{0x}D_{0x}\sigma_{0x} + g_{1x}\int \Delta D_x(\psi-\psi')\frac{d(g_{2x}\sigma_0)}{d\tau}d\tau \tag{5}$$

and for the model nonlinearity parameters respectively:

$$G_{ix} = \begin{cases} 1 & \text{for } \sigma_x < \sigma_{xc} \\[2mm] \dfrac{1-K_{ix}}{1+\dfrac{\sigma_x-\sigma_{xc}}{\sigma_{xu}-\sigma_x}e^{\left(\frac{\sigma_x-\sigma_{xc}}{\sigma_{xu}-\sigma_x}\right)}} + K_{xi} & \text{for } \sigma_x > \sigma_{xc} \end{cases} \tag{6}$$

where σ_{xu} is the ultimate tensile stress and σ_{xc} is the critical stress threshold from the linear to nonlinear region.

3. EXPERIMENTAL PROCEDURE

The material selected for the testing was a carbon/epoxy, provided by Fibredux 920-TS-5-42 prepreg sheets with a nominal weight of 0.236 Kgr/m^2. The whole fabrication process performed in the Department of Mechanics of Materials and Constructions of the Free University of Brussels (VUB). Plates of 12 layers were constructed using the hand-lay-up technique. The plates were cured in a Scholtz autoclave following the standard curing cycle proposed by CIBA GEIGY. Specimens were cut from these plates in 75°, 60° and 45° directions relative to the direction of the fibers, using a diamond wheel saw. The dimensions of each specimen specified according to the ASTM D3039-76, which is similar to the dimensions of the standard specimen, specified by the Chinese National Testing Standard GB3354-82. A polishing procedure was followed for each specimen in order to avoid surface cracks and local material heterogeneities. After the polishing procedure the specimen were tabbed using LEXAN tabbing material.

Static tests for each fiber orientation conducted in 5 specimens in an MTS-810 materials general testing machine with a Test-Art II software package, in order to obtain the ultimate tensile strength $\sigma_{u\theta}$.

The creep-recovery tests were conducted, in the Laboratory of Strength of Materials at the National Technical University of Athens. A tensile loading mode applied in each specimen, consisted of an applied tensile loading at a constant stress level σ_0 for 8 hours, followed by 8 hours of recovery. Six different stress levels σ_0 corresponding to 30%, 40%, 50% 60% 70% and 80% of the tensile rupture stress σ_u were applied. The experiments performed in a dead load creep testing machine which is facilitated by using a lever /loading arrangement with a ratio 5:1. The weights used in each test for the loading of the specimens, were carefully weighed before each testing. Strain measurements acquired with the use of a diode laser.

4. RESULTS AND DISCUSSION

The variation of the ultimate tensile strength for the carbon/epoxy specimens with respect to fiber orientation is shown in Fig-1.

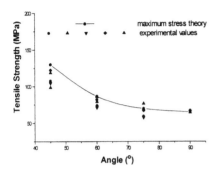

Figure 1. Tensile strength versus fiber angle for the carbon/epoxy composite.

Typical creep-recovery response for the 75° fiber orientation tested at different loading levels, is shown in Fig.-2. Nonlinear response is observed for applied stress levels of the order of 30% of the rupture stress. Isochronous curves for different fiber orientations confirm this nonlinear viscoelastic behavior, and this is shown in Fig-3. Results for the 90o specimens are also shown for the sake of comparison.

The values of go, g_1, g_2 and a_σ for each stress level, were estimated from the respective creep-recovery response occurred at that level, according to a method developed by the authors. A thorough description of the method is presented in [3].

An increase in the values of the nonlinear parameters is observed as the fiber angle decreases with an exception in the case of the shift factor a_σ.

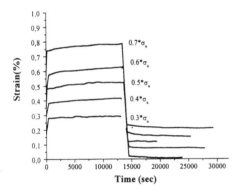

Figure 2. Creep-recovery response of the 75° carbon-epoxy composite

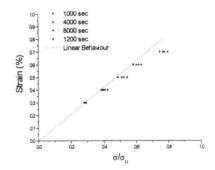

Figure 3. Isochronous curves for the carbon/epoxy composite specimen (fiber angle 75°).

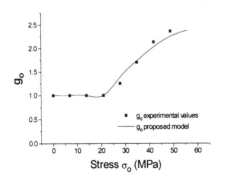

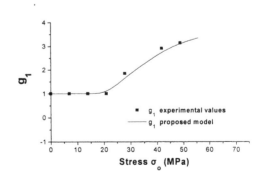

Figure 4. Comparison between the experimental values and the proposed model predictions of the nonlinear factors go and g1 as a function of stress for the 75° fiber angle specimens.

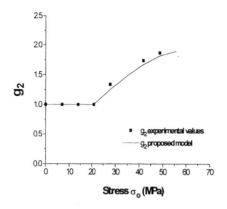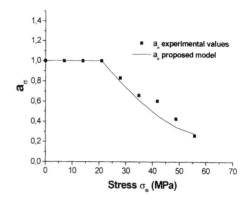

Figure 5. Comparison between the experimental values and the proposed model predictions of the nonlinear factors g_2 and a_σ as a function of stress for the 75° fiber angle specimens.

5. CONCLUSIONS

A nonlinear constitutive equation was used successfully for the viscoelastic characterization of the off axis viscoelastic behaviour of a composite material. The generic function describing each of the nonlinear parameters in the transverse direction of a composite is extended to the case of orthotropy and satisfactory predictions using the extended model can be performed. An increase in the values of the nonlinearly parameters occurs with increasing stress while this behavior is also consolidated as the fiber orientation decreases, with an exception in the case of the nonlinear parameter a_σ.

6. ACKNOWLEDGEMENTS

The help of the Associate Professor of the Laboratory of Strength of Materials at the National Technical University of Athens (NTUA) Mrs E. Kontou, in the experimental part of this study is gratefully acknowledged. Special thanks are also given to the Scientific and Technical Stuff of the Composite Systems and Adhesion Research Group (COSARGUB) of the Free University of Brussels.

REFERENCES

1. Zaoutsos, S.P., Papanicolaou, G.C. & Cardon A.H. 1998. On the Nonlinear Viscoelastic.Behaviour of Polymer Matrix Composites, *Composites Science and Technology,* 58:883-889.

2. Papanicolaou, G.C. & Zaoutsos, S.P. 1997. A Predictive Method for the Nonlinear Viscoelastic Characterization of Carbon/Epoxy Fibre-Composites, *Third Conference on Progress in Durability Analysis of Composites Systems,* Blacksburg, Virginia, U.S.A., 14-17 September.

3. Papanicolaou, G. C. & Zaoutsos, S. P. 1999. On the Prediction of the Non-Linear Viscoelastic Behaviour of Polymer Matrix Composites. *Composites Science and Technology,* 59:1311-1319.

4. Green, A. E. & Rivlin, R. S. 1957. The Mechanics of Nonlinear Materials with Memory, *Arch. Rational Mechanics and Annals,* 1:1-21.

5. Pipkin, A. C. & Rogers, T. G. 1968. A Nonlinear Integral Representation for Viscoelastic Solids, *Journal of the Mechanics and Physics of Solids,* 16:59-65

6. Schapery, R. A. 1966. A Theory of Nonlinear Thermoviscoelasticity Based on Irreversible Thermodynamics, *Proceedings of the 5th. U.S. National Congress in Applied Mechanics,* ASME, 511.

7. Schapery, R. A. 1968 On a Thermodynamic Constitutive Theory and it's Application to Various Nonlinear Materials, *Proceedings of UITAM symposium on Thermoinelasticity,* Springer-Verlag : 259-284.

8. Leaderman, H. 1943. *Elastic and Creep Properties of Filamentous Materials,* Textile Foundation, Washington.

9. Williams J. G. 1980 "Chapter 3: Time Dependent Behaviour", *Stress Analysis of Polymers,* 2nd edition, Ellis Horwood Limited, Publishers, :93-125

10. Smart, J. & Williams, J.G. 1972. A comparison of Single Integral Nonlinear Viscoelasticity Theories, *Journal of Mechanics and Physics of Solids*, 20:313-324.

11. Bernstein, B., Kearsley E.A. & Zapas L.J. 1963. A study of Stress Relaxation with Finite Strain, *Transactions of Society of Rheology*, 7:391-398.

12. Tuttle, M.E. & Pasricha A., Emery A.F. 1993 Time Dependent Behaviour of IM7/5260 Composites Subjected to Cyclic Loads and Temperatures, -Vol. 159, ASME *Mechanics of Composite Materials : Nonlinear Effects*, AMD, 159, pp. 343-357.

13. Czyz, J.A. & Szyszkowski, W. 1990. An Effective Method for Non-Linear Viscoelastic Structural Analysis, *Composites and Structures*, 37:637-646.

14. Lou, Y.C. & Schapery, R.A. 1971 Viscoelastic Characterization of a Nonlinear Fiber-Reinforced Plastic, Journal of Composite Materials, 5:208-234.

15. Tuttle M.E. & Brinson, H.F. 1986. Prediction of the long-term Creep Compliance of General Composites Laminates, *Experimental Mechanics*, 26:89-102.

Curing conditions and sample preparation for the analysis of the transverse creep behaviour of unidirectional graphite-epoxy systems

Christian Van Vossole, Pascal Bouquet & Albert H.Cardon
Free University Brussels, Belgium

ABSTRACT: An optimisation sequence for the production and preparation of testspecimen is presented. Within the context of the mechanical long term behaviour of composite materials, unidirectional carbon fiber reinforced plates are processed in order to obtain testbeams. The manipulation in the production process and storage are critical due to the sensitive measurements. The influence of environmental and damage conditions are causes of a wide dispersion of the measurements. Utmost care has to be applied yielding to representative and repeatable testresults. The paper describes the different levels on which special attention was paid for the optimisation and control of the properties of the material through the manufacturing process of solid composite material. A closer look was given at the rupture surface of the testspecimen to distinguish eventual inhomogenities which could be the source of the rupture.

1 INTRODUCTION

Durability of materials is an important feature for structural elements. The increasing number of structural applications of fiber reinforced composites demands adequate understanding of the long term behaviour. Lifetime and specifically rupture prediction of a structure submitted to mechanical loading, are of utmost importance regarding the safety requirements. Lifetime predictions of structural elements with a complex loading history need an integrated approach, taking into account the viscoelastic/viscoplastic behaviour of the polymer matrix and the damage due to the loading history.

The aim is to apply reliable accelerated test methods capable to describe long term behaviour. Therefore accelerating parameters such as stress level, pressure or temperature are applied to speed up the test. The corresponding superposition principle allows to modelize the material behaviour at a lower stress, pressure or temperature level over a longer period of time.

The fibers of the composite being almost solely elastic, the viscoelastic behaviour of the composite is mainly due to the resin. At low stress levels, we observe a linear viscoelastic behaviour on the contrary at higher stress levels a non-linear modellisation needs to be considered. For this purpose, the Schapery model can be used.

The creep behaviour of the composite is acquired from uniaxial fiber reinforced polymer matrices loaded perpendicular to the fiber direction. The stress level applied during the creep test is proportional to the ultimate stress level. A referential ultimate stress value for all testsamples had to be obtained from various production batches and samples.

Since the basic data, acquired from the creep tests at higher stress levels is essential, the knowledge of the stress level as a percentage of the tensile strength has to be determined accurately. Small differences in stress level result in big differences in time to failure.

The main purpose of this work is to keep the standard deviation of the tensile strength level of different specimen within acceptable limits in order to perform repeatable and comparable creep tests. Also the variations in time should be kept to a minimum. The solution to this critical path of testing sequence and preparation will be outlined in this paper.

2 PRODUCTION OF TEST BEAMS

Tensile tests are performed on beams with all the fibers in the transverse direction of the loading. Typical dimensions of the beams are $300 \times 17 \times 1,5 \text{mm}^3$.

These beams are cut out of bigger plates, all with identical thickness, that were cured in an autoclave, under controlled temperature and pressure conditions. The plates are laminated prepregs, type Hexcel Fiberdux 920 CX-TS-5-42

This prepreg has a fiber content of 58 weight %.

Plates of 60x30 cm² built up of 10 layers of

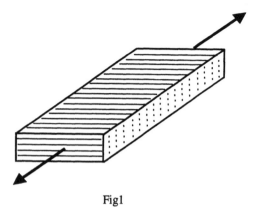

Fig1

prepreg are vacuumized for 8 hours, and then cured.

At first, the recommended cure-cycle from the producer of this prepreg was used.

3 MECHANICAL RESULTS

Beams that were cured according to the prepreg-manufacturer's cycle, were tested for their tensile strength. Averages and standard deviations are displayed in the following table.

We noticed from the experimental results that:
- Different plates: big differences in the average tensile strength
- Within one plate: big standard deviation
- Variation in time (conservation of the specimen was at room temperature and low relative humidity)

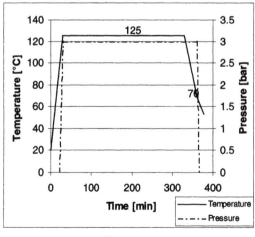

Fig2

Some factors, possible causes for these problems, had to be investigated:
- Type of epoxy groups, hardeners and additives
- Curing conditions
- Existence of fiber concentrations at fracture surface

4 CHEMICAL STRUCTURE

We were able to dissolve the prepreg resin with CHCl3 as a solvent and to separate the resin from the fibers.

The solved resin was then examined in the HNMR-apparatus to identify chemical structures. Problems arose as we weren't able to identify the hardeners nor additives but only the epoxy-groups that have the following structure:

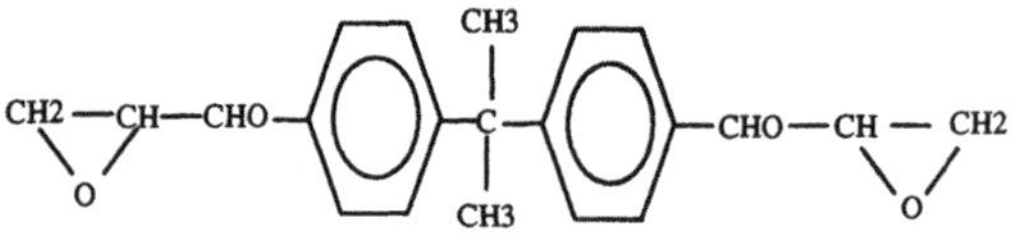

Fig3

This diglycidyl ether of bisphenol A (DGEBA) can react with amines or anhydrides.

5 FRACTURE SURFACE

When testing the specimen with the load direction transverse to the direction of the fibers, the resistance is actually provided by the resin. A strong concentration of fibers might then cause a weaker zone. The tensile strength would then be strongly dependant on the quality of the prepreg (fiber concentrations, misalignments in the prepreg).

Magnification 100x
Fig4

With the SEM (Scanning Electron Microscope) and an optical microscope a closer look was taken at the fracture surface.

Although the ten prepreg layers can still be easily recognized after the curing, no fiber concentrations could be observed at the fracture surface. Also no apparent cracks or irregularities were seen at the sawing edge.

Polishing a set of test beams didn't raise the average tensile strength level, nor did it lower the spread between different specimen.

6 CURING CONDITIONS

The epoxy - resin in the studied composite, is an amorphous polymer. To better understand the curing process, one can use the T T T (Time-Temperature-Transformation) diagram mostly developed by Gilham et. al [Ref.1,2].

Although it has been elaborated for pure amorphous polymers, it was a good guideline to improve the curing process for the composite laminate.

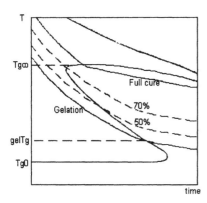

Fig5: Time-Temperature-Transformation Diagram

Since Tg is the only transition temperature for amorphous polymers, it is one of the most important material parameters. The Tg-region is a small temperature region where mechanical properties change dramatically. The material behaviour changes from a rubbery to a glassy state when going under the Tg.

During the curing of the polymer, the Tg rises, and is related to the degree of conversion. From the point where network formation starts (gelation) there is a serious increase in Tg, until full conversion at Tg∞ .

Further on, below Tg∞ , vitrication can occur, due to the higher viscosity that accompanies the network formation. Then the chemical reaction gets diffusion-controlled.

$t_{reaction} = t_{chemical\ kinetics} + t_{diffusion}$
after vitrification point $t_{diffusion} \gg t_{chemical\ kinetics}$

If we want to avoid too much exothermal reactions to occur in a short period of time, the reaction has to be diffusion controlled. This means that the cure-path should climb up the vitrification line, until Tg is reached.

To be able to compare mechanical properties from different specimen, full cure has to be achieved. A high Tg∞ will also limit the ageing-phenomena at room temperature.

A curing at a temperature above Tg∞ , will assure full cure. When there is risk for degradation, it can be achieved under Tg∞ as well.

7 DSC- MEASUREMENTS

In the DSC-oven, the heat flow of a reference pan and a sample pan (in our case 50 micro-liter) are compared during an imposed temperature programme. With temperature modulation, Cp is measured as well.

By following up these two parameters we can observe some important features:
- the temperature at which reaction starts.
- the amount of energy released once reaction begins (how high is peak).
- Tg before and after the temperature-programme.

Results:
1/Data:uncured prepreg
temperature :1st scan: -60°C to 220°C at 5°C/min
             2nd scan: 20°C to 220°C at 5°C/min

observations:
- Tg_0= -2,55°C (region: -7,98°C to 4,19°C)
- reactions begins at 105°C (high peak)
- after 1st scan : resin is partially cured
- Tg=113,28°C

2/ Data:uncured prepreg, 2 weeks out of freezer, at room temperature (scan is not displayed)
temperature: 1st scan: -60°C to 220°C at 5°C/min
             2nd scan: 20°C to 220°C at 5°C/min
observations:
- Tg_0= 0,52°C (region: -5,15°C to 6,15°C)
Even at room temperature there is some reaction going on.

3/ uncured prepreg
temperature :1st: constant at 125°C for 100 min
             2nd: scan: 50°C to 220°C at 5°C/min
             3rd: scan 50°C to 220°C at 5°C/min

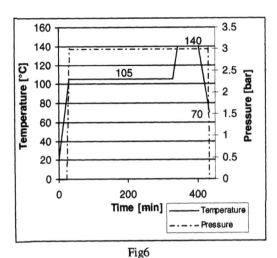

Fig6

conclusions:
- after 40 minutes reaction is finished

2nd scan: Tg has changed after 1st scan: curing after isothermal part was not fully completed

4/uncured prepreg
temperature :1st: constant at 105°C for 300 min
  2nd: scan: 50°C to 220°C at 5°C/min

conclusions:
- after 200 minutes reaction is finished
- peak is not so high
- maximum of Tg-region of the resin: 135°C (scan is not displayed)

8 IMPROVED CURE CYCLE

The conclusions of the DSC-scans let us develop a new cure cycle.

Since reaction fully starts at 105°C, a dwell is imposed at this T. From the isothermal cure at 105°C, it is observed that reaction goes on for approximately 200 minutes. Therefore, applying a margin, the dwell at 105°C in the new cycle lasts 300 minutes.

After the second scan, following isothermal cure at 105°C, the maximum of the Tg-region is 135°C. A second plateau will therefore be set at 140°C during 1 hour. Some further testing showed us that reaction was completed after this temperature programme.

9 RESULTS

DSC-samples taken from beams that were cured according to this improved cure-cycle, had a Tg of 112°C. There is still a difference with the Tg of 125°C that was reached with prepreg-specimen that were cured in the DSC itself.

The influence of the fiber content on the curing or the different curing conditions in the DSC versus the autoclave may be a cause for this difference.

The mechanical properties however, prove to be sufficient for the purpose of our tests.

Within one plate : beams tested for their tensile strength give following results:
Eg average: 73,62Mpa
  stdev: 2,04 Mpa (= 2,78%)

When we consider the averages for several cured plates (5):
  average: 74,53Mpa
  stdev: 1,43 Mpa (= 1,41%)

Also important is that there is no observable degradation in time of mechanical properties in a time span of 5 months.

10 FURTHER RESEARCH

The results and experience obtained from the preparation of testsamples will allow to proceed in the investigation of material behaviour in the long term. Since the spread of results is largely reduced, the stress level at which current creep tests are performed is more accurately defined.

The scope of our further research is to evaluate the behaviour of fibre reinforced composites under complex biaxial loading and to evaluate the damage and its influence on the lifetime after an imposed , complex, loading history.

Actual measurements clearly show the importance of a good testspecimen preparation.

11 REFERENCES

1.Gillham J.K. (1986) Formation and properties of thermosetting and high Tg polymeric materials, Polymeric Engineering and Science, vol26.

2.Aronhime M.T.,Gillham J.K. (1986)The Time-Temperature-Transformation cure diagram of thermosetting polymeric systems, Advances in Polymer Science pag.83-113

12 ACKNOWLEDGEMENTS

We would like to thank the VUB departments of Polymer chemistry of Prof. B. Van Mele, Metallurgy of Prof. J. Vereecken and the HNMR research team of Prof. R. Willem for their fruitfull cooperation.

We also want to express our gratitude to the Flemish fund for scientific research (FWO Vlaanderen) for the financial support of this research.

Applications

Recent Developments in Durability Analysis of Composite Systems, Cardon, Fukuda, Reifsnider & Verchery (eds)
© 2000 Balkema, Rotterdam, ISBN 90 5809 103 1

Durability assessment of polymeric composites for high speed civil transport

Thomas S.Gates
NASA Langley Research Center, Hampton, Va., USA

ABSTRACT: Technology development for the next generation High Speed Civil Transport (HSCT) required the construction of a large research program directed towards the durability assessment of polymeric based composites. These material systems would be used extensively in the primary and secondary structure of an HSCT and should meet the stringent requirements of a 72,000-hour design lifetime. The NASA Materials Durability Program was initiated to address this requirement and was composed of elements that addressed the development of a material durability database, predictive and accelerated test methods, and durability validation protocols. This paper will describe in broad terms these elements of the Materials Durability program and how the program elements interacted and supported each other.

1 BACKGROUND

The NASA High Speed Research (HSR) Program was conceived as a means for NASA and Industry to develop the technology necessary for the eventual design and construction of a new generation High Speed Civil Transport (HSCT) (Board [1]). The HSCT was envisioned as a 300 passenger, supersonic (Mach 2.4) aircraft capable of servicing the Pacific Rim market. The establishment of long-term durability of composite materials is one of the critical technologies required by such a revolutionary aircraft.

An economically feasible HSCT vehicle will require composite materials to help reduce the weight of the aircraft. However, previous NASA programs have shown drastic reductions in polymer composite strength and stiffness after long-term exposure (>25,000 hours) to supersonic transport service temperatures. Limited testing at reduced pressure (2 psi), representative of HSCT cruise altitudes, was more encouraging. In these tests, the specimens retained some strength after aging.

The study by Kerr and Haskins [2] recommended limiting cumulative exposure to 10,000-36,000 hours of supersonic aircraft service environment for the evaluated epoxy and polyimide materials. The epoxy, aged at 212°F, performed well, but the test was stopped at 36,000 hours. Also, this early program did not evaluate the critical mechanical properties that drive transport airframe designs. Therefore, new studies were required to evaluate modern materials under realistic HSCT conditions.

2 INTRODUCTION

The HSCT vehicle was envisioned with a design life of 72,000 flight hours and will require materials capable of withstanding a maximum temperature of 350°F during Mach 2.4 flight. The overall objective of the Materials Durability Program for HSR was to demonstrate the ability of candidate materials to withstand the harsh environment associated with a HSCT flight vehicle. It was recognized that materials used on a HSCT aircraft must withstand exposure to mechanical loads, elevated temperatures, and other environmental factors for up to 60,000 flight hours without significant degradation of mechanical properties.

In the HSR Materials Durability program, tests are being conducted to develop a material durability database for selected composite materials. Materials selected for testing either showed promise for eventual application on a HSCT or are representative of classes of materials such that they can provide validation data for analytical models of durability performance. The database includes data on the properties of these materials after short-term and long-term exposure to the HSCT mechanical, thermal, and chemical environments. Testing programs were designed to determine the degradation in material properties due to each of the elements of the HSCT environment and the extent of synergistic effects due to simultaneous exposure to multiple elements. Composite material properties of interest included static strength, elastic modulus, fatigue life, solvent resistance, creep resistance, and residual strength.

Standard test specimen configurations and procedures, conforming to the High Speed Research Composites Durability Test Guide and Recommended Test Procedures documents, were used whenever possible. Thermal-mechanical fatigue test profiles were also developed for use in mission simulation tests. Potential NDE methods for detecting manufacturing flaws and degradation due to test environments were evaluated in the course of making and testing specimens.

Materials chosen for this vehicle will experience at least one lifetime of loads, temperatures, and environments typical of the HSCT before production go-ahead is given. Considering that a 60,000-hour exposure test will require at least 7 years of testing time, it is not likely that all of the necessary information can be obtained in real-time testing. The prohibitive times and costs of testing for 60,000 hours necessitates the development of verified accelerated test methods and durability performance prediction methods. These tests should be able to provide an accurate and economical assessment of the long-term durability performance of new materials and to provide guidance for structural testing. Life, strength, and stiffness predictive models will be required for design because adequate databases for all stress/temperature/time scenarios cannot be established.

The accelerated aging methods required for the introduction of new materials later in the program should compare mechanical properties, damage modes, and physical parameters, such as weight loss, with those from specimens tested under real-time conditions. The real-time data should help validate the empirical and semi-empirical accelerated methods.

The development of accelerated aging profiles required extensive testing to define critical parameters and their interactions that drive physical aging and chemical degradation. This testing provided the insight into how these materials behaved and allowed for the development of analysis methods to predict material performance when subject to different combinations of load, temperature, and environment. Options have been explored for changing the test parameters from the expected HSCT operational values in order to reduce the test time required to achieve a desired end-state damage, but without changing the damage mechanisms identified for the operational environment.

Degradation mechanisms associated with elements of the operating environment were identified and quantified in the initial exploratory test programs. Predictive modeling strategies were selected to account for the effects of the degradation mechanisms on fatigue life, stiffness, and residual strength. The predictive models are being evaluated and refined as necessary as the program progresses by comparing predictions to results from accelerated and long-term tests.

Therefore, the specific objectives of the Materials Durability program were to:

1) Generate a limited short-term and long-term (up to 60,000 hours) durability database for selected materials.

2) Develop and verify accelerated test methods that will experimentally simulate long-term durability performance.

3) Develop and verify analytical methods that will predict long-term durability performance.

4) Develop a working set of durability evaluation protocols.

This paper will provide an overview of the progress in technology development of these four objectives by relating program issues unique to composite materials durability for HSCT. Recent developments in analytical and experimental methods will be discussed.

3 LONG-TERM TESTING, DATABASE

The test matrix developed to generate the long-term, real-time exposure data is given in table 1. The current number of exposure hours (x1000) are given in the table for each material and exposure condition.

As shown in table 1, there are currently 5 polymeric composite material (PMC) systems in long-term test. All of these PMC's utilize the same, continuous graphite, fiber system. The material abbreviations are as follows: TP- thermoplastic, PI- polyimide, -BMI bismaliemide.

The exposure conditions ranged from the simplest (isothermal with out load) to the most complex (flight profile). The flight profile condition represented the thermal and mechanical loads expected during a 4 hour "typical flight" for a wing or fuselage component. Test temperatures for each exposure condition were determined by the expected operating conditions for the material type. Maximum test temperature for any of the materials was 177°C.

Table 1. Test matrix of long-term exposure conditions for durability database. Numbers indicate actual or equivalent exposure hours (x1000) to date.

	Exposure Conditions				
Material	Isothermal w/o load	Isothermal w/ load	Thermal cycled	Isothermal fatigue	Flight profile
Gr/TP	25	15	20	-	-
Gr/PI-a	60	15	15	-	30
Gr/PI-b	25	10	10	15	10
Gr/BMI-a	60	15	15	-	30
Gr/BMI-b	60	-	15	15	10

All tests were run at laboratory atmospheric conditions. Both coupons and large panels were used in the test program.

At predetermined intervals, sets of coupons or panels were removed from test for residual property testing. These removal intervals were spaced to allow periodic assessment of degradation using both destructive and nondestructive testing. Residual mechanical properties tracked as function of exposure time included compression and tension strength and stiffness of notched and unnotched specimens. Physical property data such as weight change, surface quality, and extent of damage was also tabulated. All the measured properties were stored in an electronic database to facilitate comparisons between test materials and test conditions.

An assessment of the detailed results from the tests in table 1 are beyond the scope of this paper. However, in general, lessons learned to date from this test program include:

a) Time at temperature and maximum test temperature seems to be the most important factors in determining residual properties.

b) Damage initiation and growth can occur in the absence of mechanical load due to thermal-oxidative stability related degradation mechanisms.

c) The response to mechanical loading during aging will vary with aging time and conditions.

d) Scale-up from coupon level test results may be an issue for degradation mechanisms dependent on oxygen concentration.

e) Degradation mechanisms may be unique to material classes (eg. Thermoplastic, thermoset).

f) A full range of residual properties must be checked to ensure an adequate database.

g) NDE can be effectively used to correlate damage states with destructive testing.

4 PREDICTIVE AND ACCELERATED TEST METHODS

Early-on in the development of the Materials Durability program for HSCT, the need for research into accelerated test methods was identified. It was recognized that without the long-term data, validation of the accelerated methods was incomplete. However, it was felt that as the long-term data was collected, corrections in the accelerated test methods would take place based on periodic evaluation of success. Based on lessons learned, initial accelerated test methods were developed to help screen PMC's for long-term durability. Some of the early lessons can be found in a report published at the request of NASA by the National Research Council's National Materials Advisory Board (NMAB) (Starke [3]) on accelerated aging of future aircraft related materials.

The NMAB report identified issues related to the aging of aluminum, titanium, polymer matrix composites, and ceramic matrix composites. Suggestions were outlined for accelerated evaluation approaches and analytical methods to characterize the durability of future aircraft materials and structures throughout their service life.

As the test program continued, an Accelerated Test Methods guide was developed and updated periodically. This guide detailed both the philosophy and step-by-step methods used in the development of the accelerated tests. Table 2 provides an outline of the materials and degradation mechanisms investigated under the accelerated test program.

The current work specifically addresses issues in accelerated aging of polymer matrix composite materials suitable for high temperature applications. The general approach to development of these accelerated methods will be outlined here while a detailed description of methods may be found in Gates and Grayson [4].

4.1 Accelerated Aging

The long-term exposure of aerospace, polymeric composite materials to the use-environment will eventually result in change(s) in the original properties of the material. This process is loosely referred to as "aging". This material aging may translate to structural changes in mission-critical components which for an aerospace platform can have a potentially catastrophic effect on both the vehicle and its payload.

Verified accelerated aging methods are needed to provide guidance for materials selection and to accurately assess aging of new materials. The degradation mechanism that results in a significant loss in any important bulk physical property of the material system when exposed to environmental stress factors inside the limits of the use-environment is the critical degradation mechanism. Accelerated aging is defined as the process or processes required to accelerate a specific mechanism or mechanisms relative to a baseline aging condition; thereby resulting in the material reaching the same aged end-state as a real-

Table 2. Test matrix for accelerated tests.

	Thermal-Oxidative Stability	Thermal Stability	Hygro-Thermal	Physical Aging
Gr/PI-a	X	X	X	X
Gr/PI-b	X	X	X	X
Gr/BMI-a	-	-	X	-
Gr/BMI-b	X	X	-	-

time aged material, but in less time.

Only by understanding how each aging mechanism affects a given material system can it be determined if that aging mechanism can be properly accelerated. In the simplest case, aging is associated with a single mechanism, in which case acceleration of this mechanism will allow meaningful accelerated aging methods to be developed. More likely, the aging process involves several different mechanisms that may or may not act synergistically, complicating the problem significantly.

The highly empirical approaches taken for the majority of accelerated aging studies dictates that the primary objective of an accelerated aging method is to screen and characterize new material systems. In addition to materials screening, accelerated aging may help determine residual service life of existing structures.

The development of accelerated aging methods requires extensive testing to define critical environmental stress factors and their interactions. This testing provides insight into how materials behave and input for the development of analysis methods to predict material performance under various conditions of load, temperature, and environment. A comparison of mechanical properties, damage modes, and physical parameters; such as weight loss, changes in glass transition or fracture toughness; from accelerated testing with those from real time testing serves to validate accelerated aging methods.

4.2 Test Requirements

It is the challenge of an accelerated aging program to speed up the effects of exposure to environmental stress factors without departing from the underlying mechanisms that give rise to molecular level changes. The most important requirement to keep in mind during the development of an accelerated aging method is that it must replicate those changes that occur in the real-time testing. Thus emphasis must be put on the need to understand and reproduce degradation mechanisms associated with each accelerated aging condition.

Beyond exploratory tests, additional studies must be conducted to identify material properties that are easy to measure and that can be reliably correlated to changes in mechanical properties. These material "indicator" properties form the basis for development of more economical accelerated aging schemes for screening new materials and for evaluating the status of materials in long-term aging.

Another significant requirement for accelerated aging studies is the coupling of mechanical property data from accelerated aging of materials with data from real-time aging. The comparison with real-time data will determine the accelerating factor for any given degradation mechanism.

4.3 Degradation Mechanisms

Degradation (aging) mechanisms for PMC's are divided into three classes: chemical, physical, and mechanical. These three mechanisms may be additive or subtractive depending on the environmental and mechanical loads.

Chemical aging refers to an irreversible change in the polymer chain/network through mechanisms such as cross-linking or chain scission. Chemical degradation mechanisms include thermo-oxidative, thermal, and hydrolytic aging.

Physical aging will occur when a polymer is cooled below its T_g, and the material evolves toward thermodynamic equilibrium. This evolution is characterized by changes in the free volume, enthalpy, and entropy of the polymer and will produce measurable changes in the mechanical properties (Struik [5]).

Mechanical degradation mechanisms are irreversible processes that are observable on the macroscopic scale. These degradation mechanisms include matrix cracking, delamination, interface degradation, fiber breaks, and inelastic deformation; and thus have a direct effect on engineering properties such as bulk stiffness and strength.

4.4 Procedures

Material performance is defined by a set of indicators that measure a specific property for the material. The indicators will be different dependent upon the mechanism of interest. The suggested procedure is as follows:

a) Identify material by class (i.e. thermoplastic, thermoset).

b) Identify mechanism to evaluate (i.e. thermal stability, matrix cracking, etc.).

c) Choose an environmental stress factor for aging.

d) Conduct aging experiment within limits of the chosen environmental stress factor using established methods.

e) Perform post-aging tests with indicators sensitive to changes in material performance and compare results to unaged values.

4.5 Predictive Methods

The objective of this aspect of the program was to develop a methodology for life prediction of polymer matrix composites for HSCT applications. The approach was to understand the mechanisms of damage associated with degradation of these materials under mechanical loads and then to apply damage mechanics modeling concepts. One area of research was the cyclic loading behavior of graphite/polyimide laminates after 10,000 hours of isothermal aging at 350°F. Varna, et al. [6], and

Akshantala and Talreja [7] used the unaged and aged fatigue data along with quasi-static testing in the development of a fatigue model.

To explore the ability to predict fatigue life and remaining strength, a model was developed for composite materials subjected to cyclic mechanical and long-term thermal loads (Reifsnider and Stinchomb [8], Subramanian and Reifsnider [9]). In this approach, a representative volume was selected which was typical of the material in question. This representative volume may contain damage, such as matrix cracks, delamination, micro-buckles, or fiber fractures, but some part of it still retains the ability to carry load. It is the failure of this part of the representative volume, the so-called "critical element", which determines the fracture of the entire representative volume. The remaining strength of the critical element was calculated by using a non-linear damage evolution equation that accounts for the changing stress amplitude in the critical element. Comparing predictions with long-term test results provided evaluation and validation of the models.

Several of the individual predictive methods developed by the program were integrated into an analysis code for Integrated Strength Analysis of Aged Composites (ISAAC) which was developed at The Boeing Company. The strength analysis portion of the ISAAC code predicted laminate modulus and strength based on constituent and ply properties. The analysis was used for predicting unnotched and center notched laminates subject to axial running loads and was done on a ply-by-ply basis which provided output every time a ply was predicted to fail. Failure modes were determined and specified based on ply failures. Multiple failure criteria were provided. The effects of thermal residual stresses were also included in the analysis.

To further the development of accelerated test methods for advanced polymeric composites, a combined experimental and analytical task was initiated to investigate the effects of physical aging on graphite/polyimide composites. Recently developed novel experimental apparatus and methods were employed to test two matrix dominated loading modes, shear and transverse, for two load cases, tension and compression (Gates, et al. [10]), (Bradshaw and Brinson [11]). The tests, run over a range of sub-glass transition temperatures, provided material constants, material master curves and aging related parameters.

An analytical model based upon principles of viscoelasticity, effective time, and composite lamination theory was developed by Brinson and Gates [12] for predicting long-term tension and compression stiffness or compliance using short-term data as input.

Verification of the model through comparing predictions to long-term (1500+ hour) experiments indicated that the model worked equally as well for the tension or compression loaded cases. Comparison of the loading modes indicated that the predictive model provided more accurate long-term predictions for the shear mode as compared to the transverse mode. Parametric studies showed the usefulness of the predictive model as a tool for investigating long-term performance and temperature based acceleration.

To investigate the accelerated aging under hygrothermal environments, Nairn and Han [13] tested a graphite-polyimide composite a several temperatures and relative humidities. Measurements were taken on microcrack density and fracture toughness as a function of aging time. It was found that the toughness decreased with time and the rate was proportional to the temperature and relative humidity. The experimental results were fit with a first-order kinetics analysis and then the analysis was used to construct a hygrothermal aging master plot. This master plot could then be used to make predictions of toughness due to a variety of hygrothermal environments.

5 DURABILITY VALIDATION PROTOCOLS

A durability validation protocol is a document or set of documents that describe the step by step process for screening a candidate material for long-term durability. The objective of this task therefore, was to develop economical and definitive protocols, based on predictive and accelerated methods, to quickly assess the probable durability of a material in an HSCT service environment. These protocols made reference to the relevant long-term and accelerated test data as well as the results from the studies in predictive methods. The protocols addressed environmental stress factors associated with HSCT operation and defined key durability indicators with pass / fail criteria. The protocols will be also be used for demonstrating that material or process improvements have no adverse effects on durability. The protocols developed under the HSR Materials Durability Program include:

- Protocol for Stiffness Degradation in PMC Materials due to Physical Aging
- Solvent Sensitivity Evaluation Protocol
- Durability Screening Fatigue Spectrum Development
- Interfacial Integrity: Evaluation Protocol Development.
- Hygro-Thermal-Mechanical Fatigue: Protocol Development
- Evaluation of Protective Systems: Protocol Development
- Durability Screening Test Methods for Assessing Chemical Aging

6 SUMMARY

The NASA High Speed Research, Materials Durability Program was designed to develop the technology necessary for selection of a polymeric composite material durable enough to be considered for primary aircraft structure. The key elements of the Materials Durability Program were a database of material properties, predictive and accelerated test methods, and durability validation protocols. These elements worked together to provide the empirical and analytical tools necessary for materials selection and to ensure an in-depth understanding of material degradation mechanisms unique to super-sonic aircraft design.

REFERENCES

[1] "U.S. Supersonic Commercial Aircraft," , N. R. Council, Ed. Washington, D.C.: National Academy Press, 1997.
[2] Kerr, J. R. and Haskins, J. F., "Time-Temperature-Stress Capabilities fo Composite Materials for Advanced Supersonic Technology Application," NASA Langley Research Center, NASA Contractor Report, Contract NAS1-12308 178272, 1987.
[3] Starke, E. A. J., "Accelerated Aging of Materials and Structures, The Effects of Long-Term Elevated-Temperature Exposure," National Materials Advisory Board, National Research Council, Washington, D.C. NMAB-479, 1996.
[4] Gates, T. S. and Grayson, M. A., "On the use of Accelerated Aging Methods for Screening High Temperature Polymeric Composite Materials," presented at AIAA Structures, Dynamics and Materials Conference, St. Louis,2, AIAA99-1296 ,pp. 925-935, 1999.
[5] Struik, L. C. E., *Physical Aging in Amorphous Polymers and Other Materials*. New York: Elsevier Scientific Publishing Company, 1978.
[6] Varna, J., Akshantala, N. V., and Talreja, R., "Crack Opening Displacement and the Associated Response of Laminates with Varying Constraints," *International Journal of Damage Mechanics*, Vol. 8, pp. 174-193, 1998.
[7] Akshantala, N. V. and Talreja, R., "A Mechanistic Model for Fatigue Damage Evolution in Composite Laminates," *Mechanics of Materials*, Vol. 29, pp. 123-140, 1998.
[8] Reifsnider, K. L. and Stinchomb, W. W., "A critical element model of the residual strength and life of fatigue loaded composites," *Composite Materials: Fatigue and Fracture, ASTM STP 907*, H. T. Hahn, Ed. Philadelphia: ASTM, pp. 298-303, 1986.
[9] Subramanian, S. and Reifsnider, K. L., "A cumulative damage model to predict the fatigue life of composite laminates including the effects of a fibre-matrix interphase," *International Journal of Fatigue*, Vol. 17, pp. 343-351, 1995.
[10] Gates, T. S., Veazie, D. R., and Brinson, L. C., "Creep and Physical Aging in a Polymeric Composite: Comparison of Tension and Compression," *Journal of Composite Materials*, Vol. 31, pp. 2478-2505, 1997.
[11] Bradshaw, R. D. and Brinson, L. C., "Physical Aging in Polymers and Polymer Composites: An Analysis and Method for Time-Aging Time Superposition," *Polymer Engineering and Science*, Vol. 37, pp. 31-44, 1997.
[12] Brinson, L. C. and Gates, T. S., "Effects of Physical Aging on Long Term Creep of Polymers and Polymer Matrix Composites," *International Journal of Solids and Structures*, Vol. 32, pp. 827-846, 1995.
[13] Nairn, J. A. and Han, M.-H., "Hygrothermal Aging of Polyimide Matrix Composite Laminates," presented at International Conference on Composite Materials (ICCM 12), Paris, 1999.

Recent Developments in Durability Analysis of Composite Systems, Cardon, Fukuda, Reifsnider & Verchery (eds)
© 2000 Balkema, Rotterdam, ISBN 90 5809 103 1

Investigation into honeycomb durability through an in-flight service evaluation

J. E. Shafizadeh & J. C. Seferis
Polymeric Composites Laboratory, Department of Chemical Engineering, University of Washington, Seattle, Wash., USA

ABSTRACT: Thirteen commercial wing panels were fabricated and flown on a commercial aircraft to investigate the mechanisms of water migration through various honeycomb cores. A 12.2 J impact damage was not observed to cause damage propagation in aluminum and Korex® honeycomb materials. This was attributed to the ability of the cores to localize the impact damage. In Nomex® and fiberglass cores, a different damage propagation mechanism was observed. In these cores, the damage was not confined to the localized area around the impact. Instead, core damage was seen as far as 2.0 cm from the point of impact. This increased core damage allowed the core to retain water. The retained water helped propagate the impact damage through a freeze-thaw mechanism. Speed-tape repairs were only found to be statistically significant when water migrated through the core. Ultimately, this work provided the tools a methodology for understanding and developing damage and water tolerant honeycomb structures.

1 INTRODUCTION

Honeycomb composites are designed and manufactured to be light stiff beam structures. However, their lightweight construction also leads to problems once in-service.(Coggeshall, 1983). After manufacturing, honeycomb composite panels are designed to be 'closed' to their operating environments, but once in-service, the thin skins are susceptible to water ingression and foreign object damage. When composite facesheets are damaged, a chain reaction of events is set in motion that ultimately degrades the mechanical integrity of both the honeycomb and the composite structure.

In a high energy impact, both the facesheet and honeycomb core are damaged. Damage to the facesheet can result in matrix microcracking and fiber breakage.(Abrate, 1997) This damage is further aggravated by the presence of honeycomb core. Once the facesheet is fractured, water can easily enter the cells of the honeycomb core through cracks in the facesheet. The presence of standing water in the honeycomb core can destroy the core through a freeze-thaw mechanism.

When the facesheet is fractured, the honeycomb core is directly exposed to a set of harsh environmental conditions. At a cruising altitude of 9000 meters, the temperature inside the honeycomb core is near –40°C. At this temperature, any standing water that has accumulated within the honeycomb core will freeze. The freezing water expands and stresses the honeycomb cell walls. When the airplane lands, the water melts and the honeycomb wall relaxes. After a number of these 'freeze-thaw' cycles, the cell walls will catastrophically fail and destroy the structural integrity of the honeycomb. Along with destroying the honeycomb structure, water in the core can also delaminate the facesheet. If a honeycomb cell contains a substantial amount of water, the freezing water can expand against the facesheet and delaminate the bond between the honeycomb and the facesheet.

Facesheet delamination can also be observed on the ground. The process of repairing damaged composites can often induce damage itself. When composites are repaired, they are often heated to temperatures in excess of 100°C. Under these conditions, the water contained within the honeycomb cells of the composite part quickly vaporizes. Over 100°C, the pressure of the vaporized water can exceed the tensile strength of the bond between the facesheet and the core, resulting in delamination of the facesheet.(Garrett, Bohlmann, & Derby 1978)

As a consequence of the many problems which can occur to honeycomb sandwich structures, a number of investigations have recently been performed to try to model impact failure mechanisms and other mechanisms of damage propagation.(Abrate,1997; Cise &Lakes 1997) Through these efforts, five honeycomb composite failure modes have been identified: 1) fiber breakage in facesheet, 2) core buckling, 3) facesheet delamination, 4) core cracking and 5) matrix cracking.(Nettles & Hodge, 1990) Along with the mechanical and mathematical models developed to predict damage propagation, a number of novel studies have been conducted in an attempt to minimize honeycomb damage propagation before and after damage initiation.(Wong & Abbot, 1990) Researchers have investigated filling the honeycomb cells with foam to prevent the honeycomb from ingesting water.(Wu, Weeks, & Sun, 1995) Others have tried to prevent ingression by interleafing a sheet of PEEK within the laminate facesheet.(Weems & Fays, 1995) Some investigators have examined how to dry the honeycomb core after the core has become saturated with water under repair conditions.(Geyer, 1996) Yet despite all of the work conducted in this area, no effort has been made by academia or industry to statistically understand the mechanisms of water ingression and migration through honeycomb core as a function of in-service cycling.

This paper represents the second part of an investigation to study the scope and mechanisms of water ingression and migration through honeycomb core. In the first part of this study, the extent of honeycomb water ingression problems were explored. An investigation was conducted to determine whether water ingression was a localized problem which occasionally occurred, or whether water ingression in honeycomb sandwich structure represented a systemic composite problem. (Shafizadeh et al, submitted) To quantify the extent of water ingression and migration problems, non-destructive infrared thermographic inspections were performed on 35% of United Airlines Boeing 767's. The second part of this study focuses on a Design of Experiment (DOE) to understand how water migrates through the core once it has ingressed. For the DOE, an in-flight service evaluation was conducted on thirteen outboard fixed trailing edge panels. From these panels, the effects of core type, impact damage, and tarmac repair procedures were examined. Together, these studies represent the most comprehensive attempt to understand the in-service durability of honeycomb structures. The research performed in these two studies will aid in the understanding and design of future honeycomb composite sandwich structures.

2 EXPERIMENTAL *PROCEDURE*

2.1 *Panel Fabrication*

A total of sixteen modified Boeing 767-200 upper wing fixed trailing edge panels were fabricated and flown to study the mechanisms of water ingression and migration through honeycomb core. Upper wing fixed trailing edge panels were selected for study due to their ease of fabrication and location on the wing. As shown in Figure 1, these panels are located on the outboard section of the wing behind the rear spar.

The wing fixed trailing edge panels are lightly loaded and are manufactured with fiberglass facesheets and honeycomb core. The facesheets of the panels were fabricated from Hexcel F-155 prepreg qualified to Boeing Material Specification (BMS) 8-79 (Anon, 1989). The F-155 prepreg used was a 126°C cure, self-adhesive, epoxy based resin system which was impregnated in a 7781 style glass fabric. The panels were constructed using a vacuum bagging process where five plies of prepreg were used on the tool, or aerodynamic side of the panel.

The honeycomb core bays used in production trailing edge panels were fabricated from 3.2 mm cell size Nomex® honeycomb core with a nominal density of 50 kg/m³. In this study, the usual Nomex honeycomb material was substituted with one of four different types of honeycomb core which included aluminum, Korex®, Nomex and fiberglass based cores. The Nomex, fiberglass, Korex and aluminum cores used were all provided by Hexcel and had the following respective Hexcel designations: HRH-10-1/8 3.0, HFT-1/8-3.0, Korex-1/8-3.0, and CRIII-1/8-5052-0.001-4.5. The cell size on all honeycomb cores was 3.2 mm and all cores had a nominal density of 50 kg/m³, except the aluminum

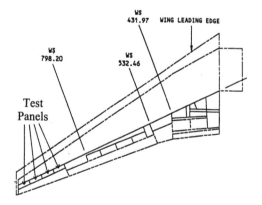

Figure 1. Diagram showing the wing location of the upper wing fixed trailing edge test panels (Anon, 1994).

core. The aluminum core had a nominal density of 75 kg/m^3 and was anodized with phosphoric acid to prevent corrosion.

In production upper wing fixed trailing edge panels, the size of the honeycomb core bays ranges from 99 to 39 cm in length and 27 to 33 cm in width. In modifying the honeycomb core bay materials, the overall size of the core bays was fixed by Boeing specification drawings. Therefore all modified core bays had to conform to the dimensions of the production core bays. In complying with this design constraint, the overall size of the core bays was divided into one, two or three smaller core bays, or sub-bays. The core sub-bays were manufactured from the different core materials listed above and spliced together with a foaming adhesive from Sovereign Specialty Chemicals (PL 685). After the bays were spliced together a 20° chamfer was machined out of the core to match the part specification drawings. The spliced core bays had the same dimension as the original production core bays. Each sub-bay was approximately 650 cm^2 in size. After fabrication, the wing panels were installed on two different United Airlines 767-200s and flown on domestic US routes for fourteen months.

2.2 Design of Experiment

In order to statistically understand and isolate the mechanisms of water migration through honeycomb core, a Design of Experiments (DOE) was constructed. The DOE was fabricated to study the interactions between different honeycomb core materials and the honeycomb operating environments as a function of core composition, damage size and limited repair procedures.

Table 1. DOE for evaluating the effects of core material, impact damage and speed-tape repair as a function of water migration and damage growth.

Core Bay	Core	Speed Tape	Impact	Plane
1	Korex	No	Yes	A
2	Nomex	No	No	A
3	Fiberglass	Yes	No	A
4	Aluminum	Yes	Yes	A
5	Aluminum	No	No	A
6	Korex	Yes	No	A
7	Nomex	Yes	Yes	A
8	Fiberglass	No	Yes	A
9	Korex	No	No	B
10	Korex	Yes	Yes	B
11	Nomex	Yes	No	B
12	Nomex	No	Yes	B
13	Fiberglass	No	No	B
14	Aluminum	No	Yes	B
15	Aluminum	Yes	No	B
16	Fiberglass	Yes	Yes	B

The DOE is outlined in Table 1. The face sheets over each core bay were intentionally damaged to allow water into the honeycomb core and create a known path for water ingression. Three holes, 1.6 mm in diameter, were drilled in the center of the aerodynamic facesheet of each core bay. The holes were drilled in the centers of three adjacent core cells forming a triangle with the centers of the holes located 3.2 mm apart.

Some honeycomb bays were also impacted to study the growth of impact damage as a function of water migration through the core. According to the Structural Repair Manual (SRM) for the Boeing 767-200, the maximum allowable damage size before a repair is necessary for a single damage site in a honeycomb core area is 5.1 cm (Anon, 1994). The panels were damaged such that all damage locations were approximately the same size. The panels were damaged with an impact drop tower. The impact energies and damage sizes for each core bay are presented in Table 2. The different impact energies and damage sizes are due to the different energy absorption characteristics of the cores.

Table 2. Impact energies and their resultant honeycomb bay damage sizes.

Core Type	Core Bay	Impact Energy	Damage Size (Diameter)
		J	cm
Nomex	7, 12	8.1	2.14
Korex	1, 10	12.2	2.09
Aluminum	4, 14	12.2	2.00
Fiberglass	8, 16	4.1	3.53

The effectiveness of speed-tape repairs to prevent water ingression through a known ingression path was also explored in the DOE. In a speed-tape repair, the damaged area is cleaned with an appropriate solvent and an aluminum foil tape is applied over the damaged area. For the modified fixed wing trailing edge panels, a square 103 cm^2 piece of speed-tape was place over the holes or damage location before the panels were installed on the aircraft. Two core bays of each construction were manufactured to insure result reproducibility.

2.3 Panel Inspection

Before and after the in-flight service evaluation, through-transmittance ultrasound (TTU) inspections were performed on the panels to measure the size of the damaged area in each honeycomb core bay. During the TTU inspections, the panels were scanned with waterjet coupled transducers, 1.91 cm in diameter. The panels were scanned at a rate of 25.4 cm/sec at a frequency of 1 MHz while being indexed in increments of 4.06 cm.

While cycling for fourteen months on the two 767 aircraft, the panels saw an average service time of 4,400 flight hours with 2 takeoffs and landings per day with an average flight time of 5 hours per take-off. After the in-service cycling, the panels were removed from the aircraft and a 420 cm² square was cut from the center of each core bay. The individual core bays were weighed and dried in a vacuum oven at 75°C under a constant pressure of 4.8 kPa for 5 days. After drying, the core bays were weighed and the difference between the initial and final weights was reported.

3 RESULTS AND DISCUSSION

3.1 *Water Absorption*

There are two basic statistical models for analyzing and interpreting Design of Experiments: factorial models and hierarchical models.(Hicks, 1982) Factorial models assume that all of the measured response variables can be analyzed independent of their source of variance. Hierarchical models assume that the response variables are interrelated with their respective sources of variance. In understanding the response of different honeycomb cores to water absorption and impact damage, a hierarchical model must be applied. The mechanisms of water migration and damage growth in honeycomb composite structures were unique to each type and density of core. Subsequently, the sources of variance had to be analyzed as a function of core type. In Table 3, the statistically significant variables that affected the water absorption characteristics of the different honeycomb cores are presented. From the analysis of variance, 73.1% of the variation in the data could be attributed to the sources outlined in Table 3.

As shown in Table 3, the locations of the individual core bays on the aircraft were not found to play a statistically significant role in water absorption. This is an important non-factor because it indicated that all core bays experienced comparable flight conditions during the fourteen-month service evaluation, independent of the individual aircraft's flight schedule or the location of the bays on the aircraft.

The absorption of water by the aluminum and Korex core bays was also found to be unaffected by a 12.2 J impact. The lack of an affect may be attributable to a few possible causes. It is first possible that the panels were not in service long enough to allow sufficient time for damage propagation. Aluminum and Korex are relatively strong web materials when compared to Nomex and fiberglass. These materials may have required a longer service time to reveal the degradative effects of continuous environmental cycling. It is also possible that while in service, water did not accumulate or remain in the core bays long enough to significantly damage the core. In Figures 2 and 3, photomicrographs of the impact damage to the aluminum and Korex core bays are shown. For these composite structures, damage to the honeycomb cores and facesheets extended only one or two cells beyond the point of impact. Outside this small damage area, the core and facesheet were unaffected by the impact, (as confirmed by optical microscopy). The localization of the impact damage prevented the creation of a volumous reservoir for the accumulation of water inside the core. By minimizing the amount of water in the cells the detrimental mechanical effects from the water continually freezing and thawing were diminished.

The localization of the impact damage may also have aided in removing water from the core. During flight, high-speed air flowing over the damaged honeycomb core bays created a negative pressure gradient inside the honeycomb core. (Giancoli,

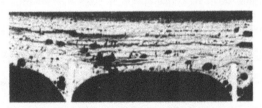

Figure 2. Photomicrograph of an aluminum core bay impacted with 12.2 J at 25X magnification.

Table 3. Statistical analysis of variance for water absorption of different honeycomb cores using a hierarchical DOE model.

Source of Variance	Core Type			
	Aluminum	Korex	Nomex	Fiberglass
Location				
Impact			XX	XX
Speed-tape				
Impact &			X	XX
Speed-tape				

XX – Significant at 1% confidence level
X – Significant at 5% confidence level

Figure 3. Photomicrograph of a Korex core bay impacted with 12.2J at 25X magnification.

1983; Shafizadeh & Seferis, 1998) This pressure gradient lowered the vapor pressure of the water or ice in the core and increased the rate of evaporation or sublimation. This effect was also enhanced at altitude. At a cruising altitude of 7,000 to 9,000 m, the vapor pressure of water is near 26 Pa and the specific humidity (grams of water/grams of dry air) is approximately 0.04%. These dry conditions further facilitate the removal of water from the core. Because all of the water in the aluminum and Korex panels was confined to a small area near the point of ingresssion, most of the water could be removed through these two mechanisms.

For the Nomex and fiberglass cores, impact damage was found to have a statistically significant effect on water ingression and migration through the sandwich structure. This was attributed to the mechanism of damage propagation through the core. When the Nomex honeycomb composite structures were impacted, both the core and facesheet sustained a great deal of damage, as shown in Figure 4. In this figure, only the area immediately surrounding the impact location is shown, however, small microcracks could be observed in the Nomex honeycomb core as far away as 1.5 cm from the impact location. This damage allowed water to migrate away from the ingression point and into the core. At altitude, the pressure gradient for water removal is largest around the point of ingression. As the distance from the impact location increased, the pressure drop from cell to cell decreased and the effective driving force for water removal decreased. The net result was a greater amount of water retained by the honeycomb core. Water retained in the core can caused freeze-thaw damage in the core, which further damaged and fractured the core, and ultimately increased the water retention volume of the core.

For the honeycomb composites manufactured with fiberglass core, similar behavior was observed. After being impacted, the facesheets of composite structures microcracked and small sections delaminated, as shown in Figure 5. But unlike the Nomex core, the facesheet damage was limited to the immediate area around the impact. Although the facesheet did not show extensive damage, the core and the skin-to-core fillets were observed to have widespread failures. From the TTU inspections, the initial diameter of the core damage was 3.5 cm, but through microscopy, small core microcracks were observed to extend as far as 2.0 cm from the center of the impact. From the photomicrographs and TTU inspections, it was concluded that the fiberglass core absorbed the brunt of the impact damage. Like the Nomex core, the fiberglass core sustained a greater amount of damage than the aluminum or Korex cores in spite of the fact that smaller impact energies were used on the fiberglass cores. The larger amount

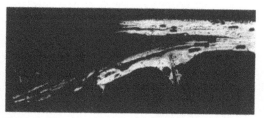

Figure 4. Photomicrograph of a Nomex core bay impacted with 8.1 J at 25X magnification.

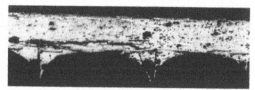

Figure 5. Photomicrograph of a fiberglass core bay impacted with 4.1 J at 25X magnification.

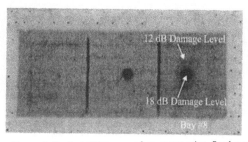

Figure 6. Typical TTU scan of an upper wing fixed panel showing 18 and 12 dB TTU damage levels after 14 months of service.

of core damage allowed a greater amount of water to be trapped within the core. The extensive damage away from the small impact center also reduced the ability of the core to dry during flight, as described earlier.

Speed-tape alone was also not found to play a significant role in the absorption of water, and alone did not have an effect when water did not accumulate in the core over time. Over an impact, speed-tape was not found to be a significant factor for the absorption of water in the aluminum or Korex cores; although, it was found to be significant in Nomex and fiberglass cores. Speed-tape did not have a significant effect when used over the aluminum and Korex core bays because, in-service, water was not found to migrate through the cores. Over impacted Nomex and fiberglass core bays, speed-tape decreased the tendency of the parts to absorb water. However, without impact, the water was limited to the drilled cells and was removed in-service. Speed-tape only mitigated water migration when water was found to migrate away from the point of ingression.

3.2 TTU Inspections

3.2.1 Gross Damage Growth

Along with the statistical analysis of water absorption, an analogous study was performed on impact damage. Through TTU inspections, the physical growth of the damage caused by drilling holes and impacting the wing panels was determined. In assessing the damage of the panels, two levels of damage were distinguished, an 18 dB threshold level and a 12 dB threshold level. The 18 dB damage level represented gross failure or delamination of the core or facesheet, while the 12 dB limit represented the 18 dB damage plus limited microcracking or absorbed water within the core. Figure 6 illustrates the differences between the two damage levels. In the far right square core bay of Figure 6, a fiberglass sub-bay is shown. In this figure, a central black damage area can be noticed. This area was defined as an 18 dB damage level. Outside this dark area, a lighter colored oblong region can also be observed. This region was defined as a 12 dB damage area.

A statistical analysis was performed on the gross damage growth for both 18 and 12 dB damage areas. The significant sources of variance at 18 dB were identical to the significant sources of variance identified in the water absorption analysis. The only difference between the two analyses was that the significant variables at the 5% level in Table 3 were significant at the 1% level in the gross damage DOE analysis. When the data was analyzed in terms of gross damage propagation at the 18 dB and 12 dB levels, 86.5% and 74.1% of the respective variance in the data could be explained.

A statistical analysis at the gross damage change at the 12 dB level revealed a slightly different set of results. In Table 4, a summary of the analyses is presented. As shown in this table, the sources of variance at the 12 dB level were only significant in the fiberglass core. This is attributed to the damage mechanism in fiberglass core. Unlike the other cores studied, when the fiberglass core was impacted, the core sustained most of the damage and little damage in the facesheet was observed, as earlier illustrated in Figure 5 and Table 2. The greater core damage allowed significant amounts of water to travel into

Table 4. Statistical analysis of gross damage growth using through transmittance ultrasound inspections with a threshold of 12 dB

Source of Variance	Core Type			
	Aluminum	Korex	Nomex	Fiberglass
Location Impact				XX
Speed-tape Impact & Speed-tape				XX

XX – Significant at 1% confidence level
X – Significant at 5% confidence level

the core and remain in the core during flight. The 12 dB damage represents the core damage front which contains water or microcracking damage, but has not caused complete failure of the core. With the other core materials, this damage front was not detected on a gross damage scale basis. Again, the presence of speed-tape over the impact was shown to reduce the growth of damage and ingression of water.

3.2.2 Percent Damage Growth

Lastly, a statistical analysis was performed on the percent damage growth. The percentage damage growth was defined as the final damage size divided by the initial damage size minus one. The minus one is a constant which does not affect the analysis of variance because the analysis is invariant to additive constants. Dividing the final damage size by the initial damage size gives additional pieces of information not included in an analysis of gross damage change. The quotient between the final and initial sizes measures the relative growth of the damage and considers the 'initial baseline damage' of panels. In analyzing the data from a percentage point of view, 89.4% and 74.4% of the respective variance in the 18 and 12 dB analyses could be explained

The 18 dB percent damage growth analysis again closely tracked the 18 dB gross damage change and water absorption analyses. At the 12 dB damage level, impact damage and impact damage coupled with speed-tape were again shown to be significant for the fiberglass core. Some differences did emerge when the 12 dB gross damage change analysis was compared to the 12 dB percent damage growth analysis shown in Table 5.

In terms of percent damage growth, impact damage was statistically significant at the 12 dB threshold level for the aluminum, Korex and Nomex cores. However, speed-tape over that impact was not found to be a significant source of variance. This implies that while speed-tape may be good for stopping gross impact damage growth, impact damage may still propagate and grow in-service. Although the overall growth may not be significant, the damage

Table 5. Statistical analysis of percent damage growth using through transmittance ultrasound inspections with a threshold of 12 dB.

Source of Variance	Core Type			
	Aluminum	Korex	Nomex	Fiberglass
Location				
Impact	XX	X	X	XX
Speed-tape				
Impact &				XX
Speed-tape				

XX – Significant at 1% confidence level
X – Significant at 5% confidence level

growth does become statistically significant when compared to the initial damage size. The 12 dB analysis also indicated that while the percent damage growth from cycling did increase from 14 months of in-service cycling, this damage was not delamination or core failure, but rather the result of some small amount of microcracking or water in the core. This damage may grow over time to facesheet delamination or core failure, but this growth was not observed in this study.

For all of the core sub-bays in the Design of Experiment, no water absorption or damage growth was noticed when only holes were drilled in the panels. This strengthens the argument for the idea that distance from point of ingression controls damage propagation through the core. When holes were simply drilled in the panel, water was confined to the area directly under the holes in the facesheet. Also, the damage was not observed to propagate on either a percent scale or on a gross damage scale. When the damage size was increased, through impacting, to extend one or two cells away from the point of water ingression, damage growth on a percent basis was observed to occur.

4 CONCLUSIONS

In this work, the mechanisms of water ingression and migration through honeycomb core were explored using an in-flight testing. The effects of impact damage and speed-tape repair on water absorption and damage propagation through the core were investigated for four different honeycomb cores and analyzed with a Design of Experiment. When fiberglass and Nomex based sandwich panels were impacted, extensive damage was seen throughout the facesheet and core. This damage allowed water to accumulate in the core and migrate through a freeze-thaw mechanism. For the aluminum and Korex cores, the impact damage was localized around the point of impact. The localization of this impact allowed the core to dry during flight. From the altitude conditions and air flowing over the impact location,

a negative pressure gradient was created inside the core. This pressure gradient provided a driving force for the removal of water around the point of impact. This effect was not seen in the Nomex or fiberglass cores due to the extent of damage. In addition, speed-tape repairs were found to be an effective repair only when water was found to migrate through the core. However, some small damage was observed to propagate through the panels despite the presence of speed-tape. From these investigations it can be shown that damage and water alone are not sufficient for the propagation of water through honeycomb core. Other factors such as core type and damage size need to be considered in evaluating the durability of honeycomb composite sandwich structures.

5 ACKNOWLEDGEMENTS

The authors would like to express their appreciation to the Hexcel Company, the Boeing Company, and United Airlines for their support and participation in the Graduate Team Certificate Program at the University of Washington. Special thanks also to J. E. Shafizadeh for continuing with the work after the formal closing of the Team Certificate Program through the Polymeric Composites Laboratory. The individual contributions of the following participant is also greatly appreciated: M. Caldwell, E. F. Chesmar, B. A. Frye, R. Geyer, C. Martin, R Popa, M. P. Thompson, W. H. Vogt and P. A. Wachter.

6 REFERENCES

Abrate, S. 1997. Localized impact on sandwich structures with laminated facings. *Applied Mechanics Reviews* 50(2): 69-82.

Anonymous 1989. *Boeing material specification 8-79 for 250°F epoxy prepreg systems*. Seattle: Boeing Commercial Airplane Group.

Anonymous 1994. Boeing 757-200 structural repair manual. Seattle: Boeing Commercial Airplane Group.

Cise, D. & R. S. Lakes 1997. Moisture ingression in honeycomb core sandwich panels: directional aspects. Journal of Material Engineering and Performance, 6(6): 732-736.

Coggeshall, R. L. 1983. Service experience with composites on Boeing commercial aircraft. 15th National SAMPE Technical Conference 15: 310-320.

Garrett, R. A., R. E. Bohlmann, & R. E. Derby 1978. Analysis and test of graphite/epoxy sandwich panels subjected to internal pressure resulting for absorbed moisture. *Advanced Composite Materials - Environemtal Effects: A Symposium*. 234-253.

Geyer B. 1996. Drying method for composite honeycomb structures, *28th International SAMPE Technical Conference*, 28: 1183-1192.

Giancoli, D. G. 1988. *Physics for Scientists and Engineers*, Englewood Cliffs: Prentice Hall.

Hicks, C. R. 1982. *Fundamental concepts in design of experiments*. Orlando: Saunders College.

Jackson, W. C. & T. K. O'Brien 1988. Water intrusion in thin-skinned composite honeycomb sandwich structures. *Journal of the American Helicopter Society* 35(4): 31-37.

Nettles A. T. & A. J. Hodge 1990. Impact testing of glass/phenolic honeycomb panels with graphite/epoxy face sheets. *35th International SAMPE Conference* 35: 1430-1440.

Shafizadeh J. E. & J. C. Seferis editors 1998. *Perception of Technological Innovation, Development and Transfer (Team Certificate Program Final Report)*, Seattle, University of Washington.

Shafizadeh, J. E., J. C. Seferis, E. F. Chesmar, & R. Geyer (Submitted). Evaluation of the in-service performance of honeycomb composite structures. *Journal of Materials* Evaluation *and Performance*.

Weems D. B. & R. E. Fay 1995. A moisture barrier for composite sandwich structure. *American Helicopter Society National Meeting*.

Wong, R. & R. Abbott 1990. Durability and damage tolerance of bonded sandwich panels. *35th International SAMPE Conference* 35: 366-380.

Wu, C. L., C. A. Weeks, & C. T. Sun 1995. Improving honeycomb-core sandwich structures for impact resistance Journal of Advanced *Materials* 26(4): 41-47.

Recent Developments in Durability Analysis of Composite Systems, Cardon, Fukuda, Reifsnider & Verchery (eds)
© 2000 Balkema, Rotterdam, ISBN 90 5809 103 1

Durability of polymer matrix composites: Some remarks by the AMAC committee on durability

D. Perreux
Laboratoire de Mécanique Appliquée R.Chaléat, UMR CNRS 6604, Besançon & Commission Durabilité de l'AMAC, France

ABSTRACT : This paper summarises the various problems discussed during a meeting organised by the AMAC committee on durability . This paper does not present solutions to problems ; rather it reports the ideas expressed at the meeting which were aimed at sharing knowledge in the field of durability.
The durability problems discussed during the meeting were mainly centred on physico-chemical and mechanical aspects. This paper deals some of these aspects.

1 INTRODUCTION

The French composite materials association (AMAC) has members from both the industrial and academic sectors. For many years, AMAC has been aware of the potential economic impact of the durability of composites, and especially of the development of the use of polymer matrix composites. To try to improve the situation, AMAC created a committee on durability in 1995, whose tasks were to study the problem and to propose ways to increase public knowledge of polymer matrix composites.

This committee has conducted a survey on durability problems in the industrial and academic sectors, has organised many meeting, two national conferences. The last one in May 1998, in Arcueil, was focused on the industrial concerns for polymer matrix composites durability. The main part of the paper is devoted to the conclusions of this last conference.

First, the results of the survey on the problem of durability is presented in the following figures. This survey is based on about 30 academic and 30 industrial answers. Thus, while it is not exhaustive, ther is sufficient number of answers to provide a basis for discussion.

The type of fibres which are studied by the academic sector and used by industries is presented in the figure 1 and 2 .

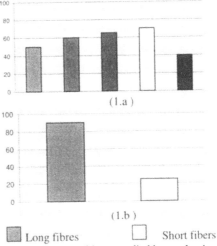

Long fibres   Short fibers

Figure 1 : (a) problems studied by academic sector (b) problems encountered by industrial sector

It is clear that here, there is agreement on the type of fibres used or studied by the industries and the academic sectors. This first point is very important for composites, a high technology industry. Progress cannot be made in the understanding of durability if there is a large gap between the materials studied by the academic sector and used by the industrial sectors.

The figure 3 is concerned by the type of problem studies by the academics sectors and the type of problems encountered by the industries.

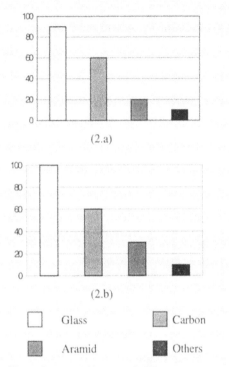

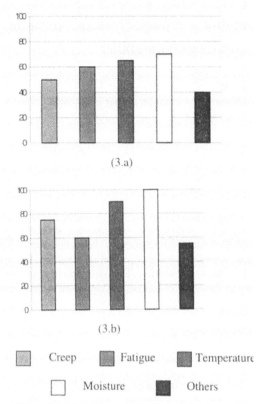

(2.a)

(2.b)

☐ Glass ▨ Carbon

▨ Aramid ■ Others

Figure 2 : (a) problems studied by academic sector
(b) problems encountered by industrial sector

(3.a)

(3.b)

▨ Creep ▨ Fatigue ▨ Temperature

☐ Moisture ■ Others

Figure 3 : (a) problem studied by academic sector
(b) problems encountered by industrial sector

Most industries are concerned with mechanical and environmental ageing problems. The academic sector studies all the problem, but proposes on party of the response. This point is not obvious on the figure 3 but it is obvious when the individual answers are analysed.

Usually, a physico-chemical scientists propose physico-chemical interpretation of the durability problems, but rarely do they propose mechanical modellings of damage which take physico-chemical phenomena into account. Mechanical scientists provide mechanical modelling but rarely considers the physico-chemical changes which occur in materials. When they do provide modellings, they often use internal variable introduced in the constitutive equations which frequently looks like a black box.

From the industrial point of view both aspects must be given more attention.

The first conclusion of this survey (and may be with a little provocative) is that the academic sector has provided a partial solution of the durability

problem, but it is not the solution required by industry. It is the only solution that the academic sector can provide due to the difficulties of the problems and due to the fact that in France but it seems to be the same away, there not enough real collaboration between physico-chemical and mechanical scientists. The AMAC committee was in large part created for this reason : in odder to improve the communication and hopefully collaboration between industries, physico-chemical and mechanical scientists.

An other interesting result is presented in Figure 4. This graphics present the proportion of various failure mode encountered in plastic and composite. This survey was performed by CETIM [1] and concern the mechanical industries.

There are three main cause of failure : the environmental ageing (moisture, temperature,…), the fatigue and mechanical stress which is a consequence of a bad design. Other causes are wear, UV effect and the creep. It must be noted that

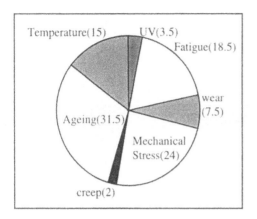

Figure 4 : Some durability problems for industries [1]

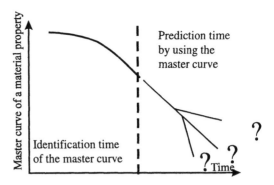

Figure 5 : Prediction confidence problems of material properties by using accelerated test.

further understanding of fatigue and the environmental ageing phenomena could reduce the risk of failure by about 60% . It is obvious that both these fields need to be intensively researched.

2 SOME REMARKS ON DURABILITY

In May 1998, the committee organised a meeting with the help of the DGA/ETCA in Arcueil (France), which focused on the industrial concerns for polymer matrix composites durability. This conference addressed only industrial concerns, and presented some problems which have not yet been solved. The presentation was divided into three groups, the first addressing the physico-chemical aspect, the second mechanical aspect and the third the coupling of phenomena.

Typical physico-chemical problems, the main point which wrer discussed and which are open questions are presented in the list below:

- To find technical solutions to problems of
  - ◆ change of colour and aspect (coating),
  - ◆ creation of micro-cracks due to moisture, photolysis and photoxidation
- Comprehension of phenomena that causes the ageing.
- How to reduce the time of tests in odder to reduce the cost.
- What artificial or accelerated tests used.

One important question which was discussed concern the degree of confidence we can have in the

results of accelerated tests. In fact lot of accelerated tests are performed over some months, and during this time, they exhibit two or more different phenomena. These behaviours are usually observed by a change of slope in a master curve. Usually, this master curve is extrapolated to perform a prediction of the behaviour for longer time. The classical question is then : are we sure that a new change of slope doesn't exist for much longer tests than the tests which were used for the identification of the master curve ? (figure 5). Of course, there is no guarantee for yes or no ! What can be done to solve these problems ?

The main type of question regarding the mechanical aspect which was discussed during the AMAC conference is summarised in the following list :

- To find a technical solution for a composite structure design which takes into account damage phenomena
  - Understand deformation, damage initiation, damage growth
  - Fatigue, creep : life time 30/40 years
  - Materials/Structure
  - Predictive models
- How processing defects affect performance
- Difficulties with standardisation
- What is the cost of the security coefficients
- Lack of feedback from experiments (confidentiality)

The main difficulties were centred around behaviour prediction. Many mechanical industries use

Computer Aided Engineering. This method requires confidence in the modelling of the materials. Usually, the time where the prediction is required is often much longer than the classical test times where the identification of the modelling parameters is performed. This last point, is one of the main difficulty of confidence in the behaviour prediction.

Another problem was the application of results obtained using specimen to determine the behaviour of a larger structure. This scale effect is still a problem for Computer Aided Engineering.

All this prediction confidence, hinder the reducing the safety factors used in the design in composite structure, as when safety factors were reduced for metal alloys. This is surely not an advantage for the economic development of composite in the industrial sector.

The last but not the least aspect was the coupling of both phenomena. It was identify that more that 40% of the failure mode is due to coupling phenomena. Here the main problem is to understand and to model the synergy of both phenomena which was already presented.

3 CONCLUSION.

The main problem on durability is the lack of understanding the   phenomena which makes it difficult to model. Many studies have already been to better understand. But can we predict the behaviour of the material over a long time? And if we can, do we know the degree of confidence of the prediction?

The AMAC committee on durability is sure that the only way to increase both of understanding and the degree of confidence in the modelling is to increase collaboration between physico-chemical   and mechanical scientists !

ACKNOWLEDGEMENTS.

This paper is based on a collective work of all the members of the AMAC committee on durability. Particular recognition is due to those who participated in the writing of the synthesis[2] of the meeting which was organised in May 1998 : *V.* Bellenger, M.L. Benzeggagh, P. Castaing, A. Chateauminois, D. Choqueuse, G. Dallemagne, F. Dal Maso, P. Davies, C. Dubois, J.P Favre, E. Ghorbel, J.M. Guillemot, F. Joubert, P. Krawczak, M.C Lafarie-Frénot, C. Limal, P. Lory, B. Mortaigne, M.F. Pays, C. Pinotti, J. Renard, A. Vautrin.

4 REFERENCE

[1] P. CASTAING AND A. LEMANSÇON : *Optimisation de la qualité et de la durabilité de pièces composites grâce à l'analyse de défaillance.* In proceeding of : Durabilité des composites à matrices organique: bilan et perspective. In arceuil, 12 may 1998. ed AMAC.

[2] V. BELLENGER, M.L. BENZEGGAGH, P. CASTAING, A. CHATEAUMINOIS, D. CHOQUEUSE, G. DALLEMAGNE, F. DAL MASO, P. DAVIES, C. DUBOIS, J.P FAVRE, E. GHORBEL, J.M. GUILLEMOT, F. JOUBERT, P. KRAWCZAK, M.C LAFARIE-FRÉNOT, C. LIMAL, P. LORY, B. MORTAIGNE, M.F. PAYS, C. PINOTTI, J. RENARD, A. VAUTRIN : *Synthése de la journée d'étude durabilité des composites à matrice organique : bilan et persperctive.* ed AMAC.

Recent Developments in Durability Analysis of Composite Systems, Cardon, Fukuda, Reifsnider & Verchery (eds)
© 2000 Balkema, Rotterdam, ISBN 90 5809 103 1

Chemical resistance of pultruded E-glass reinforced polyester composites

K. Van de Velde & P. Kiekens
Department of Textiles, Universiteit Gent, Belgium

ABSTRACT: Glass fibre reinforced composite profiles are used in an increasing number of structural components thanks to their attractive strength-to-weight ratio and good resistance to chemical attack. Such profiles
are produced via the pultrusion technique. As pultruded composites find applications in various chemical environments, a good understanding of their corrosion behaviour is necessary. In this work, the chemical resistance
of six kinds of pultruded E-glass fibre reinforced isophtalic polyesters were studied. They contain 50 w% or 60
w% glass. The composites were either unfilled or filled with calcium carbonate or kaolin clay. The mechanical,
physical, optical and structural properties of the pultruded flat profiles before and after immersion in sea water,
demineralised water, sodium hydroxide (5%) and sulphuric acid (10%) over the period of one year were analysed.

1 INTRODUCTION

Pultruded composites find applications as hand-
rails, ladders, gratings and construction profiles for
the chemical, water treatment, offshore and related
markets (Lubin 1982, Starr 1995, Martin 1978).

Compared to traditional materials, composites
are appreciated for their lower weight, good mechanical properties, electrical and thermal insulation
characteristics, lower demand for painting and maintenance and better wear, weathering and corrosion
resistance. The last mentioned characteristic is one of
the main reasons for the use of composites in chemically aggressive environments.

Although glass fibre reinforced plastics do not
corrode by an electro-chemical mechanism, they do
however degrade in other ways when exposed to
chemical environments. Chemical degradation or corrosion of a composite heavily depends upon the environment encountered, be it normal outdoor weathering, shorelines with gusting winds and salt water,
acid rain, industrial areas or immersion in some kind
of chemical fluid. The materials used and the nature
of the production process also have an important influence on the properties of the composites. Within
the world of composites one of the best mechanical
performances is obtained with the pultrusion technology where high glass volumes are used. Through
these, the corrosive properties of the pultruded composites can be reduced or influenced.

Corrosion data and corrosion design data of pultruded glass fibre reinforced plastics are sparsely
available since corrosion studies are mainly restricted
to compression moulded laminates (Springer 1984,
Sonawala 1996a, b, Harris 1984, Haarsma 1978, van
den Ende 1990, Chin 1997, Chiou 1997, Caddock
1987, Chua 1992a, b, Cai 1994, Jacquement 1998,
Renaud 1986, Folimonoff 1985, Lagrange 1990,
Castaing 1995, Shutte 1994, Bergman 1996, Gentry
1998, Van de Velde 1998).

A good understanding of the corrosion behaviour
of the pultruded composites is necessary in order to
guarantee the lifetime and the safety of the structural
components in service.

In this work, the chemical resistance of pultruded
E-glass fibre reinforced polyesters with different filler
types is studied. The mechanical, physical, optical
and structural properties of the pultruded flat profiles
before and after immersion in sea water, demineralised water, sodium hydroxide (5%) and sulphuric
acid (10%) are analysed.

2 EXPERIMENTAL

2.1 *Materials*

Flat profiles (l x t, 100 x 3 (mm)) of E-glass reinforced isopolyester were pultruded.

The polyester resin used for the pultrusions was
isophtalic polyester, a polyester of isophtalic acid,
maleïc acid anhydride and propylene glycol cross-
linked with styrene monomer. Rovings and continuous strand mats of E-glass were used as reinforcement material. Calciumcarbonate and kaolin clay

serves as a filler. Pultruded composites without filler were also made. Curing of the thermoset polyester is initiated and enhanced by peroxy catalysts. A phosphatic lubricant serves as a release agent. A polyester veil is used to cover the glass fibre reinforced plastic with a protective resin rich layer. The pultruded composites contain 50 w% and 60 w% E-glass.

An overview of the test samples can be given as follows:
- UP50z: isophtalic polyester, 50% E-glass, no filler
- UP50c: isophtalic polyester, 50% E-glass, calcium carbonate as filler
- UP50k: isophtalic polyester, 50% E-glass, kaolin clay as filler
- UP60z: isophtalic polyester, 60% E-glass, no filler
- UP60c: isophtalic polyester, 60% E-glass, calcium carbonate as filler
- UP60k: isophtalic polyester, 60% E-glass, kaolin clay as filler

2.2 Exposure environments

Strips of the pultruded composites (200 x 100 x 3 (mm)), for which the edges were coated with resin, were immersed in sea water, demineralised water, sodium hydroxide (5%) and sulphuric acid (10%) at ambient temperature. The sea water was prepared as described in ASTM D1141-90 Standard Specification for Substitute Ocean Water.

2.3 Experimental procedures

The samples were examined on their mechanical, physical, optical and structural properties after 30, 90, 180 and 360 days of immersion in accordance with ASTM C581-83 Standard Practice for Determining Chemical Resistance of Thermosetting Resins used in Glass Fiber Reinforced Structures intended for Liquid Service. Dimension and weight characterisation, visual analysis, flexural tests, tensile tests and hardness tests were done the same day after removing the sample from the test environment. Scanning Electron Microscopy (SEM) analyses were done within 48 hours. All samples were placed in an airtight polyethylene bag prior to testing.

All immersed samples were weighed on an analytical balance. Prior to weighing, the samples were dried by blotting with a paper towel.

The visual defects were classified according to ASTM D4385-84a Standard Practice for Classifying Visual Defects in Thermosetting Reinforced Plastic Pultruded Products.

The flexural properties were measured with a Monsanto T10 tensile testing machine according to ASTM D790-86 Flexural Properties of Unreinforced and Reinforced Plastics and Electrical Insulating Materials. The support span to depth ratio is 32

(L/d=32/1). The rate of cross-head motion was 5.3 mm/min. The coupon thickness determined at the time of flexural testing was used for the calculation of flexural strength and modulus. Six samples were tested for each type of composite and each exposure condition.

The tensile properties were measured on dog-bone specimens (type I) with a Zwick 1485 tensile testing machine according to ASTM D638-89 Tensile Properties of Plastics. The speed of testing was 5 mm/min. The coupon width and thickness at the time of testing was used for the calculation of tensile strength and modulus. Six samples were tested for each type of composite and each exposure condition.

Barcol hardness was measured with a Barcol-Colman impressor according to ASTM D2583-87 Standard Test Method for Indentation Hardness of Rigid Plastics by Means of a Barcol Impressor). There were 29 measurements made for each type of pultruded composite and each exposure condition.

The structure of the composite (surface and cross section) and the fracture surfaces were analysed with a scanning electron microscope (SEM) Philips 501. The samples were glued on a sample holder and coated with a gold layer of approximately 150 Å. The accelerating voltage of the instrument is 30 kV.

The failure mode was also analysed with optical microscopy (OM).

3 RESULTS AND DISCUSSION

3.1 Weights Changes

In general, the weights of the pultruded composites increase after immersion due to sorption of the fluid. During the whole period of exposure, an overall weight gain is observed pointing out that the mass loss is less than the weight gain by sorption.

For the samples and the aqueous immersion fluids under study, the first part of the curve is related to diffusion of water in the polyester matrix and penetration through surface defects. Thereafter, due to the development and presence of cracks, an increased absorption of water takes place, resulting in a weight gain. Besides the water-uptake by the polyester matrix, a water-uptake at the interfase might occur. This behaviour is seen for the samples exposed to sea water.

For the composites exposed to demineralised water the weight changes are also influenced by the leaching out of low-molecular water solvable molecules. Until about three months of exposure a weight gain take place, thereafter -between three and six months of exposure- a stabilisation of the weight is seen. After six months of exposure, another weight increase is determined. Over the whole period, the moisture gain is greater than the material loss resulting in a weight increase of the composites.

Alkaline solutions cause also chemical reactions either with the matrix as with the glass reinforcement. A saponification reaction with the polyester matrix take place. Regarding the glass fibres, the siliciumoxide layers in the network are gradually attacked by the alkaline solution. Up to three months a weight increase is determined, thereafter a little decrease followed by stabilisation or increase of the weight is seen.

Acidic aqueous solutions catalyse hydrolyses of the polyester matrix. Because of the degradation of the matrix, extra voids are arising. Through these, a migration of the solution can take place resulting in an extra sorption of fluid by the matrix and the interfase and also an attack of the glass fibres become possible. Up to one month of exposure a weight gain took place due to sorption. Thereafter a stabilisation or little decrease is determined. After three months of exposure a little increase of the weights is seen.

3.2 Flexural tests

The maximum load, the maximum stress, the strain at maximum stress, and the modulus of elasticity in bending were measured with 3-point loading test. The mean values of the testing results, the standard deviations and the coefficients of variation were calculated.

The flexural strength (maximum stress in outer fibre at moment of break) and the bending modulus against the exposure time for the various samples and per type of environment are given in Figure 1 and Figure 2.

Six kinds of pultruded composites were under study. They all consist of an isophtalic polyester matrix reinforced with 50% or 60% E-glass. The composites were either unfilled or filled with calcium carbonate or kaolin clay.

As expected the mechanical properties of the pultruded composites with the highest reinforcement level are the best. Considering the fillers, the composites containing clay or calcium carbonate are stronger than the unfilled composites.

A deterioration of the mechanical properties of the pultruded composites in flexure after exposure is observed. The reduction in strength and moduli is most obvious after one month exposure. Thereafter, considering the standard deviations, the mechanical properties are roughly maintained except for sodium hydroxide (Van de Velde 1998). The drop of the mechanical properties after one month of exposure is due to the absorption of water which can lead to plastification of the matrix as a result of interruption of Van der Waals bonds between the polymer chains.

This leads to a decrease in the matrix dominated stiffness and strength properties of the composites (Bank 1995). A slighter deterioration of the me-

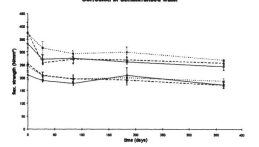

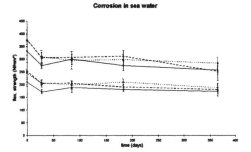

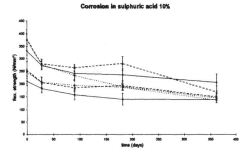

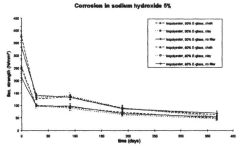

Figure 1. Flexural strength versus time of pultruded composites exposed to demineralised water, sea water, sodium hydroxide, sulphuric acid

chanical and physical properties after three, six and twelve months of exposure to demineralised water and sea water is seen. Also hydrolysis of the polyester in aqueous solutions is a well known phenome-

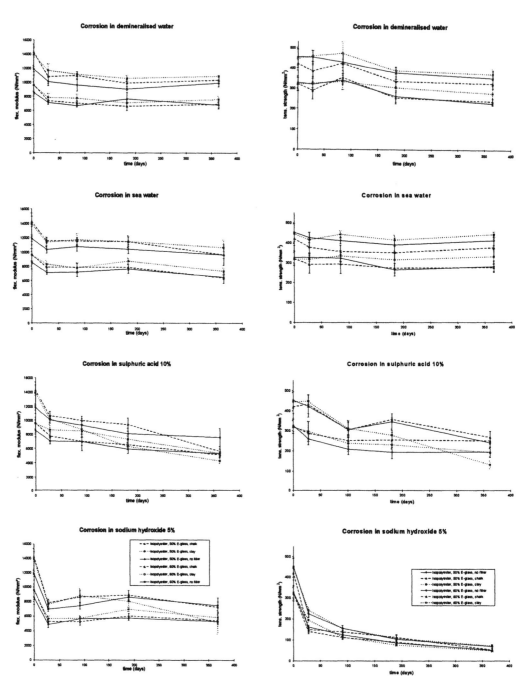

Figure 2. Flexural modulus versus time of pultruded composites exposed to demineralised water, sea water, sodium hydroxide, sulphuric acid

Figure 3. Tensile strength versus time of pultruded composites exposed to demineralised water, sea water, sodium hydroxide, sulphuric acid

non. The equilibrium of this reaction depends on the dissolution of the hydrolysis products. Low molecular components, either formed by hydrolysis reaction or already present in the composite matrix, are leached out of the matrix resulting in a deterioration of the properties. This effect is especially seen for the composites exposed to demineralised water. The extra voids in the samples are seen on the electron microscopy photographs. This effect is also seen for sulphuric acid cause of the catalytic effect of the protons to hydrolyses. Besides the influence of sulphuric acid on the isophtalic matrix, the acid can have a chemical reaction with the fillers and with the glass fibres by ion exchange. This can result in the formation of cracks at the interfase glass/matrix.

The results of the flexural tests point out that demineralised water has a more corrosive effect than sea water while sulphuric acid has even a stronger effect.

The influence of sodium hydroxide on the flexural properties of the different samples is much more severe in comparison with the other environments. For sodium hydroxide, an abrupt drop in the mechanical properties after one month of exposure is observed. In addition to the sorption effects, chemical reactions play a role. Sodium hydroxide attacks the polyester matrix, causing a saponification of the matrix. This breakdown of the polymer matrix is an irreversible reaction and leads to the formation of cracks. The siliciumoxide network in the E-glass fibre is also gradually attacked.

The fracture mechanics in flexure of the composites is a delamination between the mat reinforcement and the rovings due to the accumulation of stress between these layers.

3.3 Tensile tests

The maximum load, tensile strength, elongation at break and modulus of elasticity were measured. The tensile strength and the tensile modulus versus the exposure time is given in Figure 3 and Figure 4.

Exposure of the different pultruded composites to sea water, demineralised water, sodiumhydroxide (5%) and sulphuric acid (10%) results in a diminution of the tensile strength. This is most obvious for the alkaline environment. Attack of the glass fibres by sodiumhydroxide results, after one month exposure, in a drop of the tensile strength of 50%, and even more. After one year of exposure, 16-17% retention of the tensile strengths were observed.

Sodium hydroxide reacts with the Si-O-Si and the Si-O-R bonds of the glass fibres forming Si-O-Na and Si-OH or R-OH. The siliciumoxide layers are in this way gradually attacked. The samples exposed to sulphuric acid show a 55-78% of retention, except UP60k which show a 30% tensile strength retention. Diminution of tensile strength by acidic aqueous solutions might be due to chemical attack of the interfase and the glass fibres. Degradation of the glass fibres namely a dissociation of the Si-O-R bond (R = Na or K) can take place (Cowley 1991, van den Ende 1990, Chin 1997). For the samples exposed to demineralised water, a 70% retention of the tensile strength after one year is observed. The loss of 30% in tensile strength migth be due to the attack of the interface and the glass fibres. Like for acidic solu-

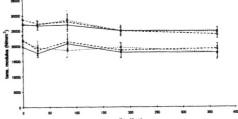

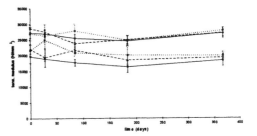

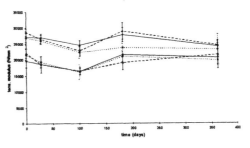

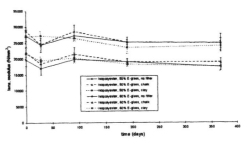

Figure 4. Tensile modulus versus time of pultruded composites exposed to demineralised water, sea water, sodium hydroxide, sulphuric acid

tions, but to a lesser extent, an ion exchange mechanism can take place. The composites immersed in sea water show a 85-100% retention. This means that the glass fibres in the composite are hardly or even not attacked by sea water.

Variations in the moduli were also mostly expressed for the samples exposed to sodium hydroxide.

3.4 Barcol Hardness

The Barcol hardness of the pultruded composites before and after exposure are given in Figure 5.

The measured values are highly spread, cause measurements can be made on the matrix, the glass reinforcement or on the filler.

The Barcol hardness of the composites filled with clay are higher than those with calcium carbonate as a filler. The unfilled composites have the lowest hardness. The higher the glass content, the higher the Barcol hardness is valid for each of them.

A global reduction of the Barcol hardness after exposure to the different environments is seen. The highest decreases are seen after one month of exposure. After longer times of exposure, increases of hardness of the composites can be seen. They are due to the higher expression and visibility of the glass mats on the surface and the erosion of the polymers. For the alkaline environment the hardness can be very low due to a very weak and degraded polyester matrix. The Barcol hardness are in accordance to the flexural strengths.

3.5 Visual observations

After exposure of the composites to sea water, demineralised water, sodium hydroxide and sulphuric acid, extra voids in the cross sections and a higher expression of the glass fibres on the surface are seen. The visual effects of sodium hydroxide on the composites are the most drastically ones. After one year immersion in sodium hydroxide the samples are unstable, glass fibres and matrix disintegrate. Large blisters are seen on the surface of the samples. The most remarkable visual observation made for the samples exposed during one year to sulphuric acid is that the glass rovings at the sides of the samples come loose. The coating veil of many samples exposed to demineralised water is swollen and spongelike. The samples exposed to sea water during a year show yellow-green algae and black fungi on their surfaces.

3.6 Scanning Electron Microscopy

With the electron microscope little cracks on the surface and in the cross sections of the composites are observed. Those cracks are the results of shrinkage and insufficient impregnation during the pultrusion process. In these weak points a migration of the exposure fluid can take place.

After exposure and depending on the type of environment, more cracks on the surface just beside a glass fibre and deeper cracks in the cross section as a

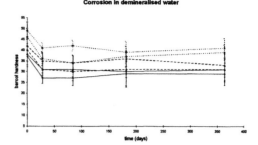

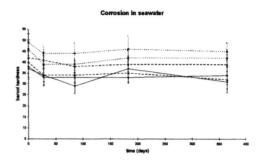

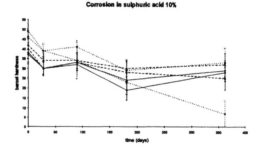

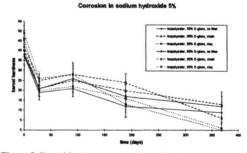

Figure 5. Barcol hardness versus time of pultruded composites exposed to demineralised water, sea water, sodium hydroxide, sulphuric acid

result of dissolution, degradation or stress-cracking due to solution sorption of matrix material and debonding between the glass fibres and the matrix material were seen. The attack of the composites is

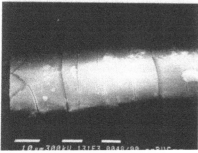

Figure 6. Micrographs of pultrudates with 50% E-glass as reinforcement and chalk as filler; respectively before corrosion, after 12 months in demineralised water, and after 12 months in sodium hydroxide - The last picture is from a corroded glass fibre in the same pultrudate (except clay as filler) after 12 months in sodium hydroxide

the worst for those exposed to sodium hydroxide. Degradation of the polyester matrix due to saponification is expressed in the formation of cracks, which are seen on the surface and in the cross section of the material. After one year of exposure, even delamination of the composite is observed. This delamination took place at the transition between the unidirectional and multidirectional reinforcement in the composites. A helicoidal degradation of the glass fibre is seen after six months of exposure. Thereafter a total disintegration of the fibre can be seen. The exposure of the pultruded composites to sulphuric acid results in the pulling out of glass fibres due to degradation of the polyester matrix and debonding between the reinforcement and the matrix. Voids, holes and debonding effects are also seen on the cross section of the composites. The most remarkable influence of demineralised water on a pultruded composite are the appaerance of little holes on the surface. Cracks along glass fibres are seen as a result of the effect of sea water. A loss of resin between the glass fibres are seen on the cross section.

The results of SEM are in agreement with the mechanical observations.

Some micrographs are shown in Figure 6.

4 CONCLUSION

Pultruded composites can be used as structural materials in various chemical environments due to the good strength-to-weight ratio and the resistance to chemical attack. Although, a good understanding of the chemical properties is necessary in order to guarantee the lifetime and the safety of the structural components in service. In this work, the chemical resistance against sea water, demineralised water, sodium hydroxide and sulphuric acid of pultruded E-glass fibre reinforced isopolyesters with different filler types is studied. Composites filled with calciumcarbonate and kaolin clay and unfilled composites containing 50% and 60% glass were analysed on their mechanical, physical, optical and structural properties.

The higher the reinforcement level, the better the mechanical properties. In each class the unfilled composites have the lowest strength. The Barcol hardness of the pultruded composites depends on the filler. The highest hardnesses are recorded for the clay filled composites, followed by the calcium carbonate filled and unfilled pultruded composites. In each class, the ones with the higher reinforcement levels are the hardest.

After exposure of the pultruded composites to demineralised water and sea water, a deterioration of the mechanical properties is observed. Particularly the flexural strength and moduli are weakened. The reduction after one month of exposure is most obvious. The diminution of the flexural strength and stiffness properties and the visual and microscopic observations point out that demineralised water and sea water have effect on the isopolyester matrix of the composites by plastification and hydrolyses. These aqueous solutions also attack the glass/matrix interfase. Demineralised water has a stronger corrosive effect than sea water. Exposure of the composites to sodium hydroxide results in abrupt diminution of tensile and flexural properties. A decline of the mechanical properties of the composites exposed to sulphuric acid is also determined. The drastic effect of sodium hydroxide is mainly due to chemical reactions either with the isopolyester matrix or with the glass reinforcement. A saponification reaction with the polyester matrix takes place which is expressed in the formation of cracks in the composite. This alkaline solution causes a gradually degradation of the siliciumoxide network of the glass fibres. In addition of the hydrolyses effect to the matrix, sulphuric acid can react with the reinforcing glass fibres by an ion exchange mechanism. This results in a diminution of the properties. It can be stated that the corrosive effect of sulphuric acid on E-glass reinforced isopolyester composites is stronger than demineralised water, but sodium hydroxide has an even greater effect. The properties of the composites are least effected by sea water.

REFERENCES

Bank, L.C., Gentry, T.R. & Barkatt, A. 1995. Accelerated Test Methods to Determine the Long-Term Behavior of FRP Composite Structures: Environmental Effects. *Journal of Reinforced Plastic Composites* 14: 559-587.

Bergman, G. 1982. Composite Performance in Corrosive Environments - A Comparison vs Metal Alternatives. *Composites Industry Conference 'Only the Strong Survive', Leipzig, 25-27 November 1996.*

Caddock, B.D., Evans, K.E. & Hull, D. 1987. The diffusion of hydrochloric acid in polyester thermosetting resins. *Journal of Materials Science* 22: 3368-3372.

Cai, L.W. & Weitsman, Y. 1994. Non-Fickian moisture diffusion in polymeric composites. *Journal of Composite Materials* 28: 130.

Castaing, Ph. & Lemoine, L. 1995. Effects of Water Absorption and Osmotic Degradation on Long-Term Behavior of Glass Fibre Reinforced Polyester. *Polymer Composites* 6(5): 349-356.

Chin, J.W., Nguyen, T. & Aouadi, K.. 1997. Effects of Environmental Exposure on Fiber-Reinforced Plastic (FRP) Materials Used in Construction. *Journal of Composites Technology & Research, JCTRER* 19(4): 205-213.

Chiou, P.L. & Bradley, W.L. 1997. Seawater Effects on Strength and Durability of Glass/Epoxy Filament-Wound Tubes as Revealed by Acoustic Emission Analysis. *Journal of Composites Technology & Research, JCTRER* 19(4): 214-221.

Chua P.S., Dai, S.R. & Piggot, M.R. 1992. Mechanical properties of the glass fibre-polyester interphase. *Journal of Materials Science* 27: 919-924.

Chua P.S. & Piggot, M.R. 1992. Mechanical properties of the glass fibre-polyester interphase. *Journal of Materials Science* 27: 925-929.

Cowley, T.W. 1991. Corrosion Resistant Fiberglass Reinforced Plastic Equipment for the Pulp and Paper Industry. *46th Annual Conference, Composites Institute, The Society of the Plastics Industry.* Session 1-C: 1-7.

Folimonoff, W. & Van De Velde, J. 1985. Etudes de la diffusion et de l'absorption de l'eau par les résines polyester insaturées. *Composites* 3: 90-93.

Gentry, T.R., Bank, L.C., Barkatt, A. & Prian, L. 1998. Accelerated Test Methods to Determine the Long-Term Behavior of Composite Highway Structures Subject to environmental Loading. *Journal of Composites Technology & Research* 20(1): 38-50.

Haarsma, J.C. 1978. Evaluation of Resin-Glass Fibre Interface under Environmental Stress. *33rd Annual Technical Conference, Reinforced Plastics/Composites Institute, The Society of the Plastics Industry.* Section 22-E: 3.

Harris, S.J., Nobel, B. & Owen, M.J. 1984. Metallographic Investigation of the Damage Caused to GRP by the Combined Action of Electrical, Mechanical and Chemical Environments. *Journal of Materials Science* 19: 1596.

Jacquement, R. & Lagrange, A. 1988. Vieillissement de stratifiées polyester/verre E et évolution de leurs caractéristiques mécaniques en milieu marin. *Composites* 4: 39.

Lagrange, A. & Jacquemet, R. 1990. Aging of E-Glass Polyester Laminates and Degradation of Mechanical Properties of Sea Water. *45th Annual Conference, Composites Institute, The Society of the Plastics Industry, Inc., 12-15 February 1990.* Session 8b: 1-7

Lubin, G. (ed) 1982. *Handbook of Composites:* 1-2. New York: Van Nostrand Reinhold Company.

Martin, J. 1978. Pultrusion - An Overview of Applications and Opportunities. *33rd Annual Technical Conference, Reinforced Plastics/Composites Institute, The Society of the Plastics Industry.* Section 8H: 1-6.

Renaud, Cl. 1986. Etude de l'influence du renfort de verre sur la résistance des stratifiés à la corrosion par les acides. *Composites* 2: 34-43.

Shutte, C.L. 1994. Environmental durability of glass-fiber composites. *Materials Science and Engineering* 13(7): 265-323.

Sonawala, S.P. & Spontak, R.J. 1996. Degradation Kinetics of Glass Reinforced Polyesters in Chemical Environment I. Aqueous Solutions. *Journal of Materials Scieince* 31: 4745-4756.

Sonawala, S.P. & Spontak, R.J. 1996. Degradation Kinetics of Glass Reinforced Polyesters in Chemical Environment II. Organic solvents. *Journal of Materials Science* 31: 4757-4765.

Technomic Publishing Co., Inc. 1984. Environmental Effects on Composite Materials 1, 2 & 3 In Springer G.S. (ed), *Technomic Publishing Co., Inc.*

Trevor F. Starr 1995. *Composites, A Profile of the Worldwide Reinforced Plastics Industry, Markets and Suppliers:* 267. England: Elsevier Sciene Ltd.

Van de Velde, K. & Kiekens, P. 1998. Chemical Resistance of E-glass Reinforced Polyester Composites made by Pultrusion. *Localised Damage 98, Computer Aided Assesment and Control. Bologna, Italy, 8-10 June 1998.*

van den Ende C.A.M. & van den Dolder. 1990. Comparison of environmental stress corrosion cracking in different glass fibre reinforced thermoset composites. *Durability of Polymer Based Composite Systems for Structural Applications, Brussels, Belgium, 27-31 August 1990:* 408-417

Recent Developments in Durability Analysis of Composite Systems, Cardon, Fukuda, Reifsnider & Verchery (eds)
© 2000 Balkema, Rotterdam, ISBN 90 5809 103 1

Monitoring and modeling the durability of polymers used for composite offshore oil transport

D. Kranbuehl, D. Hood, J. Rogozinski, A. Meyer, E. Powell, C. Higgins, C. Davis, L. Hoipkemeier, C. Ambler & C. Elko
Departments of Chemistry and Applied Science, College of William and Mary, Williamsburg, Va., USA

N. Olukcu
Departments of Chemistry and Applied Science, College of William and Mary, Williamsburg, Va., USA (Currently: Department of Petroleum Technology, University of Dokuz Eylul, Izmir, Turkey

KEYWORDS: aging, degradation, sensing, durability, dielectric properties, mechanical properties, in situ

ABSTRACT: This report focuses on the use of frequency dependent dielectric measurement sensors (FDEMS) to monitor continuously, and in situ during use the changing state of a polymeric composite pipe in the use environment. This report focuses on using FDEMS to monitor both the changes in state-health of the polyphenyl sulfide (PPS) graphite tape used for the axial armour and the polyamide PA-11 used for the composite pipes oil-gas inner lining barrier. Current life monitoring work focuses on characterizing the chemical and physical processes occurring during aging, using FDEMS and laboratory measurements of mechanical and thermodynamic properties to monitor the aging rate and state of the polymer, and then integrating the sensor output with a model for predicting the remaining service life and state of the structure. The model predictions are periodically updated through the in situ online sensing measurements. This report will also discuss work on understanding the relationship between the sensor measurement, the ionic and dipolar mobility, and the macroscopic properties of polymeric materials.

INTRODUCTION

The use of polymeric materials in extended use structures such as airplanes, bridges and pipelines, where the expected lifetimes are 20 to 40 years and where failure can be catastrophic, is rapidly expanding. As such, there is a clear need to develop in situ health monitoring capabilities. This paper describes the progress toward the development of a frequency dependent dielectric measurement sensor (FDEMS) which is capable of detecting the change in the physical and chemical state of a polymeric material in situ during use in the field environment and during process-fabrication.

Health monitoring involves monitoring the degradation of a polymeric material's performance properties. In one sense, health monitoring is the reverse of cure monitoring during fabrication.

FDEMS provides a sensitive, automated, in situ sensing technique for monitoring changes in state and properties of polymers during use as well as during intelligent processing. FDEMS in situ sensing can be designed and calibrated to monitor changes in

mechanical service life properties of polymer materials during use in the field environment as well as in processing properties. The FDEMS sensor output already has been shown and used to monitor changes in viscosity, degree of cure and T_g during cure (Kranbuehl 1997, 1989; Mijovic et al. 1995; and Parthun et al. 1992). With proper understanding of the type of polymer and the use environment, FDEMS can be used to monitor modulus, maximum load and elongation at break during use. Monitoring degradation in these performance properties is the reverse of monitoring cure during which there is a buildup in mechanical properties. The FDEMS technique has advantages over other monitoring techniques in that it is: nondestructive, accurate/reproducible, sensitive, in situ, remote and automated.

This paper will discuss recent work on the use of FDEMS to monitor aging during off shore use of flexible composite pipe designed to transport oil-gas in an offshore oil-sea environment. Current durability monitoring work focuses on characterizing the chemical and physical processes occurring during aging, determining the state of the polymer, using FDEMS sensing to monitor the aging rate, mechanical and chemical analyses, and predicting the remaining service life and health of the structure. This report will discuss work on developing a fundamental understanding of the relationship of the sensor measurement of the ionic and dipolar mobility to the macroscopic mechanical performance properties of polymeric materials. The report will describe recent work on the use of FDEMS, differential scanning calorimetry, tensile and torsional mechanical measurements to monitor aging in thermoplastics during use in salt water, oil and acidic environments. The polymers being studied are polyphenyl sulfide (PPS), graphite tape, sulfide and polyamide-11 which are used for flexible composite pipe to transport oil-gas in an offshore environment. The PPS graphite tape is used for axial windings to withstand the >100 bar internal pressures. The polyamide-11, PA-11, is used as is the inner lining barrier to contain the acidic, oil, water flow from the ocean floor up to the offshore platform. These results can be incorporated into an aging-durability model. The sensor output provides in situ online data continuously updating the current state of the polymer. Thus the model can make continuously updated predictions of the remaining service life and projected replacement date of the polymer.

BACKGROUND

Frequency dependent dielectric measurements, made over many decades of frequency, Hz-MHZ, have already been shown to be a sensitive, convenient automated means for characterizing the processing properties of thermosets and thermoplastics (Kranbuehl 1997, 1986, 1989; Mijovic et al. 1995; and Parthun et al. 1992). Using a planar wafer thin sensor, measurements can be made in situ in almost any environment. Through the frequency dependence of the impedance, this sensing technique is able to monitor changes in the molecular mobility of ions and dipoles. These changes in molecular mobility are then related to chemical and physical changes which occur during use or during processing. The FDEMS techniques have the advantage that measurements can be made both in the laboratory, in situ in the fabrication tool and in situ during use. Few laboratory measurement techniques have the advantage of being able to make measurements in a processing tool and in the field in a composite, in an adhesive bond line, of a thin film or a coating. It can be used at temperatures

exceeding 400 °C and at pressures of 60 atm, with an accuracy of 0.1% and a range in magnitude of over 10 decades. It is difficult for most other in the field techniques to attain this level of sensitivity in harsh processing environments.

At the heart of dielectric sensing is the ability to measure the changes at the molecular level in the translational mobility of ions and charge in terms of the conductivity and in the rotational mobility of dipoles in terms of a rotational time τ in the presence of a force created by an electric field. Mechanical properties reflect the response in displacement on a macroscopic level due to a mechanical force acting on the whole sample. The reason why dielectric sensing is quite sensitive is rooted in the fact that changes on the macroscopic level originate from changes in force displacement relationships on a molecular level. Indeed, it is these <u>molecular</u> changes in force-displacement relationships which dielectric sensing measures as the resin cures and ages. They are the origin of the resin's <u>macroscopic</u> changes in mechanical performance properties, during use and processing properties during fabrication.

INSTRUMENTATION

Frequency dependent complex dielectric measurements are made using an Impedance Analyzer controlled by a microcomputer (Kranbuehl 1997, 1986, and 1989). In the work discussed here, measurements at frequencies from Hz to MHZ are taken continuously throughout the entire cure process at regular intervals and converted to the complex permittivity, $\epsilon^* = \epsilon' - i\epsilon''$. The measurements are made with a geometry independent DekDyne micro sensor which has been patented and is now commercially available and a commercially available automated dielectric measurement system. This system uses commercially available impedance bridges or specially built marine environmental bridges for use on offshore oil platforms. The system permits multiplexed measurement of several sensors. The sensor itself is planar, 2.5 cm x 1.25 cm area and 5 mm thick. This sensor-bridge microcomputer assembly is able to make continuous uninterrupted measurements of both ϵ' and ϵ'' over decades in magnitude at all frequencies. The sensor is inert and has been used at temperatures exceeding 400°C and over 60 atm pressure.

RELATIONSHIP TO MACROSCOPIC PERFORMANCE AND PROCESSING PROPERTIES

The two parameter representation of the magnitude and phase relationship between the sinusoidal force, in this case electric, and the displacement-change in position of charge is the complex electric compliance $\epsilon^* = \epsilon' - i\epsilon''$. This quantity is directly analogous to the macroscopic mechanical property compliance $J^* = J' + iJ''$ or its reciprocal $G^* = G' + iG''$. For all these complex compliance-modulus quantities (ϵ^*, J^*, G^*) the real component represents energy storage and the imaginary component, energy lost. Thus measurements of the dielectric properties ϵ^* are directly analogous to mechanical measurements.

Only the difference is mechanical properties detect macroscopic displacement, while dielectric properties monitor changes in position and mobility at the molecular level. The Maxwell-Voight models are used to relate the frequency dependent macroscopic mechanical measurements of G', G'', J' and J'' to equivalent spring and viscous components. In dielectrics, the classic Debye (with its many modifications) and Einstein rotational and translational flow models (with their many slip boundary modifications) can be used to relate measurements of ϵ' and ϵ'' to conductivity σ and a relaxation time τ.

For large particles in a small molecule continuum the value of σ is equal to the product of the molecular translational mobility of the charged particles μ_i times their concentration Ni.

$$\sigma = \Sigma N_i \mu_i \qquad (1)$$

The mobility is directly proportional to the viscosity, where for spherical particles with radius r in an ideal fluid.

$$\mu_i = 6\pi r \eta \qquad (2)$$

The value of τ is proportional to the molecular rotational mobility of dipolar groups, which for spherical shaped particles can be related to viscosity from

$$\tau = \frac{8\pi r^3 \eta}{(2kT)} \qquad (3)$$

A general relation in a wide range of fluid to rubbery materials between translational mobility and viscosity of $\sigma \propto \eta^{-a}$ has been shown to be valid by many investigators. For polymers in a preglass state through this relationship, σ has been widely used to monitor viscosity during polymer resin cure (Kranbuehl 1997, 1989, Mijovic et al. 1995 and Parthun et al. 1992). As a material goes into the rubbery-glass transition state, the ionic mobility persists and can be used to monitor changes in macroscopic performance properties through laboratory calibrations which relate these changes in molecular translational mobility of charge, σ, to the macroscopic performance property of interest. Once this relation is established in the laboratory, in situ dielectric measurements of ϵ'' and thereby σ can be used to monitor with the calibration, changes in the macroscopic performance property during fabrication or during use in the field.

The general relation between τ and viscosity is also well established in the fluid state where again τ, like σ, is monitoring a molecular viscosity reflecting rotational diffusion and is proportional, in general, to the macroscopic viscosity as $\tau \propto \eta^b$. In the gel-glass transition region, it is well documented that τ monitors changes in the glass transition temperature, T_g. Exact determinations of T_g from τ can be made using a VTF or WLF type molecular model. The calibration constants relating τ and T_g are made from laboratory measurements. In the field measurements of ϵ'' dipolar and thereby τ can be used to monitor changes in T_g during cure or during use.

Finally the simple absence of change in $\epsilon''(\omega)$ indicates no change in either translational mobility of charge and rotational mobility of dipoles. This is a very strong indicator that there is no change in the performance properties of the material. Similarly, the rate of change in ϵ'' and ϵ' and its changing magnitude for complex rubbery to glassy materials, indicates change in the molecular mobility and hence change in the performance properties. The relation between these changes in molecular mobility and performance properties does not follow a single universal relationship as the molecular structure of amorphous materials is complex and theories of the structure of amorphous gel-solid materials are far from complete. Nevertheless careful laboratory measurement of the relationship between changes in ϵ'' and ϵ' due to changes at the molecular level in translational mobility of charge and rotational mobility of dipoles can be related to changes in macroscopic force-displacement properties over time for polymer materials and within given defined environments.

In summary, dielectric measurements are an extremely sensitive means to monitor changes in the state of a polymer material or resin both as it ages and during synthesis processing and cure. The reason dielectric sensing is so sensitive is the fact that changes in macroscopic force-displacement properties such as modulus, elasticity, viscosity, T_g, etc. originate from changes in mobility at the molecular level. Dielectric measurements of ϵ^* monitor molecular mobility and equally important they can be made over many orders of magnitude, in harsh environments, continuously and in situ for polymeric composites, coatings, adhesives, in virtually any environment.

LIFE MONITORING

Two materials have been monitored using dielectrics to detect changes in their performance properties during use in an offshore oil-water environment; a polyamide nylon-11 (PA-11), and polyphenyl sulfide (PPS)(Kranbuehl 1997 and 1998). The PPS material is being considered with graphite as an unidirectional tape for use as the axial wrap. It will replace steel bands and create a lighter, higher performance flexible pipe for extremely deep water and arctic environments. As the outer layer of the pipe, the PPS graphite tape is exposed to seawater. The nylon-11 on the other hand, serves as the inner fluid gas barrier and is exposed to the acidic H_2S, water, oil mixture coming up from the ground.

FDEMS sensors were embedded in both material systems by heating the polymer to its softening transition temperature, placing the sensor between 2 pieces of the polymer and encapsulating the sensor in the center with pressure. The resulting material sensor system was approximately 1 cm thick. The embedded sensor material system and mechanical test dog bones of ASTM D638 specifications were then placed in the following aging environments: Nylon-11 in 95% oil, 5% water, pH 4.6, 70°C, and 105°C; PPS in 100% simulated sea water at 90°C and 120°C. Periodically FDEMS sensor data was taken and pieces were removed for mechanical testing.

LIFE MONITORING - RESULTS

For PA-11, Figures 1-2 display the value of ϵ'' multiplied by the frequency at 100, 120, 10^3, 10^4 and 10^5 Hz versus time in the 105° and 70° oil-water acidic aging environment. The results are on a log scale. They show a rapid very large rise, 10^4, during the initial days as water diffuses into the polymer. This process occurs over 150 days at 70°, although a large fraction of this change occurs in the initial 30 days. At 105°C, the water diffusion process occurs over the initial 30 days. After water impregnation has occurred at 105° there is a gradual decline in ϵ''. This drop of ϵ'' is due to the gradual aging and approach to embrittlement of the PA-11.

Nylon Sensor 2L in 105C 5%ASTM/95% H2O

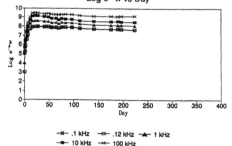

Log e"*w vs Day

Figure 1

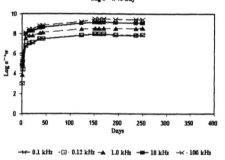

Figure 2

Earlier experimental work in our laboratory and also recently reported by others has shown that the nylon is degrading due to hydrolysis(Kranbuehl 1997, Verdu et al. 1997). This is accompanied by a sudden loss in the tensile % elongation properties of the PA-11. This drop in almost like new properties to around 30% occurs as a critical molecular weight is approached, shown schematically in Figure 3. In the past tensile properties were used to monitor the life of the PA-11. Our earlier work on measuring changes in molecular weight and the associated measurements of viscosity, size exclusion times, light scattering etc. has been shown to be a far better method to monitor the age, remaining life of PA-11 in use in an oil field and was subsequently patented as a life monitoring method for the PA-11 barrier in oil transport pipes(Kranbuehl 1997).

The use of changes in molecular weight and % elongation make it possible to calibrate the long term changes in translational mobility as measured by ϵ'' with both the amount of life remaining (through correlation with molecular weight) and the desired mechanical performance properties (through correlation with % elongation). By measuring the drop in the ionic translational mobility and the corresponding decrease in the value of ϵ'', and then correlating this change in ionic mobility and ϵ'' with the molecular weight and % elongation of samples aged in the same environment for the same amount of time, it is possible to correlate and calibrate the change in ϵ'' with the change in molecular weight and % elongation. This is displayed in Figure 4 where the decrease in $\epsilon''(t)/\epsilon''(max)$ is plotted vs molecular weight. The corresponding % elongation at each time is also plotted. In this manner, dielectric measurements of ϵ'' are able to provide continuous, in situ life monitoring in the field of the changes in % elongation performance properties. Further by knowing the initial molecular weight and the critical value molecular weight where the % elongation drops, it is possible to relate the ϵ'' correlation measurements of molecular weight to the extent of the aging and fraction of life remaining.

In the second sensor life monitoring experiment, the durability of a polyphenyl sulfide, PPS, graphite tape was examined over a 2 year period in a 90°C salt water environment. Unlike the polyamide PA-11 barrier which contains the oil-water mixture on the inner surface of the pipe, the PPS graphite tape is used as an axial armor wrapped around the outer surface of the pipe to allow the flexible pipe to be used at high pressures. Thus the PPS accelerated study was conducted in 90°C seawater with additional laboroatory accelerated aging experiments at 120°C.

A Perkin Elmer Differential Calorimeter (DSC) was used to measure the change in the heat of fusion-crystallinity and the peak in the melting temperature. As shown in Table I, the neat PPS does not display a change in the heat of crystallinity nor a change in the peak melting temperature with aging time. On the other hand, Table II shows the PPS tape does display a significant increase in its heat of fusion-crystallization over the 614 day aging period. The peak melting temperature also increases. This suggests that the interface of the PPS with the graphite tape induces some crystallization over time. In general annealing, an increase in crystallization, occurs between T_g, 88°C, and the melting temperature T_m, 280°C. The optimum annealing temperature is usually near the mean which in this case is $\approx$ 185°C. The 120°C aging temperature is closer to the 185°C mean and it correspondingly shows a larger increase in crystallinity.

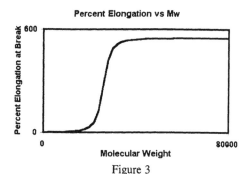

Figure 3

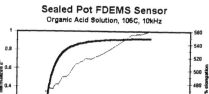

Figure 4

Table I
DSC Data Neat PPS
(samples analyzed in 1998 prior to testing 600+ day specimens)

Environment	Day	Number of Samples	ΔH (J/g)	Peak Temperature (°C)
Fresh	0	2	28.93 ± 0.69	280.6 ± 0.8
90 °C Seawater	614	3	28.94 ± 1.10	278.9 ± 1.1
120 °C Seawater	614	6	29.36 ± 0.56	278.5 ± 1.5

Table II
DSC Data PPS Graphite 6² Tapes Type A
(samples analyzed in 1998 prior to testing 600+ day specimens)

Environment	Day	Number of Samples	ΔH (J/g)	Peak Temperature (°C)
Fresh	0	1	16.65	278.0
90 °C Seawater	614	1	24.62	281.3
120 °C Seawater	614	1	29.42	281.7

417

Table III
Neat PPS 90°C Seawater

Day	Mass (g)	% Change in Mass	Thickness (mm)	% Change in Thickness
0	5.1161	0.00	3.13	0.00
10	5.1336	0.34	3.12	-0.32
24	5.1362	0.39	3.13	0.00
58	5.1370	0.41	3.15	0.64
100	5.1448	0.56	3.12	-0.32
153	5.1455	0.57	3.13	0.00
219	5.1491	0.65	3.14	0.32
275	5.1491	0.65	3.10	-0.96
301	5.1464	0.59	3.12	-0.32
400	5.1567	0.79	3.06	-2.24
500	5.1321	0.31	2.94	-6.07
741	5.1625	0.91	3.2	2.24

Table IV
Neat PPS 120°C Seawater

Day	Mass (g)	% Change in Mass	Thickness (mm)	% Change in Thickness
0	4.9773	0.00	3.13	0.00
10	5.0019	0.49	3.14	0.32
24	5.0021	0.50	3.13	0.00
58	5.0020	0.50	3.13	0.00
153	5.0082	0.62	3.13	0.00
219	5.0161	0.78	3.11	-0.64
259	5.0177	0.81	3.10	-0.96
275	5.0261	0.98	3.06	-2.24
301	5.0088	0.63	3.11	-0.64
400	5.0375	1.21	3.06	-2.24
477	4.9908	0.27	3.00	-4.15
741	5.0270	1.00	3.16	0.96

Table V
MTS Data Neat PPS 2″ Dogbones
90°C Seawater
Grip=6mPa, rate=0.224 mm/min

Day	# of tests	Max Force (N)	Displacement at Max Force (mm)	Force at Break (N)	Displacement at Break (mm)
0	5	895.43 ± 44.23	1.63 ± 0.27	668.35 ± 66.71	4.71 ± 0.66
55	3	986.85 ± 66.48	1.59 ± 0.14	754.11 ± 137.22	2.89 ± 1.18
105	3	1031.81 ± 28.87	1.48 ± 0.02	662.80 ± 24.26	3.67 ± 0.40
150	3	968.38 ± 3.62	1.40 ± 0.04	594.28 ± 7.13	3.66 ± 0.30
203	3	979.90 ± 46.81	1.46 ± 0.05	586.62 ± 51.61	3.29 ± 0.31
614	2	975.07 ± 19.53	1.54 ± 0.00	680.93 ± 42.97	3.31 ± 0.36

Table VI
MTS Data Neat PPS 2″ Dogbones
120°C Seawater
Grip=6mPa, rate=0.224 mm/min

Day	# of test	Max Force (N)	Displacement at Max Force (mm)	Force at Break (N)	Displacement at Break (mm)
0	5	895.43 ± 44.23	1.63 ± 0.27	668.35 ± 66.71	4.71 ± 0.66
203	1	891.43	1.60	609.10	5.319
614	2	1027.23 ± 1.07	1.54 ± 0.00	720.06 ± 12.36	2.93 ± 0.18

Table VII
MTS Testing PPS Graphite 6″ Tapes
90°C Seawater
Grip=15mPa, rate=1.12 mm/min

Day	Number of Samples	Load at Peak (N)	Displacement at Peak (mm)
0	10	9839.91 ± 1741.34	1.34 ±0.31
55	5	15119.51 ± 436.66	2.30 ± 0.21
104	3	14628.72 ± 250.70	3.71 ± 1.97
153	5	11625.87 ± 650.18	1.79 ± 0.13
604	2	8000.13 ± 5658.52	1.64 ± 0.66
614	2	15039.44 ± 1409.12	1.91 ± 0.18

Table VIII
MTS Testing PPS Graphite 6″ Tapes
120°C Seawater
Grip=15mPa, rate=1.12 mm/min

Day	# of samples	Load at Peak (N)	Displacement at Peak (mm)
0	10	9839.91 ± 1741.34	1.34 ±0.31
55	5	14373.09 ± 920.82	2.80 ± 1.24
104	3	13719.80 ± 912.32	2.44 ± 0.54
153	5	10343.60 ± 947.76	1.34 ± 0.01
604	1	15680.01	2.03
614	2	15708.93 ± 1043.11	2.03 ± 0.00

Tables III and IV report changes in the mass and thickness over 741 days. Within experimental error the results show less than 1% change in weight. There is no detectable change in thickness within the experimental variations. Mechanical tensile tests were made on both the neat PPS and PPS tape. Tables V and VI report the neat PPS maximum force and displacement along with the force and displacement at break. The data are for a 614 day period. Fresh samples were run both on day zero and on the 614th day. For each set of runs on fresh and aged samples there is a similar range of variation. Within the precision-variation of the measurements, there is a slight increase in the maximum load and a slight decrease in the elongation at the maximum load.

Tables VII and VIII report results of the load at peak and the displacement of the 6" inch lengths of PPS tape aged at 90°C and 120°C in seawater. Again unaged pieces were tested at day zero and on the 614th day. As in the neat PPS dog bones there is a modest increase in the load at peak. Unlike in the neat PPS, there is an increase in the elongation at the maximum load. The scatter in the data for the PPS graphite tape is greater than in the neat PPS. It should be noted that the change in the average over the 614 days is only slightly outside the scatter for the samples with identical age.

Finally torsional tests were made on the 2 inch strips of PPS tape at a frequency of 1 radian/sec over a 50°C to 160°C to 50°C temperature ramp of 3°C/min. The temperature of the maximum energy loss in the modulus G″ and the magnitude G′ at the maximum are reported in Tables IX and X for 90°C and 120°C aged tapes. Again unaged specimens were tested at day 0 and on the 614th day. For 90°C and 120°C aging, no statistically significant change is observed. All of the aged values fall within the range of results for the unaged specimens.

Finally the FDEMS output over 873 days is reported in Figure V. It shows only a slight decrease indicating a small increase in stiffness. As is the case for mechanical data, the small change occurs in the early portion of the 600+ day aging period and follows the increase in % crystallinity. The very small decrease in the FDEMS output is in complete agreement with the laboratory tests indicating both that the PPS tape is very stable in the 90° salt water environment and that the FDEMS sensor clearly monitors this stable, little to no significant change in mechanical properties over the 600 day, 90°C sea water aging period.

Table IX
Torsional Testing PPS Graphite 2″ Tapes
90°C Seawater
1 radian/second, temperature ramp = 50 °C to 160 °C to 50 °C at 3 °C/minute

Day	Number of Samples	Max of G″ (dynes/cm²) (times 10e9)	Temperature at Max G″
0	9	3.79 ± 0.91	104.5 ± 3.2
10	1	2.35	94.9
32	1	3.35	109.4
100	1	2.68	108.1
175	1	4.26	100.9
301	1	2.77	104.0
614	1	3.97	101.8

Table X
Torsional Testing PPS Graphite 2″ Tapes
120°C Seawater
1 radian/second, temperature ramp = 50 °C to 160 °C to 50 °C at 3 °C/minute

Day	Number of Samples	Max of G″ (dynes/cm²) (times 10e9)	Temperature at Max G″
0	9	3.79 ± 0.91	104.5 ± 3.2
10	1	2.15	103.5
31	1	2.62	101.8
100	1	3.18	
225	1	2.23	103.8
614	1	2.18	103.0

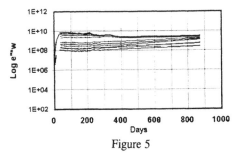

PPS Sensor in 90C Seawater

Figure 5

CONCLUSIONS

The changing state-health or durability of two polymers used for a high performance composite deep water offshore oil-gas pipe has been monitored in situ in the aging environment using dielectric (FDEMS) sensing. Dielectric properties monitor molecular mobility. This is a critical initial molecular step that causes changes in macroscopic material mechanical properties. Molecular mobility is not directly linked to molecular weight but rather the consequences of the changes in molecular weight, crystallinity as well as changes as plasticizing molecules move in and out. Both changes in molecular weight, crystallinity and plasticization effect molecular mobility first and eventually a change in the mechanical properties. The changes in dielectric molecular mobility do monitor these changes in state-health durability of the PA-11 and PPS in the composite pipe.

ACKNOWLEDGEMENTS

Support from the NSF Center of Excellence at VPI MR 912004, the North Sea Robit Offshore oil JIP and the Wellstream Corporation

REFERENCES

Kranbuehl, D., *Developments in Reinforced Plastics,* Vol. 5, Elsevier Applied Science Publishers, New York, pp. 181-204 (1986).

Kranbuehl, D., *Dielectric Spectroscopy of Polymeric Materials* ed. J. Runt, J. Fitzgerald, Am. Chem. Soc., Washington, DC 1997 pp. 303-328.

Kranbuehl, D., *Encyclopedia of Composites*, ed., Stuart M. Lee. VCH Publishers, New York, pp. 531-43 (1989).

Kranbuehl, D., *Method and Apparatus for Monitoring Nylon 11 Made from Polyomide Plastic made from Aminodecanoic Acid* #5614683, 1997, filed 1995.

Kranbuehl, D., *Progress in Durability Analysis of Composite Systems*, ed. K. Reifsnider, D. Dillard, A. Cardon, A. A. Balkema, Rotterdam 1998 pp. 11-17.

Kranbuehl, D., *Structural Health Monitoring*, ed. Fu-Kuo Chang Int. Workshop, Stanford Univ., Sept. 1997, sponsored by Air Force Scientific Research, Army Research Office, Nat. Sci. Foundation, Technomic 1997.

Mijovic, F., Bellucci, L. Nicolois, Electrochem. Soc., 142(4), 1176-1182 (1995).

Parthun,M.B., G. Johari Macromolecules 25 3254-3263) (1992).

Verdu, J., Serpe, G. Chaupert, N., *Polymer* 1997 38 1311; 1998 6-7 1375.

Prediction of the buckling of thermoplastic structural elements

J. L. Spoormaker
Faculty of Design, Engineering and Production, Delft University of Technology, Netherlands

I. D. Skrypnyk
Karpenko Physico-Mechanical Institute, Lviv, Ukraine

ABSTRACT: Plastic bottle crates are required to weight as little as possible for lower material consumption and for transportation costs. The reliability and sustainability of crates is required to be very high. It is possible to satisfy the weight requirements by applying structural optimisation. *Buckling* is an imported restrain because of the decrease in load carrying capacity. Presented is the development of methods to predict the buckling behaviour of structural elements accounting for the *non-linear time dependent behaviour of High Density Polyethylene*. The procedure for characterising the non-linear visco-elastic behaviour of polymers is shortly described. The subroutine to account for the non-linear visco-elastic behaviour in finite element methods is outlined. A few examples of the achievements with structural elements are given.

1 INTRODUCTION

We aim at predicting the buckling of a HDPE crate for beer bottles as depicted in Figure 1. Bottle crates, with (filled) bottles, are required to be stored on pallets, from several days to even weeks. The pallets are stacked one on top the other, resulting in a very high load on the bottom crate.

Figure 1. One quarter of a buckled bottle crate.

Buckling of the corner stiffening ribs of crates will cause the drop of pallets resulting in a loss the beer bottles and crates, but moreover this might injure people. A serious problem is that buckling can happen after an extended period of time. Obviously, the buckling of a crate's corner occurs after the material creeps under compressive loading.

The objective of the research is to predict using Finite Element Method (FEM), the onset of buckling of plastic bottle crates under compressive loading.

During the first stage of the investigation, the buckling behaviour of simple HDPE test specimens (strips and U-profiles) has been studied, experimentally as well as numerically, in order to obtain better understanding of creep induced buckling. The experiments and simulations with visco-elastic strips allowed us to study creep induced buckling in comparison to traditional Euler's buckling of elastic materials. The U-profile specimens have been chosen, to investigate buckling behaviour of simple elements with stiffening ribs.

2. MATERIAL MODEL

2.1 *The generalised Shapery Model*

The non-linear visco-elasticity model [6, 7] is a generalisation of the Schapery model [8]:

$$\varepsilon\big(\sigma(t),t\big) = J_0\big[\sigma(t)\big] +$$

$$\sum_i \phi_i[\sigma(t)]\int_0^t F_i(\zeta - \zeta')\frac{\partial[g_i(\sigma)]}{\partial \xi}d\xi \qquad (1)$$

The following forms for functions in equation (1) seem to be appropriate [9] for description of experimental data for many plastics:

$$J_0[\sigma] = A\cdot[\sigma + \sigma^\beta]$$

$$\phi_i[\sigma] = \exp(\gamma_i \cdot \sigma)$$

$$g_i[\sigma] = D_i \cdot \sigma^{\alpha_i}$$

$$F_i(t) = (1 - \exp(-\lambda_i t)). \qquad (2)$$

The parameters in these functions should be estimated, based on data from creep and recovery tests. For this the equation (1) can be rewritten as follows: for creep

$$\hat{\varepsilon}[\sigma_k, t] = J_0[\sigma_k] + \sum_i \phi(\sigma_k)F_i(t)g_i(\sigma_k) \qquad (3)$$

for recovery

$$\hat{\varepsilon}[\sigma_k, t] = \sum_i [F_i(\tilde{t} + t_1) - F_i(\tilde{t})]g_i(\sigma_k)\sigma_k \qquad (4)$$

$$\tilde{t} = t - t_1$$

There are strong advantages in choosing the time functions $F_i(t)$ in the form of Prony series. Firstly, this choice enables an efficient numerical scheme (Henriksen 1984) for calculation of convolution integrals. Secondly, it gives better possibilities for establishing the parameter identification procedure. This procedure is based on the idea of minimisation of the relative deviation between experimental data and model prediction. The resulting set of material parameters for HDPE is given in Table 1.

The prediction of creep and recovery behaviour of HDPE (based on the equations (1), (2) and parameter set from the Table 1) is presented in Figures 2 & 3.

2.2 Extension to 3-D formulation and Matrix formulation.

Equation (1) was extended to 3-D formulation, based on the assumptions that:

- the polymer is compressible and initially isotropic;
- the processes of change of volume and shape are uncoupled;
- the rate of viscous flow is proportional to the effective stress $\hat{\sigma}$;

Table 1. The set of model parameters for description of HDPE

	A=7.852·10⁻⁴				$\beta = 0.0$	
i	$D_{i,1}$	$\alpha_{i,1}$	$D_{i,2}$	$\alpha_{i,2}$	γ_i	λ_i
1	.1278·10⁻⁴	2.759	.259·10⁻³	1.059	-.2404·10⁻¹	10⁻¹
2	.3364·10⁻³	1.075	.591·10⁻⁵	2.872	-.1473·10⁻³	10⁻²
3	.3729·10⁻³	1.118	.967·10⁻⁵	2.740	.6727·10⁻¹	10⁻³
4	.5814·10⁻⁴	2.193	.284·10⁻⁹	6.254	.4689·10⁻¹	10⁻⁴
5	.6187·10⁻³	1.547	.675·10⁻¹¹	9.779	.9244·10⁻³	10⁻⁵
6	.4855·10⁻¹	1.637	.279·10⁻²	2.786	-3.377	10⁻⁶

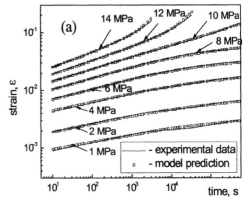

Figure 2. Experimental data [11] and model prediction for creep of HDPE

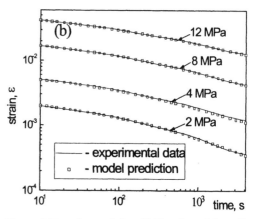

Figure 3 Experimental data [11] and model prediction for recovery of HDPE

As a result, equation (1) can be rewritten as follows:

$$\bar{\varepsilon} = [(1 + \nu_0)\mathbf{M}_D + (1 - 2\nu_0)\mathbf{M}_H]\tilde{\varphi}(\hat{\sigma})\bar{\sigma} + [(1 + \nu_1)\mathbf{M}_D + (1 - 2\nu_1)\mathbf{M}_H]\cdot \qquad (5)$$

$$\sum_{i=1}^{n}\phi_i(\hat{\sigma})\int_0^t F_i(t-\xi)\frac{\partial\big[\widetilde{g}_i(\hat{\sigma})\overline{\sigma}\big]}{\partial\xi}\cdot d\xi$$

where

$$J_0[\sigma]=\widetilde{\varphi}(\sigma)\sigma;\quad g_i[\sigma]=\widetilde{g}_i(\sigma)\sigma\ .\quad (6)$$

The matrix and vector notations are:

$$\mathbf{M}_D=\begin{cases}\dfrac{2}{3}, & \text{if }i=j\text{ and }i,j\le 3;\\[2mm]-\dfrac{1}{3}, & \text{if }i\ne j\text{ and }i,j\le 3;\\[2mm]2, & \text{if }i=j\text{ and }i,j>3;\\[2mm]0, & \text{if }i\ne j\text{ and }i,j\le 3.\end{cases}$$

$$\mathbf{M}\,_H^{i,j}\Big|_{i,j=1\dots 3}=\frac{1}{3}\qquad (7)$$

$$\overline{\sigma}=\big\{\,\sigma_{xx}\ \sigma_{yy}\ \sigma_{zz}\ \tau_{xy}\ \tau_{yz}\ \tau_{zx}\,\big\}^T;$$

$$\overline{\varepsilon}=\big\{\,\varepsilon_{xx}\ \varepsilon_{yy}\ \varepsilon_{zz}\ \gamma_{xy}\ \gamma_{yz}\ \gamma_{zx}\,\big\}^T.\qquad (8)$$

3. IMPLEMENTATION IN FEM-CODES

Since most of the FE packages based on the displacement formulation, the visco-elasticity model (1) of Kelvin-Voight type should be inverted and rewritten in the incremental form as follows:

$$\Delta\overline{\sigma}=\mathbf{L}\big(\Delta\overline{\varepsilon},\Delta t\dots\big)\qquad (9)$$

, in order to be implemented into a FEA package.

Further the main elements, necessary to derive equation (9), are given. For enough regular functions $g_i(\sigma)$ (such that $\big(\partial^2[g_i(\sigma)]/\partial t^2\big)\ll 1$) the convolution integral with the exponential kernel function can be transformed to finite form, following Henriksen scheme (Henriksen 1984):

$$\int_0^t\big(1-\exp(-\lambda_i(t-\xi))\big)\frac{\partial[g_i(\sigma)]}{\partial\xi}d\xi=g_i(\sigma)+\theta_i(t)\quad (10)$$

The hereditary integral functions $\theta_i(t)$, which can be also considered as the set of internal parameters of this model:

$$\theta_i(t)=-\int_0^t\exp(-\lambda_i(t-\xi))\frac{\partial[g_i(\sigma)]}{\partial\xi}d\xi\quad (11)$$

can be calculated recurrently as follows:

$$\theta_i(t)=\exp(-\lambda_i\Delta t)\theta_i(t-\Delta t)-\Delta\big[g_i(\sigma)\big]\Gamma_i(\Delta t)\quad (12)$$

Here, the following notation has been introduced for convenience:

$$\Gamma_i(\Delta t)=\frac{1-\exp(-\lambda_i\Delta t)}{\lambda_i\Delta t}.\qquad (13)$$

Further, the total differential of the equation (5) has to be derived and inverted to the form (9). Unfortunately, numerical scheme, based on the total differential, shows low convergence ability and often becomes unstable [12]. Therefore, similar to [13], it has been assumed that the pre-integral functions $\phi_i(\sigma)$ do not vary within a time increment. In addition only partial factorisation has been used for inversion of incremental stress-strain relation (i.e. the scheme is neither completely explicit nor implicit, but a mixed one). As a result, following incremental relation has been derived [12]:

$$\Delta\overline{\sigma}=\Big\{\big[(1+v_0)\mathbf{M}_D+(1-2v_0)\mathbf{M}_H\big]\widetilde{\varphi}(\hat{\sigma})\Big\}^{-1}\times$$
$$\Big\{\Delta\overline{\varepsilon}-\big[(1+v_1)\mathbf{M}_D+(1-2v_1)\mathbf{M}_H\big]\times\qquad (14)$$
$$\sum_{i=1}^{n}\begin{bmatrix}\big[\phi_i(\hat{\sigma})\widetilde{g}_i(\hat{\sigma})(1-\Gamma_i(\Delta t))\big]\Delta\overline{\sigma}(t-\Delta t)+\\ \phi_i(\hat{\sigma})\big(\exp(-\lambda_i\Delta t)-1\big)\widetilde{\theta}_i(t-\Delta t)\end{bmatrix}\Big\}$$

While deriving this relation it is implied that the loading history always starts from zero.

The above-derived scheme is recurrent. To calculate the stress increment, only data for the stresses field σ and internal parameters $\widetilde{\theta}_i$ from the previous step are required. For instance, the internal parameters $\widetilde{\theta}_i$ are calculated as follows:

$$\widetilde{\theta}_i(t)=\exp(-\lambda_i\Delta t)\widetilde{\theta}_i(t-\Delta t)-\Delta\big[\widetilde{g}_i(\hat{\sigma})\overline{\sigma}\big]\Gamma_i(\Delta t)\quad (15)$$

4. EXPERIMENTAL AND COMPUTATIONAL ASPECTS

4.1 *Experiments on determining the buckling of HDPE structural elements.*

Three different sets of experiments have been performed.

1. Experiments with ramp compression of strips with three different rates (5, 0.5 and 0.1 % per minute) were carried out. The maximum strain reached in these tests was 10 %.

2. Experiments were carried out with constant compressive loading. After a certain loading level (which was slightly lower, than the one necessary for instant onset of buckling) had been reached, the specimen was kept under these loading until creep buckling occurs. The time to the onset of buckling was determined.
3. Tests of ramp compression of U-profiles have been performed with a strain rate of 5% per minute until 10 % of deformation was reached.

Strips have been cut out of extruded HDPE plates and finished by milling to assure high precision and surface quality of the specimens. The cross-section of all specimens was 14.95 x 3.1 mm. The distance between clamps of the testing device will be referenced further on as to the length of the specimens. Strips with four different lengths (35, 45, 70, and 80 mm) have been tested. For the strips shorter, than 35 mm, even a small clamp misalignment drastically affects the test results. The specimens longer, than 80 mm, buckled at very low loads. This made the tests with longer strips useless because of the accuracy of the testing equipment. All tests were performed on a 10 kN Zwick testing machine with facilities for force or displacement control. The strips were clamped in the testing device. Afterwards the clamps were moved automatically to approach the zero stress level in the specimens before the test started.

As mentioned before, even a slight misalignment of the clamps can drastically influence the results of buckling tests. For instance, a small shifting (0.7 mm) of one clamp to one side, while testing a 80 mm long strip, subjected to compressive loading with a constant strain rate of 5 % per minute, leads to a drop of the buckling force by 30 % and a deviation in the post-buckling behaviour by 70 to 80 %. Therefore, special attention was paid to assure the alignment of the clamps in the testing device. As a result a good reproducibility of the experimental results was reached: for the tests with similar loading conditions and specimen length, but in different tests series, the deviation in the buckling force was less than 5 %.

The U-profiles were also manufactured from extruded HDPE plates. The wings (flanges) were bent using an electric wire bench (the bending lines were heated until bending became possible). Because of bending, some residual stresses could appear in the specimens. To avoid this, the specimens were put into a special wooden mould and heated in an oven at 90°C for half an hour. The specimens were all produced with the same width of the back wall (88mm) and of the wings (36.5 mm). The thickness of the U-profile walls was 3.1 mm. Four different lengths of specimens were chosen for tests: 256 mm,

213 mm, 170 mm and 130 mm. Again the distance between the clamps is considered as the length of the U-profiles.

Special steel clamps were produced to ensure proper clamping of the U-profile edges during testing. To verify the reproducibility of tests, at least two different specimens of the same length were used. In theory the wings of a U-profile with a perfect shape can buckle in two directions: inside and outside. In reality, the buckling mode is predefined by initial imperfections of the shape of the specimen. In the experiments both: outside and inside buckling of the wings occurred (fig. 7). For the same buckling mode the reproducibility of the experimental results was sufficient.

4.2 *FEM simulation of experimental results obtained with plastic strips.*

The mesh used for the FE modelling of the tests on strips consisted of quadrilateral shell elements of type 75 [14]. To simulate the clamping of the strip in the testing device, all degrees of freedom were restricted for the nodes at the top and the bottom of the specimens.

The deformation history was modelled by a step-wise function (100 loading steps). At each step the nodes at the top of the mesh were displaced at a distance, which caused -0.1% of additional deformation in the strip. To simulate the change of loading with time, the specimen was allowed to relax (**AUTO CREEP** option [15]) during the time increment, which relates to the strain rate, prescribed for the test simulated. The buckling of the loaded strips was simulated using the MARC facilities for buckling analysis (inverse power sweep method) [16]. In case, when the buckling problem involves material non-linearity (e.g. visco-elasticity in our case), the problem must be solved using a perturbation analysis.

This means, that at a certain moment of the loading history (at a certain increment), a linear buckling analysis is performed to estimate the eigenvector ϕ of the node displacements for the requested buckling mode.
At the next load increment, the node coordinates are modified to account for the fraction of the eigenvector:

$$X = X + \mathbf{f}^*\phi / |\phi| \qquad (16)$$

The introduced perturbation $\{\mathbf{f}^*\phi / |\phi|\}$ will grow (or diminish) in time, if the current buckling mode is stable (or unstable) under certain loading conditions. The factor $\mathbf{f}$ was found empirically. For the described model it was found to be equal to 0.5. The perturbation

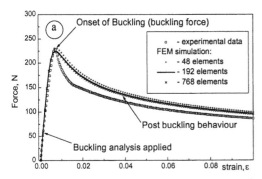

Figure 4a. Simulation of the buckling behaviour of a 70 mm long strip for different numbers of elements.

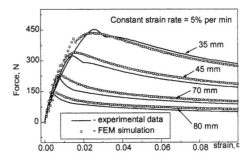

Fig. 5. FE prediction of buckling behaviour of HDPE strips with different length.

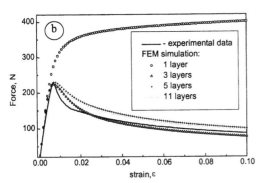

Figure 4b. Simulation of the buckling behaviour of 70 mm long strip using different numbers of layers in the shell elements.

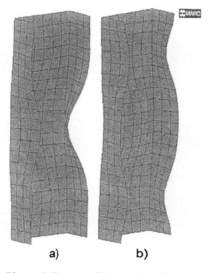

Figure 6. Two possible modes of local buckling. (a) – wings inside, (b) - wings outside.

analysis for the first buckling mode was performed by invoking the **BUCKLE INCREMENT** option [10] right after first "deformation step" (fig. 4a).

In order to verify the accuracy of FE modelling, additional calculations with different number of elements in mesh (fig. 4a) and different number of layers in shell element (fig. 4b) were performed for the 70 mm long strip loaded by strain rate 5% per minute.

For most cases (except of variant with one layer, which obviously is not functional) the accuracy of prediction of buckling force is less than 1%, while deviation between experimental data and results of simulation for post-buckling behaviour is less than 10%. Therefore, for further modelling the mesh with 48 shell elements (5 layer) was chosen.

4.3 Results of computer simulations.

The results of FE simulations of ramp compression tests with the strain rate of 5% per minute are

given in figure 6. The maximum deviation between experimental data and computer prediction is less than 10% for all the strip lengths modelled.

Figure 5. FE prediction of buckling behaviour of HDPE strips with different length.

4.4 FEM modelling of U-profile in compression.

One quarter of the U-profile was modelled in order to save calculation time. The quadrilateral, three-dimensional shell elements of type 75 [13] with 5 layers were chosen to model the U-profile. The mesh contained 72 elements and 91 nodes. It was observed in the experiments, that a small initial sag (less than 1 mm) of the back wall of the U-profile predefines a certain buckling mode: outside or inside.

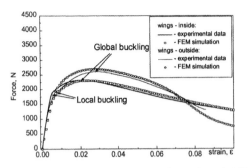

Figure 7. Comparison of experimental and FEM simulation of U-profiles for two buckling modes.

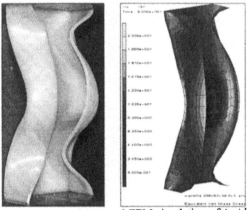

Figure 8. Experiment and FEM simulation of *inside* buckling.

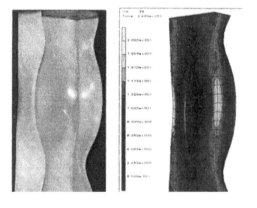

Figure 9. Experiment and FEM simulation of *outside* buckling

This observation was confirmed by FE simulation (fig. 6). In both cases good agreement (less than 6%) between the computer prediction and the test results of compressive force was obtained (fig. 7).

Because of the presence of stiffening ribs in the U-profile, buckling occurs in two steps (figures 8 & 9). First, the wings buckle that correspond to the local buckling point on fig. 4a. This leads to further buckling of the stiffening ribs.

5 CONCLUSIONS

1. The non-linear visco-elasticity model, based on tensile creep-recovery tests, can be applied for the FEM simulation of the other types of loading of HDPE structural elements.

2. A calculation schema for FEM modelling of buckling and post-buckling behaviour of the non-linear visco-elastic strips has been established.

3. The influence of model parameters (number of elements in mesh and layers in shell element) on the accuracy of FEM calculations was determined.

4. It is shown, that established calculation model enables the prediction of the buckling and post-buckling behaviour of U-profiles under ramp compression loading.

REFERENCES

1. *Samuelson L.A.* Creep Deformation and Buckling of a Column with Arbitrary Cross Section. – Report No. 10, The Aeronautical Research Institute of Sweden (FFA). – 1967.

2. *Cohen A., Arends C.B.* Creep Induced Buckling of Plastic Materials // Polym. Eng. Sci. – 1988. – **28**, No. 8, – p. 506

3. *Cohen A., Arends C.B.* Application of a Concept of Distributed Damage to Creep Induced Buckling of High Density Polyethylene Specimens // Polym. Eng. Sci. – 1988. – **28**, No. 16, – p. 1066.

4. *Rzanitsyn A.R.* Dokl. Akad. Nauk. SSSR. – 1946. – **52**, № 1.

5. *Drozdov A.D., Kolmanovskii V.B.* Stability in Viscoelasticity. – Elsevier, – 1994. – 600 p.

6. *Skrypnyk, I.D., Spoormaker, J.L.* Modelling of non-linear visco-elastic behaviour of plastic materials // Proceedings of the 5th European Conference on Advanced Materials /Euromat 97/, Materials, Functionality and Design. – 1997. – **4**. – pp. 4/491-4/495.

7. *Skrypnyk, I.D., Zweers, E.W.G.* Non-linear visco-elastic models for polymers materials. – Technical Report K345, Faculty of Industrial

Design Engineering, TU Delft, the Netherlands, –
1996. – 40 p.

8. *Schapery, R. A.* On the Characterisation of
nonlinear viscoelastic materials // J. of Polym.
Eng. Sci. – 1969. – **9.** – p. 295.

9. *Skrypnyk, I.D., Spoormaker, J.L., Kandachar,
P.V.* Modelling of the long-term visco-elastic
behaviour of polymeric materials. – Technical
Report K369, Faculty of Industrial Design
Engineering, TU Delft, the Netherlands. – 1997. –
48 p.

10. Henriksen, M 1984. *Computers & Structures.* **18:**
p.133.

11. *Beijer J.G.J., Spoormaker J.L.* Viscoelastic
behaviour of HDPE under tensile loading // Proc.
of 10th International Conference on Deformation,
Yield and Fracture of Polymers. – 1997. – pp.270
– 273.

12. *Skrypnyk, I.D., Spoormaker, J.L., Vasyljkevych,
T.O.* Modelling of the temperature dependent
visco-elasticity in plastics. – Technical Report
K381, Faculty of Design, Engineering and
Production, TU Delft, the Netherlands. – 1998. –
40 p.

13. *Lai, J.* Non-linear time-dependent deformation
behaviour of high density polyethylene. – Ph.D.
thesis, TU Delft, the Netherlands. – 1995. – 157
p.

14. MARC, Element Library, Volume B, Rev. K.7,
MARC Analysis Research Corp. – 1997.

15. MARC, Program Input, Volume C, Rev. K.7,
MARC Analysis Research Corp. – 1997.

16. MARC, Theory and User Information, Volume
A, Rev. K.7, MARC Analysis Research Corp. –
1997.

17. *Oliver, P.J.* Applications of Lie groups to
differential equations. – New York: Springer-
Verlag New York Inc., – 1986.

Durability estimation of carbon/polytetrafluoroethylene composites for hydraulic downhole motor

Kiyoshi Kemmochi, Hiroshi Takayanagi, Jun Takahashi, Hiroshi Tsuda & Hideki Nagai
– National Institute of Materials and Chemical Research, AIST, MITI Higashi, Ibaraki, Japan

Taishi Hamada *– Graduate School, Science University of Tokyo, Chiba, Japan*

Chouhachiro Nagasawa *– Kumamoto Industrial Research Institute, Japan*

Hiroshi Fukuda *– Department of Materials Science and Technology, Science University of Tokyo, Chiba, Japan*

ABSTRACT: Geothermal power energy is prospective energy resource in Japan because it is inexhaustible and clean. Utilization of geothermal energy requires to dig geothermal wells near the magma located about 3,000m below the surface of the earth. An excavator for digging geothermal wells is called downhole motor (DHM). Polymeric elastomers such as natural rubber have been used as sealing materials and stators of DHM. The life time of DHM is very short because these materials are exposed to severe geothermal environment. Therefore the development of polymeric composites with heat resistance and long-term durability has been demanded to dig geothermal wells efficiently. In the present study, long-term durability was evaluated on a Carbon/polytetrafluoroethylene (CF/PTFE) composite material which is one of the prospective materials used for DHM . We experimentally fabricated a testing machine to simulate geothermal environment. Using the testing machine creep tests with mixed-mode loading of compression and torsion were performed for 100 hours at room and elevated temperatures. The master curves to predict the lifetime of the material were obtained.

1 INTRODUCTION

At the present time, the problems of global warming are being discussed in the world. In December, 1997, the 3rd session of the conference of the parties to the united nations framework convention on climate change (COP3) was held in Kyoto. Then some regulattions to reduce the exhaust volume of CO_2 were decided. Accordingly, the devepolment of geothermal power generation is expected for a method to achieve it. The volume of CO_2 exhausted from geothermal power generation is less than that of some kinds of electric power; petroleum, coal, natural gas and so on. Concretely, it is said that is one-twenty fifth of coal and one-fourteen of petroleum.

Japan is one of the most active volcanic countries in the world. It is estimated that approximately 10 percent of the world's geothermal energy presents in Japan. As geothermal energy is one of the nation's purely indigenous energy sources, its development has been widely favored. This domestic geothermal energy is not only clean and inexpensive but also is expected to be beneficial to the local society through multipurpose utilization. Furthermore, the geothermal energy is regarded as the most promising oil-alternative energy.

Geothermal power energy is expected to be future energy resource in Japan. There are two types of geothermal power energy: shallow-seated and deep-seated. Deep-seated geothermal power energy can provide more energy than shallow-seated, however it has major difficulty that it is necessary to dig geothermal wells near the magma located about 3,000m below the surface of the earth. An excavator used to dig geothermal wells is called downhole motor (DHM). Polymeric elastomers such as natural rubber have been used as sealing materials and stators of DHM. These materials are exposed to very severe geothermal environments such as high temperature over 200°C and muddy water. Furthermore they are subjected to combined loading of compression and torsion. Thus the life time of DHM is very short. The development of polymeric composites with both heat resistance and long-term durability has been demanded to dig the well efficiently. It is also expected to establish the method to predict appropriately the life time of materials exposed to deep-seated geothermal environment.

In the present study, short carbon fiber-reinforced polytetrafluoroethylene (CF/PTFE) composite materials were used as testing materials. A testing machine, which can simulate geothermal environment, was experimentally fabricated. Using this testing machine creep tests of mixed-mode loading of compression and torsion were performed on CF/PTFE for 100 hours at room and elevated temperatures (40, 60, 80 and 100°C). The influences of both temperature and time on the deformation were discussed from the viewpoint of viscoelastic behavior of polymeric materials. From the experimental data, the master curves were obtained through the time-temperature superposition principle. The reliability of the long-term mechanical properties predicted from the master curves was verified from the viewpoint of chemical kinetics.

2 MATERIALS & EXPERIMENTAL PROCEDURES

2.1 Experimental Materials

In the present study, short carbon fiber-reinforced polytetrafluoroethylene composite materials (CF/PTFE) were used as testing materials. The geometry of the tested specimens are dog-bone shaped as shown in Fig.1.

2.2 Testing Machine

Sealing materials and stators of DHM are subjected to (1) temperature cycle from 23°C to 350°C, (2) cyclic and mixed-mode load of compression and torsion, (3) chemical attack by muddy water including chemical substances.

A testing machine which can simulate these environments was experimentally fabricated. The illustration of the testing machine is shown in Fig. 2. This machine enables to apply combined loading: compressive and torsional load. Axial force can be loaded from 30kgf to 2,000kgf with air-cylinder, and torque can be loaded up to 500kgfm with torque-actuator. This machine equips a chamber which can be heated up to 350°C. The axial deformation of the specimen is measured with an extensometer. The torsional deformation of the specimen is measured as follows. A disk with ditched side is set to the specimen as shown in Fig. 3. The ditch of the disk is cut by an interval of 1 degree. An equipped sensor read the number of the ditch so that the torsional deformation can be measured. The torsional angle can be measured up to ±75°.

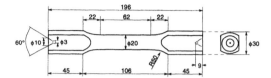

Fig.1 The tested specimen geometry

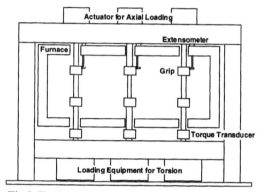

Fig.2 Testing equipment for simulating geothermal envirnment

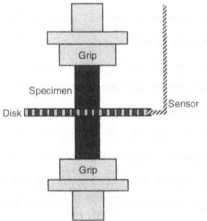

Fig.3 A method to measure the torsion angle.

2.3 Experimental Procedures

The compressive and torsional load of creep tests was set to be as follows:

compressive load: $F_{0.25} \times 0.5$ (1)

torsion load: $T_{0.25} \times 3$ (2)

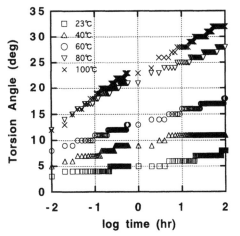

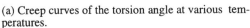

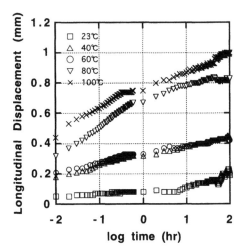

(a) Creep curves of the torsion angle at various temperatures.

(b) Creep curves of longitudinal displacement at various temperatures.

Fig.4 Experimental results of the creep tests.

where $F_{0.25}$ denotes the compressive load when the compressive strain was 0.25, and $T_{0.25}$ denotes the torsional load when the torsional strain was 0.25. The values of $F_{0.25}$ and $T_{0.25}$ were measured from monotonic compressive test and torsional test, respectively. Creep tests subjecting to combined load of compression and torsion for 100 hours were performed on CF/PTFE in air at room and elevated temperatures (40, 60, 80 and 100°C). Axial displacement and torsion angle were recorded in a microcomputer.

3 RESULTS & DISCUSSION

It is difficult to predict the long-term mechanical behavior of materials from the short-term experimental data. Establishment of accelerated test is very important to secure the long-term reliablity of advanced materials. In the present study, creep tests of combined load of compression and torsion were performed for 100 hours at room and elevated temperatures. The technique to predict the long-term mechanical properties of materials was discussed

3.1 Determination of the Creep Load Level

Monotonic compressive tests and torsional tests on CF/PTFE were performed to determine the load of creep test. Compressive modulus E, shear modulus G and load $F_{0.25}$, axial stress $\sigma_{0.25}$, torque $T_{0.25}$ and

Table.1 Mechanical properties derived from compressive test

	compressive modulus E (MPa)	compressive loading $F_{0.25}$ (N)	compressive stress $\sigma_{0.25}$ (MPa)
CF/PTFE	1850	1362.2	4.44

Table.2 Mechanical properties derived from torsional test

	shear modulus G (MPa)	torque $T_{0.25}$ (N· m)	shear stress $\tau_{0.25}$ (MPa)
CF/PTFE	528	2.06	1.34

shear stress $\tau_{0.25}$ when strain was 0.25% are shown in Table 1 and 2. The load of compression and torsion for creep tests was determined by substituting $F_{0.25}$ and $T_{0.25}$ into equation (1) and (2). The value of compressive load was 686N, and the value of torsional torque was 6.174Nm

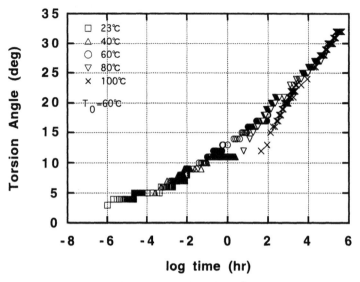

(a) Master curves related to torsion angle at $T_0=60°C$

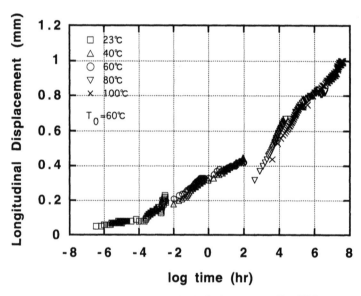

(b) Master curves related to longitudinal displacement at $T_0=60°C$.

Fig.5 Master curves

3.2 *Influence of Time and Temperature on Creep Behavior*

Creep curves for torsional and axial loading are shown in Fig. 4 (a) and (b), respectively. Both deformation by torsional and axial loading seems to be in proportional to the logarithm of time, however the relationship between temperature deformation is non-linear. The experimental dat both deformation at 80°C were close to the da 100°C.

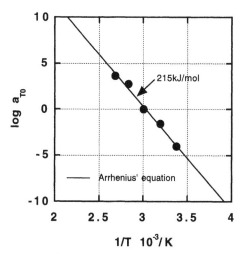

Fig.6 Arrhenius plot of time -temperature shift factors derived from master curve of the torsion angle.

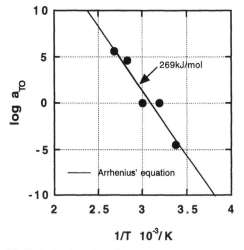

Fig.7 Arrhenius plot of time -temperature shift factors derived from master curve of the longitudinal deformation.

3.3 Master Curve

The master curves for the reference temperature $T_0=60°C$ were obtained on the basis of time-temperature superposition principle. The curves for torsional and axial loading are shown in Fig. 5(a) and (b), respectively. The creep curves of high temperature range (80 and 100°C) were shifted to the right of the reference curve, and the creep curves of low temperature range (23 and 40°C) were shifted to the left of the reference curve. The early-stage creep data at 80 and 100°C tend to deviate from the master curve. This deviation would be attributed to the different creep state. The early-stage creep data at 80 and 100°C would correspond to primary creep state and other creep data would correspond to stationary creep state. Except these area, however, smooth master curves could be obtained from the creep data for 100 hours at various temperatures.

3.4 The Validity of Experimental Shift Factor

The relationship between the time-temperature shift factor aT0 and the inverse of temperature for torsional and axial loading is shown in Fig. 6 and 7.

In both cases, the shift factors are well fit on a certain line. It can be inferred from these results that the creep deformation was controlled by the same deterioration mechanism. According to the stochastic process approach, the activation energy E_a is formulated by equation (3).

$$log\, a_{T0} = \frac{E_a}{2.303R}\left(\frac{1}{T}-\frac{1}{T_0}\right) \qquad (3)$$

where

R:	gas constant
	$=8.314\times10^{-3}$ [kJ/Kmol]
T:	testing temperature [K]
T_0:	reference temperature [K]

The estimated values of activation energy for torsional loading and axial loading are 215kJ/mol and 269kJ/mol, respectively.

4 CONCLUSION

In the present study, short carbon fiber-reinforced polytetrafluoroethylene composite materials (CF/PTFE) were used as testing materials. A testing machine which can simulate geothermal environments was experimentally fabricated. Creep tests of combined load of compression and torsion were performed on CF/PTFE for 100 hours. From the experimental results, smooth master curves for torsional and axial loading were obtained through the time-temperature superposition principle. The validity of these master curves was verified from the viewpoint of chemical kinetics. The values of activation energy for creep deformation by torsional and axial loading were quantatively evaluated.

Recent Developments in Durability Analysis of Composite Systems, Cardon, Fukuda, Reifsnider & Verchery (eds)
© 2000 Balkema, Rotterdam, ISBN 90 5809 103 1

Comparison of warp knitted and stitched non-crimp fabrics

H. Pattyn, I. Verpoest & J. Ivens
Department of Metallurgy and Materials Engineering, Katholieke Universiteit Leuven, Belgium

E. Villalon
Hexcel Fabrics, France

ABSTRACT: Warp knitting with multi-axial weft insertion is a well-known technique to produce multi-layered preforms for structural composite parts. This preform is compared with a new multi-layered, multi-axial, stitched preform produced by Hexcel fabrics. This stitched multi-axial preform offers: complete flexibility in ply orientation and stacking sequence, very fine, almost gap free plies. The preforms were processed with hand lay-up and autoclaving, using epoxy. The stitched multi-axial promotes a more homogenous fibre distribution and a better surface quality in the composite. Tensile testing was performed. In the case of carbon fibre, the normalised E-modulus was the same for both. However, in the case of glass fibre, the stitched multi-axial had a slightly higher normalised E-modulus. The normalised strength was higher for the stitched multi-axial. Impact testing revealed that the stitched multi-axial carbon is more damage tolerant. The finer fibre distribution promotes fibre bundle cracking instead of inter fibre bundle cracking.

1 INTRODUCTION

Warp knitting with multi-axial weft insertion is a well-known technique to produce tailored multi-layered preforms for structural composite parts. This material technology is also named Multi-axial (by *Karl Mayer*) or Paramax (by *Liba*). Further on, it will be referred to as 'Liba'. This Liba material is compared with a new multi-layered, multi-axial, stitched preform produced by Hexcel fabrics. Further on, it will be referred to as 'stitched multi-axial'.

Liba based composites offer high mechanical loading capacity, high stiffness, reduced weight and corrosion-chemical resistance for a reasonable cost. Typical fields of application for Liba materials include: construction of means of transport (ship – automobile – air plane), machine construction, rotor blades for wind power plants, medical devices, sport and leisure items (racquets, skis, snowboards, surf-boards).

However, the Liba technology has some limitations. Due to the machine set-up, no 0° layer can be introduced at the bottom. The preform is not gap free and the plies are still rather thick. To make a higher quality fabric, the gaps must be eliminated and the ply thickness reduced. This necessitates the use of finer yarns, which increases a lot the cost of the fabric.

2 MULTI-AXIAL PREFORMS: LIBA VERSUS STICHED MULTI-AXIAL

2.1 *Preform manufacturing*

The manufacturing technique for the Liba material is shown in Figure 1 for a typical quasi-isotropic, 4 ply preform. Weft yarns are laid down, oriented layer by layer. At both sides, a continuous chain of hooks keeps the yarns straight and well in place. These chains transport the yarn layer assembly to the knitting unit, where the layers are bonded together into the final stable preform. This process realises flat filament winding in a continuous way.

The manufacturing process for the stitched multi-axial by Hexcel is shown in Figure 2. Heavy tows (12k up to 80k) are spread, combined into tape and laid down, oriented layer by layer. A belt transports the lay-up to the stitching unit, which assembles the lay-up into the final preform.

2.2 *Advantages of the stitched multi-axial preform*

Compared to Liba, the stitched multi-axial offers complete flexibility in ply orientation and stacking sequence, overcoming the limitations posed by the side chains with hooks. This technique allows the creation of a multi-axial preform with very fine plies, starting from heavy tows. The stitched multi-axial is almost gap free, see Figure 3 and Figure 4.

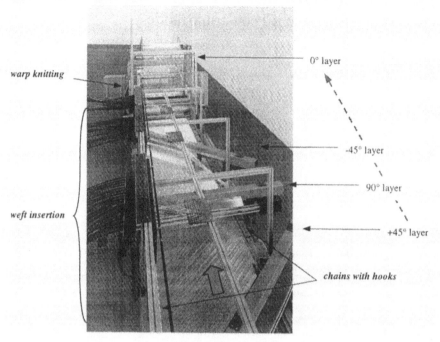

Figure 1. Warp knitting with multi-axial weft insertion *(by Karl Mayer)*

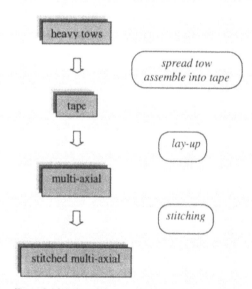

Figure 2. Manufacturing process for stitched multi-axial

Figure 3. Liba glass preform

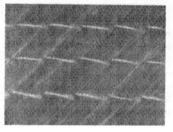

Figure 4. Stitched glass multi-axial preform

3 COMPOSITE PROCESSING

Hand lay-up and autoclaving was used to process the multi-axial preforms. This technique is known to be well suited for the production of good quality thin laminates. The curing cycle is shown in Figure 5.

The number of resin films (epoxy resin film 250 gr/m², by Hexcel) was chosen in order to obtain approximately Vf = 40% without resin being squeezed out.

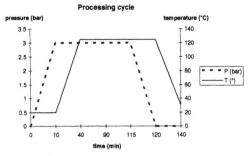

Figure 5. Processing cycle

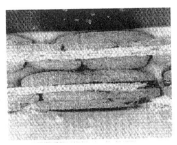

Figure 7. Liba carbon (+/- 15x)

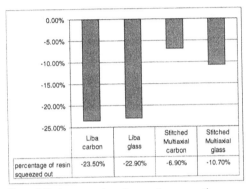

	Liba carbon	Liba glass	Stitched Multiaxial carbon	Stitched Multiaxial glass
percentage of resin squeezed out	-23.50%	-22.90%	-6.90%	-10.70%

Figure 6. Resin squeezed out for autoclave processing

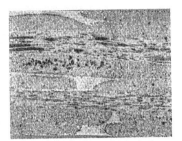

Figure 8. Stitched multi-axial carbon (+/- 15x

In Figure 6, it can be seen that the amount of resin squeezed out is 2 to 3 times higher for the Liba material. The gaps between the yarns of the Liba material act as "channels" for the resin. This higher amount of squeezed out resin results in a higher loss of resin and indicates that the in-plane permeability is higher for the Liba than for the stitched multi-axial.

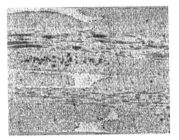

Figure 9. Liba glass (+/- 35x)

4 FIBRE DISTRIBUTION IN THE COMPOSITE

The processing conditions were the same for all preforms. The fibre distribution in the composite is shown in Figure 7, Figure 8, Figure 9 and Figure 10. The black zones in Figure 9 are surface defects, caused by small glass filaments during polishing.

The photographs show that the fibre distribution is much more homogeneous for the stitched multi-axial than for the Liba.

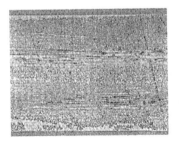

Figure 10. Stitched multi-axial glass (+/-35x)

5 SURFACE QUALITY

The surface of the stitched multi-axial samples is smoother compared to the Liba samples. The roughness profile was measured mechanically with a Taylor-Hobson apparatus (see Figure 11 for the results).

It should be noted that the plates were produced using a Teflon foil on top and bottom of the lay-up, what does not guarantee best surface quality. Using

Table 1: Elastic properties

Glass-Epoxy			E (GPa)				nu			
	Angle	Vf	exp	dev	theor	exp/theor	exp	dev	theor	exp/theor
Liba	0°	51.2%	16.2	0.6	17.63	92%	0.30	0.03	0.30	102%
	90°	51.2%	17.4	1.3	17.63	99%	0.33	0.01	0.30	110%
	67.5°	51.2%	17.6	0.5	17.42	101%	0.32	0.03	0.31	103%
Stitched multi-axial	0°	45.0%	16.4	0.3	15.63	105%	0.33	0.00	0.30	112%
	90°	45.0%	16.8	0.7	15.63	107%	0.35	0.02	0.30	116%
	67.5°	45.0%	15.2	0.1	15.44	99%	0.35	0.02	0.31	114%

Carbon-Epoxy			E (GPa)				nu			
	Angle	Vf	exp	dev	theor	exp/theor	exp	dev	theor	exp/theor
Liba	0°	52.6%	41.7	1.3	42.34	99%	0.34	0.01	0.27	124%
	90°	52.6%	43.3	2.4	42.34	102%	0.32	0.05	0.27	117%
	67.5°	52.6%	43.9	1.1	40.99	107%	0.37	0.05	0.29	126%
Stitched multi-axial	0°	40.8%	34.6	1.7	33.25	104%	0.35	0.03	0.28	127%
	90°	40.8%	33.2	1.7	33.25	100%	0.34	0.02	0.28	123%
	67.5°	40.8%	34.4	1.1	32.22	107%	0.34	0.02	0.30	114%

Table 2: Strength properties

Glass-Epoxy			S (MPa)			
	Angle	Vf	exp	dev	theor (LPF)	exp/theor
Liba	0°	51.2%	319	15	334	96%
	90°	51.2%	359	13	334	108%
	67.5°	51.2%	301	4	330	91%
Stitched multi-axial	0°	45.0%	320	5	256	125%
	90°	45.0%	361	15	256	141%
	67.5°	45.0%	329	15	175	188%

Carbon-Epoxy			S (MPa)			
	Angle	Vf	exp	dev	theor (LPF)	exp/theor
Liba	0°	52.6%	548	27	585	94%
	90°	52.6%	573	30	585	98%
	67.5°	52.6%	354	15	560	63%
Stitched multi-axial	0°	40.8%	547	25	476	115%
	90°	40.8%	538	16	476	113%
	67.5°	40.8%	394	20	468	84%

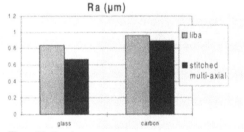

Figure 11: Roughness measurement

6 MECHANICAL IN-PLANE PERFORMANCE

6.1 Tensile testing

Tensile tests were performed for at least 5 specimens in each direction according to ASTM D3039M-93 to evaluate the in plane performance. Both a longitudinal and transverse extensometer were used. The test data were compared with theoretical values based on classical laminate plate theory, in order to compensate for the differences in Vf, see Table 1 and Table 2.

6.2 Interpretation

The theoretical values should be higher than the experimental ones, but this is not always the case. This is probably due to some deviations on the used fibre/matrix properties. The theoretical strength properties (last ply failure) are less reliable than the

special foils might allow a significantly better surface quality, but under the same processing conditions, the more homogenous fibre distribution promotes a better surface quality and might also be beneficial for the surface finishing.

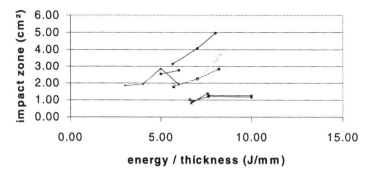

Figure 12: Impact zone versus energy/thickness

theoretical elastic properties.

Due to the straight alignment of the fibres, both materials have a high in-plane performance. For both the stitched multi-axial and the Liba, the elastic behaviour is more or less isotropic, while the strength behaviour is anistropic (more pronounced for carbon). For glass-epoxy, the elastic performance is slightly better for the stitched multi-axial than for Liba (compare 97% with 104 %); for carbon-epoxy, there is no difference. The strength performance is systematically better for the multi-axial than for Liba: for glass epoxy: Liba = 98% compared to multi-axial = 151%, for carbon epoxy : Liba = 85% compared to multi-axial = 104%.

7 DAMAGE RESISTANCE

7.1 Drop weight impact testing

Drop weight impact testing was used to characterise the resistance to damage development. The damaged zone was used as an evaluation parameter. For the glass-epoxy specimens, the damaged zone could be observed visually, as these were semi-transparent. For the carbon-epoxy specimens, the damaged zone was determined using ultrasonic C-scanning, see for example Figure 13. The results are summarised in Figure 12.

Figure 13: Damaged zone in liba carbon-epoxy specimen

7.2 Interpretation

The glass variations can absorb much more ergy than the carbon variations. Taking into acc the fibre volume fraction, the stitched multi-a glass performs slightly better than the Liba gl The stitched multi-axial carbon performs sigi cantly better compared to the Liba carbon. The ference between the pet stitching and aramid sti ing is small. Comparing the stitched multi-axial 3mmx3mm aramid stitching with 5mmx5mm c not reveal a major difference.

After impact, the Liba specimens show l꜒ cracks among the fibre bundles whereas for stitched multi-axial specimens, cracks among th꜒ bre bundles are smaller. The damaged zone is π constrained for the stitched multi-axial. In the pact zone, the stitched multi-axial specimens sl more fragmentation of the fibre bundle. Finer f distribution promotes fibre bundle cracking instead of inter fibre bundle cracking, so that the stitched multi-axial can absorb, locally, more energy and is more damage tolerant.

8 CONCLUSIONS

1. Stitched multi-axial preforms based on tape offer the following advantages:
  - complete flexibility in ply orientation and stacking sequence
  - very fine plies
  - gap free
  - cost effective

2. The in-plane permeability is lower for the stitched multi-axial than for the Liba. Stitching has a big influence on the through-the-thickness permeability.

3. The fibre distribution in the composite is more homogeneous for the stitched multi-axial compared to Liba. This promotes:
  - better surface quality
  - higher strength
  - higher damage resistance

There was no major difference in performance found for the pet stitch yarn compared to the aramid stitch yarn and also not for the 3mmx3mm stitch spacing compared to the 5mmx5mm.

Recent Developments in Durability Analysis of Composite Systems, Cardon, Fukuda, Reifsnider & Verchery (eds)
© 2000 Balkema, Rotterdam, ISBN 90 5809 103 1

Predicting the durability of PMR-15 composites aged at elevated temperatures

Kenneth J. Bowles
Polymers Branch, Materials Division, National Aeronautics and Space Administration, Glenn Research Center, Cleveland, Ohio, USA

ABSTRACT: Earlier work, which reported relationships between the compression properties and elevated temperature aging times and weight losses, (1) also pointed out the apparent influence of a surface layer formation and growth on the retention of compression properties during extended aging times. Since that time, studies have been directed toward evaluating the growth of the surface layer. This layer was found to change in its composition and features as the aging temperature changed. Microcracks and small voids initiated and advanced inward at all temperatures. Visible oxidation at the surface occurred only at the temperatures above 260 °C. Relationships between layer thickness and aging time and temperature were evaluated and empirically formulated. Then the compression properties were graphically related to the surface layer thickness with excellent correlation.

1. INTRODUCTION

Programs are underway at the Glenn Research Center to develop advanced propulsion systems for 21st century aircraft. To do this, researchers must develop predictive models for propulsion components made from polymer matrix composites that describe their durability under extreme ambient conditions. This paper is focused toward developing an engineering-based description of the thermal and mechanical durability of T650-35 graphite-fabric-reinforced PMR-15 (polyimide) composites at temperatures from 204 to 343 °C. When compression properties of T650-35 fabric/PMR-15 composites were measured at 204, 260, 288, 316, and 343 °C in a previous study, (Bowles 1997), aging times reached 26,300 hr for specimens aged at 204 °C. During aging at elevated temperatures, however, the surfaces of the specimens oxidized, and a damaged surface layer forms as the material lost weight. This layer was found to play an important role in the degradation of composite's compression properties. For the current study, particular attention was given to those chemically induced physical changes that have the most influence of the degradation of compression properties. Specimens were evaluated according to the composite's (1) thermal oxidative stability (2) compression properties, and (3) microstructural changes.

2. MATERIALS

The material that was studied was PMR-15 reinforced with T650-35, 24 by 23, eight-harness-satin-weave graphite fiber fabric. The aged specimens measured about 11 by 9 cm in length and width by either 4, 8 or 20 plies in thickness (nominally 1.27, 2.77 and 6.73 mm respectively). These dimensions were chosen to provide nominal cut-edge to total-surface-area percentages of 3, 5 and 12 percent, where the total surface area consisted of both cut and molded surfaces. The molded surfaces were those that were in contact with the metal mold or vacuum bag during the curing process. The materials were processed at the GE Aircraft Engine Inc., in Evendale, Ohio, USA.

3. TESTING

The specimens were aged in air-circulating ovens at temperatures of 204, 260, 288, 316 and 343 °C and an air flow maintained at 100 cm³/min. The laminates were removed periodically, allowed to cool in a desiccator, weighed, and either returned to the oven or permanently removed for testing. The aging time was considered completed when the weight loss exceeded 10 percent.

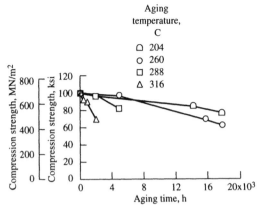

Figure 1. Compression strength of T650–35/PMR–15 composite specimens as a function of aging time at various temperatures. Number of plies, 20.

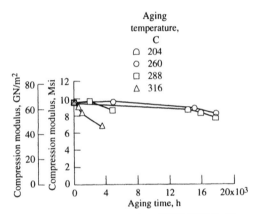

Figure 2. Compression modulus of T650–35/PMR–15 composite specimens as a function of aging time at various temperatures. Number of plies, 20.

All specimens were conditioned at 125 °C for 16 hr before compression tests were performed. All compression tests were performed as specified in ASTM *Test Method for Compressive Properties of Rigid Plastics* (ASTM D-695M), with a cross-head speed of 1.2 mm/min, a temperature of 23.3 °C, and a relative humidity of 50 percent. No end tabs were used. Strain was measured with an extensometer, and moduli were measured using strains and loads at 500 and 1500 microstrain. Surface layer thickness were measured from photomicrographs of sectioned specimens using differential interference contrast to accentuate the visibly undamaged core. The final polishing medium was 0.05 mm colloidal silica emulsion.

4. RESULTS

Selected specimens were removed from the aging ovens for compression testing at different times during the aging periods. Figures 1-2 (Bowles, Roberts and Kamvouris, 1995) show strengths and moduli, respectively, of the 20- ply specimens plotted against aging time. When the ordinate variable is aging time, all the relationships will appear to be separate linear curves with a different slope for each temperature. However, the data from the 204 and 260 °C tests appear to be almost identical. When percent weight loss is the independent variable, (Fig. 3) all the data except those for the specimens aged at 204 °C appear to collapse onto a single curve with a relationship $\ln S_C = 4.614\text{-}10.259\times10^{-2}\,w$, where S_C is the compression strength in MN/m^2 and w is the percent weight loss. (Details are given in Bowles et al. 1995). Neither of these two relationships, percent weight loss or aging time, produce one weight loss curve that accommodates the data at all the temperatures that were studied. The data from Figure 1 indicate that the PMR-15 composite material will not retain its strength very long at

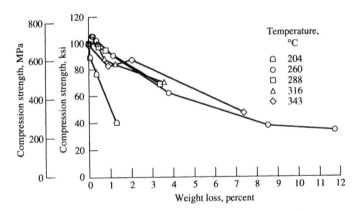

Figure 3. Compression strength of T650-35/PMR-15 composite specimens as a function of percent weight loss at various temperatures.

442

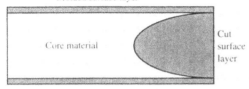

Figure 4. Microcrack and oxidation layer growth during isothermal aging of T650–35/PMR–15 composite specimens.

temperatures over 260 °C. The initial moduli values appear to be retained for longer periods of time at the lower temperatures (Fig. 2).

Figure 4 shows a schematic of the surface damage growth during the aging period. The depth of cut surface damage increased with increasing specimen thickness. Because of this, weight loss data cannot be compared for specimens of different thicknesses. Also, weight losses from cut surfaces exceeded those from the molded surface (Bowles & Meyer 1986, and Bowles & Kamvouris 1995).

Two types of surface degradation occur in these composites. Aging at the higher temperatures (316 to 343 °C) produces a light-colored surface layer that grows inward and voids and microcracks initiate and grow within the layer, as in Figure 5a. The light color is attributed to the formation of solid oxidation products at the higher temperatures. During aging at the lower temperatures (Fig. 5b), specimens show the same advance of voids and microcracks into the surface, but the oxidized light band of matrix material is not visible. The two degradation mechanisms that are operating during isothermal aging are surface oxidation and bulk thermal degradation. Results of compression testing of composite layers that were machined parallel to the molded surface layer show that after aging was completed at 204 °C for 26,300 hr the compression strength of the visibly damaged layer was one half that of the visibly less damaged central core material (Bowles McCorkle and Inghram, 1998). This leads one to believe that the growth of the cracked surface layer contributes to the degradation of the mechanical properties of PMR-15 composite material. It was observed that the central core also experienced a substantial loss in compression strength.

Figure 6 shows the relationship between the thickness of the resin-rich surface damage layer and the aging time at all temperatures. The relationships appear to be linear at all five temperatures, with slower growth rates at the lower temperatures. The data from the two lower temperature tests indicate what may be an initial fast rate of growth and then a slower steady rate after 1000 hr of aging. This may be normal scatter, however. One item of interest is that these linear curves appear similar to the compression strength curves in Figure 1.

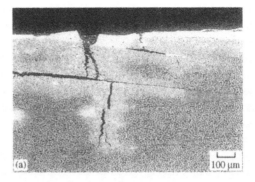

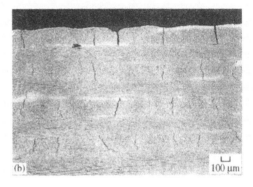

Figure 5. Surface oxidation of T650–35/PMR–15 composite specimens aged in air. (a) 1000 h at 316 °C. (b) 10 000 h at 204 °C.

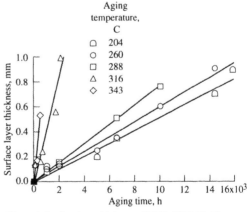

Figure 6. Surface layer thickness of T650–35/PMR–15 composite specimens as a function of aging time at various temperatures.

Figure 7 presents compression strength, plotted as a function of layer thickness of the composite at various temperatures. Each figure contains data for one specimen thickness: 1.50, 2.77 or 6.78 mm (4, 8, or 20 plies re-

443

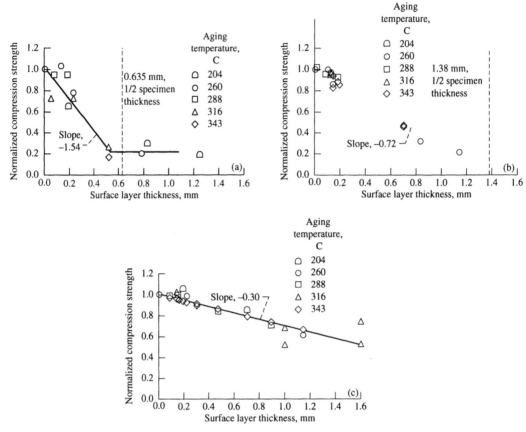

Figure 7. Compression strength of T650–35/PMR–15 composite specimens as a function of surface layer thickness at various temperatures. (a) Number of plies, 4. (b) Number of plies, 8. (c) Number of plies, 20.

spectively). All the data for the two thicker specimens (Figs. 7b, c) fall on one curve. The compression strength of the inner, crack free core material specimens that were machined parallel to the molded surfaces of a large specimen decreased in strength by a considerable amount (as much as 35 percent) (Bowles, McCorkle and Inghram, 1998). The measured strength of the core material was close to that of an aged specimen with surface degradation and the same initial specimen thickness. These data indicate that the formation and growth of the surface layer does not significantly reduce the compression properties of 8- and 20-ply fabric reinforced composites. Figure 7a presents the data for the 4 ply specimens. The scatter in the data appears to be greater than that of the 8 and 20 ply specimens. Two important items stand out. The initial half thickness of the 4-ply specimens is nominally 0.635 mm. When the surface layer thickness reaches this value, the entire cross section of the specimen is damaged and consists of surface layer material. The second observation is that the normalized compression strength seems

to reach a minimum at 20 to 25 percent. The surface layer thickness values to the right of the half-thickness line are taken from Figure 6. The data in Figure 7b extend to almost 1.2 mm of surface layer thickness. The half-thickness is 1.38 mm. It appears that when the entire cross section is damaged material, the residual compression strength is 20 percent when normalized. It would be logical to assume from the data in Figure 7a that the strength would remain at this level with increasing surface layer thicknesses. Figure 8 shows the moduli as a function of the layer thickness the modulus data shown in Figure 8a are grouped into two separate curves. In Figures 8b, c the modulus data collapse onto one linear curve. No lower bounds are observed at the half-thicknesses in these figures.

Estimates of structural lifetimes for PMR-15 composites reinforced with T650-35 fabric, as shown in Figure 9, can be made using Figure 6 in conjunction with a plot like those shown in Figure 7. Accelerated testing may be possible as shown in Figure 10, which is an adjustment of Figure 7b. Once a straight line relationship can be assumed

444

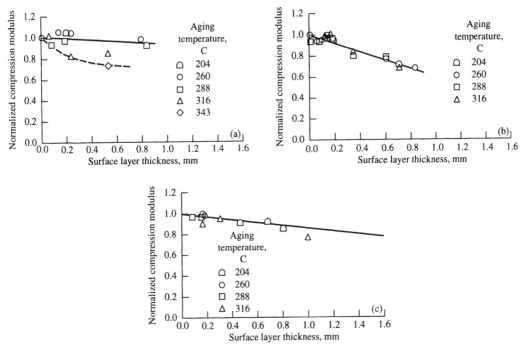

Figure 8. Compression modulus of T650–35/PMR–15 composite specimens as a function of surface layer thickness at various temperatures. (a) Number of plies, 4. (b) Number of plies, 8. (c) Number of plies, 20.

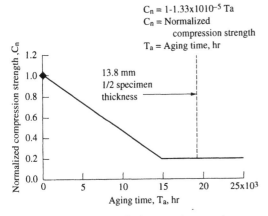

Figure 9. Composite normalized compression strength as a function of aging time at 288 °C. Number of plies, 8.

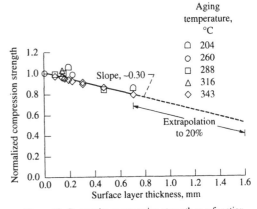

Figure 10. Composite compression strength as a function of surface layer thickness at various temperatures. Number of plies, 20.

with a minimum amount of data points, it can be extrapolated to the desired duration or minimum strength. This should give a reasonably close estimation.

5. SUMMARY AND CONCLUSIONS.

The results of this study indicate that simple, linear relationships exist between the compression properties of the T650-35 graphite fiber fabric/PMR-15 composites and the depth of the surface layer that develops and grows during periods of aging at elevated temperatures. The buildup of the surface layer is indicative of the physical condition of the fabric reinforced PMR-15 composites at all temperatures that were studied. However, although the surface layer is indicative of the decrease in strength, the central core volume is the main contributor.

Specimen thickness is a significant factor in compression properties during such periods of exposure. It is apparent from viewing Figures 7b, c that the influence of the surface layer diminishes as the composite thickness increases and the core strength becomes more important. The strength data in Figure 7a, from the 4-ply specimens aged at 316 and 343 °C, are below those measured at the other three temperatures. As noted earlier, the surface layer for a specimen aged at 316 °C had a compression strength about half that of the core material after aging at 204 °C for 26,300 hr. Thus, for specimens that had a significant amount of oxidative attack in the surface layer, thinner specimens should show lower strengths than those aged at lower temperatures. The minimum normalized compression strength that is attained, when the complete cross section of the specimen is composed of surface layer material, appears to be around 20 to 25 percent. That is what we see in Figure 8.

These results can be used to produce empirical relationships between aging time and compression properties for T650-35/PMR-15 composites of varying thicknesses. There is also the potential for estimating composite lifetimes using a shorter, accelerated test procedure.

REFERENCES

Bowles, K.J and Meyers, A., Specimen Geometry Effects on Graphite/PMR-15 Composites During Thermal Oxidative Aging," *32st International SAMPE Symposium and Exhibition*, (1986), 1285.

Bowles, K.J., Roberts, G.D., and Kamvouris, J.E. 1995. Long-term Isothermal Aging Effects on Carbon Fabric-reinforced PMR-15 Composites: Compression Strength. Also ASTM STP-1302, High Temperature and Environmental Effects on Polymeric Composites; 2nd Volume. American Society for Testing and Materials, 1997, pp. 175–190.

Bowles, K.J., and Kamvouris, J.E., 1995. Penetration of carbon-fabric-reinforced composites by edge cracks during thermal aging. *J. Advanced Materials* 26(2):2–11.

Bowles, K.J., McCorkle, L., and Inghram, L., "Comparison of Graphite Fabric Reinforced PMR-15 and Avimid N Composites After Long Term Isothermal Aging at Various Temperatures, NASA TM—1998-107529, February 1998.

Recent Developments in Durability Analysis of Composite Systems, Cardon, Fukuda, Reifsnider & Verchery (eds)
© 2000 Balkema, Rotterdam, ISBN 90 5809 103 1

Ultimate bearing capacity of cement composite sections under long term physico-chemical exposure

T.Z. Błaszczyński
Poznań University of Technology, Poland

ABSTRACT: Bearing capacity of transverse concrete composite sections subjected to bending and eccentric compression with relation to oil influence as the physico-chemical exposure is evaluated. Two oil conditions are considered: nonoiled (dry) and oiled after 12 months of influencing. Ultimate bearing capacity of reinforced sections influenced by crude oil products is defined numerically by the limit state interaction diagrams. The analysis is carried out for rectangular section with tension and compression reinforcement. In the case of oiled condition bearing capacity of a section decreases up to 50%, when compared with the nonoiled state.

1 INTRODUCTION

The cement composite (like RC) sections bearing capacity is presented in many papers. In most cases it concerns an analysis of bearing capacity without taking into account the service conditions of the element. Properties of concrete, such as strength and deformability depend on various environmental influence, e.g. temperature, humidity or aggressive agents. The bearing capacity of RC sections therefore depends not only on the strength and deformation parameters of concrete, but is also influenced by environmental factors.

The article presents an evaluation of crude oil products like mineral oils influence on bearing capacity of RC sections subjected to bending or compression with bending. This main issue is solved using limit state interaction diagrams for the sections. Interaction diagrams have been established for oiled and nonoiled state of concrete.

2 CRUDE OIL PRODUCTS (AS PHYSICO-CHEMICAL AGENTS) INFLUENCE ON CONCRETE STRENGTH AND DEFORMABILITY

Long term laboratory experiments have been conducted to assess the changes of physico-mechanical characteristics of oil contaminated concrete. The compressive strength was determined from 100 mm cubes for concrete type B-25 (29,8 MPa) as most commonly used for industrial RC structures in Poland, the average 28 day compressive strength of concrete is shown in brackets. The water-cement ratio was 0,59 and aggregate-cement ratio was 6,70.

Concrete was impregnated with the most commonly used industrial oils of different kinematic viscosities namely turbine oil TU-20 (81 mm²/s), machine oil M-40 (211 mm²/s) and hydraulic oil H-70 (383 mm²/s). These oils have low neutralisation numbers with values between 0,05 and 0,75 mg KOH/g. The oils was first applied to concrete 2 months after casting, subsequently the specimens were examined every 4 or 12 months during total period 72 months. The control specimens (samples) were additionally examined after 28 days and 2 months.

The results of the experiments on concrete compressive strength f_c are presented on figure 1.

The results clearly show, that as a result of the influence of the oils used, the significant decrease of f_c

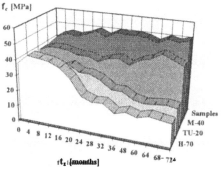

Fig 1. Variation of concrete B-25 compressive strength during the period of exposure to H-70, TU-20 and M-40 oils.

with oil contamination. Comparison of the influence which the oils had on the compressive strength shows that oils TU-20 and H-70 decrease f_c the most.

Contamination of concrete by hydrocarbons gives an almost new material, which behaves differently. The results of the stress (σ) - strain (ε) relation in dry state and after 12-months of oiling by mineral oil TU-20 for concrete B-25 in function of the longitudinal strains are different (Fig. 2.). The non-linear behaviour of strength and strain variations depends on the contents of hydrocarbon and its type. It can be noticed that the strain ε_R, corresponding to the maximum stress, is lower for oil saturated concrete then for dry concrete. On the other hand, the maximum limit strain ε_m is correspondingly higher.

3 THE METHOD OF SOLUTION

The influence of mineral oils on the bearing capacity of RC sections was assessed by the limit state interaction diagrams.

The following assumptions were adopted in calculations of the bearing capacity:
- plane sections remain plane,
- strains of tension and compression reinforcement are the same as strains of the surrounding concrete,
- for section under axial compression the concrete compressive strain is limited to ε_R,
- for sections compressed in part, the limit concrete strain is assumed to be ε_m,
- the tensile strength of concrete is neglected,
- the σ-ε relation for concrete follows the Saenz function (fig. 3.a.),
- the σ-ε relation for steel is based on the following formulae (fig. 3.b.),

$$\sigma(\varepsilon) = E_a \cdot \varepsilon_a \text{ for } \varepsilon_a < \varepsilon_o$$
and
$$\sigma(\varepsilon) = f_{ak} \text{ for } \varepsilon_o \leq \varepsilon_a \leq \varepsilon_k ,$$

Limit state interaction diagrams for concrete sections have been defined using the equilibrium conditions at limit bearing capacity state.

The (M, N) co-ordinates of the interaction diagram have been defined from equilibrium conditions, taking advantage of Gauss quadratic formula in numerical integration. According to graph showing the strains, presented in fig. 4, appropriate ranges of co-ordinates were adopted for the limit state diagram curves.

The (M, N) co-ordinates were approximated using n-th order polynomial. The order of the polynomial was chosen in such a way, that the discrepancy in the approximation did not exceed 0,1%.

The numerical analysis of the bearing capacity of a section using the limit state interaction diagrams was

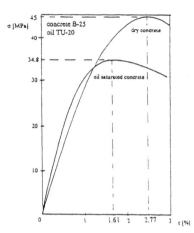

Fig 2. σ-ε diagrams for dry and oiled by oil TU-20 concrete B-25.

a/ for concrete

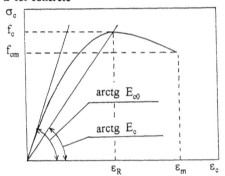

b/ for steel

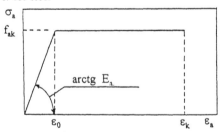

Fig 3. σ - ε diagrams for concrete and for steel

carried out for nonoiled and contaminated state. In order to compare the influence of oiling on the bearing capacity of RC section, presented in the form of limit state interaction diagram, the discrepancy R_w was introduced between the bearing capacity areas defined for concrete in dry and oiled state (fig. 5.). It follows that,

$$R_w = \frac{r_2 - r_1}{r_1} 100$$

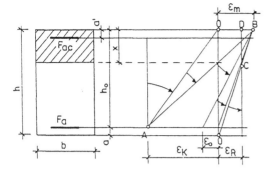

Fig. 4. Strains in concrete section at bearing capacity.

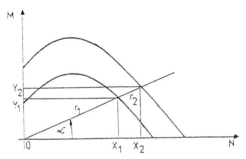

Fig. 5. Discrepancy between the interaction curves.

4 NUMERICAL ANALYSIS

The numerical analysis of RC section bearing capacity was carried out for oiled and nonoiled states.

The coefficients of the Saenz function were find by the numerical calculation using APROKS software. In order to find the values of the limit strains ε_m, the ratio $\varepsilon_R/\varepsilon_m$ was taken to be 0,75 for nonoiled concrete and 0,5 for oiled concrete according to.

Therefore (from fig. 2.):

– for oiled B-25 concrete (after 12 months of oiling)

$$\varepsilon_m = \frac{\varepsilon_R}{0,9} = \frac{0,00161}{0,5} = 0,00322$$

– for nonoiled B-25 concrete (in the same age)

$$\varepsilon_m = \frac{\varepsilon_R}{0,9} = \frac{0,00277}{0,75} = 0,00369$$

Rectangular concrete section with tension and compression reinforcement was examined by numerical analysis. Geometrical parameters of the section are given in fig. 6. The analysis was carried out for two extreme values of total reinforcement percentage ($\mu_{max} = 6\%$, $\mu_{min} = 0,4\%$) and for two classes of reinforcing steel ($f_{ak} = 220$ and 360 MPa).

An example diagram of the limit state interaction curves for both states is shown in fig. 6. This figure refers to the section with steel of $f_{ak} = 360$ MPa and a maximum percentage of reinforcement.

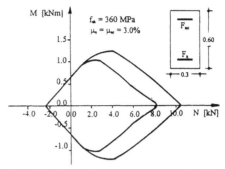

Fig. 6. Limit state interaction diagram for RC section with symmetrical reinforcement and to extreme cases of oiling in concrete B-25.

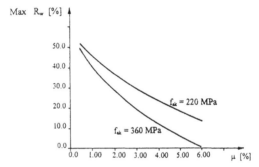

Fig. 7. Discrepancy between the limit state interaction diagrams for section with symmetrical reinforcement of $f_{ak} = 220$ MPa at two extreme percentages of steel area.

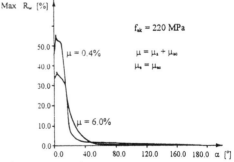

Fig. 8. The maximum discrepancy values in concrete B-25 section with symmetrical reinforcement, subjected to bending as a function of the total reinforcement percentage.

The distribution of discrepancy R_w (in function of angle α - as shown in fig. 5.) between the areas of bearing capacity of oiled and nonoiled states, for steel with $f_{ak} = 220$ MPa and two extreme percentages of reinforcement is presented in fig. 7. As come from fig. 5. angle $\alpha = 0°$ refers to axial compression, $\alpha = 90°$ refers to bending and $\alpha = 180°$ refers to an axial tension of the section.

In fig. 8. the discrepancy R_w is shown as a function of the total percentage of reinforcement in the section subjected to bending ($\alpha = 90°$) with symmetrical tension and compression reinforcement.

5 CONCLUSIONS

As the presented numerical analysis proves, the limit state interaction diagrams are determined by oiling in concrete. The loss of bearing capacity of RC section using concrete in mineral oil environment extends up to 50%, for high contribution of shear force and the lower percentage of reinforcement. The discrepancy rises also with the lowering of steel strength.

Important lowering of bearing capacity of RC oiled sections points to the necessity of taking this fact into account in the design and asses process.

REFERENCES

Nurmaganbetov, E.K. 1991. Evaluation of RC bars strength subjected to bending in normal cross-sections (in Russian). *Beton i Żelezobeton*, 3: 18-19.

Basista, M. and Mames J. 1981. A study of eccentrically compressed concrete sections (in Polish). Arch. Inż. Ląd., vol. 27, No 2: 233-247.

Sulimowski, Z. 1984. Vector representation of concrete sections bearing capacity (in Polish). Arch. Inż. Ląd., vol. 30, No 2-3: 515-533.

Błaszczyński, T. 1994. Durability analysis of RC structures exposed to a physico-chemical environment. *Proceedings of the Third International Conference on Global Trends in Structural Engineering*, Singapore: 67-70.

Saenz, L.P. 1964. Equation for the stress-strain curve of concrete. *Journal of ACI*, vol. 61, No. 9: 1227-1239.

EUROCODE 2, *Design of concrete structures*, vol. 1, Polish version, ENV 1992-1-1, 1991.

Godycki-Ćwirko, T. 1982. *Concrete mechanics*. Arkady, Warszawa.

Kozaczewski, J. 1987. *The influence of mineral oils on some physico-mechanical concrete features*. PhD Thesis, Poznań University of Technology, Poznań.

Recent Developments in Durability Analysis of Composite Systems, Cardon, Fukuda, Reifsnider & Verchery (eds)
© 2000 Balkema, Rotterdam, ISBN 90 5809 103 1

Hybrid composites with nonlinear behaviour for reinforcing of concrete

V.Tamuzs
Institute of Polymer Mechanics, University of Latvia, Riga, Latvia

R.Tepfers
Chalmers University of Technology, Göteborg, Sweden

ABSTRACT: Composite rods as reinforcement of concrete are used in small amount during the last decade. The nonlinear ductile behavior of such rods at extreme loading level is desirable for safety reasons. Such nonlinear behavior can be obtained by designing the appropriate hybrid composites. Nonlinearity of hybrid composites is caused by extended damage accumulation (fragmentation of brittle fibers) before the final failure. In the paper the nonlinear hybrid composite deformation and failure is analyzed theoreticaly and experimentaly as well. The failure of concrete beam reinforced by hybrid composite rods is investigated.

1 UP-TO-DATE SITUATION IN PRODUCTION, CONSUMPTION AND APPLICATIONS OF ADVANCED COMPOSITES

Commercial production of advanced fibers for composites started substantially in the early 1970s. At that time various reinforcement fibres were developed, namely boron, carbon from viscose and PAN, aramid, alumina, silica, etc. One of the typical sort of fibers are carbon fiber. Since specific elastic modulus and specific tensile strength of carbon fibres from PAN and mesophase pitch are very high, these fibres have gained an advantageous position as reinforcements of advanced composite materials. For the structural parts of aircraft, spacecraft, sporting goods carbon fibres have gained a majority share. According Matsui (1993) current world production capacity of carbon fibers in 1993 was about 10,000 tons per year (50% Japan, 30% USA and 15% Europe). Consumption was 6500 tons per year (35% USA, 23% Japan, 20% Europe and 23% the other countries). In 1986 to 87 aeronautic use counted for 42-50% versus sporting goods 40-37%. The world-wide recession and a cutback of military spending in the USA and Europe hit the aircraft industry, resulting in aeronautic use going down to 20% and the share of sporting goods increasing up to 55% in 1993 (Matsui (1993)). In early nineties the application of fiber composites in infrastructure started to increase. The main consumption of advanced composites in civil engineering is the repairing and retrofit of old and damaged concrete structures by external reinforcement. It can be estimated (Ueda, pers. comm.) that in Japan in late nineties the amount of carbon composites used in concrete industry equaled 300 tons per year. The dynamics of composite amount used in civil engineering in Swiss is given by Meier (1996) Figure 1. Besides of external reinforcement also some attempts have been done to replace the traditional concrete steel bar reinforcement by composite bar reinforcements.

2 COMPOSITE BARS AS REINFORCEMENT FOR CONCRETE

At present some comercial available products of fiber reinforced plastic (FRP) bars as reinforcement of concrete can be mentioned: CFCC (Tokyo Rope Carbon Fiber Composite Cable), Arapree - aramid reinforced composite bars, Technora-aramid reinforced bars, unidirectional glass fiber composite rods etc.

The main advantage of composite reinforcement consists in its resistance to corrosion and consequent durability in contrary to steel. The light weight and

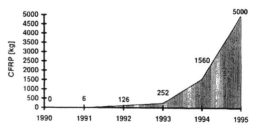

Figure 1. Use of CFRP-Laminates for strengthening purposes of concrete in Switzerland.

transparence to electromagnetic waves also can be desirable in specific cases.

The main disadvantages are the follow:
1 high material price (carbon and aramid)
2 creep (aramid fibers)
3 sensibility to alkaline attack (glass fibers)
4 low ultimate strain (especialy carbon fibers)
5 low modulus (aramid and glass fibers)
6 linear behavior till the failure without extended plasticity.

Some of these undesirable features can be reduced by using the hybrid composite rods: carbon + glass or carbon + aramide fibers: a) increase of modulus comparably with aramide or glass composite, b) reduction of price comparably with carbon composite rods, c) appearance of nonlinear "ductile" behaviour at high stress level.

Ductility of concrete structures is necessary due to safety requirements to avoid brittle fracture. In traditional steel reinforced concrete structures the ductility is guaranteed by ductile behaviour of cold worked reinforcing steel.

Classical stress strain diagram for steel is shown on Figure 2. A fictitious yield stress is stipulated as the stress $\sigma_{0.2}$ at which the permanent elongation after removing the load is 0.2%. Further the steel must have a strain at maximum load of at least 3.0%. Moreover there should be a load increase between the $\sigma_{0.2}$ and the stress at maximum σ_{su}. The relation between $\sigma_{su}/\sigma_{0.2}$ should be at least somewhere in between 1.05 and 1.10 according to code requirements in different countries. It is desirable to about reproduce for FRP this stress-strain relationship.

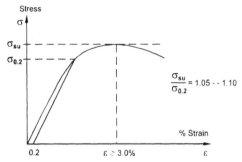

Figure 2. Stress-strain diagram for cold worked steel.

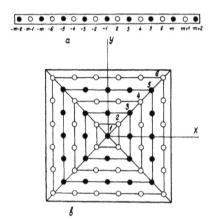

Figure 3. Schematic of unidirectional hybrid composite (a) single layer and (b) multilayer.

3 ESTIMATION OF MODULUS AND NONLINEARITY OF HYBRID COMPOSITES

Let us analyze the behaviour of unidirectional hybrid composite, where two sorts of fibers with different modulus, ultimate elongation and strength properties are mixed.

The initial modulus E_c of hybrid composite with great accuracy can be expressed by simple rule of mixture

$$E_c = E_{Hf}\mu_H + E_{Lf}\mu_L + E_m(1 - \mu_H - \mu_L) \qquad (1)$$

where indices H and L denote the high and low modulus fibers, m = matrix; f = fibers; and μ_i is the volume fraction of fibers.

Let ultimate deformation $\varepsilon_H < \varepsilon_L$. At $\varepsilon = \varepsilon_H$ the H fibers start to break transferring the load

$$N = \mu_H E_H \varepsilon_H + \mu_L E_L \varepsilon_H$$

to L fibers. It is necessary that stress in L fibers after breaking process of H fibers and nonlinear deformation don't exceeds the ultimate strength of L fibers allowing to load slightly increase (up to 20%)

analogicaly as it is in Figure 2 for steel.

Therefore

$$\mu_H E_H \varepsilon_H + \mu_L E_L \varepsilon_H = 0.8 E_L \varepsilon_L \mu_L \qquad (2)$$

So the desirable volume fraction ratio μ_H / μ_L depends on the ratio of fiber moduli and fiber strength:

$$\frac{\mu_H}{\mu_L} = \frac{E_L}{E_H}\left(\frac{0.8\varepsilon_L}{\varepsilon_H} - 1\right) = 0.8\frac{\sigma_L}{\sigma_H} - \frac{E_L}{E_H} \qquad (3)$$

Of course it is very rough estimation and more detailed analysis of nonlinear behaviour of hybrid composite is based on the statistical model of fiber fragmentation in hybrid composites Gutans & Tamuzs (1987).

The main assumption used in Gutans & Tamuzs (1987). was that hybrid composite has been formed by well mixtured fibers of low and high elongation (high modulus (*HM*) and low modulus (*LM*) respectively) Figure 3.

For fracture process prediction it was necessary to know besides of elastic constants the fiber strength statistical distribution for both fibers and their inef-

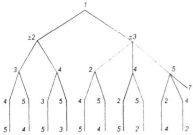

Figure 4. Tree graph T pattern of probable fiber failure sequence.

fective lengths l_H and l_L. It was assumed, that strength distribution follows the Weibul's rule for both types of fibers. Stress concentration coefficients in fibers around the defects have been calculated in Gutans (1985) and for particular case referred in Gutans & Tamuzs (1987).

The breaking sequence of HM and LM fibers mixed regularly as shown on Figure 3 can be very different and depends on strength and stiffness properties of fibers. The sequence of fiber breakage can be illustrated by tree graph Figure 4 Gutans & Tamuzs (1987).

The numbers on the graph denote the corresponding fiber failure. The odd and even numbers refer, respectively, to the HM and LM fibers. Assuming that the HM fiber fails first it corresponds to the root of the tree. Further the n - 1 lines on the tree show the different fiber failure sequence. The extreme right-hand side yields the sequence 1, 3, 5 ... i.e total HM fiber breaking before LM fibers start to break. In Gutans & Tamuzs (1987) the algorithm of the most probable failure sequence seeking was formulated and particular numerical results were obtained. For our aims it is important to note, that the damage level of H fibers obtained in hybrid composite was 20 times more high than the damage level calculated in regular H fibers composite (8.3% and 0.47% respectively). The result obtained is quite evident because of different ultimate elongation of fibers in hybrid composite. The ultimate damage level can be even 2-3 times more high depending on the ratio of properties of both fibers. It means that in the ideally organized (mixtured) hybrid composite the high modulus

fibers will split into fragments with mean length equal approximately 5-10 "ineffective lengths" i.e. 100-500 diameters of fibers.

The reducing of modulus because of arosed damage can be calculated by two approaches. The first aproach consists in taking the virgin composite modulus as an initial one and calculating the reduction of modulus as a result of arising of small cracks in transverse isotropic body Grushetsky et al. (1982). Another possibility consists in taking the matrix modulus as initial and calculating of the modulus of material by reinforcing the matrix with the system of straight parallel fiber fragments Maksimov & Kochetkov (1982).

However in practise such sofisticate approach is rather useless. The Weibul's statistics for fibers are not known and what is even more important the mixture of fibers in real composite rod is far away from that on Figure 3. The rod consists on rather irregularly mixtured bundles of both sorts of fibers. Strongly speaking the new theoretical model of fracture, damage and nonlinear stress strain behaviour in such composite should be elaborated. For practical needs the elementary formula (3) can be used, which corresponds to extreme right-hand branch of fiber failure tree graph Figure 4.

4 EXPERIMENTAL RESULTS

Two kinds of hybrid unidirectional composites were produced: carbon-aramid-polyester and carbon - glass-polyester Apinis et al. (1998). The raw materials used for specimens were "Toray" carbon fiber, T400HB-6K-40D produced in Japan; aramid fiber ZSVM-300, produced in Russia; E-glass fiber produced in Valmiera, Latvia; and polyester resin NORPOL440M850, made in Norway. The characteristics of materials are given in table 1.

The columns E_{ref}, σ^*_{ref} ε^*_{ref} are values of modulus, ultimate stress and strain of fibers according the data in literature. The columns E, σ^*, ε^* are data obtained on direct measurement of flat composite coupons with known fiber content. These data were used in our calculations.

Carbon-aramid rods were made of T400HB carbon fiber and Russia-made ZSVM - 300 aramid fiber.

Table 1

Type of fiber	E	E_{ref}	σ^*	σ^*_{ref}	ε^*	ε^*_{ref}
	GPa	GPa	GPa	GPa	%	%
E-glass	75.0	72-75	1.84	1.25-3.8	2.64	3-4.8
Aramid ZSVM	106	115-135	2.70	3.8-4.2	3.29	3-4
Carbon fiber T400HB	244	250	2.93	4.49	1.21	1.8

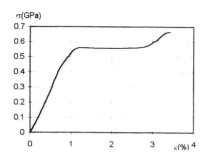

Figure 5. Carbon-aramid hybrid composite σ - ε diagram.

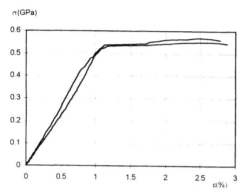

Figure 6. Carbon-glass hybrid composite σ - ε diagram.

The required proportion of the fiber kinds was found using (3), which yields: 76% aramid and 24% carbon. Total volume fraction of fibers in the rod was μ_{tot} = 0.36. Cross section of rod was 195 mm² (diameter 16 mm).

Figure 5 shows σ - ε curve obtained on carbon-aramid specimen in tension and it is seen, that desirable ductility is achieved.

According to calculations, the initial modulus of the material is E = 51.7 GPa. Experiments give E = 52 GPa. The calculated value of the modulus in the final stage of loading is 22.4 GPa (the modulus of the aramid fiber used decreases as strain grows so that just before rupture it is about 75% of the initial value; this effect is taken into account here). Experimental value is $\approx$ 20 GPa. Therefore, the experiment confirms that the wanted hybrid effect was attained. The ultimate load capacity was 17 tons. It is worth to note that this nonlinear effect was achieved only on specimens with relatively large diameter, (16 mm) and absent on thin unidirectional hybrid specimens. In such specimens the failure onset was observed just after the first break of carbon fiber bundle. Increasing the specimens cross-section decreases a ratio of the fiber bundle cross-section to the total specimen cross-section and improves homogeneity of fibers mixing.

Glass fiber is much cheaper than aramid fiber, which makes prospective its using for concrete reinforcing very attractive and motivates research in hybrid rods with E-glass fibers instead of aramid. Calculations using (3) show that such a material must contain 83% glass fiber and 17% carbon fiber. Hybrid composite specimens containing 80% E-glass fiber and 20% T400HB carbon fiber were made and tested. Total volume fraction of the fibers was μ_{tot} = 0.47. The tension tests were carried out at a constant loading rate of 20.8 N/s. Figure 6 shows the results of these experiments.

It is seen that also this hybrid composite has a plateau. 3% strain was almost reached in these first experiments and the stress growth before the rupture was about 3% to 5%.

The creep tests of hybrid composites are still in progress, but for carbon-glass hybrid during 3 months no creep was observed under load equal 0.7 of the static strength.

5 FLEXURAL ROTATIONAL CAPACITY CONCRETE BEAM TEST

It is of interest to study the pseudoductile behavior in concrete of developed hybrid CFRP + GFRP rod. Measuring the flexural rotational capacity of concrete beam with hybrid rods as tensile reinforcement can do the study.

Flexural rotational capacity of a concrete beam depends on many factors. Among these are, yield capacity of the reinforcement, the ultimate elongation and load increase between the load when yielding starts and ultimate failure. Further the bond of the reinforcement to concrete has importance, because to good bond does not allow reinforcement parts between the cracks to be engaged in yielding. The rotational capacity is also dependent on the performance and deformations in the concrete compressive zone. Thus the whole flexural rotational capacity problem is very complex, and the test reported father is only the preliminary one.

Rotational capacity tests were performed using four point bending. The scheme of beam and layout of displacement gauges for measurement the deflection are shown in Figure 7.

The beam had 2 hybrid composite rods (HCR) as tensile reinforcement.

The detailed analysis of results is presented in (Apinis et al., in press). Here we shortly note only, that measuring of deflections and strain in compressed zone allow to estimate the strain level in HCR in concrete beam, assuming the linear strain distribution across the beam.

In the Figure 8 the results of estimated strain level of HCR in concrete beam are summarized. It is seen that all methods (by using curvature of deformed

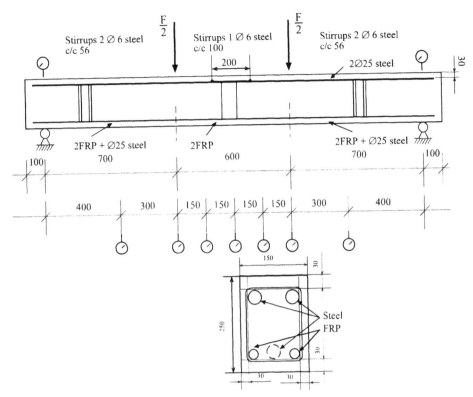

Figure 7. Beam reinforcements and the measuring devices.

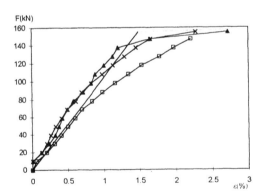

Figure 8. Calculated tensile strain of hybrid composite rods in central part of beam:
———▲——— Result of calculation using measured compression strain and curvature of beam;

———×——— Result of calculation using inclination angle and curvature of beam;

———□——— Result of calculation using measured compression strain and neutral axis position from the equilibrium of longitudinal stresses;

——— ..... Idealized HCR stress strain curve according Figure 6 and equilibrium of external and internal moments in beam.

beam, inclination angles of beam ends and compression strain in the intact part of concrete) lead to the similar nonlinear HCR deformation curve, showing the maximum strain in HCR up to 2.5% After testing the HCR were extracted from concrete beam and polyester matrix was burned out at temperature $400°C$ during 4h. The carbon fiber fragmentation was revealed only at the ends of working part of beam near the loading points, where the rod failed due to some stress concentration. Therefore for more detailed and comprehensive analysis of hybrid composite behaviour as reinforcement of concrete the performing of additional tests is necessary.

6 CONCLUSIONS

1. It is possible by appropriate choose of hybrid composite fibers to obtain a nonlinear pseudo-ductile behaviour of hybrid composite rods till the ultimate strain 3%.
2. The fiber mix proportion, mixing uniformity, size of fiber bundles, resin impregnation procedure and size effect of rod are very important to achieve a desirable nonlinear performance of hybrid rods.

3. Carbon-glass composite rods having carbon fibers up 20% of total fiber amount don't reveal the tensile creep deformation at stress level 0.7 of static strength.

4. The ductile behaviour of hybrid rods in concrete structures should be investigated more completely.

REFERENCES

Apinis, R. et al. 1998. Ductility of hybrid fiber composite Reinforcement FRP for concrete. *Proceedings of ECCM-8 Naples* 2: 89-96.

Apinis, R. et al., in press. Rotation capacity of concrete beams with ductile hybrid CFRP + GFRP and elastic CFRP tensile reinforcement. *In Proceedings IRACC-99 Baltimore USA. November 1999.*

Grushecky, I. et al. 1982. Change in stiffness of unidirectionally oriented fibrous composite material due to crushing of fibers. *Mechanics of Composite Materials* 18(2): 139-144.

Gutans, J. 1985. Analysis of stress concentration in rupture of fibers in a hybrid composites. *Mechanics of Composite Materials* 21(2): 171-175.

Gutans, J. & Tamuzs, V. 1987. Strength probability of unidirectional hybrid composites. *Theoretical and Applied Fracture Mechanics* 7: 193-200.

Maksimov, R & Kochetkov, V. 1982. Description of the deformation of a hybrid composite with allowance for the effect of fibre fracture. *Mechanics of Composite Materials* 18(2): 157-162.

Matsui, J. 1993. PAN based carbon fibres. *In Proceedings ECCM-6 (Euro-Japanese colloquium on ceramic fibres).* 38 - 54.

Meier, U. 1996. Composites for structural repair and retrofitting. In *Fiber Composites in Infrastructure, ICCI 96 Tucson*: 1203-1216.

Ueda, T. - pers. comm.

Recent Developments in Durability Analysis of Composite Systems, Cardon, Fukuda, Reifsnider & Verchery (eds)
© 2000 Balkema, Rotterdam, ISBN 90 5809 103 1

Freeze-thaw durability of polymer matrix composites in infrastructure

K.N.E.Verghese, M.R.Morrell, M.R.Horne & J.J.Lesko
Department of Engineering Science and Mechanics, Virginia Polytechnic Institute and State University,
Blacksburg, Va., USA

J.Haramis
Department of Civil Engineering, Virginia Polytechnic Institute and State University, Blacksburg, Va., USA

ABSTRACT: The present paper discusses the effects of aging and freeze-thaw damage in polymer matrix composites. Differential scanning caloriemetry (DSC) data obtained showed that the neat resin, although saturated, did not contain any freezable traces of water, whereas the composite system did show traces of freezable water. Scanning electron and optical micrographs taken on composite cross-sections indicate clearly the presence of both interfacial cracking upon saturation along with some cracks in the resin rich areas of the pultruded component. This led the authors to believe that water tends to reside in these locations. The effect of the combination of fatigue, saturation and freeze-thaw (100 cycles) on the mechanical behavior of stitched $[90/0]_{3S}$ vinyl ester E-glass composite was also examined.

1 INTRODUCTION

Fiber reinforced polymer (FRP) composite materials are under serious consideration for use and investigation for civil infrastructure within the U.S. The high strength and stiffness to weight ratios along with their flexibility in designing specific structural characteristics make them attractive for various structural applications. In addition, the serviceability and functional service life of a composite superstructure such as a bridge may be greater than those built using conventional materials. An uncertainty that plagues the community in its attempt to routinely implement FRP in highway structures is the proof of their environmental durability. A comprehensive database of information does not exist as of today nor does a fundamental understanding of the physics of the problem at hand.

One such environmental condition that remains an unknown is that of freeze-thaw resistance. Limited research has been conducted in this area, primarily with regards to aerospace applications[1]. When considering civil infrastructure applications, Gomez and Casto[2] have reported mechanical data on freeze-thaw tests conducted on insopthalic polyester and vinylester pultruded glass reinforced composites. These specimens were aged in accordance with ASTM C666 (namely, 40°F to 0°F followed by a

hold at 0°F and a ramp up to 40°F followed by a hold) while submerged in 2% sodium chloride and water. Specimens were removed after every 50 cycles and tested in flexure mode. The results clearly indicated reduction in flexure strength and modulus after 300 cycles.

Lord and Dutta[3] were according to the authors one of the first to highlight the importance of cracks in the matrix and fiber-matrix interface as being the cause of the damage in composite materials. When these cracks form beyond a certain critical size and density, they coalesce to form macroscopic matrix cracks. This then increases the diffusion of water into the system so that now condensed matter could settle within these cracks and result in crack propagation as well the formation of micro as well macro level ply delamination during the expansion of water undergoing a liquid-solid phase transition.

In Kevlar fabric laminates used by Allred[4] and subjected to 2 hour temperature cycles from −20°F to 125°F, ultimate tensile strength of the laminate was found to decrease by 23% after 360 cycles and by 63% after 1170 cycles.

The work done thus far has not elucidated the controlling mechanism(s) or kinetics to explain the process of freeze-thaw damage. Thus, the primary goal of the present work is to therefore examine the occurrence, extent and mechanisms controlling freeze-thaw degradation in composite materials. The problem has been broken down to examine the response of each of the constituents (fiber, matrix and interphase/face) to the presence of moisture under cyclic thermal conditions. Differential scanning calorimetry (DSC) will serve as an initial technique to study the freeze-thaw process in an effort to identify the nature and presence of freezable water within each constituent. Subsequent work will examine the nature of freeze-thaw damage within the composites and their effects on residual properties.

2 MATERIAL SYSTEM AND EXPERIMENTAL PROCEDURE

A commercial vinylester was used as the neat, unreinforced resin. This resin is produced by the Dow Chemical Company and marketed under the brand name Derakane 441-400. Styrene is added as the co-monomer and helps reduce the room temperature viscosity of the resin. Benzoyl peroxide (BPO) is added to the system and the resin undergoes a rapid, free radically initiated addition polymerization to form the crosslinked network.

The glass transition temperature, T_g for the pure resin is around 135°C (from DSC) and that of the composite is around 126°C. The lower T_g for the composite can be attributed to fact that the pultrudable resin contains fillers and other additives such as mold and air release agents that may plasticize the resin. The composite is a $(0°/90°)_{3s}$, glass fiber, vinyl ester system, pultruded at Hexcel Corporation using the continuous resin injection technique.

A Perkin Elmer, Pyris 1 differential scanning calorimeter was used to perform the freeze-thaw experiments. Liquid nitrogen was used to achieve the cryogenic temperatures needed for the experiment. The base line of the instrument was constructed on a daily basis to ensure experimental accuracy. Regular aluminum pans were used for the single cycle experiments whereas special Perkin Elmer, pressure pans had to be used in the cyclic experiments. These are pans with special threaded lids to prevent leakage due to pressure build up inside the pans leading to subsequent opening of the seal. In the case of the cyclic experiments a few drops of water was dispensed into the pan in addition to the saturated specimen in order to investigate the role of this free-water. Circular disks of the material were punched out in the case of the resin and machined in the case of the composite before moisture saturation and DSC testing. A temperature cycle of -150°C to +50°C was used for the single cycle experiment and –18°C to +4°C for the cyclic experiment. In both cases a rate of 5°C/min was used and the melt endotherm was monitored owing to its greater reliability as explained by McKenna.[5] This thaw rate is clearly much greater than that observed in our terrestrial environment (5°F/hr.).[6] In each case two samples were examined and a dwell time of about 3 minutes was given to the sample at the extreme temperatures in order to maintain thermal stability in the system.

Quasi-static tensile tests were performed on the composite samples using an Instron test frame at a crosshead speed of 2 mm/min. R=0.1 fatigue testes were conducted on an MTS servo-hydraulic machine to determine the fatigue life of the material at 17% of its ultimate strength. Next, three different levels were chosen to correlate with 0.03% (61 cycles), 0.73% (1500 cycles) and 40% (81814 cycles) of the material's life. Once the specimens were fatigued to each of these levels, measurements of crack density were made from X-ray radiographs. The specimens were then placed into a water bath at 65°C and allowed to saturate. No edge coating procedure was adopted for these specimens. Once the specimens were saturated, they were tested for residual strength, stiffness and strain to failure. These tests were also repeated after 1 year to examine the effects of long-term hygrothermal aging. Freeze-thaw experiments were also carried out on the composite specimens under a known profile determined from ASTM C666 for 100 cycles. Residual properties were once again assessed.

3 RESULTS AND DISCUSSION

3.1 Thermal analysis and microscopy

Heat flow measurements as obtained by DSC, under the above stated thaw conditions for a neat, un-reinforced resin sample, subjected to a single cycle, showed no endotherm. Again, this resin had been saturated in a water bath at 65°C and had an equilibrium moisture content of 0.9% by weight (Table 1). The lack of a melt endotherm indicates the absence of freezable water in the pure resin. This is not surprising considering the fact that the water essentially resides in the free volume of the resin and this free volume size is of the order of about 6-20 Å. According to the Thompson's equation this void size is too small thermodynamically for water to freeze. In addition, Verghese et.al[7] have reported that specific interactions in the form of hydrogen bonding exist in the presence of water for this vinyl ester resin. These specific interactions bind the water to the resin backbone inhibiting crystallization.

The melt endotherm region in a similar heat flow experiment that was carried out on a saturated (at 65°C) E-glass/Derakane 441-400 composite sample is shown in Figure 2. The existence of a weak melt endotherm is clear, indicating the presence of freezable water. From this melt endotherm region the enthalpy under the peak can be estimated. Using the enthalpy of fusion of water (333 J/g), the amount of freezable water can be ascertained (Table 1). A supression in the freezing point is seen to exist and this is conceivable as the degree of supercooling is a function of the crystal size (diameter in the case of a spherical crystal).

The melt endotherm portion of several freeze-thaw cycles is shown in Figure 3. Again, the composite specimen was saturated in a 65°C water bath before testing. In this experiment, a few drops of water were dispensed into the pan in addition to the saturated specimen in order to investigate the role of free versus bound water. It is clear that certain features of the curve change as freeze-thaw cycles proceed. Initially, a shoulder appears at the lower temperatures and is attributed to the water contained inside the saturated specimen. The position of the peak seems to be shifted to around -4°C as opposed to -6°C in the previous case and this is because these samples were saturated in distilled water as opposed to tap water. The second more prominent peak is associated, due to its shear magnitude, to the melting of the free water. After 5 cycles this low-end shoulder disappears. We speculate that this is due to the accumulation of damage in the composite that then opens up the size of these spaces and allows the water to freeze in an increased free-water like manner. This however needs to be inspected with further testing. This increase in damage can be associated with fiber matrix interface failure as well as preexisting flaws within the composite. A scanning electron micrograph of a saturated glass/vinyl ester composite (EXTREN™)[8] clearly shows the presence of interfacial cracks (figure 4).

The sizes of these cracks along with that shown in Figure 5, which is an optical micrograph, are large enough to harbor freezable water. The crack in Figure 5 is estimated to be about 20 microns x 250 microns in dimension and is formed in the resin rich regions of the "as received" material detailed in the experimental section. This processing flaw is due to the large cure shrinkage associated with the vinyl ester resin (~6% by volume) and the mismatch in

Table 1. Saturation and DSC data obtained on both the neat resin as well as composite

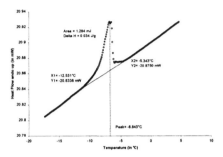

Sample	Number	Maximum Moisture Content (wt.%)	Amount of Water, W1 (mg)	Area under Endotherm (mJ)	T_m (°C)	Freezable Water, W2 (mg)
Derakane 441-400	1	0.9	0.2	None	—	None
	2	0.87	0.2	None	—	None
Composite	1	0.53	0.1	1.05	-6.34	0.038
	2	0.55	0.1	1.51	-6.48	0.045
	3	0.56	0.1	1.28	-6.83	0.039

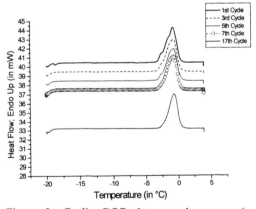

Figure 2. Melt endotherm region for the composite specimen. Indicated here are both the position of the melt endotherm at –6.8°C and the area under the peak

Figure 3. Cyclic DSC data on the composite specimen. The pan contained both the saturated specimen as well as a few drops of free-water.

coefficient of thermal expansion between glass fibers and vinyl ester[9]. The authors therefore have sufficient reason to believe that water can reside in these areas in the composite and that these regions are large enough to facilitate the freezing of water during aging.

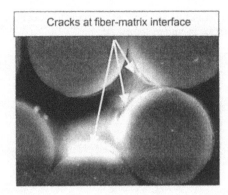

Figure 4. Scanning electron micrograph of fresh water aged material at X 3000 magnification

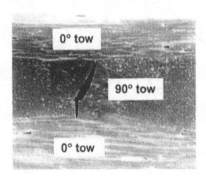

Figure 5. Optical micrograph of the "as received" pultruded cross ply laminate (E-glass/Derakane 441-400) viewed from the transverse to the pull direction

3.2 Mechanical Testing

The average tensile strength of the virgin composite was found to be 49.8 ksi. Both stiffness as well as strength were seen to reduce with increasing fatigue cycles. Shown in figure 6 is the relationship between crack density as measured from X-ray radiography and fatigue cycles.

Surprisingly, no clear correlation between damage content and diffusivity was seen to exist. This is clear from figure 7 which shows the moisture uptake curves for the different composites.

The strength of the composite was, however, strongly affected by the presence of moisture. Figure 8 shows a plot of normalized strength as a function of crack density for both the dry and saturated composites. A similar plot of stiffness showed a weak correlation, indicating that moisture does not have profound effect.

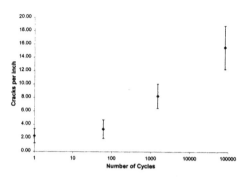

Figure 6. Fatigue cycles versus crack density

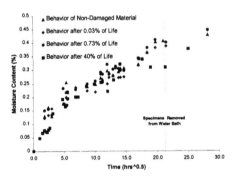

Figure 7. Water uptake of damaged vinyl ester E-glass composite

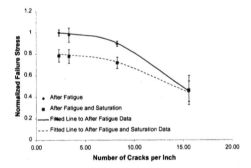

Figure 8. Normalized residual strength versus crack density

The fits in figure 8 were derived from Blikstad[10] and have the following forms, for the dry but fatigued samples:

$$\frac{\sigma_{fat}(d)}{\sigma_{ult}} = 1 - 0.00343 \left(\frac{d}{d_{unfat}} \right)^{2.67} \qquad (1)$$

and for the fatigued and saturated samples:

$$\frac{\sigma_{sat+fat}(d)}{\sigma_{ult}} = 0.8 - 0.00343\left(\frac{d}{d_{unfat}}\right)^{2.45} \quad (2)$$

where, σ_{ult} is the ultimate tensile strength, σ_{fat} is the residual strength after fatigue, $\sigma_{sat+fat}$ is the residual strength after fatigue and saturation and d is the crack density.

Specimens that had been removed from the 65°C bath upon saturation and immersed in a room temperature (28°C) water container for storage purposes, showed an extreme reduction in strength when removed and tested after a year. This is illustrated in figure 9.

Similar trends were seen to exist with strain to failure, however stiffness showed only minor changes. Due to a limited number of samples, X-ray radiography could not be performed as a result of which, crack density data is not available for these conditions.

Freeze-thaw aging of both dry as well as saturated specimens were conducted in a Blue-M conditioning chamber. A thermal cycle in accordance with ASTM C666 was used and programmed into the chamber's controller. Figure 10 is a picture of the unloaded fixture that was used to house the specimens. Each fixture can hold 8 specimens. As mentioned in the caption, every other slot is a blind slot in order to house the specimen and water. The other slots are through slots that help increase the efficiency of heat transfer. This is crucial to the success of the experiment failing which, large temperature overshoots can occur as the chamber tries to maintain the desired heating/cooling rate during the liquid-solid phase transition. Temperature inside each fixture was monitored with the help of an embedded thermocouple.

Figure 10. Picture of the unloaded fixture used house specimens during freeze-thaw aging. Eve other slot is a blind slot

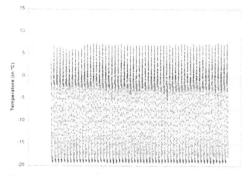

Figure 11. Temperature response over 100 freez thaw cycles inside the specimen fixture

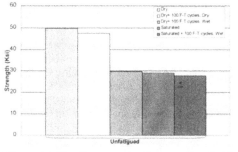

Figure 12. Tensile strength of composites that hav been freeze-thaw (F-T) aged both dry and wet

Figure 11 shows the thermocouple trace over 1(cycles. The response seems to be very consistent.

In order to understand the individual and combine effects of temperature, saturation, and fatigue c freeze-thaw durability, aging tests in which specimens aged under one or two of the above

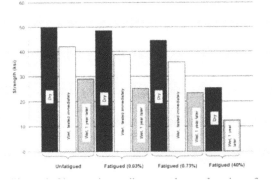

mentioned scenarios was examined. These tests included freeze-thaw conditioning on a range of samples from unfatigued, dry materials to high fatigue damaged, saturated materials. Figure 12 shows strength data for unfatigued composite specimens that were subjected to several different conditioning scenarios.

In the above plot, "Dry" indicates virgin material and "Dry + 100 F-T cycles, Dry" stands for virgin material that was subjected to a 100 freeze-thaw cycles without the presence of water, namely thermal cycling only. "Dry + 100 F-T cycles, Wet" stands for samples that have been aged in the freeze-thaw fixtures surrounded with water as shown in figure 10. The presence of water in the fixture does seem to have had an influence during the aging process of dry samples.

4 CONCLUSIONS

It is virtually impossible to freeze water in a highly crosslinked amorphous polymer like vinyl ester. This is in part due to the geometric space constraints and to hydrogen bonding that further impedes the process. In the composite system, however, the crack dimensions are large enough to facilitate the freezability of water. The authors believe that it is this mechanism of freezing and the associated volume increase during the transition that leads to the propagation of cracks and the accumulation of damage.

Samples that had been fatigued showed drops in both strength as well as stiffness. X-ray radiography proved to be a powerful technique to count cracks in the sample. This resulted in the establishment of a clear relationship between strength (figure 8) as well as stiffness and crack density. Surprisingly, no evident relationship between diffusivity and crack density was seen to exist as seen in figure 7.

The influence of both fatigue and moisture aging were seen to be significant as seen in figure 9. Under low cycle fatigue conditions, the influence of water was to lower strength and was found to be large, whereas under high cycle fatigue conditions, this effect was smaller and the reduction in properties were mainly due to the fatigue process itself. This indicated that a mechanism of relieving residual stresses by water was the prime reason for the drop in tensile strength under low cycle fatigue. However, it is the combination of both fatigue (namely transverse cracking and longitudinal splitting) as well as moisture damage (namely at the fiber-matrix interface) that was the cause for the

further reduction in properties under high cycle fatigue conditions. In addition to the above, significant changes in strength were observed in samples that had been aged in water for an additional year. This could be due to both continued fiber-matrix interfacial damage as well fiber stress corrosion damage that results in the deterioration of the reinforcement itself.

Effects of preliminary freeze-thaw aging on the strength of the unfatigued specimens were seen to exist (figure 12), however, the authors believe that 100 cycles may be low. In addition, the authors believe that the opening of existing cracks via the addition of load will be crucial to the synergism of the environment. Existing cracks would then be opened facilitating the transport of water deep into the crack tips and the subsequent freezing and thawing of that water. Figure 12 does show differences between samples that are labeled "Dry + 100 F/T, Dry" and "Dry + 100 F/T, Wet". This could be because the latter specimens were placed inside the freeze-thaw fixtures during aging, as a result of which water could have been absorbed by the specimens.

5 FUTURE WORK

Presently, work is underway to subject specimens to 300 freeze-thaw cycles. The authors feel that 100 cycles may have been too little, keeping in mind that the samples were not aged under stress which would have kept the cracks open for further damage. Work is also under way to begin aging another batch of glass fiber/vinyl ester composites under stress. A four point bend freeze-thaw fixture has been designed along the lines of that shown in figure 10, to facilitate pre-stressing of the composite specimens during freeze-thaw aging. The dimensions of the support and loading pins have been designed to subject the specimen to 6000 micro-strain in the constant moment region. The choice of this strain was based on the location of the "knee" in the tensile stress-strain curve. Finally, since our preliminary work showed an encouraging correlation between crack density and mechanical properties, the authors are presently also exploring the use of a non-destructive acousto-ultrasonic technique to evaluate the damage state of the composite.

Traditional ultrasonic nondestructive evaluation (NDE) techniques typically deal with imaging of discontinuities such as cracks. They require an additional step of interpreting the affect of the discontinuity on the performance of the material to be useful for monitoring damage development. Acousto-Ultrasonics (AU) is an ultrasonic NDE technique useful for quantifying small distributed changes not easily detected with traditional ultrasonic techniques.[11] It is not an imaging technique. It returns results related to the ability of the interrogated material to transfer mechanical energy. In this set-up, the transducers are spaced in a manner (typically, far apart) so that there is no direct or minimally reflected energy transmission between them and the primary transmission is by plate wave propagation. AU results are calculated from the energy spectral distribution of the AU signal by moment analysis. The most useful AU parameters tend to be the area under the curve (zeroth order moment) and the centroidal frequency (ratio of the first and zeroth moments). The area is directly related to the amount of energy transfer along the specimen. Shifts in the centroidal frequency also indicate changes in the materials ability to propagate particular modes. Typically, the spectral content of the AU signal has several relatively discrete frequency peaks with a few dominant ones. Each peak corresponds to a different mode or order of plate wave propagation. The modes are typically categorized as symmetric (extensional) and anti-symmetric (flexural). The order of a particular mode indicates the complexity of that type of deformation. Since each peak represents energy propagating with a different type of deformation, it possibly can be sensitive to different types of degradation. The authors feel that this technique will facilitate the close monitoring of damage as a function of aging, whether it be hygrothermal and/or freeze-thaw.

6 ACKNOWLEDGEMENTS

The authors would like to acknowledge the financial support of the National Science Foundation Center for High Performance Polymeric Adhesives and Composites under contract number, DMR-9120004, and the NSF CAREER award program through Civil and Mechanical Systems. The authors would also like to thank the Hexcel Corporation for supplying the composite material under the NIST-ATP proposal.

7 REFERENCES

1) Mitra, Dutta and Hansen, "Thermal Cycling Studies of a Cross-plied P100 Graphite Fiber Reinforced 6061 Aluminum Composite Laminate", *Journal of Materials Science*, 26, November 1991

2) Gomez, J. P. and Casto, B., "Freeze-Thaw Durability of Composite Materials", *Virginia Transportation Research Council*, 1996

3) Lord, H. W., and Dutta, P. K, "On the Design of Polymeric Composite Structures for Cold Regions Applications", *Journal of Reinforced Plastics and Composites*, 7, 1988

4) Allred, R. E., "The Effects of Temperature and Moisture Content on the Flexural Response of Kevlar/Epoxy Laminates: Part II [45/0/90] Filament Orientation", *Environmental Effects on Composite Materials*, Volume 2, George Springer, editor, Lancaster, PA: Technomic Publishing Company, Inc., 1984

5) McKenna, G, and Catheryn, L. J., "On the Anomalous Freezing and Melting of Solvent Crystals in Swollen Gels of Natural Rubber", *Rubber Chemistry and Technology*, 64, 1990

6 Larson, T., Cady, P., Franzen, M. and Reed, J., "A Critical Review of Literature Treating Methods of Identifying Aggregates Subject to Destructive Volume Change when Frozen in Concrete and a Proposed Program of Research", *Highway Research Board Special Report 80*, pp. 30, 1964

7) Verghese, K. N. E, Hayes, M. D., Garcia, K., Carrier, C., Wood, J., and Lesko, J. J., "Effects of Matrix Chemistry on the Short Term Hygrothermal Aging of Vinyl Ester Matrix and Composites", *Journal of Composite Materials,* In Press

8) McBagonluri, F, Garcia, K, Hayes, M. D, Verghese, K. N. E, and Lesko, J. J, "Characterization of Fatigue and Combined Environment on Durability Performance of Glass/Vinyl Ester Composite for Infrastructure Applications", *Accepted for publication in the International Journal of Fatigue*

9) Phifer, S., " Quasi-Static and fatigue Evaluation of Pultruded Vinyl Ester/E-Glass Composites", *Masters Thesis, Virginia Polytechnic Institute and State University*, January 1999

10) Blikstad, M., Sjöblom, W., Johannesson, T.R., (1984), "Long-Term Moisture Absorption in Graphite/Epoxy Angle-Ply Laminates", *Journal of Composite Materials*, Vol. 18, pp. 107-121.

11) Kiernan, M.T. and J. C. Duke, Jr., "A Physical Model for the Acousto-Ultrasonic Method," NASA Grant NAG3-172, October 1990.

Recent Developments in Durability Analysis of Composite Systems, Cardon, Fukuda, Reifsnider & Verchery (eds)
© 2000 Balkema, Rotterdam, ISBN 90 5809 103 1

Durability of titanium-graphite hybrid laminates

D.A. Burianek & S.M. Spearing
Technology Laboratory for Advanced Composites, Massachusetts Institute of Technology, Cambridge, Mass., USA

ABSTRACT: Experimental results are presented for elevated temperature (177°C), tension-compression fatigue tests performed on titanium-graphite (TiGr) hybrid laminates containing open holes. The key damage modes were identified as 0° ply splitting, facesheet cracking, and facesheet delamination. The growth of damage during fatigue resulted in large reductions of the local stiffness adjacent to the hole. This stiffness reduction was predominantly due to delamination of the facesheet from the polymer matrix composite core and showed a significant temperature dependence. Using a simple one-dimensional model it was shown that the delamination growth rates depend on the applied loading according to a "Paris" law, with an exponent of 1.9 on the cyclic strain energy release rate. This observation was substantiated by observations of delaminations growing in isolation in specimens containing facesheet seams.

1. INTRODUCTION

Titanium-graphite (TiGr) fiber metal hybrid laminates (FMHL) offer the potential to achieve high levels of durability at low structural mass fractions for elevated temperature airframe and engine applications. These materials consist of polymer matrix composite (PMC) plies interspersed with titanium foil layers. The use of titanium for the outer plies protects the PMC core from environmental effects such as oxidation and moisture ingress as well as potentially providing improved impact resistance and bearing properties. The concept for TiGr builds upon previously developed FMHL materials such as ARALL (aramid fiber - aluminum) and GLARE (glass fiber - aluminum) which have been found to have excellent fatigue resistance and damage tolerance, particularly compared to non-hybrid composites under impact loading [Bucci 1989, Gunnink 1988 , Wilson 1991].

In order to assess the suitability of titanium graphite FHML materials for their target applications it is essential to obtain durability test data under service environments. Previous research on TiGr laminates [Miller 1994, Li 1996, Li 1998a] has shown that the material can achieve superior fatigue resistance at room temperature.

However there is little data regarding the mechanical performance of TiGr at elevated temperatures. In addition to the need for obtaining test data the high cost of long duration testing, makes it desirable to develop models for the material response that are capable of guiding accelerated testing and of underpinning durability assessments out to service lifetimes.

The approach taken in the present work was to conduct elevated temperature (177°C), tension-compression loading tests on a particular TiGr laminate. The key damage mechanisms and failure modes were identified and quantified, preparatory to constructing mechanism-based models for these aspects of durability. Open hole specimens were chosen for initial testing as being generically representative of structural details, and therefore a likely site for damage initiation in fatigue loading conditions. Subsequent tests were conducted on specimens containing seams in the facesheets in order to isolate facesheet delamination and to obtain quantitative data for its growth. The particular laminate investigated in this work consisted of a graphite fiber-reinforced laminated PMC core with titanium foils as the outer plies. A schematic of this TiGr laminate configuration is shown in Figure 1.

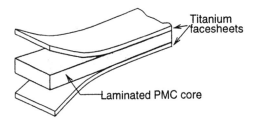

Figure 1: Schematic of a TiGr laminate.

Table 1. Elastic Properties of TiGr constituents and laminate.

	Ti 15-3	PMC [0°]	TiGr [Ti/0/90/0_2]$_s$
E_1 (GPa)	107	162	122
E_2 (MPa)	112	6.9	61
G_{12} (MPa)	41	4.5	11
υ_{12}	0.33	0.35	0.162

2. EXPERIMENTAL PROCEDURE

This section provides a brief description of the specimens and procedures used in the test program. A complete description is available elsewhere [Burianek 1998a].

The TiGr laminate used in this study consisted of Ti exterior layers with a polymer matrix composite core. The exterior plies were 127 μm thick titanium 15V-3Cr-3Al-3Sn (Ti-15-3) foil and the polymer matrix composite (PMC) core consisted of graphite/polyimide matrix plies with a nominal ply thickness of 137 μm. [Ti/0/90/0_2]$_s$, a high strength laminate architecture was used throughout. Material properties for the constituent materials and for the laminate are given in Table 1. The constituent properties were supplied by The Boeing Company [Li 1996b] and the laminate properties were calculated using classical laminated plate theory (CLPT). The laminates were manufactured in the form of plates and machined to 305 mm x 38 mm rectangular coupons by water jet cutting at The Boeing Company. A 6.4 mm diameter hole was drilled in the center of each specimen. Some unnotched specimens were subsequently tested with a seam in one of the facesheets, orientated perpendicular to loading axis. This represents a

practical concern since titanium foil is typically supplied in 610 mm wide strips, thus seams will inevitably be present in large built up structures. For the purposes of the present work the seams acted as a means of initiating and growing delaminations in a controlled manner.

The mechanical tests were carried out using a servohydraulic testing machine with a maximum capacity of 222 kN which was equipped with a digital controller. A custom-designed temperature cabinet and hydraulic grips with water-cooled faces were used for the elevated temperature tests. An anti-buckling guide was used to prevent overall specimen buckling during the compression portion of the load cycle. To monitor the stiffness loss of the specimens during cycling, a high temperature extensometer was used. The extensometer had a gauge length of 50 mm and was mounted axially on the specimen surface symmetrically across the hole. The details of the temperature cabinet, antibuckling guide and extensometer are described more fully elsewhere [Burianek 1998a].

The fatigue tests were performed with maximum cyclic stresses ranging from 10% to 60% of the open hole tension (OHT) strength. The ratio of minimum load to maximum load (R-ratio) was equal to -0.2 except where otherwise stated. The waveforms were sinusoidal with frequencies in the range 6-15 Hz. Cycling was interrupted at predetermined intervals to measure the tangent stiffness at the mean cyclic load.

Selected specimens were sacrificed at predetermined cycle counts and sectioned and polished so that optical microscopy could be carried out to track damage accumulation. Micrographs were taken of the cross-sectioned specimens and then the procedure was repeated at a different cross-section in order to generate a three dimensional views of the evolving damage state [Burianek 1998a, Burianek 1998b].

3. EXPERIMENTAL RESULTS

3.1 Open Hole Tension Tests

Open hole tension tests were performed to establish a baseline for the fatigue tests. The specimens were tested at 177° C, with the anti-buckling guide installed. The tests were run in displacement control using a single ramp at a loading rate of 0.04 mm/s. The load vs. local displacement (as measured by the extensometer) curve was linear until just before final fracture.

The four OHT strengths ranged from 963 MPa to 1123 MPa with an average of 1033 MPa. Figure 2 shows a photograph of a fractured specimen. Fracture occurred along the narrowest ligament of the specimen, with very little fiber pullout or delamination. The relatively small amount of visible damage is consistent with the nearly linear elastic load deflection curve.

Figure 2. Fractured OHT specimen. Note the relatively "clean" fracture surface.

3.2 Fatigue Damage Progression

By contrast with the observation of damage in monotonically loaded open hole tension specimens, the specimens loaded in fatigue developed far more damage prior to failure. Figure 3 shows a specimen that failed in fatigue loading. The fracture surfaces are very fibrous, with extensive delamination. The sequence of sub-critical damage evolution, determined by direct observation and cross-sectioning and microscopy, is documented fully elsewhere [Burianek 1998a&b].

Figure 3. Fatigue-fractured open hole specimen. Note the very fibrous fracture surface and extensive delamination.

The principal damage modes were identified as cracking and delamination of the facesheets and splitting, transverse ply cracking and delamination within the PMC core. These damage modes propagated in a coupled manner during the early part of the life of the coupon. Figure 4 shows a schematic of damage in a specimen cycled to 30,000 cycles at 177° C, R= 0.1, and σ_{max}=30% OHT strength. The Ti facesheet cracks extend approximately one third of the way across the facesheet ligaments and are accompanied by splitting, facesheet delamination and transverse ply cracking within the PMC core.

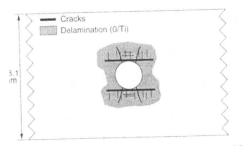

Figure 4: Schematic of damage after 30,000 cycles.

At higher numbers of cycles the mode of combined damage propagation changes. When the facesheet cracks, and accompanying delamination, reach the edge of the specimen, subsequent delamination propagation occurs axially, parallel to the loading direction. This is accompanied by secondary splitting of the 0° plies away from the hole, transverse ply cracking growing in from the free edges and delamination at the interior 0/90 interfaces. A sketch of the damage state after 100,000 cycles is shown in Figure 5.

Close inspection of the delamination fracture surfaces revealed that polymer matrix material and a few fibers remained attached to the titanium facesheets over much of the fracture area. Figure 6 illustrates this observation. The tendency for delaminations to propagate within the PMC plies indicates that the delamination growth is likely to be controlled by the delamination behavior of the

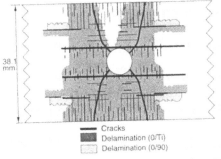

Figure 5. Schematic of damage after 100,000 cycles

composite plies rather than the polymer matrix/titanium interface.

3.3 Mechanical Response

The end point for the fatigue tests depended strongly on the load level applied. At the highest cyclic load levels, 60% OHT, the specimens failed in the manner shown in figure 3 after less than 50,000 cycles, with separation of the specimens into two pieces, albeit with considerable frictional load transfer due to fiber and bundle pullout. At intermediate load levels, between 30% and 45% OHT strength, the tests were stopped when the extensive delamination caused the specimen to wedge in the anti-buckling guide as the titanium facesheets peeled away from the core. At lower load levels specimens could be cycled to one million cycles, arbitrarily defined as "run out", without severe damage accumulating.

Figure 7 shows the progressive stiffness reduction as a function of cycles for these tests. As expected, the rate of stiffness loss increases as the maximum stress increases, indicating a more rapid accumulation of damage. It should be noted that these large stiffness reductions reflect the local damage around the hole, rather than the overall specimen compliance.

In order to provide insight into the role of temperature in determining the durability of TiGr a single specimen was cycled at room temperature (18°C), at 30% of the elevated temperature open hole tensile strength and with an R-ratio of -0.2. In figure 8 the stiffness reduction in this specimen is compared with that of a specimen tested at identical load levels at 177°C. The specimen tested at elevated temperature reaches a 50% stiffness reduction nearly ten times faster than the room temperature specimen, indicating that temperature plays an important role in the rate of damage propagation.

4. DISCUSSION AND ANALYSIS

The experimental observations of damage modes indicate that the durability of TiGr is governed initially by multiple interacting damage modes, namely: Ti cracking, PMC splitting and transverse ply cracking and delamination. Some insight into the factors affecting these damage modes have been presented in previous work [Burianek 1998b]. However, in the mechanical test results presented here it is apparent that "failure" at least at the coupon level, is dominated by the development of extensive delamination, principally of the Ti facesheets from the PMC core.

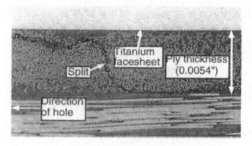

Figure 6. Cross section of fatigue damage in a TiGr. Laminate. Note that the delamination at the upper left is within the PMC plies.

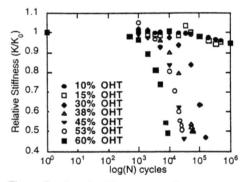

Figure 7. Local stiffness vs. cycles at seven stress levels. Stiffness is normalized by the initial value. Note the tendency for increasing rate of stiffness reduction with stress level.

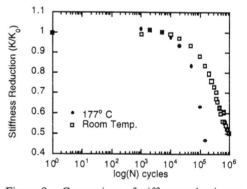

Figure 8. Comparison of stiffness reduction at room temperature and 177°C. Maximum stress was 30% of the OHT strength. Note the accelerated stiffness reduction at the higher temperature.

468

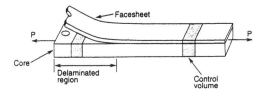

Figure 9. Schematic of strain energy release rate calculation

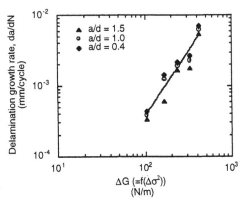

Figure 10. Delamination growth rate as a function of cyclic strain energy release rate.

The delamination of the facesheets in fatigue is assumed to be governed by the cyclic strain energy release rate as is the case for delamination in non-hybrid PMC's [Trethewey 1988]. By assuming that the delamination propagates as a planar crack spanning the width of the specimen the strain energy stored in the specimen can be simply calculated for a control volume ahead of and behind the delamination front, as shown in Figure 9.

The difference between the two calculated strain energies can be equated to the area of delamination surface created, which gives dU/dA, the strain energy release rate, G, available to propagate the delamination, as given in Equation 1. A full derivation is given in reference [7]. A similar approach has been applied to other materials [Ashby 1985].

$$G = \frac{dU}{dA} = \frac{\sigma_L^2}{2WE_L}\left[1 - \frac{E_L}{E_C}\frac{t_L}{t_c}\right] \quad [1]$$

where: E_L and E_C are the Young's modulus of the laminate and PMC core respectively, σ_L is the applied stress, W is the specimen width and t_t and t_C are the thickness of the laminate and PMC core respectively.

hence:

$$\Delta G = \sigma_{max}^2 * f(E, t, W) - \sigma_{min}^2 * f(E, t, W) \quad [2]$$

where: $f(E,t,W)$ is a function of material properties (see equation 1) and σ_{max}, σ_{min} are the applied load at maximum and minimum of cycle.

Combination of the calculations of strain energy release rate with the measurements of damage growth, as indicated by stiffness reduction, allows the data to be presented in the form of a crack growth rate - cyclic strain energy release rate relationship. The slopes of the each of the stiffness reduction curves in figure 7 were measured at three levels of stiffness reduction,

corresponding to three damage extents, and reduced via a calibration curve [Burinek 1998a]. The delamination growth rate can then be plotted as a function of the cyclic strain energy release rate, as shown in figure 10. It is apparent that a power law of the form:

$$da/dN \propto (\Delta G)^{m/2} \quad [3]$$

holds over the range of data shown, with the exponent m taking a value of 3.8. This exponent is somewhat lower than is usually found for delamination of composites, which typically fall in the range 6-10 [Trethewey 1988].

In order to verify that the delamination growth dependence inferred from figure 10 is representative of that for facesheet delamination in TiGr, specimens with facesheet seams were tested. In these specimens delamination, without other damage modes present, originated from the seams. Growth could easily be measured from the edge of the specimen using direct observation and edge replicates, as shown in Figure 11.

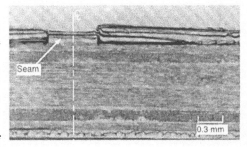

Compared to the open hole specimens relatively little secondary damage was evident.

Cyclic strain energy release rates were calculated using equation [1] and [2] for the load levels used in the tests. The crack growth rate data for tests conducted at room temperature and 177 °C are presented in figure 12.

There are two interesting features of the data in figure 12. Firstly, a power law fit to the 177°C data has an exponent, m, of 3.8, which corresponds closely to the value found for the homogenized treatment of damage growing from notches in figure 10. Secondly the increase in temperature from 20°C to 177° results in a significant acceleration in crack growth. This latter observation is consistent with the increase in stiffness loss at elevated temperatures shown in figure 8, and further reinforces the conclusion that delamination is likely to be the damage mode most responsible for limiting the durability of the TiGr laminates examined in the present study. The effect of temperature in increasing the delamination growth rate suggests that temperature may be an effective means of conducting accelerated testing for this material system.

It should be noted that the strong influence of delamination on the stiffness response and the visible deterioration of the coupons used in this study is partially a result of the finite scale of the test coupons. The very high relative stiffness reduction shown in figure 7 results, in part, from the extent of the delamination being of the same order as the gauge length of the extensometer.

The highest delamination growth rates only occur after the facesheet cracks reach the edge of the specimen. These circumstances may not be relevant to conditions in the intended structural applications since the fraction of free edge present in a large aerospace vehicle is significantly less than in the small coupons tested in the present work. This would suggest that the actual durability of TiGr may be considerably greater than would be inferred from the experiments presented in this work. Further analytical and experimental work is required to confirm this supposition. The development of models to permit the scaling of coupon level test data up to the structural level is an important ongoing activity to address this issue.

5. SUMMARY

Elevated temperature tension-compression fatigue tests were conducted on open hole TiGr 2-6-2 specimens. Ultimate failure either occurred by fracture at higher load levels or extensive delamination at intermediate load levels. In all cases, failure was preceded by a significant reduction in stiffness, reflecting considerable damage accumulation. The key subcritical damage modes were identified as facesheet cracking, 0° ply splitting and delamination.

Simple models for the degradation in stiffness and the strain energy release rate associated with delamination growth were formulated, and their combination suggests that facesheet delamination growth rate has a power law dependence on the strain energy release rate amplitude. Temperature was identified as a significant accelerating factor for facesheet delamination and therefore the stiffness reduction in open hole tension specimens.

Further work is required to characterize these damage modes and in particular, to relate the behavior of the test coupons to the envisioned application in aerospace structures.

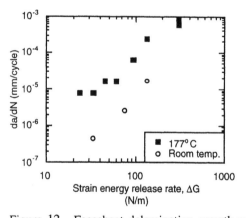

Figure 12. Facesheet delamination growth rate vs. cyclic strain energy release rate amplitude for two temperatures. Note the increased growth rate at the higher temperature.

ACKNOWLEDGMENTS

The authors would like to acknowledge the work and support provided by the HSCT structures group at The Boeing Company, especially Ron Zabora, Ed Li, Bill Westre, Antonio Rufin, and Matthew Miller. In addition, we would like to thank Todd Harrison, Barbara Huppe, and Julio

Rodriguez for their contribution and participation in this project.

REFERENCES

Ashby, M. F., K. Easterling, R. Harrysson, & S. Maiti 1985. The Fracture and Toughness of Woods. *Proc. Royal Soc., London.* A398:265-280.

Bucci, R., L. Mueller, L. Vogelesang, & J. Gunnink, 1989. ARALL Laminates. *Aluminum Alloys-Contemporary Research and Applications*, 31: 295-322.

Burianek, D. A. 1998a. Fatigue Damage in Titanium Graphite Hybrid Laminates. S. M. Thesis, Massachusetts Institute of Technology.

Burianek, D. A. & S. M. Spearing, 1998b Durability of Titanium/Graphite Hybrid Laminates. AIAA paper 98-1959, Presented at AIAA SDM meeting, Long Beach.

Burianek, D. A. & S. M. Spearing, 1998c. Fatigue Induced Delamination in Titanium-Graphite Hybrid Laminates. American Society for Composites, 13th Technical Conference, Baltimore, MD.

Gunnink, J.W. 1988. Design Studies of Primary Aircraft Structures in ARALL Laminates. *Journal of Aircraft*: 1023-1032.

Li, E., The Boeing Company, 1996a. Personal Communication.

Li, E. Residual Strength Study of Fatigued Open-Hole Titanium-Graphite Hybrid Composite Laminates. in The 39th AIAA/ASME/ASCE/AHS/ASC Structures, Structural Dynamics, and Materials Conference, 1998. AIAA-98-1960.

Li, E. and W.S. Johnson, 1996b. An Evaluation of Hybrid Titanium Composite Laminates for Room Temperature Fatigue. in *The Tenth International Conference on Composite Materials*.

Miller, J.L., D.J. Prograr, W.S. Johnson, & T.L.S. Clair 1994. Preliminary Evaluation of Hybrid Titanium Composite Laminates. technical memorandum, NASA Langley Research Center.

Trethewey, B. R., J. W. Gillespie & L. A. Carlsson 1988. Mode II Cyclic Delamination Growth. *Journal of Composite Materials*, 22: 459-483.

Wilson, C. D. & D. A. Wilson 1991. Effective Crack Lengths by Compliance Measurement for ARALL-2 Laminates, *Composite Materials: Fatigue and Fracture*, ASTM STP 1110: 791-805.

Recent Developments in Durability Analysis of Composite Systems, Cardon, Fukuda, Reifsnider & Verchery (eds)
© 2000 Balkema, Rotterdam, ISBN 90 5809 103 1

Durability of stressed E glass fibre in alkaline medium

Grzegorz Świt
Kielce University of Technology, Poland

ABSTRACT: The topic of this paper is analysis of E glass fibre bunch under constant static load and in alkaline environment condition. The incipient of cracking process of tested specimens was detected using acoustic emission and the influence of the alkaline medium on stress corrosion damaged fibre is shown on scanning micrograph.

Keywords: Glass fibre, stress corrosion, acoustic emission,

1 INTRODUCTION

For the last years, the composite materials have been used often in civil engineering. They can be used both as the main element of frequently the engineering structure and as the element reinforcing existing constructions (Focacci 1998, Tasnon 1996).

The limits in using composites as a reinforced material are caused by the fact that E glass fibres are not resistant to stress corrosion. There have been many papers publicised about stress corrosion fibre as an element of composite with polymer matrix in acid medium (Kumosa 1997, Chandler 1984, Das 1991, Golaski 1995). However, there is no information about behaviour of these fibres in alkaline medium.

That is why I decided to establish the influence of alkaline medium concentration and the level of load on the stress corrosion effect. This is important, when, if we want to predict the durability of reinforced elements under combine influenced of stress and alkaline environment.

2 EXPERYMENTAL PROCEDURE

The tests were made on bunches of E glass fibres - ER-3005 produced by Glass Works in Krosno (Poland). Bunches have been taken randomly without eliminating damaged fibre. The experiment was made on rowing of boroaluminosilicate glass which contained less than 1% alkaline with text rowing 1200 g/km and the nominal diameter of a single fibre equal $\phi = 10$ μm. The number of fibres in a bunch was approximately 6000. The surface of rowing was covered by glycidoepoxysilicanian preparation. The nominal length of specimens was equal $l_{av} = 275$ mm, and the cross section was equal $A_N = 0.49$ mm^2. The tests have been made in alkaline medium Ca(OH)$_2$ with concentration from pH12 to pH8. In the first stage of tese experiments the strength using the UTS 20 testing machine. The medium strength of a bunch of fibres was equal $P_{av} = 154$ N.

In corrosion tests, loads in the range from 0.5 to 0.7 of mean strength were applied. To detect the beginning of fracture, the acoustic emission (AE) AET 204 and MISTRAS 2001 processors were used. To evaluate the stress corrosion, damages of the specimen surfaces were analysed with a scanning electron micrograph after failures.

3 INVESTIGATION OF CRACK CORROSION

Stress corrosion in alkaline medium is important when glass fibres are used as a filling material of composite to reinforce concrete and concrete elements of structures, because the concentrations mentioned above can be encountered in concrete. The concentration pH12 occurs on surface of non-corroded concrete while pH8 is observed in corroded concrete.

The stress corrosion cracking produces stress waves, which can be the detected acoustic emission. When analysing the AE events vs. time plots during stress corrosion cracking three intervals in damage intensity can be selected:

1. strong process AE signals produced by the fracture of the fibre with unadvantageous orientation and friction between the fibres.
2. stable increase of AE signals, which may be the results of incidental fibres cracking or corrosion process
3. sudden increase of AE signals as the results of gross fibre breaking.

The plots mentioned above are show on figure 1.

For further analysis it was assumed that the number of AE events equals the number of fractured fibres in a bunch. Analysing the plot given above it can be noticed that the fast process of fibre bunch destruction takes place when 30% of all fibres were fractured.

The service time is indicated by the point of intersection of line s1 and s2.(as it is show Fig.1)

The influence of load on durability is shown Fig.2

While comparing these plots it can be noticed that 30% of fractured fibres limited service time in corrosive environment. The number of fractured fibres at final fracture of bunches depends neither on concentration on the level of load.

The damages of fibre surface, which took place as a result of stress corrosion, were examined with the help of scanning electron micrograph.

The examples of alkaline concentration on stress development load for 0.7 P_{av} and different alkaline concentration after fracture are shown on Fig.3

While comparing the pictures given above, three types of fibre corrosion may be observed:
1. a large amount of surface elongated pits which are not connected with each other causes significant cavity of the area of fibres - fig. 3a
2. wide, deep pits with irregular edges, but not

connected with each other - fig.3b
3. wide, deep, irregular pits connected with each other; a type of corrosion which is the sum of mechanisms mentioned above – fig.3c

The observed stress corrosion damages probably take place in accordance with the equation of the chemical process indicated below:

$$Ca(OH)_2 + SiO_2 \rightarrow CaSiO_3 + H_2O$$

To estimate if the fibre damage is a result of combined influence of stress and environment, unloaded fibres were immersed for 7 days in bath with concentration from pH12 to pH8.

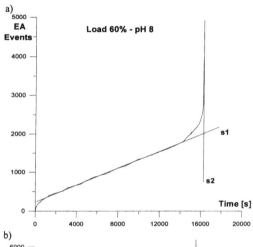
a)

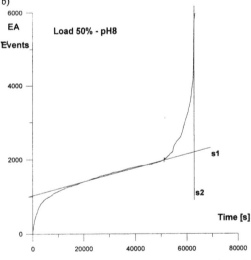
b)

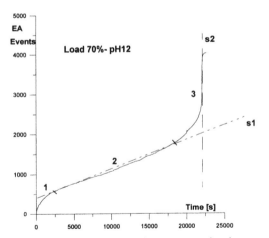

Fig.1. Characteristic periods, during stress corrosion destruction of fibre bunch

Fig.2. Comparison of plots: (AE) events (number) vs. time in dependence on load for pH8
a) load: 0.6 P_{av} - pH8
b) load: 0.5 P_{av} - pH8

474

a) Scanning electron micrograph of surfaces of tested fibre in concentration pH12

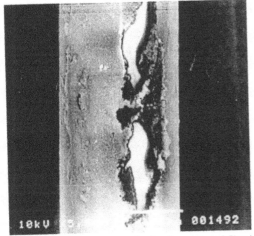

b) Scanning electron micrograph of surfaces of tested fibre in concentration pH10

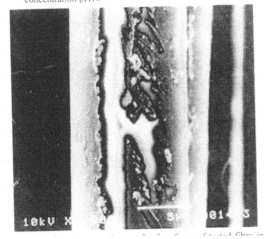

concentration pH8

Fig.3. The influence of alkaline concentration on stress corrosion damage of fibre under constant load 0.7 P_{av}.

Fig.4. Scanning electron micrograph of unloaded E-glass fibres immersed in solution of $Ca(OH)_2$.

Next the fibres surface were examined with the help of scanning micrograph. The results is shown on Fig.4.

In this picture there are no signals of corrosion damages on fibres surface, that leads to a conclusion that the environment surfaces aren't a sufficient condition to initiate corrosion of fibres.

4 CONCLUSION

While analysing of fibre corrosion in dependent on alkaline concentration and load, it can be concluded, that:

a) stress corrosion fracture of fibre bunches takes place when about 30% of all fibres are broked,

b) the most intensive damage of stress corrosion process occurs in alkaline medium with concentration pH8

c) the damages of unloaded fibres in alkaline environment (pH8-pH12) have not been observed

5 REFERENCES

FOCACCI F., NANNI A., FARINA F., SERRA P., CANNETI C. 1998. Repair and Rehabilitation of an Existing RC Structure using CFRP Sheets. *ECCM-8*: 51-58.Woodgead Publishing Ltd.,

TASNON M., MISSIHOUN M., GERIN G. 1996. Lajoie, restorarion of bulding facades with composite materials. *Advanced composites materials in bridges and structures:* 589-593 Montreal-Quebec,

KUMOSA M., QIU Q. 1997. Corrosion of E-glass fibers in acidic environments. *Composites Science and Technology.* 57: 497-507

CHANDLER H., JONES R. 1984. Strength loss in „E" glass fibres treated in stronng solutions of mineral acids. *Journal of Materials Science.* 19: 3849-3854

DAS B., TUCKER B., WATSON J. 1991. Acid corrosion analysis of fibre glass. *Journal of Materials Science.* 26: 6606-6612

GOŁASKI L., RANACHOWSKI J. 1995.Service life prediction of glass reinforced plastic under stress corrosion condition by acoustic emission. 101-108

Author index